Age-Differentiated Work Systems

Christopher Marc Schlick
Ekkehart Frieling · Jürgen Wegge
Editors

Age-Differentiated Work Systems

Springer

Editors
Christopher Marc Schlick
Institute of Industrial Engineering
and Ergonomics
RWTH Aachen University
Aachen
Germany

Jürgen Wegge
Institute of Work and Organizational
Psychology
TU Dresden
Dresden
Germany

Ekkehart Frieling
Institute of Work and Organizational
Psychology
University of Kassel
Kassel
Germany

ISBN 978-3-642-35056-6 ISBN 978-3-642-35057-3 (eBook)
DOI 10.1007/978-3-642-35057-3
Springer Heidelberg New York Dordrecht London

Library of Congress Control Number: 2012955995

Printed on acid-free paper

Springer is part of Springer Science+Business Media (www.springer.com)

Contents

Contributors

Prof. Dr. Peter Angerer is full professor of Occupational and Social Medicine at the Heinrich Heine University Düsseldorf (M.D. 1984, habilitation 2003). His research interests are in the field of coronary risk factors at work, psychosocial health, and health services research, e-mail: Peter.Angerer@uni-duesseldorf.de

Dr. Katrin Arning obtained her Ph.D. degree in Psychology, evaluating ICT trainings for older users. She is currently a post doc at the chair of Communication Science at the Human–Computer Interaction Center at RWTH Aachen University. Her research interests focus on HCI usability evaluations, technology acceptance, and user diversity, e-mail: arning@comm.rwth-aachen.de

Prof. Dr. Gisa Aschersleben is full professor of Developmental Psychology at Saarland University, Saarbrücken (Ph.D. 1993, habilitation 1999). Her research interests are in the field of early cognitive development, action control, theory of mind, and cognitive aging. She has published two monographs, two edited books, two special issues, 75 journal articles, and 34 book chapters related to these topics, e-mail: aschersleben@mx.uni-saarland.de

Dipl.-Wirtsch.-Ing. Marcel Becker was research assistant from 2002 to 2008 and is doctoral candidate at the ifab-Institute of Human and Industrial Engineering of the Karlsruhe Institute of Technology (KIT; formerly University of Karlsruhe). He is now working in the industry. His research interests concern the integration of workforce and resource planning, personnel-centered simulation, and demographic development

Dr. Max Bierwirth studied Business Administration and Economics in Mannheim and Frankfurt and majored in Industrial Business Operations and Logistics. From 2008 to 2012 he worked as research associate at Institute of Ergonomics at the Technische Universität Darmstadt. The focus of his research and his doctoral thesis (2012) is the management of ergonomics activities and processes in manufacturing companies, e-mail: bierwirth@iad.tu-darmstadt.de

Dipl.-Ing. Kerstin Börner graduated in Systems Engineering. She is working as research assistant at the Chair for Ergonomics at the Institute of Industrial Management and Factory Systems at the Chemnitz University of Technology since 2006. From 2010 to 2012 she attended the research project focused on age-differentiated work systems, e-mail: kerstin.boerner@mb.tu-chemnitz.de

Prof. Dr. Ralph Bruder is Director of the Institute of Ergonomics of the Technische Universität Darmstadt (IAD). From 2003 until 2006 he headed the Zollverein School of Management and Design as president and managing director. Professor Bruder is President of the German Society of Ergonomics (Gesellschaft für Arbeitswissenschaft GfA). His research interests include evaluation and design of work systems in general as well as innovative product design, e-mail: bruder@iad.tu-darmstadt.de

Dipl.-Ing. Jennifer Bützler graduated in Mechanical Engineering with the Dipl.-Ing. degree from RWTH Aachen University in 2009. Since then, she is a research assistant and Ph.D. candidate at the Institute of Industrial Engineering and Ergonomics at RWTH Aachen University. Her research focuses on ergonomics and human–computer interaction, e-mail: j.buetzler@iaw.rwth-aachen.de

Dr. Stefan Diestel is a scientist of Work and Organizational Psychology at the Leibniz-Research Centre for Working Environment and Human Factors at the Technical University of Dortmund. His research focuses on burnout development, self-control at the workplace, emotional labor, and employee absenteeism, e-mail: diestel@ifado.de

Dr. José Alonso Enríquez Díaz is employed at a well-known automotive manufacturer in the Department of Production Systems. He completed his first diploma in Mechanical Engineering at the University Nacional de San Agustin in Peru and holds a second diploma with focus on manufacturing and work psychology from the University of Kassel, where he was engaged as a scientific staff member. In 2012 he finished his Ph.D. on the ergonomic evaluation of production systems for the component assembly in the automotive industry, e-mail: alonso.enriquez@volkswagen.de

Prof. Dr. Michael Falkenstein is Professor for Psychology at the Technische Universität Dortmund and head of the project group "Ageing" and of the research initiative "elderly workers" at the Leibniz Research Centre for Human Factors and Working Environment (IfADo) in Dortmund (diplomas in electrical engineering and psychology, MD in medicine in 1978, and habilitation in psychology in 1996). His main research topics are electrophysiological measures of cognitive functions in young and old subjects, and interventions to ameliorate age-related cognitive changes in seniors and older workers, e-mail: falkenstein@ifado.de

Prof. Dr. Ekkehart Frieling was a full professor of Work and Industrial Psychology at Kassel University (Dr. phil. 1974, habilitation 1979). His main research topics are: industrial psychology in the automotive industry, task analysis, job analysis methods, work design, and redesign, e-mail: ekkehart.frieling@t-online.de

Dipl.-Volkswirt. Jan Fries is a researcher at the Department of Labour Economics, Human Resources, and Social Policy at the Centre for European Economic Research (ZEW). He studied economics at the University of Freiburg and the Free University Berlin, and started to work on his dissertation in 2009. His research focusses on topics in personnel economics and the evaluation of active labor market policy, e-mail: fries@zew.de

Dr. Kathrin Gimpel is head of the Personnel Development at the Corporate Center of Human Resources of the ThyssenKrupp AG in Essen/Germany. In 2006 she obtained a doctoral degree at the Karlsruhe Institute of Technology (KIT) (formerly Karlsruhe University). Having studied business engineering with corporate planning as a specialty, she was employed as a scientific assistant at the Ergonomics Department of the local Institute for Industrial Production (IIP). In research and in practice, she focused on ageing-appropriate qualification, diversity management, working in virtual structures, employability, and innovative working-time design, e-mail: kathrin.gimpel@thyssenkrupp.com

Prof. Dr. Jürgen Glaser is full professor of Applied Psychology at the University of Innsbruck (Ph.D. 1997, habilitation 2004). His research interests are in the field of work analysis and design, occupational health, and human service work, e-mail: juergen.glaser@uibk.ac.at

Prof. Dr. Ben Godde is professor of Neuroscience at the Jacobs Center on Lifelong Learning and Institutional Development of the Jacobs University Bremen (Ph.D. 1996) and co-speaker of the research center AgeAct—Interaction of Aging Processes. He is interested in mechanisms of adult cortical plasticity and its relationship to learning processes, particularly in the sensory system. A special focus of his research is on healthy aging. Ben Godde also published numerous journal articles in national and international journals as well as book chapters, e-mail: b.godde@jacobs-university.de

Prof. Dr. Christian Göbel is professor of economics with a specialization on labor market theory and statistics, labor and social policy at the University of Applied Labour Studies of the Federal Employment Agency, Mannheim (HdBA) and research associate at the Centre for European Economic Research (ZEW). He received his Ph.D. from the Université Catholique de Louvain in 2007. From 2006 to 2011, he was a researcher at the department of "Labour economics, human resources, and social policy" at ZEW. His research focusses on labor economics, policy evaluation and applied econometrics, e-mail: christian.goebel@arbeitsagentur.de

Dr. Anna Grube worked between 2007 and 2008 as research assistant at the Department of Work and Organizational Psychology, University of Würzburg, Germany, and 2008–2010 at the Department of Organizational and Business Psychology, University of Münster, Germany. She finished her Ph.D. thesis on age differences in explicit work values at the University of Münster, and now works in a large public ICT service provider organisation in Southern Germany, e-mail: anna.grube@uni-muenster.de

Dr. Melanie Hahn is research associate in the project group "Ageing" at the Leibniz Research Centre for Human Factors and Working Environment (IfADo) in Dortmund (Ph.D. 2008). Her main research interests concern the neurophysiology of cognitive functions in aging, especially executive processes and their influence on driving performance in young and old subjects, e-mail: hahn@ifado.de

Prof. Dr. Mathias Hegele is assistant professor of Sport Psychology and Motor Learning at the Justus Liebig University Gießen. He earned his Ph.D. in Experimental Psychology in 2009 after having been a doctoral student in the SPP since 2005. His research interests primarily pertain to the visuomotor control of manual actions, the effects of tool use on perception and action planning, and the role of the sensorimotor system in higher-level cognitive processes such as prediction and categorization (embodied cognition), e-mail: mathias.hegele@sport.uni-giessen.de

MD Barbara Heiden is a researcher at the Institute and Outpatient Clinic for Occupational, Social and Environmental Medicine, Ludwig-Maximilians-University Munich. Her research interests are psychosocial stressors, work-related musculoskeletal disorders and participatory interventions in the workplace, e-mail: barbara.heiden@med.lmu.de

Prof. Dr. Guido Hertel was Professor of Work and Organizational Psychology at the University of Würzburg, Germany (2004–2008), and since 2008 holds a chair in Organizational and Business Psychology at the University of Münster, Germany (Ph.D. 1995, habilitation 2002). In addition to demographic changes in organizations (editor of three special issues on this issue), his research interests are in the field of teamwork, electronic human resource management, negotiation and conflict management, and scientist-practitioner gap. His research has been published in Journal of Personality and Social Psychology, Journal of Management, Research Policy, Human Resource Management Review, Journal of Managerial Psychology, and European Journal of Work and Organizational Psychology (among others). Guido Hertel also works as consultant for organizations, e-mail: ghertel@uni-muenster.de

Prof. Dr. Herbert Heuer is Professor of Work Psychology and Experimental Psychology at IfADo—Leibniz Research Centre for Working Environment and Human Factors (Dr.rer.nat. 1978, habilitation 1984). His research interests are in the field of human performance (movement—perception—action), work-related variations of performance, work load, and occupational health, e-mail: heuer@ifado.de

Dr. Nicole Jochems studied computer science at RWTH Aachen University. From April 2005 since September 2009 she was a research assistant at the Institute of Industrial Engineering and Ergonomics at RWTH Aachen University. In 2010 she received the Dr.-Ing. degree from RWTH Aachen University. Since 2010 she is deputy head of division Ergonomics and Human–Machine-Systems. Her research interests include human–computer interaction, design of age-differentiated and age-robust user interfaces, and ergonomics, e-mail: n.jochems@iaw.rwth-aachen.de

Dipl.-Psych. Franziska Jungmann is a research assistant and doctoral student at the Institute of Work and Organizational Psychology at the Technical University of Dresden. She did her undergraduate studies in psychology at the Technical University of Dresden, Germany, and the University of British Columbia, Vancouver, Canada. Her research interests include age diverse teams, leadership of an aging work force, psychological stressors at work, trainings, and occupational health psychology, e-mail: jungmann@psychologie.tu-dresden.de

PD Dr. Dorothee Karl is Head of the Communal Policies of Employment at the Jobcenter Mannheim and coach for the support of work ability. She obtained a doctorate and qualified for professorship at the Karlsruhe Institute of Technology (KIT) (formerly Karlsruhe University), where she was employed as a scientific assistant and project manager in charge of demographic change, work ability, lifelong learning, aging-appropriate work design, and diversity in both research and practice. She completed a course in psychology at Mannheim, e-mail: dorothee.karl64@gmx.de

Prof. Dr. Karsten Kluth received his Dr.-Ing. degree in Ergonomics in 1996 from Siegen University. Thereafter, he was involved in large-scale laboratory and field research in the evaluation of the working conditions in a chain of grocery stores. Subsequently, he finished his habilitation thesis in 2000 on the basis of the development and evaluation of an ergonomically optimized scanner checkout. After several years of field and laboratory research in the field of working in deep cold at Darmstadt University, in 2008 he became Professor and Head of the Ergonomics Division in the Department of Mechanical Engineering at Siegen University, e-mail: kluth@ergonomie.uni-siegen.de

Prof. Dr. Peter Knauth is former head of the Ergonomics Department in the Institute for Industrial Production (IIP) and full professor at the Karlsruhe Institute of Technology (KIT) (formerly Karlsruhe University). His research focuses on working hours, human resource management (e.g. ageing workforces, diversity management, work-life balance) and new forms of organisation, e-mail: peter.knauth@web.de

M.Sc. Psych. Daniel Kotzab is research assistant of Work and Organisational Psychology at the University of Kassel. His research interests are in the field of work motivation, work satisfaction, and ergonomics. He has published five conference articles, e-mail: kotzab@uni-kassel.de

Dr. Stefan Krumm is a post-doctoral researcher at the Department of Organizational and Business Psychology, University of Muenster, Germany. He finished his Ph.D. 2005 at the University of Marburg, Germany. Before he joined the University of Muenster, he worked as a visiting scholar at the Educational Testing Service, Princeton, and as a senior consultant at a management consultancy. His research interests include virtual teamwork, the relevance of cognitive abilities in applied settings, and age-related differences in work values, e-mail: stefankrumm@uni-muenster.de

Prof. Dr. Kurt Landau was full professor of Ergonomics at Technische Universität Darmstadt and Head of the Institute of Ergonomics from 1996 to 2005 (Ph.D. 1978). His research interests include work and stress analysis as well as the ergonomic design of human machine interfaces, e-mail: landau@iad.tu-darmstadt.de

Dr. Bianca Leitner-Mai is a biologist. She is a research assistant at the Chair for Ergonomics at the Institute of Industrial Management and Factory Systems at the Chemnitz University of Technology since 2010 and is focused on age-differentiated work systems, e-mail: bianca.leitner-mai@mb.tu-chemnitz.de

Dipl.-Psych. Susanne Christina Liebermann is a doctoral student at the Institute of Work and Organizational Psychology at the Technical University of Dresden. She did her undergraduate studies in psychology at the University in Mannheim, Germany. Her research interests include processes of group identification, employability of elderly workers, retirement, and health, e-mail: liebermann@psychologie.tu-dresden.de

M. A. Mot. Dorothee Müglich studied motology at Philipps-Universität Marburg and holds a bachelor degree in physical therapy. She is a research associate at Institute of Ergonomics at the Technische Universität Darmstadt with the main focus on performance assessments. Her research interests are in the field of human performance and occupational health, e-mail: d.mueglich@iad.tu-darmstadt.de

Dr. Andreas Müller is a researcher in industrial and organizational psychology at the Institute of Occupational and Social Medicine at the Heinrich Heine University Düsseldorf, Germany (Ph.D. 2007). His research interests are in the areas of work design, employee well-being, aging at work and action regulation, e-mail: andreas.mueller@uni-duesseldorf.de

Prof. Dr. Jochen Müsseler is a full professor of Psychology at the RWTH Aachen University and Head of the Department of Work and Cognitive Psychology. He received his Ph.D. in psychology at the University of Bielefeld in 1986 and his postdoctoral (habilitation) degree in psychology at the University of Munich in 1995. His research interests are in the field of perception and action, selective attention, dual-task performance, stimulus-response compatibility and cognitive ergonomics, e-mail: muesseler@psych.rwth-aachen.de

Dr. Susanne Mütze-Niewöhner received her doctoral degree from Faculty of Mechanical Engineering at RWTH Aachen University. She heads the department of work organization at Institute of Industrial Engineering and Ergonomics, which focuses on analyzing, evaluating and designing of work systems and processes. From 2009 to 2010 she coordinated the Priority Program 1184—Age-differentiated Work Systems, e-mail: s.muetze@iaw.rwth-aachen.de

Dr. Mario Penzkofer studied Industrial Engineering at the University of Siegen. Since 2006, he has been working as a research assistant at the Ergonomics Division. In 2012 he received the Dr.-Ing. degree from the University of Siegen. Within the scope of his doctoral thesis he supervised the project "Age and Working in the Cold", funded by the German Research Foundation. In the course of the project he carried out extensive field studies to evaluate the effects of various cold exposures on human beings, e-mail: penzkofer@ergonomie.uni-siegen.de

Dr. Anja Philipp is post doctoral research scientist in Educational Psychology at the University of Frankfurt (Ph.D. 2010). Her research interests are in the field of self-regulation strategies of teachers and emotion regulation in the workplace, the development of self-regulation strategies over a teacher's career and occupational health, e-mail: philipp@paed.psych.uni-frankfurt.de

Dipl.-Wirtsch.-Ing. Holger Rademacher studied business engineering with mechanical engineering as technical field in Darmstadt. From 2005 to 2011 he worked as research assistant at the Institute of Ergonomics at the Technische Universität Darmstadt (IAD). At the IAD he worked in several industry projects related to production ergonomics research and was particularly engaged in the research project "Assistance System for Age Differentiated Work Design and Employee Assignment". His doctoral thesis (in preparation) focuses on the analysis of physical work exposures and employee capabilities as basis for an age-differentiated work design, e-mail: Rademacher@gesamtmetall.de

Dr. Miya K. Rand is a motor control specialist at IfADo—Leibniz Research Centre for Working Environment and Human Factors (Ph.D. 1993). Her research interests are mechanisms behind changes in movement control due to adaptation, learning, aging, and neurological disorders, e-mail: rand@ifado.de

Dipl.-Psych. Cornelia Rauschenbach is Ph.D. student at the University of Münster, Germany. From 2008 to 2012 she has worked as research assistant at the Department of Organizational and Business Psychology, University of Münster. Currently, she works at the German Federal Employment Agency. Her research interests are in the field of demographic change and occupational well-being, e-mail: rauschenbach@uni-muenster.de

Eva-Maria Reuter M.Sc. is research assistant in Human Performance at the Jacobs Center on Lifelong Learning and Institutional Development of the Jacobs University Bremen. Her research interest is in the field of motor control and sensory perception with a specific focus on the age-related differences of somatosensory processing and its neurophysiological correlates, e-mail: e.reuter@jacobs-university.de

Dipl.-Psych. Birgit Claudia Ries has worked as a scientist of Work and Organizational Psychology at the Leibniz-Research Centre for Working Environment and Human Factors at the Technical University of Dortmund; currently in further education as psychological psychotherapist at the Ruhr-University of Bochum. Her research focused on age diversity within work groups, e-mail: BC.Ries@web.de

Dipl.-Wirtsch.-Ing. Christian Scherf is a research assistant at the Chair for Ergonomics at the Institute of Industrial Management and Factory Systems at the Faculty of Mechanical Engineering at the Chemnitz University of Technology since 2007. From 2010 to 2012 he attended the research project focused on age-differentiated work systems and age-simulation, e-mail: christian.scherf@mb.tu-chemnitz.de

Prof. Dr. Christopher M. Schlick is full professor of Mechanical Engineering at RWTH Aachen University and Head of the Institute of Industrial Engineering and Ergonomics (Ph.D. 1999, habilitation 2004). He is a member of the management board of the Fraunhofer Institute for Communication, Information Processing and Ergonomics (FKIE). His research interests include modeling and simulation of work and business systems as well as the ergonomic design of human-machine interfaces, e-mail: c.schlick@iaw.rwth-aachen.de

Prof. Dr. Klaus-Helmut Schmidt is Professor of Work and Organisational Psychology at the Leibniz Research Centre for Working Environment and Human Factors at the Technical University of Dortmund. His research interests are in the field of self-control, stress and strain at work, and withdrawal behavior, e-mail: schmidtkh@ifado.de

Prof. Dr. Heinz Schüpbach is Head of the School of Applied Psychology of the University of Applied Sciences and Arts Northwestern Switzerland as well as honorary professor of Work and Organizational Psychology at the University of Freiburg (Ph.D. 1985, habilitation 1993), e-mail: heinz.schuepbach@fhnw.ch

Dipl.-Psych. Michael Sengpiel works as research assistant and doctoral student at the professorship of Engineering Psychology/ Cognitive Ergonomics at the Humboldt University Berlin. His research interests are in the field of human–computer interaction, usability engineering, user centered design, computer literacy and design for older adults, e-mail: michael.sengpiel@psychologie.hu-berlin.de

Dr. Meir Shemla is a postdoctoral researcher at the Institute of Work and Organisational Psychology at the Technical University of Dresden. His research covers topics such as diversity in organisational teams, team leadership, emotions, and values, e-mail: shemla@psychologie.tu-dresden.de

Dr. Andrea Sinn-Behrendt is a research associate at Institute of Ergonomics at the Technische Universität Darmstadt since October 2000. Her profession is medical practitioner and she completed her post-graduate education in internal and occupational medicine. Her research interests concern work-related musculoskeletal disorders and rehabilitation, e-mail: sinn@iad.tu-darmstadt.de

Dipl.-Psych. Malte Sönksen is research assistant at Professorship of Engineering Psychology/Cognitive Ergonomics at Humboldt-Universität zu Berlin. His interests are in the field of usability engineering, especially user research and requirements analysis, e-mail: malte.soenksen@psychologie.hu-berlin.de

Prof. Dr. Birgit Spanner-Ulmer held the Chair for Ergonomics at the Institute of Industrial Management and Factory Systems at the Chemnitz University of Technology from 2004 to 2012. The main research and teaching topics were related to product development and product manufacturing. The professorship's activities focus on the topics of ergonomic product design, ergonomic process design and the creation of innovative working environments, which are characterized by interaction between human, technology and organization, e-mail: awi@tu-chemnitz.de

Prof. Dr. Christian Stamov-Roßnagel is Professor of Organisational Psychology at Jacobs University Bremen (Ph.D. 1994, habilitation 2001). His research interests include work-related learning, work motivation, and innovative work behaviour. He also works as a consultant to companies such as Bosch, Michelin, and Vodafone and is Academic Director of Jacobs University's Next Generation Leadership executive education programme, e-mail: c.stamovrossnagel@jacobs-university.de

Prof. Dr. Helmut Strasser has been Head of the Ergonomics Division at Siegen University until retirement in 2007. Topics of main interest were developing and applying work-physiological methods. Main topics of applied ergonomics field research in the last years were analysis, assessment and design of industrial working conditions and workplaces in the self-service area. For two 6-year terms he was a member of the Executive Committee of the German Ergonomics Society (GfA), and from 2007 to 2009 he served as President of GfA. From 2004 to 2008, he acted as main reviewer of the German Research Foundation. Helmut Strasser edited several books and contributed to ergonomic textbooks. He published more than 400 papers in peer-reviewed journals and Conference Proceedings. In 2006, he received the IEA Fellow award, e-mail: h.strasser@aws.mb.uni-siegen.de

M.A. Alina Sytch is research assistant of Work and Organisational Psychology at the University of Kassel. Her research interests are in the field of work motivation, healthy leadership, and teamwork. She has published four conference articles, e-mail: a.sytch@uni-kassel.de

Dipl.-Psych. Markus M. Thielgen is research assistant and Ph.D.-student at the Department of Organizational and Business Psychology, University of Münster, Germany. His research interests are in the field of implicit, explicit and work-related motives, as well as person-environment fit in organizations, with a special focus on age differences, e-mail: Markus.Thielgen@uni-muenster.de

Dr. Sebastian Vetter studied psychology at the Universities of Maastricht, Netherlands, and Aachen, Germany. He received the Dipl.-Psych. degree from RWTH Aachen University in 2007. Between 2007 and 2012 he worked as research assitsant at the Institute of Industrial Engineering and Ergonomics at RWTH Aachen University. In 2012 he received the Dr.-phil. degree from RWTH Aachen University. His research interests include ergonomics and human-computer interaction, e-mail: s.vetter@iaw.rwth-aachen.de

Solveig Vieluf M.Sc. is research assistant in Human Performance at the Jacobs Center on Lifelong Learning and Institutional Development of the Jacobs University Bremen. Her research interest is in the field of motor control and motor learning with a focus on age- and expertise-related differences in fine motor control as well as its neurophysiological correlates, e-mail: s.vieluf@jacobs-university.de

Prof. Dr. Claudia Voelcker-Rehage is full professor of Human Performance at the Jacobs Center on Lifelong Learning and Institutional Development of the Jacobs University Bremen (Ph.D. 2002). Her research interests are in the field of motor learning and motor control across the lifespan and the plastic-adaptive capacities of older persons as well as the interaction between sensorimotor and cognitive performance. She has published numerous journal articles in national and international journals and book chapters related to these topics, e-mail: c.voelcker-rehage@jacobs-university.de

Dipl.-Wirtsch.-Ing. Martin Waldherr was from 2006 until 2012 research assistant at the ifab-Institute of Human and Industrial Engineering of the Karlsruhe Institute of Technology (KIT; formerly University of Karlsruhe). He is now working in industry. His research interests were in the field of industrial engineering, simulation, and demographic development

Prof. Dr. Hartmut Wandke is full professor of Engineering Psychology/Cognitive Ergonomics at Humboldt Universität zu Berlin (Ph.D. 1979, habilitation 1988). His research interests are in the field of human–computer interaction, usability engineering, human–machine-systems, user centred design, e-mail: hartmut. wandke@psychologie.hu-berlin.de

Prof. Dr. Jürgen Wegge is full professor of Work and Organisational Psychology at the Technical University of Dresden (Ph.D. 1994, habilitation 2003). His research interests are in the field of work motivation, leadership, occupational health and excellence in organizations. He has published three books, three special issues, 40 journal articles and 66 book chapters related to these topics, e-mail: wegge@psychologie.tu-dresden.de

Dr. Matthias Weigl is a researcher in industrial and organizational psychology at the Institute and Outpatient Clinic for Occupational, Social, and Environmental Medicine of Munich University, Germany (Ph.D. 2006). His research interests are in the areas of work design, employee well-being, performance, and quality in health care organizations, e-mail: matthias.weigl@med.lmu.de

Prof. Dr. Nele Wild-Wall is full professor of Psychological Research Methods and Diagnostics at the Rhein-Waal University of Applied Science (Ph.D. 2004). Her research interests concern the neurophysiology of cognitive aging, attentional and motor processes. She has published 20 journal articles and book chapters related to these topics, e-mail: nele.wild-wall@hochschule-rhein-waal.de

Prof. Dr. Gert Zülch was from 1985 until 2012 full professor and director of the ifab-Institute of Human and Industrial Engineering of the Karlsruhe Institute of Technology (KIT; formerly University of Karlsruhe). His research interests were in the field of communication ergonomics in production, computer supported design of assembly and disassembly systems, work structuring, personnel-centered simulation, modeling, and organization of production and service companies and strategies of production planning and control, e-mail: gert.zuelch@kit.edu

Prof. Dr. Thomas Zwick is full professor of Human Resource Management and Organization at the University of Würzburg (Ph.D. 1998 University of Maastricht, Habilitation 2006 University of Zurich). He was, at the time of the research presented in this book, professor for human resource management at Ludwig-Maximilians University, Munich (LMU) and researcher and later research associate at the Centre for European Economic Research, Mannheim (ZEW). His main fields of interests are microeconomic and micro econometric labor market analyses, qualification research and personnel economics, e-mail: thomas.zwick@uni-wuerzburg.de

Age-Differentiated Work Systems: Introduction and Overview to a Six-Year Research Program in Germany

Christopher M. Schlick, Ekkehart Frieling and Jürgen Wegge

The disproportionate aging of the populations in many nations around the world is a unique occurrence in the history of humankind. It has a major impact on the working population, and thus on the age structures in companies. This is because such aging of the population usually leads to disproportionate aging of employees in organizations, while ever fewer young people are available in the employment/work sector. Furthermore, the percentage of individuals of working age declines in the population as a whole.

Fertility and mortality have a major influence on the developments referred to as demographic change. The fertility rate states the average number of live births per woman aged between 15 and 44. It is also referred to as the birth rate. The mortality rate expresses the number of deaths in relation to the overall population. A decrease in fertility leads to a declining percentage of younger people in the population as a whole and in the working population, while declining mortality is associated with growth in the group of older people and older employees. Life expectancy is often used in place of the mortality rate as a measurement of mortality. The development is inversely proportional—a decrease in the mortality rate is associated with higher life expectancy.

C. M. Schlick (✉)
Institute of Industrial Engineering and Ergonomics, RWTH Aachen University, Aachen, Germany
e-mail: c.schlick@iaw.rwth-aachen.de

E. Frieling
Work and Industrial Psychology, University of Kassel, Kassel, Germany
e-mail: ekkehart.frieling@t-online.de

J. Wegge
Institute of Work, Organizational and Social Psychology, Technical University of Dresden, Dresden, Germany
e-mail: wegge@psychologie.tu-dresden.de

C. M. Schlick et al. (eds.), *Age-Differentiated Work Systems*,
DOI: 10.1007/978-3-642-35057-3_1, © Springer-Verlag Berlin Heidelberg 2013

In the light of demographic change, it is becoming increasingly important to develop and use the potential of older employees. In the future, fostering, preserving, and making appropriate use of the knowledge, skills and abilities of older employees will be a major objective for any company or firm. The focus will be on topics such as age-specific motivation and skill development as well as designing tasks, equipment and tools ergonomically, and establishing appropriate working hours for older and aging employees.

Demographic Developments and Their Effects on the Working Population

Demographic Developments in Germany

It is expected that the current population of just under 82.4 million in Germany will decrease to between 65 and 75 million by 2050 (Eisenmenger et al. 2006).

The fertility rate in Germany has declined almost continuously since peaking during the baby boom of the mid-1960s. In fact, the mortality rate has exceeded the birth rate since the beginning of the 1970s. In 2009, life expectancy in Germany was 77.8 years for men and 82.8 years for women, slightly above the EU average. Compared with 1993, life expectancy at birth has increased by five years for men and by 3.5 years for women (European Commission 2010).

The development of life expectancy and the decline in fertility for the period from 1995 to 2050 will be almost diametrically opposed linear processes. This means that there will be a long-term change in the age structure in Germany. The number of people over 65 will rise from 16 million in 2008 to over 22 million in 2030 to around 23 million by 2050. At that point in time, every third inhabitant of Germany will be 65 or older (Eisenmenger et al. 2006).

If these demographic developments continue, fewer people will live in Germany in the future, and the percentage of older people in the population as a whole and in the working population will increase significantly. The dependency ratio, which expresses the ratio of people of working age to those of non-working age, is an important indicator of the development of the working population. Since the 1960s, the age dependency ratio has constantly increased. Forecasts show that it will increase further in the coming decades. This means that the number of workers in Germany will decrease from 41.9 million in 2000 to 29.6 million in 2050 (Cologne Institute for Economic Research 2005). A rapidly declining number of people of working age can be expected as early as 2020. This will have a particularly strong impact on the number of young skilled workers available. Increasing "aging of society" will have the following effects on the age structure of the employment and working sector: By 2020, the average age of people of working age will have already risen by 2.2 years in Germany. While the highest percentage of labor potential consisted of people aged between 30 and 45 in 2000,

50–60 year-olds will make up the largest group in 2020. The aging process described in this book will primarily occur during the next 10–15 years.

International Demographic Developments

In contrast to the population in Germany, the global population will continue to grow in the future. The latter, which currently stands at seven billion, is forecast to rise to 9.6 billion by 2050 (DSW 2011). Almost all of this population growth will occur in developing countries. Asia is the continent where this development will be most evident. Huge growth can be expected in India in particular. It is expected that India will have a population of over 1.6 billion by 2050, thus replacing China as the country with the largest population (DSW 2011).

Figure 1 shows the fertility rate as a major factor in population growth for Europe, the United States and Asia. The countries with the highest fertility rates worldwide are in Africa and western Asia (United Nations 2004). Unlike other developed regions, the population aging process in the U.S. is not as pronounced. This is due to a higher fertility rate and slightly lower life expectancy—particularly in comparison with European countries. According to a recent report of the European Commission (2010), the fertility rate in Europe has been below 1.5 children per woman since 1995.

Most of the countries with a very high life expectancy in global terms are found in Europe (United Nations 2004, see Fig. 2). In 2009, life expectancy at birth in EU countries was 76.4 years for men and 82.4 years for women in 2009. In the U.S., life expectancy at birth was 75.9 years for women and 80.9 years for men in 2011. Life expectancy is thus in the top third worldwide. Life expectancy at birth in Asia was 68.9 years from 2005 to 2010.

The United Nations forecasts that fertility and mortality rates in the developing regions will evolve similarly to those of the industrialized nations, albeit at a slightly later stage (see Figs. 1 and 2) so that all countries and regions worldwide will be affected by population aging in the end (United Nations 2004).

Fig. 1 Development of the fertility rate between 1950 and 2010 in Europe, the United States and Asia. Data according to United Nations Department of Economic and Social Affairs, Population Division (2011). World Population Prospects: The 2010 Revision

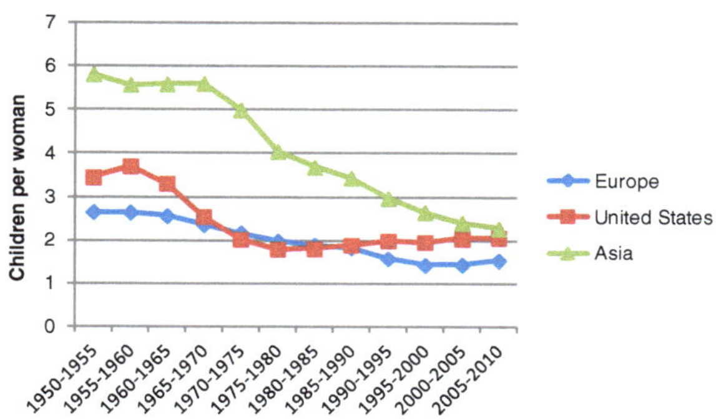

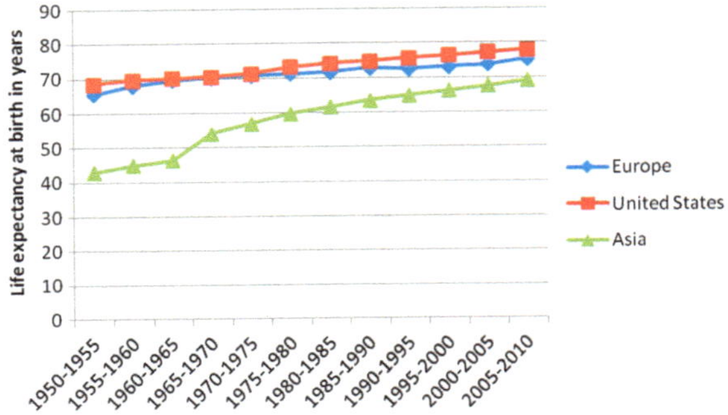

Fig. 2 Development of the life expectancy at birth between 1950 and 2010 in Europe, the United States and Asia. Data according to United Nations, Department of Economic and Social Affairs, Population Division (2011). World Population Prospects: The 2010 Revision

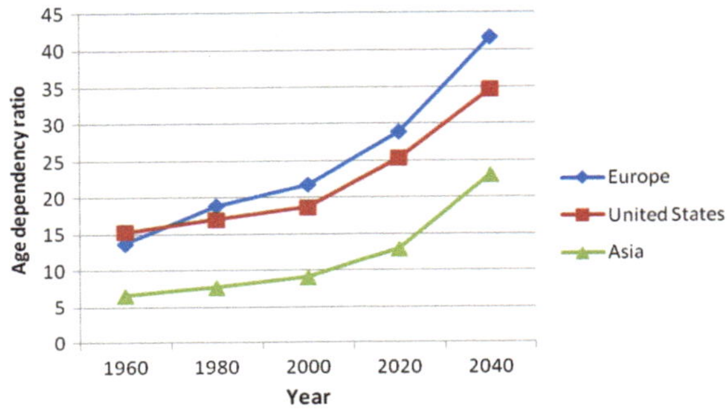

Fig. 3 Development of the age dependency ratio (ratio of population aged 65+ per 100 population 15–64) between 1960 and 2040 in Europe, the United States and Asia. Data according to United Nations, Department of Economic and Social Affairs, Population Division (2011). World Population Prospects: The 2010 Revision

The development of age distribution among the global population in the past decades shows that the percentage of younger people in the population as a whole is declining (the share of those aged between 0 and 14 will decrease by 10 % between 2000 and 2050), while the group of older people is constantly growing (the share of people over 65 will increase to 15.9 % between 2000 and 2050).

These developments have an impact on the previously mentioned dependency ratio (Fig. 3). An increase in the age dependency ratio can be observed not only in developed regions such as Europe and the U.S., but also—albeit at a lower level—in developing regions like Asia. In 2040, only slightly more than two people from the group of people of working age will provide for one person from the group of people over 65 (United Nations 2011). A high percentage of children among the overall population leads to a high dependency ratio in developing countries, while the increasing number of old people causes this ratio to rise in developed regions.

Implications for Hiring Persons, Designing Human Work, and Conducting Research in the Field of Ergonomics

Forecasts indicate that the population aging process in industrialized nations will accelerate during the first decades of the 21st century. This means that there will be ever fewer younger workers in the employment system, while older workers will become increasingly important for companies' economic and social development. Simultaneously, there is a trend for people to retire early. The combination of both developments means that companies have significant difficulties in meeting their needs for qualified, motivated and capable persons. In the future, many companies will face the challenge of fostering, preserving and making even more effective use of the knowledge, skills and abilities of their increasingly older employees under the existing competitive market conditions.

In this context, the fact that demographic change is a global phenomenon cannot be ignored. While the aging process will initially accelerate in the industrialized nations, in the long-term it will be felt worldwide.

In geographical terms, the regionally diverging parameters and effects of demographic change should not be discounted either. In other words, the population aging process must be viewed differently for different regions. Regional economic strength and job supply have an impact on these developments. For example, the trend is for young people to migrate from structurally weak regions. The average age of the population rises, while the number of potential employees declines. As a result, there is a greater focus on older employees in the recruitment of new staff.

Apart from the challenge facing companies of fostering, preserving and making appropriate use of the knowledge, skills and abilities of older employees, the complementary task of improving the quality of working life of younger persons through preventative, prospective work design must also be taken into account. Successfully designing work so that it is suited to older or aging employees requires knowledge of (1) the employee and the development of his/her organismic, perceptive, cognitive, sensorimotor and motivational "sub-systems" and knowledge of the effects of stress and strain resulting from (2) specific work tasks, (3) the equipment and tools used at work and (4) the work environment in terms of human performance and reliability. The following questions arise as a direct result:

- How should tasks be structured in an age/aging-appropriate way so that they support the use of individual optimization and compensation strategies as regards performance and health?
- What principles and rules must be used when designing work equipment and tools so that they can be used effectively, efficiently, safely, and with an appropriate level of strain, by older and younger users?
- How should the work environment be designed so that the intensity of environmental factors does not cause harm and, in the case of older people who may not be able to withstand as much stress, does not lead to unacceptable levels of strain?
- How can companies deploy employees appropriately in terms of age and aging in order to keep them in the work systems longer and promote intergenerational teamwork?

The aim of the six-year Priority Program 1184—Age-differentiated Work Systems was to answer these questions.

Age-Differentiated Work Systems

The Priority Program 1184—Age-differentiated Work Systems was launched by a pluridisciplinary team of researchers and funded by the German Research Foundation (DFG). The program began in fall 2005 and was set up for a six-year period. During these six years, a total of 19 universities and non-university research institutions from all over the Federal Republic of Germany were involved in the program. The research findings of 17 project teams are presented in this edited volume of the same name in condensed form. Project teams from different work-related disciplines, such as occupational psychology and ergonomics, worked on different yet interrelated relevant research tasks.

The priority program's research objective was to develop models and methods that companies could use to design and optimize their work systems and offer their employees age/aging-appropriate working and learning conditions. The aim was that the models and methods would have a sound theoretical basis in terms of their scientific derivation and form of explanation. Furthermore, they were to be empirically sound with regard to the effectiveness of the interventions based on them. Over 40 laboratory experiments involving 2,000 participants and 50 field studies involving over 25,000 employees were conducted under the priority program. The idea was that the findings from this research would help companies detect and avoid errors in their assessment of the capabilities and motivation of older workers and create objectively good working conditions.

In order to achieve the main research objective, there was continuous coordination and regular discussions between the individual projects. Findings were presented and discussed, and cross-project and cross-disciplinary cooperation networks were established at regular meetings. Apart from discussion and cooperation with research groups/initiatives at the national level, as well as with various organizations (e.g., automobile manufacturing, fiscal authorities, financial services, and schools), a large number of activities was undertaken to make the concepts, methods and results of the priority program also available on an international level. These included writing journal articles (e.g., for the Journal of Applied Psychology, Journal of Motor Behavior, and Journal of Industrial Ergonomics), publishing special issues [Occupational Ergonomics (2010); Zeitschrift für Arbeitswissenschaft (2006, 2010); Zeitschrift für Personalpsychologie (2009); Wirtschaftspsychologie (2008); Journal of Managerial Psychology (in press)], and conducting sessions at international conferences [(e.g., the International Congress of Psychology 2008; European Association of Work and Organizational Psychology 2007, 2009, 2011; Conference on Applied Human Factors and Ergonomics 2010; International Symposium of Human Factors in Organizational Design and Management 2011; and World Congress of the International Ergonomics Association 2012]. Over the six-year program period, the wide range of the research areas combined with various theoretical and methodological approaches improved the generalizability of the research findings and their transfer to and application in corporate practice.

The priority program focused on the design and evaluation of age-differentiated work systems. In the following section, the concepts "age-differentiation" and "work system" will be explained in greater detail as well as in relation to each other and discussed in the context of the objectives of the priority program.

Conceptual Framework

Since the system approach is a general one, the work system concept does not initially imply a particular level for viewing work processes—in other words, it can refer equally to an individual job or to an entire company. However, it generally refers to the former level and to the operation and movement of tools and at machines involved in a job. Depending on the question posed, the observed structure of the work system can be differentiated in various ways. However, it always includes a person and a task as a minimum requirement (Rohmert 1983). In general, a work system can be described using the following elements: person, task, equipment and objects. Furthermore, physical and social environmental factors must be considered. A work system can also be linked with other systems through input–output relationships. This provides a structured scheme for describing most work structures systematically.

The priority program viewed the work system elements listed above as part of a whole and analyzed their interdependencies. In addition, the program observed the diverse interactions between individuals, focusing on factors such as strategies for regulating emotions, motivation and actions. Person-centered structuring of age-differentiated work systems implies that the opportunities and challenges for an individual's capabilities and motivation to change with age are considered. At first sight, the demand for age-differentiated work systems is highly ambivalent, since it tends to imply that older people in particular need "different" working conditions from younger people due to changing performance and strain profiles. "Different" often means tasks with greater levels of mechanization or automation in terms of ergonomics or tasks that place lower physical, perceptive or cognitive demands on workers. In this context, the term "age-differentiated work systems" could imply a certain stigmatization of old age due to the general differentiation that it makes between the performance and well-being of younger and older workers. Obviously, this should be rigorously avoided in practice, firstly because objective findings partly contradict this assumption and secondly because it lowers the acceptance of interventions derived from methodology. For example, more recent research shows that there is no systematic relationship between a person's calendar age and his/her work performance (Ng and Feldman 2008; Wegge et al. 2008). In recent years, there have been significant changes in the explanatory and forecasting models that deal with notions of aging people's performance and strain. While earlier models presumed that the only changes would be for the worse (deficit models), that is, they assumed that a constant process of decline starts from the age of 28 (Naegle 2004), the compensation or competence model is now used to describe changes in older adults. These models assume that behavior, experience and strategies can compensate for the changes caused by age and that older adults have a command of a qualitatively more differentiated performance spectrum than younger people (Astor et al. 2006; Ng and Feldman 2008; Martin and Kliegel 2010). The assumption of a "differential age", which assumes different developments in human functions during the aging process, has increasingly replaced the hypotheses derived from the deficit model. Information-processing subsystems are assumed to have multiple resources that can be used differently in term of quality. Aging processes are understood differentially and in a far more complex manner in terms of their development (Czaja 1997; Brandstädter 2007). In this way, differing inter-individual aging processes can be described and explained. The extent of the changes can vary. In addition, they can occur at different times and proceed in various directions (Birren and Schaie 2006; Maintz 2002). Hence, it cannot be assumed that performance generally declines with increasing age, but rather that changes occur in mental and physical abilities. On one hand, this variability is caused by the fact that performance differs both among individuals, that is, between different people (inter-individually), and in a single person (intra-individually). For example, it can depend on a person's level of training or current state of health. On the other hand, increasing variability, particularly in older people, also results from the growing risk of deteriorating health. In practice, it is often difficult to differentiate between "normal" (primary) and "pathological" (secondary) aging.

An initial conclusion from the above observations is that the rules for designing work systems cannot apply universally, but rather must take intra- and inter-individual differences into account. Further analysis shows that while certain measures for designing work systems can have a significant positive impact on younger workers, the effects on older workers can differ in terms of effectiveness and even in terms of direction. However, ergonomic working conditions specially developed for older workers can often be of benefit to younger workers too, and can help improve well-being. In this context, one speaks of a design-for-all approach.

The priority program researchers examined selected work system elements in terms of both their age-differentiated impact and their interactions. In this context, the focus was not on "engineered" age-differentiated design, but on the type and extent of the impact of work-related factors on changes in performance and motivation. The idea was that appropriate work design measures should be used to adapt technical, organizational and social conditions to people so that companies can provide harmless, feasible, bearable and unimpaired working conditions and make individual development possible (Luczak and Volpert 1987). Various empirical studies have shown that it is not feasible to have standardized "optimum" work structures and processes for all workers (Zink 1978; Triebe 1980, 1981). In this respect, the research on work systems was age-differentiated, for example when the objective was to answer questions about distributing tasks among workers in terms of stress and strain (such as which forms of work are particularly suited to younger people and older workers respectively) or when the research involved the age-differentiated design of work processes and equipment (such as which organizational, technical and social conditions support younger and older workers' performance, motivation, and health). The findings can be used to adapt existing work structures and processes retroactively to the requirements of human work. This involves a correcting or corrective type of work design. The findings can also be applied when designing new (differential) work systems. The aim of this preventative strategy is to avoid damage and impairment to health from the outset and to take ergonomic, technical, organizational, and economic requirements into account in the planning and design phase (see Ulich 2005).

The priority program's analysis of a work system's individual elements and their interaction was age-differentiated, and in some cases it also used other criteria such as skills (analysis level). This part of the priority program involved an individual approach that can be used to develop design measures that can prevent existing or future work-related declines in performance and impairments to health. The individual findings and solutions were subsequently summarized in the work design process as an overall solution (synthesis). At the synthesis level, the research projects derived criteria, principles and rules for designing work structures and processes. This derivation followed the preventative work design strategy. It can be described as aging-appropriate work design that takes work-relevant skills and capabilities and the resulting ergonomic requirements into account. Aging employees are not the only ones affected by such measures; designing work systems in terms of prevention can help to forestall a loss of work and employment capacity among younger workers too. The aim is to maintain and further develop employees' perceptive, motor, motivational, emotional and social-communicative skills throughout their entire working life.

The overall objective was that the priority program's findings would provide a valid, methodologically sound foundation for the integrated design of aging-appropriate work systems. Such systems are age-differentiated or age-specific where economic, technical or organizational constraints and limitations prevent the implementation of ergonomic working conditions for all groups of employees, where workers' performance has already been curtailed, or where improvements in efficiency, effectiveness, and occupational health and safety only result from the inclusion of age-specific factors. However, cohort effects must be taken into account when applying the priority program's findings to real-life work systems. These effects are closely related to the analytical methods used (Birren and Schaie 2006; Schaie and Hertzog 1983). In the priority program, both longitudinal and cross-sectional studies were conducted; in the latter, cultural and social differences between the cohorts deserve special consideration when interpreting the findings.

Designing age-differentiated work systems—taking into account the work system's individual elements—not only involves hardware and software ergonomics, but the design of tasks and specific work time models, mixed age groups, teamwork, and aspects concerning age-specific motivation and competence development. The focus here is on differential work design that goes beyond an inclusion of inter-individual differences and stipulates the simultaneous availability of various structures that individual workers can choose from. In terms of age-differentiated work system design, this implies the individualization of tasks, equipment and compensation strategies—provided that the design measures affect younger and older workers differently. The priority program's primary research topic involved the question of what type and scope of age-differentiation the design measures would require. To what extent and in what form should work systems be designed in terms of age-differentiation? Which design measures are particularly suited to younger and older workers respectively? What are the limits of age-differentiated design and when should it be supplemented or replaced with "age-robust" design in the sense of "design for all"? The individual chapters in this book address the debate over age-differentiated and aging-appropriate design of work systems, answer the questions raised at various levels of observation, and consolidate the findings in the form of "lessons learned".

Structure of the Priority Program

Design and intervention strategies at the micro- and macro-ergonomic level are essential to the development of sustainable concepts for age-differentiated work systems in companies. As a result, the 18 subprojects integrated over the course of the program period utilized both of these levels.

The focus of the subprojects ranged from economic analyses of productivity among older employees in different sectors to analyses of stress and strain at the workplace level to observations of the impact of working in cold conditions on older workers' thermoregulation. The subprojects' research topics and questions were grouped based on the level model by Luczak and Volpert (1987). Seven levels were differentiated:

1. Sectors and Value Networks
2. Enterprises and Companies
3. Cooperation in Work Groups
4. Holistic Activities and Work Forms
5. Tasks and Workplaces
6. Sensorimotor Control of Tools
7. Autonomous Organismic Systems and the Work Environment

In some cases, the projects could clearly be assigned to a particular level, while in others they addressed the overlaps or junctions between two levels.

1. *Sectors and Value Networks*

Productivity, employability, and corporate age structure mutually influence each other. At the first level, these interdependencies were examined from an overarching economic perspective on different sectors and the underlying value networks. Among other things, the effects of different staff age structures on productivity combined with various types of work systems, human resources measures, and corporate structures were estimated empirically on the basis of micro-econometric procedures (see chapter Age-Differentiated Work Systems Enhance Productivity and Retention of Old Employees, Zwick et al.).

2. *Enterprises and Companies*

The focus at the second level was on researching how workers could be deployed in companies in an age-appropriate way in order to keep them in the business processes for a longer time.

Under certain circumstances, the strain of an identical stress on older workers is greater than that on younger workers. The design of working hours plays a particular role here. The second level focused on the impact of different working-time models on aging workers in companies. It took into account various corporate variables such as different shift systems and approaches to work design, for example as regards the division of labor. Both short-term and long-term effects at the corporate and individual level were examined.

Furthermore, the analyses concentrated on sensitizing senior managers and on developing relevant aging-appropriate human resources deployment strategies, taking into account age-appropriate performance and stress. Staff qualifications and the way that human resources are deployed are major variables in age-appropriate work system optimization (see chapters Development and Evaluation of Working-Time Models for the Ageing Workforce: Lessons Learned from the KRONOS Research Project, Knauth et al.; Effects of an Ageing Workforce on the Performance of Assembly Systems, Zülch et al.).

3. *Cooperation in Work Groups*

When examining the deployment of humans in work systems, the different abilities and skills must be taken into account. Equally, social and communication needs must be considered in terms of organizational aspects of cooperative work. Social needs and individual performance are closely connected. Examples include situations when team work or group work leads to a positive working atmosphere, thus simplifying cooperation processes or improving performance. The priority program concentrated on cooperation between people in work groups. The focus was on age diversity in the work groups and its impact on performance and health (see chapter Age Diversity and Team Effectiveness, Ries et al.).

4. *Holistic Activities and Work Forms*

In order to be able to design work systems in an age-differentiated way, the person-centered design process should begin with an investigation of which strategies the workers for whom the system is being designed use to regulate their emotions, motivation and actions. Hence, the fourth level involved subprojects that investigated, among other things, the effects of emotion regulation on health. However, the extent to which workers are prepared to exploit their abilities depends on their motivation. Motivation, which is also referred to as drive regulation, is physiologically determined by the stimulation level of organs or organ systems. Psychologically, it is determined by attitudes to performance and motives such as needs, interests, intentions or convictions (Schlick et al. 2010). During the priority program, age differences in drive regulation were examined in terms of career-related motives, job satisfaction, emotional resilience and the use of control strategies in pursuing career goals.

At the fourth level, the research on emotion and drive regulation was extended to include findings on action regulation. Psychological gerontology shows that the individual level of function can be maintained if available resources are deployed effectively using selection, optimization and compensation (SOC) action strategies (Baltes and Baltes 1990). Another important theoretical development is the Socioemotional Selectivity Theory (SST, Carstensen 2006) which predicts that older workers de-prioritize goals related to growth and advancement and instead emphasize social and affective values. For the priority program, the focus was not only on examining the link between SOC strategies and the health and performance of older workers but also on the age-related differences in work motivation and control strategies while pursuing career goals (see chapters Age Differences in Motivation and Stress at Work, Hertel et al.; Age-Related Differences in the Emotion Regulation of Teachers in the Classroom, Philipp and Schüpbach; and Successful Aging Strategies in Nursing: The Example of Selective Optimization with Compensation, Müller et al.).

5. *Tasks and Workplaces*

Tasks and the conditions under which they are carried out influence capabilities and motivation. At the task and workplace levels, the priority program primarily examined the influence on capabilities. This was supplemented by analyses of physical and mental stress and strain. In this context, the research projects focused on the aging-appropriate design of work systems in automobile assembly and on system design in the realm of vehicle driving. In the former area, the effects of "Takt"-driven, short-cycle assembly work on employees was examined, with age taken into account. Specific load situations were analyzed. Combined with the skills profiles, these situations justify age-differentiated work design in the relevant production systems. The research projects aimed to provide fundamental insights into a balanced, age-differentiated load profile and the strain related to it in order to maintain older workers' capacity for employment and work. The specific research on driving included an examination of the effects of complex visual-motoric dual tasks, such as driving with an additional secondary task, in terms of age-related differences. Here the aim was to provide recommendations for aging-appropriate design of traffic situations and technical aids. In addition, recommendations can be derived for the optimum design of driving training for older drivers (see chapters Assembly Tasks in the Automotive Industry: A Challenge for Older Employees, Frieling et al.; Capability Related Stress Analysis to Support Design of Work Systems, Rademacher et al.; Field Study of Age-Critical Assembly Processes in the Automotive Industry, Börner et al.; Age-Related Differences in Critical Driving Situations: The Influence of Dual-Task Situations, S-R Compatibility and Driving Expertise, Aschersleben et al.; and Age-Related Changes of Neural Control Processes and their Significance for Driving Performance, Hahn et al.).

6. *Sensorimotor Control of Tools*

The fundamental principles of elementary physical (e.g., movement coordination) and mental (e.g., memory capacity) functions were examined at the sixth level. This level of the priority program consisted of subprojects concerning various aspects of human information processing, particularly perception, cognition, action selection and action execution in the interaction between humans and machines.

The focus was on the analysis, evaluation and age-differentiated design of interactive systems, as well as on the development of training concepts for teaching and supporting older users.

In addition, age-related changes and their influence on dexterity, and transformed movements that occur, for example, in the use of certain hand-held tools, were examined at the action execution level. The analyses focused on the influence of age-differentiated factors on movement coordination and motor performance, as well as on the learning ability of older workers in particular (see chapters Integrating Training, Instruction and Design into Universal User Interfaces, Sengpiel et al.; Ergonomic Design of Human–Computer Interfaces for Aging Users, Schlick et al.; Age-Related Variations in the Control of Electronic Tools, Heuer et al.; and Influence of Age and Expertise on Manual Dexterity in the Work Context: The Bremen-Hand-Study@Jacobs, Voelcker-Rehage et al.).

7. *Autonomous Organismic Systems and the Work Environment*

The seventh level examined fundamental thermodynamic variables of the work environment and its effects on autonomic body functions such as thermoregulation. The focus was on measuring and evaluating climate as regards work in cold conditions. On average, the analysis was conducted for temperatures below -20° C for the entire workday. Since energy metabolism declines with age due to the decrease in the basal metabolic rate as part of the total metabolic rate, people's ability to protect themselves from hypothermia automatically deteriorates with age. Hence, guidelines for the ergonomic design of work in cold conditions were developed during the priority program in order to provide preventative industrial safety for all age groups and to safeguard the effectiveness and efficiency of the work (see chapter Physiological Responses of Two Male Age Groups to Working in Deep Cold and Subjectively Experienced Stress and Strain, Kluth et al.).

Structure of the Book

In the following section, the seven levels for examining work processes outlined above (cf. Luczak and Volpert 1987) serve as a classification system for the priority program's individual projects and determine the sequence of the individual chapters in terms of content. The chapters start with the macro-ergonomic level

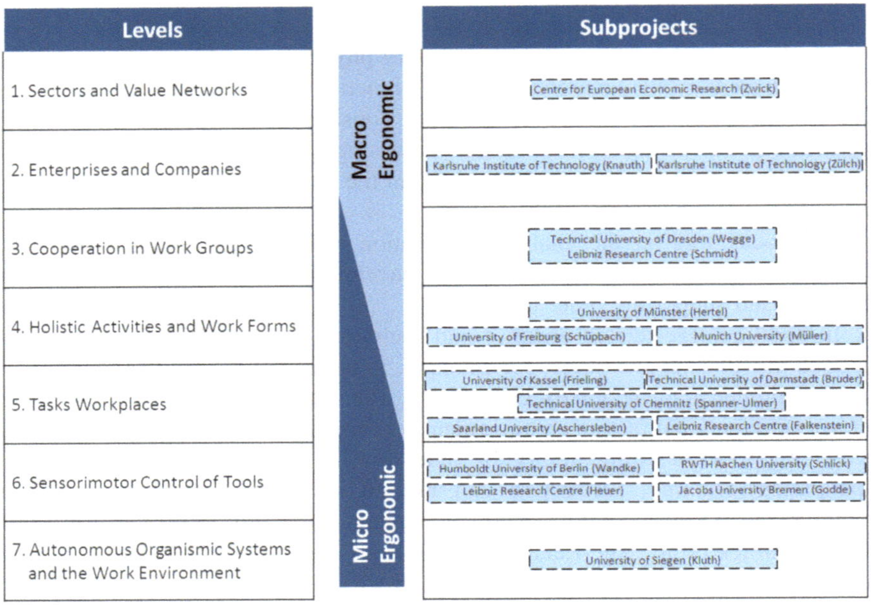

Fig. 4 Levels of the Priority Program 1184 and structure of the edited volume. On the right hand side the research institutions as well as the principal investigators are listed

and then move to the micro-ergonomic level. The levels of the priority program and structure of the edited volume are shown in Fig. 4.

1. The research project conducted by the Centre for European Economic Research and LMU Munich (headed by Prof. Thomas Zwick and Prof. Bernhard Boockmann) formed part of the first level of the priority program. The aim of the project was to examine work systems from the perspective of work organization and economics. The chapter written by Thomas Zwick, Christian Göbel and Jan Fries analyzes the relationship between specific human resources measures targeted at older employees (SMOE) and establishment outcomes. More specifically, the authors examine the relationship between SMOE and the relative productivity of older employees in comparison with younger employees and the job stability of older employees. The studies in this chapter are based on a large, representative panel data set. This allows the authors to analyze the causal relationship between workforce age and establishment outcomes. Zwick, Göbel and Fries analyze the correlation between these measures and the relative productivity and tenure of older employees. They find that several SMOE increase the relative productivity of older employees. Another important finding is that on average, old employees are not less productive than younger employees. SMOE could also be used to lengthen tenure and postpone the retirement decision. The authors identify human resources

measures and employment and training strategies that can keep valuable knowledge in the organization and help avoid skills shortages.

2. The second level starts with the project conducted by the Karlsruhe Institute of Technology (headed by Professor Peter Knauth). The chapter written by Peter Knauth, Dorothee Karl and Kathrin Gimpel reports the results of the KRONOS research project, in which working-time models were developed, implemented, and evaluated for an aging workforce in six German companies from the automobile, steel, pharmaceutical, and chemical industries. It presents a total of ten subprojects in each of the following working-time design fields: (1) number of work hours per day, week or year, (2) adequate rest periods, (3) shift work, (4) employee influence on working-time design, (5) long-term time accounts, and (6) beginning and ending of the daily work-time. An important result was that there is no such thing as an "ideal" working-time model for an aging workforce, firstly because there is a large inter-individual variance in employees' work ability and work time preferences, and secondly because companies pursue different objectives with regard to operating hours and adaptation to demographic change. This is why working-time models need to be tailor-made.
3. The focus of the project at the Karlsruhe Institute of Technology (headed by Professor Gert Zülch) was on discrete-event simulation of human resources structures for aging-robust design of manufacturing systems. The authors Gert Zülch, Martin Waldherr and Marcel Becker present the results of a four-year research project on this issue. The project's main question concerned which effects would influence the overall performance of complex assembly systems. This also raises the question of the extent to which innovative approaches can be used in the specific design of an aging-robust work system. These issues were treated in a simulation-based approach in order to analyze and restructure existing assembly systems and to plan new assembly systems such as pulsed lines. The authors also present recommendations for planning aging-robust assembly systems.
4. Age diversity is increasing in group work as a result of demographic developments. The project conducted by the TU Dresden and the Leibniz Research Center for Working Environment and Human Factors (headed by Professor Jürgen Wegge and Professor Klaus-Helmut Schmidt) formed part of the third level of the priority program. The project focused on the impact of age diversity in work groups as a determinant of innovation, team effectiveness, and health. Due to the continuous increase of age differences in the workforce, human resources are called upon to find strategies for managing age-diverse teams successfully. However, age diversity in teams may result in both advantages and disadvantages, with the latter being more prevalent. The ADIGU project focused on identifying prerequisites for the effectiveness of age-diverse teams, such as task complexity, salience, and appreciation of age differences. The authors Birgit Claudia Ries, Stefan Diestel, Meir Shemla, Susanne Christina Liebermann, Franziska Jungmann, Jürgen Wegge and Klaus-Helmut Schmidt developed a new model that describes mediators and moderators explaining the connection between age diversity and team effectiveness. In order to test the

central assumptions of this model, the researchers collected data from over 745 natural work teams with 8,848 employees in different fields (e.g., car production, administration). A new training course for supervisors was designed and evaluated on the basis of these findings. The chapter describes the proposed model and training course, summarizes the main results, and outlines future research and practical implications.

5. The project conducted by the University of Münster (headed by Professor Guido Hertel) opens the fourth level. In this project age-related differences in work motivation and in control strategies while pursuing career goals were examined. In an increasingly aging workforce, age differences in motivation and stress at work provide important guidelines for age-differentiated human resources management and job design. The authors Guido Hertel, Markus Thielgen, Cornelia Rauschenbach, Anna Grube, Christian Stamov-Roßnagel and Stefan Krumm report the results of a six-year research effort exploring systematic age differences in workers' implicit and explicit motives, experiences with stress, job satisfaction, and control strategies at work. The results revealed that quantitative and qualitative differences between older and younger worker exist. The differences include a higher subjective importance assigned to autonomy and generativity values, a better fit between work values and task characteristics, higher congruency between implicit and explicit motives at work, more negative reactions to the lack of person-environment fit and motivational incongruencies, lower stress levels, higher job satisfaction, and more effective control strategies at work. These results not only underline the need for age-differentiated human resources management, they also document significant strengths and resources among older workers.
6. The research work conducted by the University of Freiburg (headed by Professor Heinz Schüpbach) investigated to what extent teachers' emotion regulation in stressful teaching situations is of importance in terms of health and which age-specific variables must be taken into account. Teaching can be emotionally demanding and being able to regulate one's emotions successfully is a resource that contributes to teachers' well-being and health. The authors Anja Philipp and Heinz Schüpbach show that particularly teachers in the last phase of their career—but also their less experienced colleagues—seem to struggle with the emotional demands of teaching. Three studies provide indications of which emotion regulation strategies help teachers to remain well and healthy. The studies also indicate how to design classroom interventions on emotion regulation that are beneficial to health.
7. The chapter by the Institute and Polyclinic for Occupational, Social and Environmental Medicine at LMU Munich (headed by Dr. Andreas Müller) investigates the importance of autonomy in tasks as a condition for older employees' health and performance. The project added a new research and application field to the priority program by focusing on the care sector. The future efficiency of healthcare provision in Western countries is closely related to the challenge of enabling healthcare employees to remain healthy and capable in their jobs. The authors' research aimed to investigate in particular

how the interplay between successful aging strategies and working conditions affects nurses' well-being and ability to work. Their findings confirm that nurses apply successful aging strategies in terms of selection, optimization and compensation (SOC) (Baltes and Baltes 1990). The authors Andreas Müller, Barbara Heiden, Matthias Weigl, Jürgen Glaser and Peter Angerer found that SOC in nursing significantly contributes to nurses'—and particularly older nurses'—ability to work. Moreover, in aging nurses and nurses with musculoskeletal impairments, the autonomy to schedule work, make their own decisions, or choose methods—i.e., job control—substantially enhances the effectiveness of SOC. The findings foster the theoretical development of occupational lifespan models. Furthermore, the results provide a foundation for the development of SOC-specific interventions that aim to maintain nurses' well-being and ability to work.

In general, five chapters/research projects addressed questions at the fifth level. These chapters can be divided into two specific subcategories. The first three texts focus on the design of industrial work systems, using the example of automobile assembly. The other two chapters deal with engineering psychology and concentrate on designing interactions between vehicle drivers and technology.

8. The project by the University of Kassel (headed by Professor Ekkehart Frieling) focuses on the impact of assembly tasks on an aging workforce. The authors Ekkehart Frieling, José Alonso Enríquez Díaz, Daniel Kotzab and Alina Sytch examined various tasks in equipment assembly and final vehicle assembly at two international automobile companies. In a longitudinal study, two assembly groups are compared to each other at three intervals (approximately two years apart). Additionally, in one of the companies two more assembly groups were surveyed every two years. These two groups are organized according to the Chaku–Chaku principle. Their job is to assemble handbrake levers and exhaust systems. All study samples are assessed using the same survey methods: video recordings of relevant body postures, standardized questionnaires, and on-site expert evaluations. The results demonstrate that physical complaints increase rather than decrease despite ergonomic work design. The intensification of the work and the avoidance of non-added value tasks induce more strain on the employees. However, the four samples also reveal differences among the work structures. Well-organized teamwork helps temper the impact of the high physical strain; this is of particular importance to the aging persons.
9. The project conducted by Technische Universität Darmstadt (headed by Professor Ralph Bruder) also examined assembly jobs in the automobile industry. The aim was to develop an assistance system for age-differentiated work design. The main purpose of the authors Ralph Bruder, Kurt Landau, Holger Rademacher, Andrea Sinn-Behrendt, Max Bierwirth and Dorothee Müglich was to assess the age-related changes of relevant physical capabilities among workers in automotive and supplier companies. The researchers assessed exposures at over 200 workplaces and examined the capabilities of

106 workers aged 20–63. Several capability tests were developed in order to assess industry-relevant physical capabilities. The results of the capability tests show statistically significant differences between younger (20–35 years) and older workers (45–63 years) in terms of handling materials manually and maintaining awkward body postures. These types of exposures were also found to be the most frequent types in workplace assessments in the automobile industry. Tasks conducted in a vehicle's interior in particular combine higher load weights and awkward body postures. Ergonomic work design can help to reduce such critical exposures. As higher grades of limitations were also found among younger workers, a more differentiated view on the individual worker's physical capabilities is needed.

10. The research work conducted by Chemnitz University of Technology (headed by Professor Birgit Spanner-Ulmer) provided the study on assembly jobs in the automobile industry with ergonomic analyses. The researchers investigated whether and in what form age-differentiated process model components can help reduce the strain that results from age-related changes in performance. The chapter by Kerstin Börner, Christian Scherf, Bianca Leitner-Mai and Birgit Spanner-Ulmer focuses on the question of whether older employees experience more strain than younger workers when both are exposed to the same level of stress. For this purpose, two field studies were conducted in an automobile supply company and an automobile company in order to analyze age-differentiated strain experienced by healthy assembly workers. Subjective and objective data were collected over the entire eight-hour morning shift. The data comprised motion recordings, measurements of heart rate, and several questionnaires. In field study A, 23 female workers aged 27–57 were observed at single workplaces. In field study B, 31 male workers aged 21–60 were monitored at an assembly line. Based on these data, it can be stated that older assembly workers do not experience more strain than younger workers. Nevertheless, the results show significant differences in assembly times, pulse differences, and questionnaire responses.
11. The interface between humans and technology was expanded to include vehicle driving in the research project conducted at Saarland University (headed by Professor Gisa Aschersleben and Professor Jochen Müsseler). The aim of the project was to analyze age-related differences in the applied context of critical driving situations. In the chapter by Gisa Aschersleben, Katrin Arning and Jochen Müsseler, the authors replicated findings obtained by Müsseler et al. (2009) that show the typically observed advantage of spatially compatible responses is reversed for dangerous situations in natural scenes. In accordance with the findings reported in the literature, a decreased performance of older drivers was identified. However, spatial compatibility effects were not increased with age either in single or dual-task situations. Moreover, driving expertise moderated the effects of age in driving performance and resulted in reduced vulnerability to interfering stimuli. As a result of these findings, the authors suggest that older drivers should regularly attend driving

classes, particularly if they drive shorter distances or less frequently than before.

12. Driving was also the topic of the project conducted by the Leibniz Research Center for Working Environment and Human Factors at the TU Dortmund University (headed by Professor Michael Falkenstein and Professor Nele Wild-Wall). As a main source of mobility for elderly people, driving is a very complex task that requires visual, motor, and cognitive skills. These skills are subject to specific age-related changes. In a series of three experiments, the authors Melanie Hahn, Nele Wild-Wall and Michael Falkenstein were able to show that older people's tracking performance was particularly hampered when they also had to respond manually to a relevant secondary stimulus. This motor interference could not be minimized by changing the response modality of the second task from manual to verbal response. Furthermore, the authors were able to demonstrate that older people seem to have problems in differentiating between relevant and irrelevant stimuli. This may affect efficient response to critical and significant stimuli in the traffic environment, even if they occur after irrelevant stimuli, and especially when the interval is short. The findings may have important implications for the relationship between age-related cognitive changes and driving performance. Moreover, they can be helpful in the development of age-adapted training programs and technical innovations in the driver's environment.
13. The project conducted by HU Berlin (headed by Professor Hartmut Wandke) forms the start of the sixth level. It focuses on the question of which training and equipment-design measures can support older people in the use of computer-based interactive systems. The chapter by Michael Sengpiel, Malte Sönksen and Hartmut Wandke summarizes the findings from seven experiments conducted with a simulated ticket vending machine. The experiments investigated various training techniques such as error-guided training, worked examples, and model-based training, compared with a basic training program, as well as instruction and design aspects, by comparing the effects of brief video instructions and a wizard redesign of the graphical user interface for older ($N = 247$, mean age of 68 years) and younger ($N = 142$, mean age of 25 years) participants. They show that older users can learn to use interactive systems with degrees of effectiveness and efficiency approaching those of young users provided a careful universal design is combined with minimal instruction and training endeavors. These endeavors seem justified given the chasm between current use and potential benefits of interactive technology for an aging society.
14. The project conducted by the Institute of Industrial Engineering and Ergonomics at RWTH Aachen University (headed by Professor Christopher M. Schlick) builds directly on the research work at HU Berlin. The aim was to develop models and methods for an age-differentiated adaptation of human–computer interfaces and to evaluate them ergonomically using prototype user interfaces in a series of experiments conducted under laboratory conditions. The chapter by Christopher M. Schlick, Sebastian Vetter, Jennifer Bützler,

Nicole Jochems and Susanne Mütze-Niewöhner summarizes the findings from seven experiments with a total of 445 participants aged between 20 and 77 years. The authors focused on analyzing the stages of visual perception, cognition, and response execution regarding effective, efficient, and satisfying human–computer interaction on an adequate level of mental workload. They investigated human–computer interaction on the basis of a self-developed project management application software. This was done for two reasons: Firstly, computer-aided project management is important in many manufacturing and service industries, and secondly, older workers are often responsible for the coordination and supervision of large projects due to their extensive experience and excellent communication skills. For most, although not all, of the studies, the results reveal that age-induced differences in human–computer interaction can best be compensated by an ergonomic "design for all." In some cases, however, an age-differentiated graphical representation and manipulation of complex informational structures is favorable for project planning, scheduling and schedule management/control.

15. The project by the Leibniz Research Center for Working Environment and Human Factors at the TU Dortmund University (headed by Professor Herbert Heuer) conducted research on the age dependence of adaptations to new visual-motor transformations in which the execution and observation of movements are spatially separated (e.g., use of a computer mouse, minimally invasive surgery). The difficulties that older people have with these transformations may result in part from generalized slowing, but also in part from reduced learning capabilities. The chapter by Herbert Heuer, Mathias Hegele and Miya Kato Rand reports the results of a series of experiments in which the authors explored the nature of the latter kind of change. In all of these experiments, participants learned novel visual-motor transformations. Different tests served to assess implicit and explicit knowledge that was acquired from learning. These were tests of after-effects (implicit knowledge), judgments of visually presented amplitudes and directions on "correct movements" (explicit knowledge), and tests of adaptive shifts (combination of explicit and implicit knowledge). The experiments revealed that explicit knowledge and adaptive shifts decline with age, while after-effects remain unaffected (sometimes they increase). Countermeasures against age-related changes seem to require special efforts for improving explicit learning at an older working age.
16. The project at Jacobs University Bremen (headed by Professor Claudia Voelcker-Rehage and Professor Ben Godde) examined the importance of manual dexterity for the employability of older employees. Tasks which require a high level of manual dexterity such as the use of precision tools (e.g., forceps, tweezers, and screwdrivers), the assembly of computer boards, or the soldering of electronic devices require much less physical strength than those involving the whole body, such as lifting or carrying, and thus provide a potential field of work for older employees. The authors Claudia Voelcker-Rehage, Eva-Maria Reuter, Solveig Vieluf and Ben Godde

investigated age-related changes in manual dexterity during working age at the behavioral and neurophysiological level with a special focus on precision grip control and tactile performance. Furthermore, they were interested in whether work-related expertise arising from continuous and elaborated use might delay age-related changes. The results revealed that manual dexterity already declines throughout the working lifespan. However, expertise and training might counteract decline in finger force control and haptic perception. The authors conclude that the transition from physically demanding jobs to jobs requiring manual dexterity can be successfully implemented. However, systematic task-related training is required to counteract age-related decline in manual dexterity that might hinder such horizontal career tracks.

17. The working environment was the primary topic of the research project carried out at the University of Siegen (headed by Professor Helmut Strasser and Professor Karsten Kluth). Order-picking in deep cold-storage depots means that employees must work in temperatures of around −24 °C for the entire work day. The chapter by Karsten Kluth, Mario Penzkofer and Helmut Strasser focuses on the question of whether or not age-related organization of work times and breaks is necessary. In order to assess possible age-dependent physiological effects, 30 male subjects were classified into two age groups (20–35-year-olds and 40–65-year-olds). In tests conducted over the entire work day, heart rate, skin surface temperature, blood pressure and body core temperature were recorded. The heart rate readings indicated a high physiological strain for both age groups. As a result of the deteriorating ability of heat generation with advancing age, substantial decreases in the body core temperature were recorded among the older employees. Age-related differences in skin temperature and blood pressure could not be detected. In terms of physiological strain, appropriately adapted work-time break regimes have to be provided for older employees to ensure their ability to work over the long-term.

This edited volume *Age-differentiated Work Systems* thus provides a final report on the six-year priority program of the same name and presents selected research findings in condensed form. The book is aimed in particular at students and company practitioners. The idea is that it will serve both as a reference book and overview of the current state of research in ergonomics, occupational psychology and related disciplines. It provides new models, methods, and procedures for analyzing and designing age-differentiated work systems with the aim of supporting social actors in their decisions on labor and employment policies.

Further objectives of the edited volume were to provide a pluridisciplinary compilation of the extensive information acquired over the six-year program period, to illustrate the range of the research field, and to convey an integrated understanding of age-differentiated work systems to readers. Those interested in further information will find relevant references, which provide access to the specialized literature found in journals and proceedings. In addition, the standardized structure of the individual chapters should help readers view the contents quickly and filter information where necessary. To this end, the main findings from the various subprojects are summarized at the start of each chapter in order to provide readers with a compressed overview.

Although this volume reflects the current state of research in the above-mentioned specialized areas and is believed to be the first interdisciplinary program of its kind in basic research, it cannot claim to cover this wide and comprehensive field in its entirety. Instead, the idea is to stimulate academic discourse and provide motivation for further research. Apart from describing the main findings, all of the published chapters point to research deficits or further research questions that should be tackled as part of additional individual projects or similar interdisciplinary research programs.

On behalf of all those involved, the editors would like to thank the German Research Foundation for funding Priority Program 1184. We are particularly grateful to Dr. Andreas Engelke and Dr. Anne Brüggemann for their superb support in all phases of the entire project. They made a major contribution to the successful conclusion of this interdisciplinary initiative through their deep commitment to the development and supervision of the research program. The editors wish to repeat at this occasion their thanks to Dr. Markus Buch, Dr. Susanne Mütze-Niewöhner, Dr. Katharina Hasenau and Dr. Nicole Jochems for their outstanding support of the priority program and the excellent coordination of the research groups. We would also like to thank the companies, organizations and factories, as well as the employees and works councils involved in the priority program, for providing information and profound insights into their work practices.

References

Astor M, Koch C, Klose G, Reimann F, Rochhold S, Stemann M (2006) Zu alt, um Neues zu lernen? Chancen und Grenzen des gemeinsamen Lernens von älteren und jüngeren Mitarbeitern. In: QUEM-Materialien der Arbeitsgemeinschaft Betriebliche Weiterbildungsforschung e.V. (AWBF), vol 77, pp 1–165

Baltes PB, Baltes MM (eds) (1990) Successful aging: perspectives from the behavioral sciences. Cambridge University Press, New York

Birren JE, Schaie KW (2006) Handbook of psychology of aging, 6th edn. Academic Press, London

Brandtstädter J (2007) Das flexible Selbst. Selbstentwicklung zwischen Zielbindung und Ablösung. Elsevier, Heidelberg

Carstensen LL (2006) The influence of a sense of time on human development. Science 312:1913–1915

Czaja SJ (1997) Computer technology and the older adult. In: Helander M, Landauer TK, Prabhu P (eds.) Handbook of human-computer interaction. Elsevier Science, New York, pp 797–812

DSW—Deutsche Stiftung Weltbevölkerung (2011) DSW-Datenreport 2011. Soziale und demographische Daten zur Weltbevölkerung

Eisenmenger M, Pötsch O, Sommer B (2006) Koordinierte Bevölkerungsvorausberechnung, Annahmen und Ergebnisse
European Commission (2010) Commission staff working document: demography report
Luczak H, Volpert W (1987) Arbeitswissenschaft. Kerndefinition—Gegenstandskatalog—Forschungsgebiete. RKW-Verlag, Eschborn
Maintz G (2002) Leistungsfähigkeit älterer Arbeitsnehmer. Abschied vom Defizitmodell. Fehlzeiten-Report 2002. Forschungsbericht, Springer, Berlin
Martin M, Kliegel M (2010) Psychologische Grundlagen der Gerontologie, 3. komplett überarbeitete Auflage. In: Tesch-Römer C, Wahl HW, Weyerer S, Zank S (eds) Grundriss Gerontologie: Band 3. Kohlhammer, Stuttgart
Müsseler J, Aschersleben G, Arning K, Proctor R (2009) Reversed effects of spatial compatibility in natural scenes. Am J Psychol 122:325–336
Naegele G (2004) Zwischen Arbeit und Rente: gesellschaftliche Chancen und Risisken älterer Arbeitnehmer, 2nd edn. Maro-Verlag
Ng TWH, Feldman DC (2008) The relationship of age to ten dimensions of job performance. J Appl Psychol 93:392–423
Rohmert W (1983) Formen menschlicher Arbeit. In: Rohmert W, Rutenfranz J (eds) Praktische Arbeitsphysiologie. Georg Thieme Verlag, Stuttgart New York
Schaie KW, Hertzog C (1983) Fourteen-year cohort-sequential analyses of adult intellectual development. Dev Psychol 19:531–543
Schlick C, Bruder R, Luczak H (2010) Arbeitswissenschaft. Springer, Berlin
Triebe JK (1980) Untersuchungen zum Lernprozess während des Erwerbs der Grundqualifikation (Montage eines kompletten Motors). Arbeits- und sozialpsychologische Untersuchungen von Arbeitsstrukturen im Bereich der Aggregatefertigung der Volkswagenwerk AG. HA 80-019. BMFT, Bonn
Triebe JK (1981) Aspekte beruflichen Handelns und Lernens. Eine Feld-und Längsschnittuntersuchung zu ausgewählten Merkmalen der Struktur und Genese von Handlungsstrategien bei einer Montagetätigkeit. Unpublished Disseration, University of Bern
Ulich E (2005) Arbeitspsychologie, 6 Auflage. Schäffer-Poeschel Verlag, Stuttgart
United Nations, Department of Economic and Social Affairs (2004). World population to 2300. New York
United Nations, Department of Economic and Social Affairs, Population Division (2011) World population prospects: the 2010 revision
Wegge J, Roth C, Neubach B, Schmidt KH, Kanfer R (2008) Age and gender diversity as determinants of performance and health in public organization: the role of task complexity and group size. J Appl Psychol 93:1301–1313
Zink KJ (1978) Zur Begründung einer zielgruppenspezifischen Organisationsentwicklung. Zeitschrift für Arbeitswissenschaft, vol 32, pp 42–48

Published Special Issues

Deller J, Hertel G (guest editors) (2009) Demographic change in work organizations. Sonderheft der Zeitschrift für Personalpsychologie
Frieling E (guest editor) (2006) Altersgerechte Arbeitsgestaltung. Zeitschrift für Arbeitswissenschaft
Hertel G, van der Heijden B, De Lange A, Deller J (guest editors) (2013a) Facilitating age diversity in organizations—part I: challenging popular misbeliefs. Spec issue J Manag Psychol (in press)
Hertel G, van der Heijden B, De Lange A, Deller J (guest editors) (2013b) Facilitating age diversity in organizations—part II: managing perceptions and interactions. Spec issue J Manag Psychol (in press)

Schlick C (guest editor) (2010) Age-differentiated work systems. Occupational Ergonomics
Schlick C, Mütze-Niewöhner S (guest editors) (2010) Altersdifferenzierte Arbeitssysteme. Zeitschrift für Arbeitswissenschaft
Wegge J, Frieling E, Schmidt KH (guest editors) (2008) Alter und Arbeit. Sonderheft Wirtschaftspsychologie

Age-Differentiated Work Systems Enhance Productivity and Retention of Old Employees

Thomas Zwick, Christian Göbel and Jan Fries

Lessons Learned

The rapid aging of the workforce in Germany threatens to undermine the competitiveness of enterprises mainly in two respects. The productivity of enterprises might decrease when relative productivity of old employees is lower than productivity of younger employees, and when employers cannot avoid an increase in the share of old employees. In addition, enterprises might lose valuable knowledge and skills when large cohorts of well-educated employees of the baby boom generation retire and have to be replaced by employees from smaller cohorts of labor market entrants.

This chapter reviews recent findings on the relationship between **S**pecific Human Resource **M**easures that are targeted on **O**ld **E**mployees (SMOE) and establishment outcomes. More specifically, we study the relationship between SMOE and the relative productivity of older employees in comparison to younger employees and job stability of old employees. The studies summarized below put into question some traditional, widely accepted and frequently used paradigms. In the existing literature, old employees are often described less motivated, less productive or reluctant to learn. In addition, most previous studies concentrate on

T. Zwick (✉)
Faculty of Economics and Management, University of Würzburg, Würzburg, Germany
e-mail: thomas.zwick@uni-wuerzburg.de

C. Göbel
University of Applied Labour Studies of the Federal Employment Agency (HdBA), Mannheim, Germany
e-mail: Christian.Goebel@arbeitsagentur.de

J. Fries
Research Department Labour Markets, Human Resources and Social Policy, Centre for European Economic Research (ZEW), Mannheim, Germany
e-mail: fries@zew.de

C. M. Schlick et al. (eds.), *Age-Differentiated Work Systems*,
DOI: 10.1007/978-3-642-35057-3_2,

early retirement instead of ways to retain old employees. On the basis of extensive and detailed German linked employer-employee panel data, our project results show that there is no negative relationship between workforce age and establishment outcomes. We also find that SMOE increase the relative productivity of old employees and that some of them may be helpful for retaining productive old employees longer.

These are the most important lessons learned:

- Old employees are on average not less productive than younger employees; there are large variances between enterprises, however.
- The application of age-specific workplaces, specific jobs for old employees, and mixed-age working teams in establishments is associated with a significantly higher relative productivity of old employees.
- The relative productivity of old employees in establishments that apply reduced working time and training for old employees is not significantly different from establishments without these measures in our data set. Reduced working time has been mainly used to implement early retirement, and training of old workers frequently does not take into account age-specific learning motivation.
- The use of mixed-age working teams is associated with significantly longer employment durations of employees aged 50 and above. By contrast, an age-specific part-time program (*Altersteilzeit*) significantly reduces employment durations of older workers. We do not find a relationship between employment duration and training measures and other policies, such as reduced work requirements or specific equipment of workplaces for older employees.

Introduction

Germany experienced a strong increase in workforce age and this trend is likely to continue in the future (Göbel and Zwick 2009). If older employees are structurally less productive than younger employees, then higher shares of older employees in the workforce might reduce productivity of concerned establishments and lower their competitiveness. On the other hand, retirement of the relatively large cohorts of highly educated baby-boomers and the labor market entry of relatively small cohorts of young people might lead to an undersupply of skilled employees and the loss of specific knowledge (Frosch et al. 2011). In order to assess the importance of the threats that are caused by an aging workforce, in a first step we compute the relative productivity of old employees. In other words, we investigate how productivity of establishments changes when the share of old employees increases.

The age composition of employees in the establishments changes significantly from year to year. It should also be noted, that the age structure cannot entirely be determined by the management because Germany has strict labor protection laws especially for old employees. We find that establishments apply different employment strategies in order to avoid strong aging. One group of establishments is mainly hiring younger employees while others mainly dismiss their old employees. Strong variation in the age structure allows us to identify the impact of a change in the age structure of the employees on establishment productivity. We also consider that the effect of age on productivity might be different between economic sectors.

Enterprises have the possibility to respond to an expected reduction in establishment productivity when their workforce ages. They can, for example, use human resource management measures especially targeted on old employees. These measures can be important components of age-differentiated work systems. In order to study which human resource management measures are effective, we analyze the correlation between these measures and the relative productivity of old employees. SMOE might also be used for an increase in the tenure of old employees and a postponement of the retirement decision. We identify effective personnel measures that can keep valuable knowledge in the establishment and help to avoid skills shortages by prolonging the tenure of old employees.

Background

The impact of the age structure on establishment productivity is at the center of a lively scientific debate. At the core of this debate is the question whether the decline in for example cognitive functioning, executive or memory capability measured, muscle strength and sight documented in many medical and psychological studies (Skirbekk 2008; Ng and Feldman 2008) leads to a measurable decline of average productivity of old employees. There are at least two arguments against this hypothesis. First, establishments might lay off old employees if they observe a decline in productivity. Second, productivity is not so much driven by peak or short period performance frequently measured in medical studies, but by long period performance. In addition, declining performance dimensions may be substituted by experiential knowledge (so-called crystallized intelligence, Kanfer and Ackerman 2004) and specific human capital or be compensated by mixed-age working teams, where members are selected according to their specific and complementary knowledge and skills.

Skirbekk (2008) reviews 14 chapters on the relationship between the age of employees and firm productivity, starting from the chapter by Hellerstein and Neumark (1995). Between 2008 and 2011, about the same number of studies appeared, covering different countries. One reason for the renewed interest in this subject is the availability of suitable representative panel data and a paradigm shift. Early chapters almost unanimously (in Skirbekk's literature review 13 out of 14 chapters) found that the relation between workforce age and productivity is hump-shaped (or describes an inverse "*U*"). In other words, young and old employees have a lower productivity than mid-aged employees. Consequently, the early literature had a focus on the question, at which age the decline in productivity starts.

The widely held believe that old employees lead to a reduction in productivity at the firm level first was put into question by the seminal work by Patrick Aubert and Bruno Crépon. In a series of chapters, they argue theoretically and show empirically that previous results have been driven by the endogeneity of the age structure in establishment productivity estimation (see for example Aubert and Crépon 2006). In other words, productivity and age structure are frequently jointly driven by third, mostly unobserved factors. One example is that establishments in declining industries see their productivity erode while they cannot afford to hire new employees and the existing workforce slowly ages. Another example is that establishments that are about to introduce an innovation can increase their productivity and at the same time hire young employees. In both cases, a cross-sectional analysis or a simple panel analysis that does not take endogeneity of the age structure into account leads to biased results for the effect of the age structure on productivity. Aubert and Crépon (2006) show how to mitigate this estimation problem and demonstrate that its correction mainly leads to an increase of the measured impact of the share of older employees on establishment productivity. In other words, in their estimation approach, the age productivity profile is like an inverse "*L*"—young employees contribute less than the average employee to establishment productivity but from a certain age onwards, the productivity contribution of employees in different age groups is stable.

The finding that the correction of endogeneity in age-productivity estimations leads to an increase in the relative productivity impact of old employees has been replicated several times since 2006. In 2011, two major international conferences in Louvain-La-Neuve and St. Gallen have been explicitly devoted to the relationship between aging workforces and productivity. Contributions to these conferences discuss estimation issues and differences between regions, economic sectors and age-productivity as well as age-earnings profiles. Many of the contributions to these conferences appeared or will appear in special issues of well-known international journals—De Economist, volume 159 (2) in 2011 and Labour Economics, forthcoming in 2012. Both special issues will certainly have influence on the discussion of age-productivity profiles and their estimation.

Besides the question whether old employees are as productive as younger employees, many chapters discuss the question whether specific personnel measures for old employees increase their productivity. The first insight from this literature is that old employees are on average not less motivated than younger employees, however, different measures are effective in motivating old employees (Stamov-Roßnagel and Hertel 2010).They stress that old employees mainly want to adopt their working environment to fit their (partly reduced) resources as well as possible. Younger people primarily strive for gains in, for example, earnings, status or employment security; older people, however, often focus on maintenance, returns from prior investment, and the prevention of losses. Stamov-Roßnagel and Hertel (2010) argue that interest in tasks that involve acquiring new skills, knowledge or career opportunities decrease with age. On average, motives like autonomy, positive relationships with colleagues and supervisors, and self-realization increase in importance during the life cycle. There are at least two reasons for that. First, many old employees have already achieved goals that motivate younger employees, like employment security, promotion or a high income level. Another variant of this argument is that old employees enjoy better institutional rules like higher employment security, hiring subsidies or longer unemployment payments that also change their motivation (Boockmann et al. 2012b). Second, old employees do not want to compete for example in a promotion tournament in ability areas they have clear disadvantages in (Kanfer and Ackerman 2004). These authors stress that motivation for activities that mainly demand fluid cognitive abilities becomes less interesting because the ability to learn radically new things declines with age. Activities that are mainly based on the development of crystallized cognitive abilities remain attractive, however, because the experience of old employees is an advantage for such activities.

The most pervasive personnel measure specifically aimed at old employees is to select the most able and best fitting employees and dismiss less productive employees (Howard 1988; Frosch et al. 2011). Strict labor market protection for old employees in Germany does not allow selective dismissal of old employees or makes this strategy expensive. Strategic human resource management measures that directly tackle disadvantages of old employees therefore provide alternatives for early retirement, old age working time reductions or dismissals. These measures are based on the insight that old and younger employees have complementary competencies and capabilities (Boockmann and Zwick 2004; Skirbekk 2008). Specific measures might also increase the chances to retain old employees. Establishments that face a shortage of young skilled workers, or establishments where a large cohort of employees approaches retirement age, might have an interest in applying specific measures for old employees in order to lower quits. Moreover, employment duration of old employees is likely to depend on relative productivity of old employees. Therefore, specific measures that target an enhancement of older workers' productivity are likely to prolong employment, too. Therefore it may make sense to offer SMOE, even if this may be perceived as discrimination.

Data and Empirical Strategy

All analyses presented in this chapter use linked employer-employee panel data sets (LIAB) of the Institute for Employment Research, Nuremberg (IAB). On the establishment level, the LIAB uses the survey data of the IAB establishment panel. This panel entails questions on value added, investment, industrial relations, sector, average employee characteristics and expectations of managers. The establishment data are linked to administrative records on employees, by the means of a common identifier. The employee data set is based on official data of the IAB employment register. Yearly information on wages, qualification, gender, tenure, and age can therefore be linked to the employer data. Altogether, our version of the LIAB covers almost 7 million employees and more than 8,500 establishments.

The data set therefore has the advantage that information on establishments from all economic sectors and the structure of their employees can be captured over time. We restrict our analysis to profit oriented enterprises with easily measureable output (and therefore exclude public administration enterprises, agriculture, banks and insurances). In addition, the individual information such as wage and age is drawn from administrative sources and is therefore practically free of measurement errors. One disadvantage is that there are problems with item non-response for several variables on the establishment level such as sales or innovations. In some chapters, we imputed missing observations of important variables, when values were reported before and after the missing observation. Dates for tenure and experience were reported only after January 1st 1975 in West Germany and after January 1st 1990 in East Germany. This means that more than 10 % of the West German and more than 25 % of the East German employees have censored values for tenure and experience. We account for censoring by multiply imputing their values. Yearly imputation of the values for tenure could lead to excess variance in these variables and therefore, for each employee, only the first value for tenure is imputed and following observations are derived from the initial value, accordingly; for each additional year the employee stays in the same establishment we update the value for tenure by adding one year to the value of the last year.

The panel data are provided in several versions, see Jacobebbinghaus (2008) for details. The so-called cross-sectional version used by Göbel and Zwick (2009, 2012a, 2012b) and Frosch et al. (2011), provides one observation per year for establishment characteristics and virtually all employees of the observed establishments on June 30th of the respective year. The so-called longitudinal version uses individual spell data. This means that employment spells and their characteristics such as wage and duration are completely known for all employees who worked in one of the included establishments. This version of the LIAB is used by Boockmann et al. (2012a). The spell data in the LIAB allow us to reconstruct employment spells that last for more than 25 years. The exact duration of an employment relationship can be observed with daily accuracy. One of the disadvantages of the longitudinal version of the LIAB is that there is no information on retirement. However, establishments that apply specific measures for old employees are mainly interested in employment duration, for which analysis the longitudinal version of the LIAB is well suited. We draw a sample of workers who are employed on January 1st 2002 and who are more than 40 years old, correcting for potential stock-sample bias. These individuals are followed in the data until they leave the establishment permanently or reach official retirement age at 65.

In Göbel and Zwick (2009, 2012a), we estimate the impact of the workforce age structure and its changes on establishment productivity. We argue that cross-section estimations of the relationship between the establishment age-structure and productivity are likely to give biased results. Assume that the productivity-contribution of employees aged 20–30 years is lower than that of the prime age workers (those aged 30–40 years). Moreover, the relative productivity of the 40–50 years old employees and the 50–60 years old employees is on the same

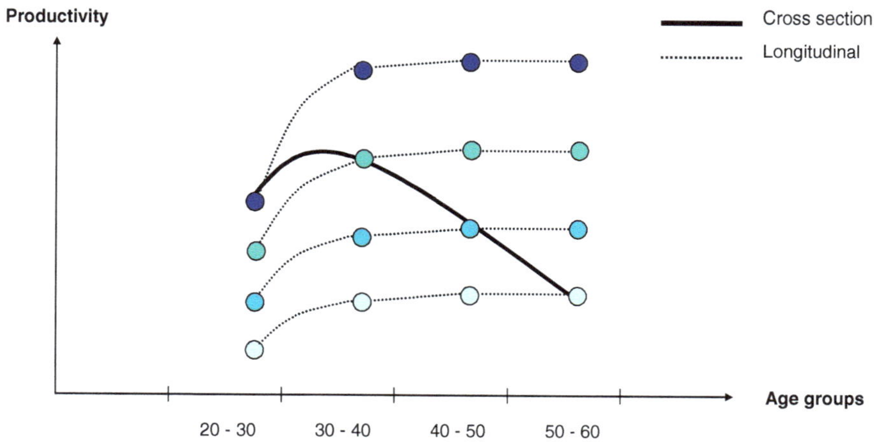

Fig. 1 Longitudinal vs. cross-sectional observations of age and productivity

level as relative productivity of the 30–40 years old employees. This means that the relation between age share and productivity can be represented as shown in the lowest line in Fig. 1. Furthermore, let us assume that there is an exogenous technical improvement induced by learning and innovations that increases total establishment productivity uniformly over time for all age groups. The described upwards shift would for example occur when later birth cohorts are better educated or when they work with newer, more productive technologies or machines. The shift could apply for birth cohorts at the level of employees and for cohorts of establishments that are founded at the same moment in time. In other words, the technical improvement does not particularly favor a certain age group, but shifts the productivity of all age groups upwards, as shown in the second lowest line in comparison to the lowest line in Fig. 1. We finally assume that there are no age composition effects on productivity, i.e. the productivity of employees of a certain age group does not depend on the share distributions of the other age groups. In this case, the lowest line of Fig. 1 represents the true age productivity profile of the employees of one cohort. This profile represents what researchers would obtain if they could directly observe the development of the individual productivity contributions in this cohort for 40 years. Note that this is the approach for example proposed by Schaie and Willis in the Seattle Longitudinal Study (Schaie and Willis 2002) to study individual aging processes. The second lowest line is the analogous age-productivity profile for another cohort that enters the labor market 10 years later. From the longitudinal perspective, observing age and productivity over time, we obtain several trajectories of age-productivity curves in an inverted *L*-shape.

Now suppose that we compute age-productivity profiles derived from cross-sectional data, observing differences in age and productivity between our four cohorts at one point in time. This approach cannot take into account the productivity trend over time because technological shocks or productivity changes based on innovations usually cannot be observed in cross sectional data sets. The cross-section approach results in an inverted U-shaped age-productivity profile as indicated by the bold line in Fig. 1.

In our empirical approach, we imitate the longitudinal and cross-sectional estimation approaches depicted in Fig. 1. First, we calculate an age-productivity pattern in a pooled ordinary least squares (OLS) approach that is driven by cross-sectional variation in the data. Then we take into account unobservable time-invariant heterogeneity between establishments (probably induced by unobservable technological shocks over time) in a longitudinal approach. Our preferred specification exploits changes in productivity and the age structure over time within the same firms and takes into account that third factors such as the business cycle might have an impact on both productivity and the age structure (which would lead to endogeneity). In addition, our data allow us to separate cohort effects from age effects. The simultaneous identification of age-, cohort- and calendar-effects is often awkward in empirical models. The fundamental issue is that at the individual level, year of birth plus age is equal to calendar time. This renders it impossible to separate the effects of birth-year, age, and calendar time at the individual level. This equality also can lead to estimation problems (collinearity) when analyzing establishment data. Here, the share of employees with a specific year of birth is equal to the corresponding share of employees with the corresponding age in years. Therefore, additional requirements have to be met for the simultaneous identification of cohort-, age- and calendar-effects with establishment data. First, the chosen time windows for age groups and cohort groups are not allowed to completely overlap. Complete overlap would lead to perfect collinearity and would render any estimation impossible. This implies that one has to decide on the width of the time windows, which can put constraints on the empirical specification. In empirical applications, time windows typically comprise several years. This requirement often does not put serious constraints on the specification. In order to analyze the sensitivity of the results, however, it is advisable to test different specifications of the window widths. Second, sufficient variation over the possible age and cohort shares is required. The use of panel data at the establishment level is certainly helpful to provide this variation, since there is typically ample variation over age and cohort composition within the enterprises over time (compare Göbel and Zwick 2009). This variation is the outcome of an aging workforce combined with employees that quit and enter the establishments in the course of the observation period.

We estimate a structural so-called Cobb-Douglas production function. Productivity of the establishments is measured by value added (sales minus input) per head. We explain productivity by capital per head, the share of employees in certain age groups, and other establishment characteristics. We include a broad range of relevant establishment-specific information besides capital and age shares. It is especially important to control for age related variables such as tenure (Haegeland et al. 1999; Daveri and Maliranta 2007), qualification (van Ours and Stoeldraijer 2011), the age of the firm (Cardoso et al. 2011), and the birth cohorts of the employees. Otherwise, the age of the employees might capture part of the influence of these factors on productivity, which could lead to biased results. We use age classes in five-year brackets and only report the coefficients of employees between 20 and 60 years of age. The estimates for the other age classes are summarized in a separate variable, but are not reported because they are likely to be dominated by unobserved characteristics of employees at the fringes of the age distribution—very young employees and very old employees are usually specific individuals. In addition, they represent only a small fraction of the population of all employees (in the year 2005, only 3.5 % of the employees are younger than 20 years and only 3.8 % are older than 60 years old, OECD 2005). In order to allow for persistence in the level of productivity, we specify a dynamic model where the production level p of an establishment j of one year t is allowed to be a function of its past values $p_{j,t-k}$.

We consider various ways of estimating the production function. In a first step, we estimate pooled OLS of the Cobb-Douglas production function without the lagged dependent variables. However, the OLS estimates are likely to be biased because of endogeneity since the value added and the age structure are determined simultaneously. Successful establishments, for example, recruit more workers, and job entrants tend to be younger than those who leave the enterprise (Frosch et al. 2011). In addition, the variation between the establishments is likely to drive the results, and we can only observe part of the heterogeneity between establishments. For example, establishments with good industrial relations might be able to bind their employees longer, while they enjoy a higher productivity. Since we cannot control for all establishment characteristics, estimates from OLS estimation of the age-productivity profile are likely to be biased.

Therefore, we switch from a cross-sectional (between) estimation to a longitudinal (within) estimation in a next step and hereby control unobservable differences between enterprises (so-called unobservable time-invariant heterogeneity). In order to control for endogeneity, we apply so-called dynamic General Method of Moments (GMM) methods and use lagged values of the explanatory variable to instrument contemporary values (Arellano and Bover 1995; Blundell and Bond 1998). In order to find the correctly specified model, we start with moment conditions that require relatively mild assumptions and augment the set of instruments systematically. The validity of the additional instruments is tested by means of the standard Sargan/Hansen test for over-identifying restrictions. We also apply a test for serial correlation in the disturbance term in order to check whether the dynamic specification of the model is correct (for details see Göbel and Zwick 2009). Finally, we use the model with the smallest number of instruments and lags that satisfies all necessary test statistics.

In Göbel and Zwick (2012b), we use the age and establishment productivity profiles described above and split the establishments into groups with and without age-specific personnel measures. This allows us to test the hypothesis that the correlations between specific personnel measures and the relative productivity of different age groups are confined to the group of old employees. One exception is age-mixed teams because also younger employees might profit from the presence of older employees with complementary skills (Wegge et al. 2008, see chapter Age

Diversity and Team Effectiveness, Ries et al.). One disadvantage of our data set is that we only know whether an establishment uses SMOE or not, we do not know the effort or pervasiveness of the measure and which employee groups are affected.

In order to estimate the effect of human resource management measures on job durations of older employees, in Boockmann et al. (2012a) we implement a hazard rate model. In the existing empirical literature on employment duration there are two distinct types of models. One strand of the literature focuses on the duration of employment or tenure (van den Berg and van der Klaauw 2001). This literature emphasizes the correct empirical implementation of duration dependence and discusses how transitions out of employment depend on tenure. It is common in this literature not to pay particular attention to age effects; most studies focus on younger workers and exclude workers aged above 50 or 55. Another strand of the literature focuses on the analysis of employment duration of older workers or retirement decisions. It emphasizes age effects, but does not take duration dependence into account (for a recent application we refer to Hanel and Riphahn 2012). Blau (1994) and Blau and Riphahn (1999) derive their empirical duration model from a theoretical model of labor force transitions, without considering the interaction of age and duration dependence. However, the sampling probabilities, transition rates and survival probabilities during the observation period typically depend on both duration dependence as well as age dependence. Therefore, an appropriate empirical model should take both effects into account.

We specify a duration model where duration dependence, age dependence and the explanatory part enter multiplicatively. This specification is an extension of the proportional hazard model. In our specification, the transition rates are allowed to shift proportionally with the explanatory variables and age. Note that in this empirical model, survival in employment does not only depend on elapsed duration but also very flexibly on years of age. Our specification includes years of age as the baseline hazard; age effects are estimated in annual intervals for each year of age beyond 40. We can, thus, account for spikes in the probability of leaving the job, such as early retirement age. Identification of this model has been proven by Imbens (1994). This study extends existing results for identification of a parametric duration model by the simultaneous consideration of tenure and age effects.

We use the longitudinal version of the LIAB data to construct employment spells. Since we cannot observe the start of employment for spells before 1975 in Western Germany and 1991 in Eastern Germany, we have to take left-censoring into account. In order to limit the impact of left censoring, we only analyze the employment duration for Western Germany. Since we have a large sample, we use only transitions out of employment that occurred in 2002, the year in which we observe the application of specific measures for older employees. More precisely, we use a stock sample of all workers who are employed at January 1st, 2002 and who are older than 40 years, correcting for stock-sampling bias appropriately. We analyze transitions out of employment during the year 2002. Employment spells that continue beyond the year 2002 are right-censored at the end of the year and we adjust the likelihood accordingly. Since we have a large number of employment spells (59,099 spells from

300 establishments), the sample provides many transitions out of employment for all considered years of age and employment durations.

We exploit the rich information of the LIAB data and alternatively include employer characteristics or establishment fixed effects, as well as a set of employee characteristics. We then split our sample between establishments that use or use not particular measures and compare the age profile of the job exit probability (implied by our estimated baseline hazards) between these groups.

In Zwick (2011), the "Berufliche **We**iterbildung als Bestandteil **L**ebenslangen **L**ernens—Continuing Training as Part of Lifelong Learning" (WeLL) data set is used. This data set is also based on establishments observed in the IAB establishment panel (like the LIAB data set). About 150 establishments that indicated that they invested in training have been chosen from this data set. In addition, more than 6,000 individual answers on training behavior and socio-demographic information from employees in these establishments have been added to the establishment characteristics from the years 2007 and 2008.

Results

The aim of our Cobb-Douglas production function estimates is to show the relationship between the age shares and establishment productivity. We therefore mainly present these results here and refer the reader interested in the other coefficients, tests and sample details to the original chapters. In Göbel and Zwick (2009), the impact of different age groups on establishment productivity is first calculated for a pooled cross-section of establishments. The typical inverse U-shape of the age-productivity profile found in numerous studies can be replicated (see Fig. 2)—only the age groups older than 50 years are significantly less productive than the reference group 35–40 years (their confidence intervals do not overlap the zero line). Please note that the age group 35–40 years has no confidence intervals because it is the reference group. In addition, the value of the coefficient has to be normalized to zero. The influence of the different age groups depends on additional explanatory variables included. When we do not include decisive variables such as tenure, experience and cohorts, especially the impact of young age groups is much more negative than in the estimation presented here with the full range of explanatory variables (not shown here).

In a next step, endogeneity and unobserved time-invariable heterogeneity are taken into account by estimating the impact of changes in age groups on changes in productivity using GMM methods. All validity tests for the difference-GMM specification are fine. The age-productivity profile is an inverse L-shape and again, the included explanatory variables have a sizable impact on the results (see Fig. 3). Göbel and Zwick (2012a) show that there are no significant differences in age-productivity profiles between the manufacturing and services sectors. Even the metal manufacturing sector has a similar age-productivity profile, even though one

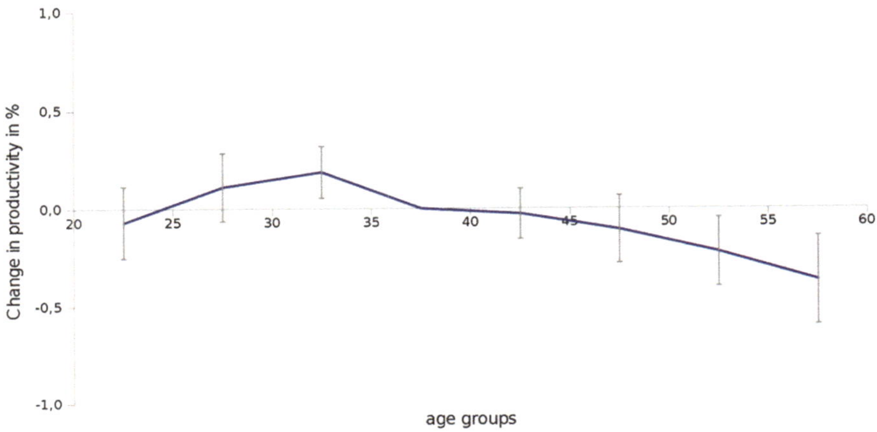

Fig. 2 Pooled OLS estimates of the age-productivity profile (*Note* 95 % confidence intervals are indicated by bars)

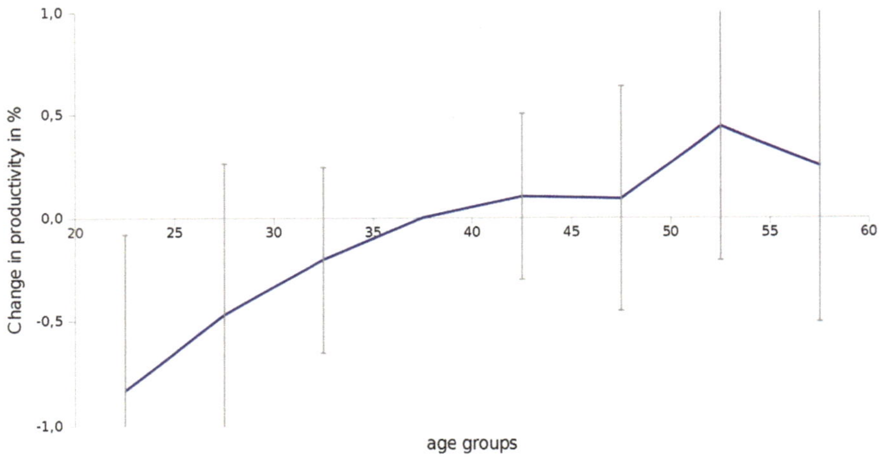

Fig. 3 Dynamic difference-GMM estimates of age-productivity profiles (*Note* 95 % confidence intervals are indicated by bars)

might assume that old employees have larger difficulties to keep up their productivity contribution in this sector.

The confidence intervals in our age-productivity profiles are relatively large, however. There are several possible explanations for the relatively imprecise estimation of age-productivity profiles. First, due to unobserved differences between enterprises or measurement errors, it could be that the specification does not capture the true correlations between age and productivity well. However, given the quality of the data, our rigorous specification and over-identification tests

Table 1.1 Percentage of enterprises that offer specific measures for old employees (SMOE)

	Share (%)
At least one SMOE	27.8
Specific equipment of workplaces	1.8
Reduced working time	14.5
Age specific jobs	3.9
Mixed-age working teams	11.1
Training for old employees	10.4

Note Source IAB establishment panel, wave 2002

and the flexible specification, we believe that this explanation seems not very likely. Second, there is indeed a large variance between enterprises—some enterprises actually can increase their productivity when the share of older employees increases and others see their productivity decline. The second interpretation can be tested by finding significant differences between these groups of enterprises and at the same time smaller variances for both groups.

Therefore, we re-estimate the age-productivity profiles separately for establishments with and without specific personnel measures for old employees in Göbel and Zwick (2012b). We find important differences between the age-productivity profiles depending on the application of different measures. This finding suggests that real variation in age-productivity profiles between establishments is one of the causes for the large standard errors of our estimates. The shares of establishments that use these specific measures are displayed in Table 1.

We find for three out of five measures that they have a positive impact on the relative productivity of old employees—the bands indicating the standard errors do not overlap here. In all cases, the variance bands for the separate enterprise samples are narrower than in the full sample. Establishments with a high relative productivity of old employees apply the following measures: specific workplaces for old employees, age-specific jobs for old employees, and mixed-age work teams (also see chapter Age Diversity and Team Effectiveness, Ries et al.), see Fig. 4. Enterprises with mixed-age work teams also have a relatively high productivity of their younger employees.

Göbel and Zwick (2012b) also find that there are two measures that are not related to relative productivity of old employees: reduced working time for old employees and training of old employees. An important reason for the finding of reduced working time for old employees is that this measure has been used by about 90 % as the so-called block model. This means that instead of flexibly reducing their working time in old age, employees work full time with reduced income for a certain time span (usually two years) and fully retire earlier. This form of hidden early retirement proved to be attractive for establishments until recently because it was subsidized publicly. Zwick (2011) shows that training of old employees is frequently not effective because enterprises do not take age-specific differences in training motivation into account. The consequence is that although old training participants get the same resources and comparable contents

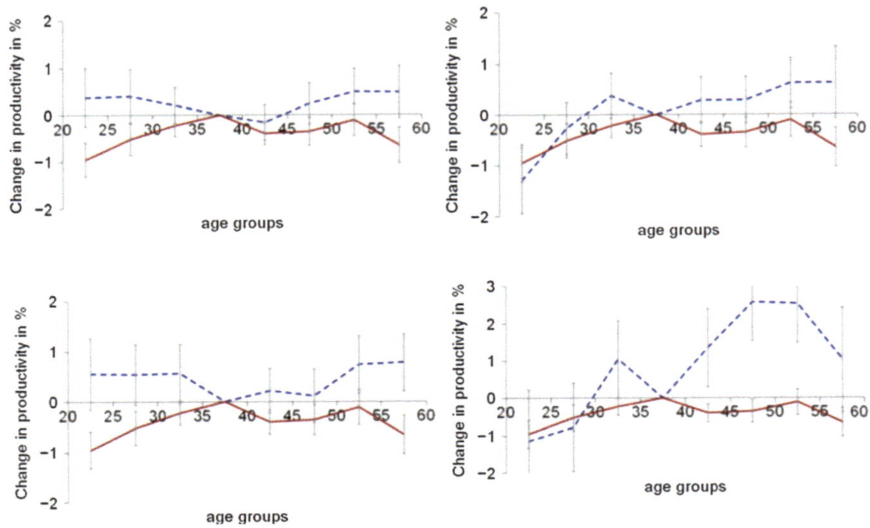

Fig. 4 Dynamic difference-GMM estimates of age-productivity profiles for different samples of enterprise. (*Note* From *left above* to *right below*: *Blue dashed lines*: enterprises with any personnel measure, enterprises with age specific jobs for old employees, enterprises with age mixed work teams, and enterprises with specific workplaces for old employees. *Red solid lines*: always enterprises without any measures. Standard errors are indicated by *bars*.)

and training forms, the self-assessed training efficiency with respect to relevant variables such improvement of career chances, higher income, higher flexibility, workplace security or higher productivity is significantly lower than for younger employees.

According to the hypotheses by Roßnagel-Stamov and Hertel (2010) and Kanfer and Ackerman (2004) on shifts in motivation by old employees described above, some training contents and methods are more effective than other training measures. In a direct comparison of the effectiveness of training methods, Zwick (2011) shows that, for old employees, formal seminars are less effective than training on the job and self-managed learning. The interpretation of this finding is that seminars frequently do not deliver knowledge that can easily and immediately be used at the workplace and training periods in seminars are not flexible enough. In addition, training in information and communication technology and in technical contents is less effective than training in management and communication skills. These results demonstrate that old employees are not keen on acquiring knowledge that is demanding with respect to fluid skills. They instead prefer training contents that can be easier absorbed by crystallized abilities. Management and communication skills in addition deliver immediate value-added for the quality of the workplace and communication with colleagues or team members.

Frosch et al. (2011) investigate how establishments achieve workforce rejuvenation—from the inflow of employees who are younger than their workforce or from outflows of relatively older employees. In addition, it is explored whether certain staffing patterns are more likely to appear under different economic regimes. The analysis of linked employer-employee data shows that most of the establishments covered rejuvenate by inflows of younger workers. Half of the establishments also use the outflow of older workers, however. Furthermore, in growing establishments, the workforce is more likely to become more age-heterogeneous. Moreover, in times of workforce decline, rejuvenation is primarily caused by outflows of older workers, and this is occurring regardless of the dominance regime (dominant firms have well-functioning internal labor markets and are able to attract and retain workers with high levels of human capital, dominated firms have less attractive career opportunities and lose valuable and self-trained workers to dominant firms). Further subdividing establishments into growth regimes according to whether they are dominant or dominated employers finally reveals that this phenomenon only proves true for dominated establishments. In contrast, more dominant establishments rejuvenate through the inflow of younger workers even in times of high external labor demand. Finally, the chapter does not find evidence for the hypothesis that a youth-centered human resource strategy (always) fosters innovation.

Now we present the results of Boockmann et al. (2012a). We focus on the effect of specific measures for older employees on the job exit probability at each year of age beyond 40, comparing between establishments that use or do not use particular measures for older workers: mixed-age working teams, age-specific part-time, reduced work requirements, a special subsidized part-time scheme, specific training for older employees, participation of older workers in general training, and age-specific equipment of workplaces. The main question behind our analysis is whether older workers stay longer with their employers if employers use one or several of theses policies.

As we can see in Fig. 5, based on our preferred specification with establishment-level fixed effects, age does not play a dominant role for transitions out of employment until the age of 55 for most of the measures. Only after the age of 58 is there a sharp increase in the effect of age on the transition rate out of current employment. Because job mobility is low at more advanced ages, this sharp increase is likely to be caused by labour force exits of the older workers concerned rather than by job-to-job transitions. Emphasis is on the shape of the duration profile rather than on its level. Indeed, the level information is absorbed by the establishment fixed effects.

The use of mixed-age work teams seems to reduce transition rates of older workers substantially in most of the age groups (upper left graph). Even at ages 45 and 52, transition rates are significantly lower in establishments using mixed-age work teams. Starting at age 55, the differences in transition rates are quite large in magnitude and are further increasing with age.

In the upper right part of the graph, we show the transition profiles of establishments that do and those that do not offer age-specific part-time. As mentioned above, there is evidence that a large part of establishments that use this measure apply the so called "block-model" for reduced working time. This is basically a

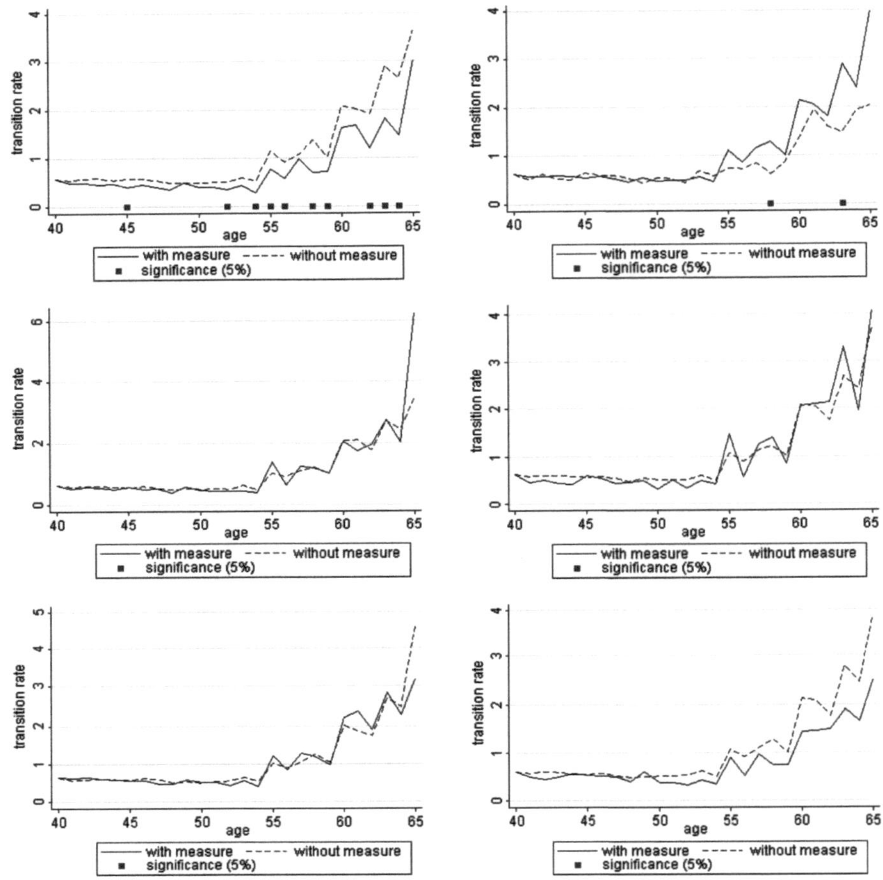

Fig. 5 Baseline hazards depending on age for the job exit probability. (*Note Solid lines* display estimates for establishments that apply a specific measure for old employees, *dashed lines* display estimates for establishment without this measure. From *left above* to *right below* mixed-age working teams, age-specific part-time, reduced work requirements, specific training for older employees, participation of older workers in general training, age-specific equipment of workplaces)

subsidized early-retirement scheme, and we would expect workers to quit earlier in establishments that apply this measure. Although the worker is effectively retired during half of the time within this scheme, he or she counts as being employed in our data. This underestimates the effective exit rates during the scheme. In spite of this underestimation, the results show that the application of age-specific part-time tends to reduce employment duration. However, the difference is only significant at particular ages. Results from other specifications show that job duration for younger age groups may be more stable in establishments using the part-time scheme.

Reduced work requirements for older workers do not influence employment duration of older workers; for none of the ages between 41 and 65 are there any significant differences between establishments applying or not applying this measure. The same applies to age-specific training and participation in standard training. In both cases, we would expect the measures to decrease transition rates. However, the figures show that exit rates have the same magnitudes irrespective of whether training is provided. Finally, the transition profile of establishments that have age-specific equipment of workplaces also does not indicate that this measure delays job exits.

Overall, we find that, among the measures considered, only aged-mixed teams are associated with lower job exit rates of older employees. This effect complements existing findings on the productivity effects of this particular measure. Age-specific part-time is related to shorter employment durations of older workers. This result is in line with descriptive information from the literature on the use of age-specific part-time as a means of early retirement. For other measures, such as training, reduced work requirements and specific equipment, we do not find a relation to employment duration. These results do not necessarily have a causal interpretation. Yet they show that one needs to evaluate each of the measures separately if one wishes to address the highly important question of how to achieve longer employment durations of older workers.

Discussion

A core result of all studies described in this chapter is that many negative associations with old employees are a result of correlations between old employees and other (frequently unobserved) characteristics that reduce performance. When these other characteristics are taken into account in addition to age, the relationship between age and performance is much more positive and the impression that age has a negative effect on establishment performance vanishes. These are some examples of mechanisms that negatively bias our perception of the performance of old employees:

- they are on average less educated,
- they belong to a cohort that was less productive already when their members were younger because they had to work with older vintages of technology,
- they are more likely to work in declining industries and establishments.

Moreover, one should keep in mind that age is also associated with characteristics that are positively correlated with performance. Prime examples are tenure and experience.

Besides taking into account individual and enterprise characteristics that are correlated both with the share of old employees and productivity or innovativeness, we have learnt another important lesson from this research. Establishments can influence their age structure, and the way they rejuvenate their workforce is correlated with their productivity. Highly productive and innovative enterprises with a high wage level for example have access to the best job candidates even in periods when less dominant firms do not find enough suited young applicants to fill their vacancies. Therefore we have also to be aware of the so-called reversed causality when looking at the relationship between age structure and employer performance—many firms might change their age structure after they have been successful or made an important invention and not the other way around.

Another important choice variable for enterprises is the selective use of early retirement or dismissals of old employees whose labor productivity is too low for their wage. The share of employees who have been dismissed against their will instead of voluntarily resigning from the job increases with age, and therefore the relatively high productivity of old employees may be a consequence of selectivity.

Also personnel measures aimed specifically at old employees are profit maximizing choices of managers and not random events. This means that we should not generalize positive effects of certain personnel measures on employee tenure and performance of establishments that have not used them so far. Probably their effects are smaller or their costs are higher for this group of establishments. In addition, personnel measures aimed at old employees might be part of a bundle of measures or be applied more frequently in enterprises with a specific work climate. An example is an establishment with strong internal labor markets that reserve many vacancies for promotion tournaments of insiders. This enterprise needs long-term incentives in order to motivate and retain its best employees and especially those who could not be promoted in a tournament. A widely used and efficient personnel measure to reach both goals is seniority wages where young employees get earnings below and long tenured old employees get earnings above their productivity (Zwick 2012). Seniority wages are associated with a longer tenure for old employees but also with a higher qualification level of the workforce and they are found more frequently in certain sectors than in others. A separate look at specific measures for old employees without taking into account the age earnings profile and internal labor markets would lead to biased results.

Outlook

So far, the potential selectivity of old employees especially beyond the age of 60 has not been taken into account in the literature. We know, however, that selectivity currently strongly increases after the age of 60 and we also find that only the more productive old employees stay. After the abolition of public subsidies for early retirement and since recently many establishments are keen on retaining their old employees in order to avoid skill gaps, we expect that the share of employees older than 60 years to increase dramatically during the next years. The change of selectivity might strongly affect the impact of old employees on productivity and innovation performance (and of course retention of old employees). When establishments are forced to keep low performing old employees, specific personnel measures for old employees might gain importance for the competitiveness of enterprises. So far, mainly establishment characteristics and their personnel measures have been analyzed in order to explain which enterprise can retain old employees. In order to obtain a complete picture, also individual characteristics should be included and compared with the weight of enterprise characteristics. Finally, the research so far is concentrated on employees until the age of 65 or before retirement. Almost ten percent of those who have been officially retired still work, however. Also the share of the so-called 'silver workers' is likely to increase in the future because people over 65 get healthier and the level of old age pensions guaranteed by enterprises and the state is bound to decrease further. Therefore, also this group of employees should be included in future research.

Göbel and Zwick (2012a) and Frosch et al. (2011) show that practitioners and policy makers should be aware that, despite all the efforts made in recent research, our knowledge about the interplay between workforce age, personnel measures and innovative performance remains very limited, especially due to the methodological caveats most studies so far experience. In this context, research on workforce age and innovation is still severely hampered by the lack of comprehensive innovation and human resource management data. This calls for the creation of a longitudinal dataset that includes reliable innovation indicators, such as patenting activity, detailed R&D expenditure or other innovation and personnel management activities as well as information on workers, their qualifications and previous careers. Combining existing linked employer-employee datasets with official and reliable patenting statistics or more detailed questionnaires on personnel measures and their intensity would provide the opportunity to study the career courses of workers and innovation processes and the impact of age-differentiated personnel measures on productivity on a methodologically and conceptually sound basis. Also multi-level approaches where case study evidence is combined with representative survey data seems promising for this research field.

Acknowledgments This project has been partly funded by the German Research Foundation (grant ZW172/1-2).We thank Katharina Frosch and especially Bernhard Boockmann for valuable advice and their input throughout the project.

References

Arellano M, Bover O (1995) Another look at the instrumental variable estimation of error-components models. J Econometrics 68:29–51

Aubert P, Crepon B (2006) Age, wage and productivity: firm-level evidence. Discussion paper INSEE, Paris

Blau DM (1994) Labor force dynamics of older men. Econometrica 62:117–156

Blau DM, Riphahn RT (1999) Labor force transitions of older married couples in Germany. Labour Econ 6:229–251

Blundell R, Bond S (1998) Initial conditions and moment restrictions in dynamic panel data models. J Econometrics 87:115–143

Boockmann B, Zwick T (2004) Betriebliche Determinanten der Beschäftigung älterer Arbeitnehmer. Zeitschrift für ArbeitsmarktForschung/J Labour Market Res 37:53–63
Boockmann B, Fries J, Göbel C (2012a) Specific measures for older employees and late career employment. ZEW discussion paper 12-059, Mannheim
Boockmann B, Zwick T, Ammermüller A, Maier M (2012b) Do hiring subsidies reduce unemployment among older workers? Evidence from two natural experiments. J Eur Econ Assoc 10:735–764
Cardoso A, Guimaraes P, Varejao J (2011) Are older workers worthy of their pay? An empirical investigation of age-productivity and age-wage nexuses. De Economist 159:95–112
Daveri F, Maliranta M (2007) Age, seniority and labour costs. Econ Policy 49:118–175
Frosch K, Göbel C, Zwick T (2011) Separating wheat and chaff: age-specific staffing strategies and innovative performance at the firm level. Zeitschrift für ArbeitsmarktForschung/J Labour Market Res 44:321–338
Göbel C, Zwick T (2009) Age and productivity—evidence from linked employer employee data. ZEW discussion paper 09–024, Mannheim
Göbel C, Zwick T (2012a) Age and productivity: sector differences. De Economist 160:35–57
Göbel C, Zwick T (2012b) Are personnel measures effective in increasing productivity of old workers? Labour Econ, forthcoming. http://dx.doi.org/10.1016/j.labeco.2012.11.005
Haegeland T, Klette T, Salvanes K (1999) Declining returns to education in Norway? Comparing estimates across cohorts, sectors and over time. Scand J Econ 101:555–576
Hanel B, Riphahn R (2012) The timing of retirement-New evidence from Swiss female workers. Labour Econ 19:718–728
Hellerstein J, Neumark D (1995) Are earnings profiles steeper than productivity profile? evidence from Israeli firm-level data. J Human Resour 30:89–112
Howard A (1988) Who reaches for the golden handshake? Acad Manage Executive 2:133–144
Imbens G (1994) Transition models in a non-stationary environment. Rev Econ Stat 76:703–720
Jacobebbinghaus P (2008) LIAB Datenhandbuch, Version 3.0, FDZ Datenreport 3/2008, Nuremberg
Kanfer R, Ackerman P (2004) Aging, adult development and work motivation. Acad Manag Rev 29:440–458
Ng T, Feldman D (2008) The relationship of age to ten dimensions of job performance. J Appl Psychol 93:392–423
OECD (2005) Ageing and employment policies—Germany, Paris
Schaie W, Willis S (2002) Adult development and aging, 5th edn. Prentice-Hall, New York
Skirbekk V (2008) Age and productivity capacity: descriptions, causes and policy options. Ageing Horiz 8:4–12
Stamov-Roßnagel C, Hertel G (2010) Older workers′ motivation: against the myth of general decline. Manag Decis 48:894–906
van den Berg GJ, van der Klaauw B (2001) Combining micro and macro unemployment duration data. J Econometrics 102:271–309
van Ours J, Stoeldraijer L (2011) Age, wage and productivity. De Economist 159:113–138
Wegge J, Roth C, Neubach B, Schmidt K, Kanfer R (2008) Age and gender diversity as determinants of performance and health in a public organization: the role of task complexity and groups size. J Appl Psychol 93:1301–1313
Zwick T (2011) Why training older employees is less effective. ZEW discussion paper 011–046, Mannheim
Zwick T (2012) Consequences of seniority wages on the employment structure. Ind Labor Relat Rev 65:108–125

Development and Evaluation of Working-Time Models for the Ageing Workforce: Lessons Learned from the KRONOS Research Project

Peter Knauth, Dorothee Karl and Kathrin Gimpel

Lessons Learned

German companies need to develop strategies to maintain the work ability and health of their ageing workforces to the retirement age of 67. Next to a variety of fields like management, qualification, work organisation, and health promotion, working-time designs that meet the requirements of ageing workers in their various life phases are of particular importance.

Under ten sub-projects implemented in six German companies from the automobile, steel, pharmaceutical, and chemical industries, ageing-appropriate working-time models (e.g. part-time work, short breaks, ageing-appropriate shift rotas, long-term time accounts) were examined and/or newly developed, introduced, and evaluated in the KRONOS research project. This chapter presents a total of ten sub-projects in each of the following working-time design fields:

- number of working hours per day, week or year,
- adequate rest periods,
- shift work,
- employee influence on working-time design,

P. Knauth (✉)
Karlsruhe Institute of Technology (KIT) (formerly Karlsruhe University),
Karlsruhe, Germany
e-mail: peter.knauth@web.de

D. Karl
Communal Policies of Employment, Jobcenter Mannheim, Mannheim, Germany
e-mail: dorothee.karl64@gmx.de

K. Gimpel
Corporate Center of Human Resources, ThyssenKrupp AG, Essen, Germany
e-mail: kathrin.gimpel@thyssenkrupp.com

C. M. Schlick et al. (eds.), *Age-Differentiated Work Systems*,
DOI: 10.1007/978-3-642-35057-3_3, © Springer-Verlag Berlin Heidelberg 2013

- long-term time accounts,
- beginning and ending of the daily working time.

Specifically, the following recommendations have emerged from KRONOS: uniformly reducing the daily working hours of all older employees is not recommended as personal health and work ability vary very widely among individuals. However, older employees who do hard work should take more frequent breaks. Shift systems based on fast forward rotation have a more favourable effect on work ability and health than traditional shift schedules that rotate backwards on a weekly basis. It is also recommended reduce the number of night shifts per person and year. The study shows, for example, that night-shift manning levels may be reduced even in partly automated processes by rescheduling certain activities from the night to the morning and afternoon shifts. Morning shifts should not begin before 06:00 a.m. Results also suggest that working-time choice and sabbatical options should be offered more widely.

An important result was, that there is no such thing as an "ideal" working-time model for an ageing workforce. First, there is a great inter-individual scatter in the work ability and the working time preferences of the employees and second companies follow different objectives with regard to operating hours and adaptation to demographic change. This is why working-time models need to be tailor-made.

Essential factors for the success of ageing-appropriate working-time models include:

- giving consideration to ergonomic recommendations,
- giving employees an opportunity to influence the design of their working time,
- ensuring favourable framework conditions (e.g. managerial attitudes, adequate participatory strategies of introduction), and
- integrating all constituent projects into a corporate master strategy to deal with demographic change.

Aim of the KRONOS Research Project

The conceptual question of the KRONOS project is: How can we maintain, support, and improve the health and work ability of our ageing workforce?

Only a comprehensive strategy of multi-dimensional interventions can help to maintain or improve the work ability of the ageing workforce. One important intervention within this strategy is the design and implementation of suitable working-time models.

The design of a working-time model has to take the following parameters into consideration:

- duration of the working time
  - (per day, week, month, year, phase of worklife),

- position of the working time
  - (time of day, weekday, season, phase of worklife),
- start and end of shifts
  - (fixed, staggered, sliding, variable),
- flexibility of the working-time arrangement
  - (no adaptation possible, planned change between different models, short-term adaptations),
- autonomy or control over arranging one's working hours.

Only 27 % of employed and 8 % of the self employed people have the classical working day, 07:00/08:00 a.m. to 05:00/06:00 p.m. from Monday to Friday (3rd European survey on working conditions). The number of unfavourable working-time arrangements has increased and the question is: Are there suitable working-time models for the ageing workforce? Therefore, the aim of the KRONOS research project was to study the effects of various working-time models on the ageing workforce and to deduce practical recommendations from the results.

Potential Negative Effects of Inadequate Working-Time Models

There are a number of potential negative effects of inadequate working-time models due to, for example,

- night-and shift work,
- long working hours,
- intensified work stress without adequate rest periods,
- inadequate compatibility of work with private life, and
- the transition from work to retirement.

Shift workers do not live in harmony either with their 'biological clock' (the diurnal rhythm of physiological functions) or the life rhythms of their social environment. This may result in a number of complaints (e.g. sleep or appetite disturbances)' social life and performance impairments and, in the long run, gastro-intestinal or cardiovascular diseases (Knauth 2007a). These problems escalate in many shift workers as they grow older e.g. more sleep problems (Bosch and De lange 1987; Foret et al. 1981; Härmä 1995; Moneta et al. 1996; Smith and Mason 2001; Tepas et al. 1993), enhanced risk of cardiovascular disease (Knutsson et al. 1986). On the other hand, there are shift workers who display no sign of major problems even at a relatively old age.

There are a number of preventive and compensatory measures designed to mitigate the psychological, social, and health problems of shift workers, one of the most effective being the design of shift rotas on the basis of ergonomic findings. When shift schedules for ageing workforces are designed on the basis of the results of recent studies (e.g. Härmä et al. 2006; Knauth et al. 2009a), not only older but also younger shift workers stand to benefit.

Important design parameters include working-time scheduling (shift work) and duration. Overlong daily working hours as well as working weeks that are longer than average may have a negative impact on, for instance, sleep quality, alertness, health, and social life, and they may cause accidents (Knauth 2007b; Wagstaff and Sigstad Lie 2011; Wirtz 2010). A combination of long working hours, a heavy workload, and unfavourable influences from the working environment is particularly critical. As it may be supposed that older employees have more problems dealing with such stress combinations, it has been suggested to reduce the working time (per day, per week, or per year) of older people in general. This misguided theory is based on the assumption that the work ability of all older employees is inferior to that of younger workers.

There are other problems next to shift work and overlong working hours, namely the increasing intensity of work and the narrowing scope of action (Ilmarinen 2006). In those cases where the causes of the problem cannot be eliminated, providing especially older employees with the option of taking an adequate number of breaks when needed appears to make sense (Ilmarinen 1999).

Another problem that has been discussed frequently in the past (Costa et al. 2004; Giebel et al. 2004; Janßen and Nachreiner 2004; Kandolin et al. 2001; Knauth 2007c) is the impossibility of reconciling one's professional and private life in the face of unfavourable work scheduling (e.g. flexible working hours based solely on the interests of the company). However, there are insufficient adequate working-time models being practised in the field and, more importantly, there are still too many managers who believe that the employees should be left to solve the problem on their own (Freier 2005). While child care is one of the key interests among younger employees, older employees are more interested in looking after relations. At all events, reconciling one's professional and private life can be done only where employees are given more influence on their daily, weekly, or yearly working time.

Similarly, retirement may be managed in various ways. Available options include the block model, gradual withdrawal from professional life, and working 100 % to retirement age. The block model of flexible retirement (e.g. working 100 % for 2.5 years and 0 % for another 2.5 years until retirement age), which has been generally used in Germany in the last years, does not make sense from the ergonomic point of view. More desirable alternatives include sabbaticals for regeneration during the worklife or a gradual withdrawal from it. Retirement from worklife is one of the most significant and largest transitions to occur during the life course. Withdrawing gradually and stepwise from worklife helps preparing for changes in private life, for example, in income, health services, functional capacity, hobbies and friends as well as experience sharing with a successor in the company.

Research Hypotheses and Methods

The key research hypotheses of the project may be summarised as follows:

The effects of age-differentiated working-time models on the health, work ability, and job satisfaction of an ageing workforce as well as on corporate performance will be all the more positive

- the more opportunities employees have to influence the design of their working time in the different phases of their life,
- the more ergonomic recommendations on working-time design are taken into account at an early stage, and
- the more favourable the prevailing framework conditions are (e.g. management attitudes towards older employees, ergonomic workplace design, in-service training for all age groups, working conditions that encourage learning, adequate strategies for launching new working-time models).

Specific research hypotheses for each sub-project are presented in Knauth et al. (2009a).

Various working-time models were developed, implemented, and evaluated using a mix of different methods (questionnaire surveys, interviews, working-group sessions, workshops, software development) (Table 1). For further information, see Knauth et al. (2009a).

In six companies from different industries, a number of sub-projects were realised whose objectives will be briefly described in the following.

The investigation in **Company 1** (automobile industry) aimed to establish the extent to which employees and managers are motivated to work part-time themselves and/or tolerate part-time work by colleagues and employees. Another objective was to think about a strategy that might help to enlarge the scope of part-time work. A questionnaire study was realised with 374 employees.

One of the sub-projects implemented in **Company 2** (chemical industry) similarly aimed to identify opportunities to extend the scope of part-time work. Thirteen employees were interviewed. Another sub-project in the same company was designed to analyse the experience of non-pay scale employees with long-term time accounts in the last two years. Ten non-pay scale employees were interviewed. Under the third and largest sub-project in the Company, medical examinations as well as a newly-developed questionnaire were used to establish the work ability of 981 employees working under three different working-time systems (3×12 h, 4×12 h and daytime work). All participants were examined by company physicians to get objective medical data [e.g. Framingham and PROCAM risk factors as well as many relevant ICDs, details cf. Knauth et al. (2009a)]. The questionnaire included questions referring to the type of shiftwork, lifestyle factors, social factors, and the Work Ability Index (WAI).

The objective of the first sub-project in **Company 3** (automobile industry) was to develop a holistic method of evaluating shift schedules on the basis of ergonomic recommendations and the employees' views. 104 shiftworkers participated

Table 1 Overview of the methods used in the various sub-projects

Methods	Company 1	Company 2			Company 3		Company 4	Company 5	Company 6	Total
	Part-time work	*Working time and health*	*Part-time in shift work*	*Long-term time accounts*	*Shift rota evaluation*	*Ageing-appropriate shift rota*	*Short breaks and rescheduled shift-change times*	*Long-term time accounts*	*Working-time choice*	
Working-group sessions	6	8	4	2	5	3	10	5	2	45
Information meetings	2	–	1	1	20	1	5	10	1	41
Interviews	8	–	13	10	14	13	–	9	–	67
Surveys	374	981	–	–	104	–	391	54	–	1904
Medical examinations	–	981	–	–	–	–	–	–	–	981
Workshops	–	–	–	–	1	2	2	–	–	5
Measurements	–	–	–	–	–	–	2395 Alertness tests, 5544 Tiredness scales, 5544 Physical complaints, 2395 Sleep problems			15878

in a questionnaire study. Among other things, interviewees were asked to assess the importance of the major shift-system characteristics (e.g. number of successive days off, number of successive night shifts, number of days off after the last night shift, shift duration, start of the morning shift, forward shift rotation) under the aspects of family, leisure and health. A scale of five ratings ranging from "entirely unimportant" to "very important" was used. Results were afterwards used to weight ergonomic criteria (deduced from the literature cf. Knauth and Hornberger 2003) for the design of shift systems (Knauth et al. 2009b).The new "multi-perspective IT evaluation tool for shift schedules" that emerged from these investigations has been described by Knauth et al. (2009b). The second sub-project in Company 3 was to check whether and, if so, how the individual stresses caused by night-time work can be reduced in a partly automated production process by reducing night-shift manning levels.

Based on the theory that is better to have several short breaks than a single long one, and that a good break schedule is even more important for older than for younger workers where hard work is concerned (Ilmarinen 1999), one of the sub-projects in **Company 4** (steel industry) was designed to optimise the existing unfavourable break schedule and investigate the effect of breaks on younger and older employees. The objective of another sub-project in Company 4 was to study the impact of specific shift-change times on quality of sleep, tiredness, and reaction time. Another item of investigation was to reduce tiredness in the morning shift by moving the start of the shift from 05:30 a.m. to 06:00 a.m. so as to allow more time for sleeping. In Company 4 the following methods were used:

- vigilance test (Walter Reed palm-held psychomotor vigilance test, cf. Thorne et al. 2005),
- visual analogue scale for subjective drowsiness scaling (cf. Kiesswetter 1988),
- scale to identify physical complaints (cf. Corlett and Bishop 1976)
- sleep questionnaire (in-house development).
- The numbers of measurements are shown in Table 1.

The sub-project in **Company 5** (pharmaceutical industry) was mounted to develop and implement a long-term time account concept for the workforce and evaluate a one-year test phase.

In **Company 6**, (chemical industry) the authors introduced a working-time choice model (Knauth et al. 2003) ten years ago which may be regarded as a model of the future for ageing workforces (Knauth 2009). The purpose of the project was to conduct a follow-up survey to establish the track record of the model over several years. After various working-group sessions and an information meeting, it was concluded that a survey would not be appropriate because the introduction of short-time work had caused considerable irritation within the plant. The works' council was afraid that only a minority of the workers would be willing to participate in a questionnaire study.

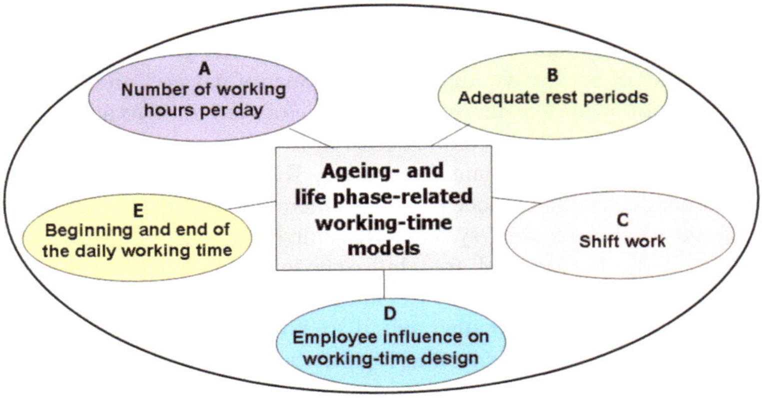

Fig. 1 Design fields for ageing- and life phase-related working-time models

Results

In the following, the main findings will be summarised in seven recommendations relating to various aspects of working-time design shown in Fig. 1. Each recommendation will be illustrated by selected results. The remaining findings have been published in the research report (Knauth et al. 2009a)

Number of Working Hours per Day

Recommendation 1 (Part-Time Work)

> Uniformly reducing the daily working hours of all older employees is not recommended as personal health and work ability vary very widely among individuals in the older age groups.

Broken down not by age but by the number of years which an individual spent working in shifts, the WAI totals shown in Fig. 2 display extensive inter-personal scatter. If all or most of the experienced shift workers would have a bad WAI, we would recommend to reduce the number of working hours per day. However, the group with the prolonged shift experience contained persons with a good as well as others with a bad WAI. In addition to this intra-individual scatter of subjective health and performance there is also an intra-individual variability of affect at work (Grube and Hertel 2008). In the 'more than ten years' category in Fig. 2, the scatter would probably be even wider without selection (healthy worker effect). Obviously, the solutions needed for people with a good WAI are different from

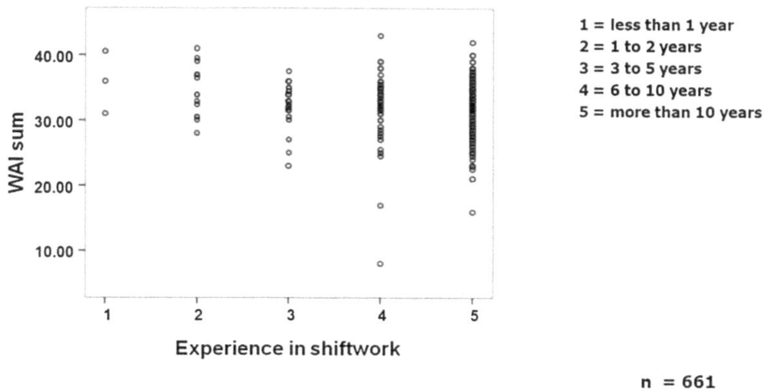

Fig. 2 WAI total versus shift biography (Karl et al. 2007)

those for persons whose WAI is bad (see also Ilmarinen 1999; Ilmarinen and Tempel 2002).

Adequate Regeneration Periods

Recommendation 2 (Breaks)

> Older employees with a heavy workload should be given additional breaks.

In a steel mill, 92 shift workers assessed the regeneration value of each break after it was over with help of a visual analogue scale ranging from "very restful/ refreshing" to "not at all restful/refreshing". Shift workers above 40 years of age rated the regeneration value of breaks significantly ($p < 0.05$) more badly than those below 40. This appears to underpin the demand for more breaks for older employees, e.g. published by Jansen and Haas (1991). They refer to a survey carried out by Shepherd and Walker in the metal industry: Economic losses per working shift dependent on the work strain, the breaks and the age of the workers. Under the condition "very heavy load without breaks" there were more annual production losses per working shift in the group of workers over the age of 45 years (7.55 %), than among the workers under 45 years (6.77 %). These losses decreased when "serial breaks" were allowed (<45 years 3.64 % and $\geq$45 years 3.51 %).

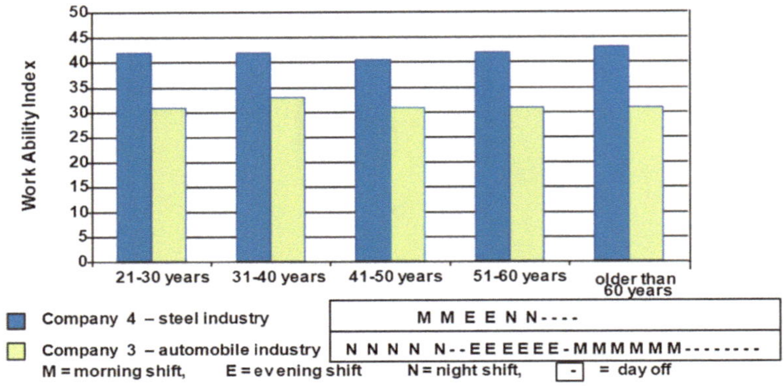

Fig. 3 Comparison of WAI averages under an ergonomically favourable (Company 4, N=391) and an unfavourable shift system (Company 3, N=104) broken down by age group (Knauth et al. 2009a)

Working-Time Scheduling/Shift Work

Recommendation 3 (Shift Rota Design)

> Shift systems designed to conform to recent ergonomic recommendations (e.g. fast forward rotation) have a more favourable effect on the WAI than traditional shift systems that rotate backwards on a weekly basis.

The lower part of Fig. 3 shows two shift schedules. The upper shift system, which rotates forward quickly, was introduced by the authors in Company 4 (steel industry) twelve years ago (Gördes et al. 2004). This shift system conforms to ergonomic recommendations regarding shift rota design. Conversely, the lower shift system implemented in Company 3, which rotates backwards slowly, does not comply with these recommendations.

Under the quickly rotating shift system, average work ability totals were significantly ($p < 0.05$) superior in all age groups to those of the traditional shift system. The data are based on cross-sectional studies. Unfortunately, longitudinal studies over many years and including all possible factors influencing the work ability could not be realised.

The working-time system is not the only factor by which the WAI is influenced negatively (e.g. strenuous physical work) or positively (e.g. adequate behaviour of the management towards older employees) (Tuomi et al. 1997). When comparing the shift systems in Companies 3 and 4 (Fig. 3), the fact that the condition of the working environment in the steel industry is markedly inferior to that in the automobile industry must be taken into consideration. Nevertheless, the WAI figures of Company 4 are significantly superior. This finding seems to underpin the importance of age-appropriate shift systems.

Week/Team	Mon	Tues	Wed	Thur	Fri	Sat	Sun
1	M	M	E	E			
2			M	M	E		
3	N	N			M	M	
4	E	E	N	N	N		
5	M	M	E	E			
6			M	M	E		
7	N	N			M	M	
8	E	E	N	N	N		
9	M	M	M	M	M	M	
10	E	E	E	E	E		

M = morning shift E = evening shift N = night shift ■ = day off

Fig. 4 Example of a discontinuous shift system involving 10 sub-groups and a one-third reduction of the night shift manning level

The fact that the WAI figures did not deteriorate in the older age groups is probably due to selection effects (healthy worker effect).

Recommendation 4 (Reduced Night-Shift Manning Levels)

> As the night shift is the most critical of all shifts as far as sleep quality, tiredness, performance, and health are concerned, it is recommended to reduce the number of night shifts per person and year. Attempts should be made to reduce night-shift manning levels by rescheduling certain activities from the night to the morning and afternoon shifts.

In a partly automated engine manufacturing plant belonging to Company 3 (automobile industry), a concept was developed to reduce night-shift manning levels. The first step taken was to identify any activities that could be rescheduled from the night shift to the morning or afternoon shifts. Mathematical simulations, which were carried out by the company, but could not be published, demonstrated that the manning level of the night shift could indeed be reduced by one third.

In the second step, we organised a workshop in which we cooperated with shift workers on developing three alternative discontinuous shift systems that largely conformed to ergonomic guidelines. Figure 4 shows one of these models.

If, for example, ten persons work on Mondays in each of the sub-groups shown in Fig. 4, the morning and afternoon shifts will be manned by 30 persons and the night shift by no more than 20. While this concept was developed during the duration of the KRONOS project, its implementation was deferred beyond the end of the project.

Employee Influence on Working-Time Design

Recommendation 5 (Working-Time Choice)

> Models that permit employees to switch between various weekly or annual work schedules in the course of their active life (working-time choice) are meaningful and attractive not only for older workers but also for younger employees who might want more leisure time.

In Company 6, shift workers working under a continuous shift system may choose between the following three weekly work schedules: 33.6/35.0/37.5 h/week (Knauth et al. 2003). If, to develop a continuous shift system, we divide the 168 h of work per week by 5 teams, we arrive at an average working week of 33.6 h in the basic shift schedule. However, a shift worker wishing to reach the full 37.5 h stipulated in the collective agreement for the average working week, which would give him his full pay, will have to work twenty-two extra shifts per year. If he opts for part-time, he receives less pay. A working week of 33.6 h requires no extra shifts, while part-time work of 35.0 h per week calls for fourteen extra shifts per year.

In the pilot year, it was mainly older shift workers suffering from gastrointestinal or sleep problems who opted for part-time work. Later on, however, younger shift workers interested in having more leisure time similarly opted for part-time work under the continuous shift system.

At the end of the pilot year, even full-time workers were more satisfied with the working-time regulation than before even though they had not taken advantage of the option to choose their own working time. The mere chance of switching to part-time should a personal need arise in the future had a positive effect on satisfaction (Knauth 2009).

Next to working-time choice, there are numerous other options for granting shift workers greater influence on the design of their working hours, of which only a few will be named below by way of example (Knauth et al. 2009c):

- time-autonomous groups,
- holiday planning by shift workers,
- 'personalised' duty rotas (Gauderer and Knauth 2004),
- variable working hours by group consultation,
- functional flexitime with three working-hour categories:
  - core time (all are present),
  - functional flexitime (hours directly before or after the core time in which at least one member of a group is present to handle customer contacts),
  - individual flexitime (hours beyond the aforementioned period in which workers may come and go as they please),
- time windows,

- short-and long-term time accounts.
- The last-named option was investigated under the KRONOS project.

Recommendation 6 (Long-Term Time Accounts)

From the ergonomic point of view, sabbaticals that promote regeneration are better than models under which employees work very many hours of overtime so that they may retire early later on, when their health may already be impaired. Options to pay into and withdraw from long-term time accounts must be tailored to the needs of the target groups.

Long-term time accounts are one option of adapting working-time designs to individual life phases. Thus, for example, long-term time accounts permit taking sabbaticals or retiring early without loss of income. The advantages of long-term time accounts extend not only to employees but also to employers as they may serve to enhance the attractiveness of the jobs offered, secure the loyalty of qualified employees and maintain their performance (if, for example, sabbaticals are used for purposes of regeneration or professional development).

A long-term time account concept was developed, implemented, and evaluated in two pilot departments of Company 5 (pharmaceutical industry), a manufacturing and a laboratory operation. The long-term time account was more attractive for the workers younger than 40 years than for the workers older than 40 years. On a scale from "entirely true" (1.0) to "not at all true" (5.0) the mean value for the younger workers was 1.76 and for the older ones 3.64.

Figure 5 shows sample results of a survey addressing desired options of withdrawing from long-term time accounts. Most interviewees thought that early

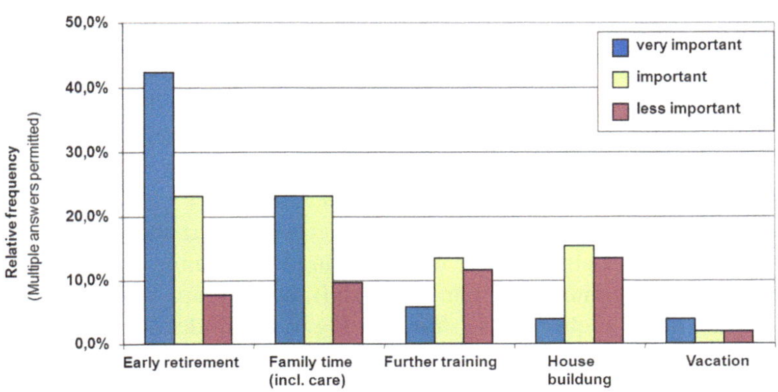

Fig. 5 Desired withdrawal options from long-term time accounts (N=54) (Knauth et al. 2009a)

retirement was either 'important' or 'very important'. Family time followed in second place.

As far as sabbaticals are concerned, most interviewees thought that relatively short leave periods such as four weeks made sense.

86 % of the full-time workers stated that they were able to "pay" overtime into their long-term time account, whereas 53 % of the part-time workers answered in the same way. 92 % of the under 40 years old workers wanted to use the long-term time account for an early retirement in contrast to only 70 % of the over 40 years old workers. More of the under 40 year old workers (64 %) than of the over 40 years old workers (27 %) wanted to use the long-term time account for time with their family. As the questions about objectives, options, limitations, and information levels revealed significant differences ($p < 0.05$) between full-and part-time workers as well as between younger and older employees ($p < 0.05$), it will be important to tailor long-term time accounts to the needs of the respective target groups and involve the persons concerned in the development of the concept.

Additional experiences with long-term time accounts and sabbaticals in four companies have been described by Zimmerman (1999, 2003).

Beginning and End of the Daily Working Time

Recommendation 7 (Start of the Morning Shift)

> The morning shift should not begin before 06:00 a.m. as an earlier start will have a negative impact on the quality of sleep before the morning shift as well as on tiredness and reaction time during the first hours of the shift.

This recommendation applies to average workforce members, not to extreme 'early birds'.

In Company 4, the tiredness of shift workers was measured on a visual analogue scale (Kiesswetter 1988) from wide awake (=1) to very tired (=9). In Fig. 6 the data of the control group and the pilot group were combined. Before the intervention of the pilot group (shifting the start of the morning shift to 06:00 a.m.) both groups started the morning shift at 05:20 a.m.

The average duration of sleep of the pilot group (5.6 h) and the control group (5.2 h) did not differ significantly when starting the morning shift at 05:20 a.m. However, at the beginning of the morning shift, both groups were almost as tired as at the end of the night shift as shown in Fig. 6 (more details see Knauth et al. 2009a).

Reaction-time measurements at the beginning and end of a shift produced comparable results (Knauth et al. 2009a).

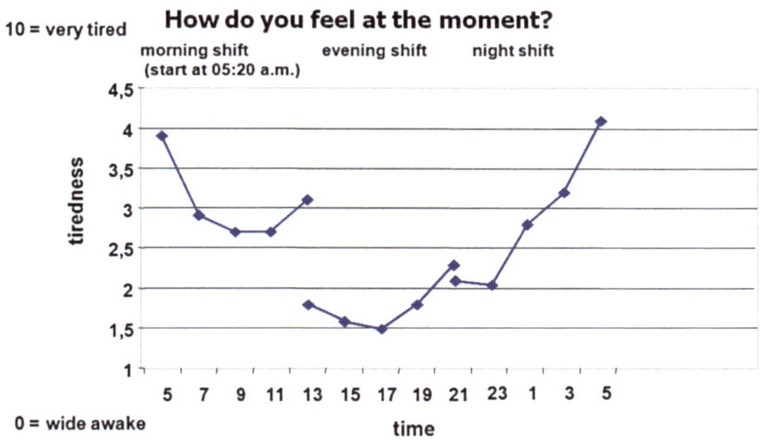

Fig. 6 Results of 4,796 measurements of subjective tiredness in steel mill (Karl et al. 2006)

Discussion

One important conclusion to emerge from the KRONOS research project is that there is no such thing as an 'ideal' working-time model for an ageing workforce.

The inter-personal scatter of work ability and health among employees increases with their age. This being so, you may find persons with a good Work Ability Index as well as others with a middling or even a bad WAI among shift workers as well.

Moreover, employees' working-time preferences change inter-as well as intra-individually during their active life, depending on, for example, their marital status, the age of their children, their non-professional activities, health, and need for money. We recommend, therefore, creating personalised solutions for employees, such as the option to choose their weekly or annual working time. In addition, other options to influence their working-time design might be offered, such as the joint planning of holidays by the members of a group, involvement in the development of a new shift system (which is required in German law to conform to recent ergonomic findings), sabbaticals, and agreements in time-autonomous groups about modifying shift-change times on short notice.

On the corporate side, too, various requirements apply to working-time and operational design. What has stood the test of time is tailor-made working-time models that reflect corporate objectives as well as the employees' wishes and the findings of ergonomics. The objective should always be to create a win–win situation for both employers and employees. An employer striving for greater working-time flexibility should give his employees a chance to influence their working-time design. Thus, for example, Kandolin et al. (2001) proved that combining corporate and individual factors in the design of working hours has a positive effect on the well-being of individuals and the conformability of their working and family life.

Working-time design for an ageing workforce should be embedded in a corporate master strategy for demographic change which covers management, work organisation, qualification, and health promotion as fields of action (Knauth et al. 2009c). In this context, the attitude and behaviour of the management are of particular importance. A Finnish longitudinal study found that the work ability of persons who were highly satisfied with the behaviour of their superiors was 3.6 times better than that of persons dissatisfied with it (Ilmarinen and Tempel 2002).

Outlook

Although there are a number of cross-sectional studies on potential negative effects of different dimensions of unfavourable working-time arrangements of younger and older workers, well-controlled longitudinal studies on the effects of age are missing.

Two trends, i.e. first the ageing workforce and second the increasing unsocial and irregular working hours, mean an increasing occupational and safety risk. There are studies, showing that individual flexibility may alleviate the negative effects of the company-based flexibility in working hours on subjective health, safety and the interference of the working hours with social well-being (Costa et al. 2004; Bohle et al. 2011; Kecklund et al. 2011; Takahashi et al. 2011; Viitasalo et al. 2008). Therefore Costa et al. (2004) suggest:

- "To develop methods for the evaluation of the impact of flexible working hours on workers' conditions, focusing more on human-centered outcome variables on health and well-being (long term effects, more objective assessments), preferably longitudinal studies with adequate time intervals according to the different effects monitored.
- To analyze better the interaction between work load, flexible working hours and work organization, and to support interventions studies with careful control of the effects (…).
- To develop guidelines on non-impairing or positive arrangements of flexible working hours according to ergonomic principles through participatory planning and publishing "best practice" examples showing that this is possible."

Besides research needed referring to flexible working hours we do not know enough about optimal arrangements of breaks for older workers and optimal arrangements of stepwise retirement.

Last but not least more fixed time windows for learning have to be provided for the ageing workers (e.g. specific days within a shift system). Although everybody agrees that lifelong learning is necessary, adult education decreases significantly in the oldest age-group of working people (Ilmarinen 2006). However, not only tight schedules at the workplace but also an insufficient support by the superior are obstacles to vocational training of older workers.

Acknowledgments The "Deutsche Forschungsgemeinschaft" (DFG) has supported the research project SPP 1184, GZKN 170/8-1. We owe special thanks to our colleagues PD Dr. Sonia Hornberger, Dr. Patrick C. Gauderer, Dr. Matthias Rott, and Dr. Christine Watrinet, who helped us to realize many subprojects. We would also like to thank almost 2000 workers and other representatives of the involved companies, who participated in the studies or organised e.g. meetings, interviews, questionnaire studies, workshops and medical examinations.

References

Bohle P, Willaby H, Quinlan M (2011) Flexible work in call centers: effects on working hours, work-life conflict, and health. Shiftwok International Newsletter, 41

Bosch LHM, De Lange WAM (1987) Shiftwork in health care. Ergonomics 30:773–791

Corlett EN, Bishop RP (1976) A technique for measuring postural discomfort. Ergonomics 9:175–182

Costa G, Åkerstedt T, Nachreiner F, Balteri F, Carvalhais M, Gadbois C, Gärtner J, Grzech-Sukalo H, Härmä M, Kandolin I, Sartori S, Silvéro J (2004) Flexible working hours, health, and well-being in Europe: some considerations from SALTSA project. Chronobiol Int 21(6):831–844

Foret J, Besimon G, Benoit O, Vieux N (1981) Qualitiy of sleep as a function of age and shiftwork. In: Reinberg A, Vieux N, Andlauer P (eds) Night and shiftwork. Biological and social aspects. Pergomon Press, Oxford, New York, Toronto, Sydney, Paris, Frankfurt am Main, 149–154

Freier K (2005) Work Life Balance Zielgruppenanalyse am Beispiel eines deutschen Automobilkonzerns. [Target group analysis referring to Work life balance in a German automotive company]. Peter Lang, Frankfurt, Berlin, Bern, Bruxelles, New York, Oxford, Wien

Gauderer PC, Knauth P (2004) Pilot study with individualized duty rotas in public transport. Le Travail Humain 67(1):87–100

Giebel O, Janßen D, Schomann C, Nachreiner F (2004) A new approach for evaluating flexible working hours. Chronobiol Int 21(6):1015–1024

Gördes C, Höfer K, Meiler RC (2004) Arbeitszeitflexibilität und Bündnis für die Arbeit. In: Kroll D, Neumann H (eds) Neue Wege der Organisation. [Flexibility of working time and alliance for work. New ways of organisation], vol 6. Gabler/GWV Fachverlag GmbH, Wiesbaden, Kap, pp 88–111

Grube A, Hertel G (2008) Altersbedingte Unterschiede in Arbeitsmotivation, Arbeitszufriedenheit und emotionalem Erleben während der Arbeit. [Age-related differences in work motivation, job satisfaction and emotio006Eal experiences at work.]. Wirtschaftspsychologie 10(3):18–29

Härmä M (1995) Ageing, physical fitness and shiftwork tolerance. Appl Ergon 27:25–29

Härmä M, Hakola T, Kandolin I, Sallinen M, Virkkola J, Bonnefond A, Mutanen P (2006) A controlled intervention study on effects of a very rapidly forward rotating shift system on sleep-wakefulness and well-being among young and elderly shift workers. Int J Psychophysiol 59:70–79

Ilmarinen J (1999) Ageing workers in the European Union—status and promotion of work ability, employablity and employment. Finnish Institute of Occupational Health, Ministry of Social Affaires and Health, Ministry of Labour, Helsiniki

Ilmarinen J (2006) Towards a longer worklife! Ageing and the quality of worklife in the European Union. Finnish Institute of Occupational Health, Ministry of Social Affaires and Health, Ministry of Labour, Helsiniki

Ilmarinen J, Tempel J (2002) Arbeitsfähigkeit 2010—Was können wir tun, damit Sie gesund bleiben? [Work ability 2010—What can we do that you stay healthy?]. VSA-Verlag, Hamburg

Jansen G, Haas J (1991) (eds) Kompendium der Arbeitsmedizin [Compendium of occupational medicine]. TÜV Rheinland, Köln

Janßen D, Nachreiner F (2004). Health and psychosocial effects of flexible working hours. Rev Saúde Pública, 38(Suppl.):11–18

Kandolin I, Hämä M, Toivanen M (2001) Flexible working hours and well-being in Finland. J Human Ergol 30:35–40

Karl D, Elmerich K, Knauth P (2007) Alter ist nicht alles! Der Einsatz des WAI bei Schichtarbeitern. [Age is not everything! The application of the WAI in a group of shiftworkers.] In: baua Bundesanstalt für Arbeitsschutz und Arbeitsmedizin (eds) Why WAI? Der Work Ability Index im Einsatz für Arbeitsfähigkeit und Prävention—Erfahrungsberichte aus der Praxis. [Why WAI? The Work Ability Index used for work ability and prevention—reports based on experiences in practice.] Lausitzer Druck-und Verlagshaus: 77–81

Karl D, Knauth P, Elmerich K, Rott, M, Watrinet C (2006) Teilprojekt KRONOS. Lebensarbeitszeitmodelle: Chancen für Unternehmen und Mitarbeiter. Subproject KRONOS. Lifelong working-time models: opportunities for enterprises and employees. Z Arb Wiss 4(60):256–264

Kecklund G, Åkerstedt T, Oxenstierna G, Westerlund A, Theorell T (2011) Work time control, shift work and disturbed sleep. Shiftwork International Newsletter 98

Kiesswetter E (1988) Das circadiane und adaptive Verhalten psychischer und physischer Funktionen bei experimenteller Schichtarbeit. [The circadian and adaptive behavior of psychic and physical functions during experimental shiftwork.] Verlag Peter Lang Frankfurt am Main, Bern, New York, Paris

Knauth P (2007a) Schichtarbeit. [Shiftwork.] In: Letzel S, Nowak D (eds) Handbuch für Arbeitsmedizin. [Handbook for occupational medicine.] Ecomed Medizin, Hüthig-Jehle-Rehm GmbH, 5. Erg. Lfg. 12/07:B IV-2, 1–29

Knauth P (2007b) Extended work periods. Industrial Health 1(45):125–136 Available from: http://www.jstage.jst.go.jp/article/indhealth/45/1/45_125/_article plus Erratum in following issue. Available from: http://www.jstage.jst.go.jp/article/indhealth/45/2/E1/_errata/brief (21.05.2007)

Knauth P (2007c) Kombination von Flexibilisierung und Individualisierung der Arbeitszeit. [Combination of transition to flexible and individual working-times] In: Dilger A, Gerlach I, Schneider H (eds) Betriebliche Familienpolitik [Managerial family policy.] Verlag für Sozialwissenschaften, Wiesbaden, pp. 141–158

Knauth P (2009) Wahlarbeitszeit—ein Zukunftsmodell. [Optional working-time—a model of the future.] Z Arb Wiss, 63(2):123–125

Knauth P, Hornberger S (2003) Preventive and compensatory measures for shift workers. Occup Med 53:109–116

Knauth P, Hornberger S, Scheuermann G (2003) Wahlarbeitszeit für Schichtarbeiter. [Optional working-time for shiftworkers.] In: Landau K (eds) Good Practice, Ergonomie und Arbeitsgestaltung. [Good practice, ergonomics and industrial engineering.] Ergonomie Verlag, Stuttgart, pp. 387–395

Knauth P, Karl D, Elmerich K (2009a) Lebensarbeitszeitmodelle. Chancen und Risiken für das Unternehmen und die Mitarbeiter. [Lifelong working-time models. Opportunities and risks for the enterprises and the employees.] Universitätsverlag Karlsruhe, http://digbib.ubka.uni-karlsruhe.de/volltexte/1000009932 (08.12.2009)

Knauth P, Gauderer PC, Elmerich K, Karl D (2009b) Multi-perspective IT evaluation tool for shift schedules. In: Schlick CM (ed) Industrial engineering and ergonomics, vision, concepts, methods and tools, Festschrift in Honor of Professor Holger Luczak. Springer, Berlin, Heidelberg, pp 211–219

Knauth P, Elmerich K, Karl D (2009c) Risikofaktor demografischer Wandel—Generationenvielfalt als Unternehmensstrategie. [Risk factor demographic change—diversity of generations as a corporate strategy.]. Symposion Publishing GmbH, Düsseldorf

Knutsson A, Åkerstedt T, Johnson B, Orth-Gomer K (1986) Increased risk of ischemic heart disease in shift workers. Lancet 2:86–92

Moneta GB, Leclerc A, Chastan J, Tran T, Goldberg M (1996) Time trends of sleep disorder in relation to night work. A study of sequential 1-year prevalences within the GAZL cohort. J Clin Epid 49:1133–1141

Smith L, Mason C (2001) Age and the subjective experience of shift work. J Human Ergol 30:307–313

Takahashi M, Iwasaki K, Sasaki T, Kubo T, Mori I, Otsuka Y (2011) Fatigue recovery, daytime sleepiness and depressive symptoms in a working population: the role of worktime control. Shiftwork International Newsletter, 160

Tepas D, Duchon JC, Gersten AH (1993) Shiftwork and the older worker. Exp Ageing Res 19:295–320

Thorne DR, Johnson DE, Redmond DP, Sing HC, Belenky G, Shapiro JM (2005) The Walter reed palm-held psychomotor vigilance test. Behav Res Methods 37(1):111–118

Tuomi K, Ilmarinen J, Seitsamo J, Huuhtanen P, Martikainen R, Nygard CH, Klockars M (1997) Summary of the Finish research project (1981–1992) to promote health and work ability of ageing workers. Scand J Work Environ Health 23(1):66–71

Viitasalo K, Kuosa E, Laitinen J, Härmä M (2008) Effects of shift rotation and the flexibility of a shift system on daytime alertness and cardiovascular risk factors. Scand J Work Environ Health 34(3):198–205

Wagstaff AS, Sigstad Lie JA (2011) Shift and night work and long working hours—a systematic review of safety implications. Scand J Work Environ Health 37(3):173–185

Wirtz A (2010) Gesundheitliche und soziale Auswirkungen langer Arbeitszeiten. [Health and social effects of long working hours.] Bundesanstalt für Arbeitsschutz und Arbeitsmedizin, Dortmund, Berlin, Dresden

Zimmermann E (1999) Alternativen zur Entberuflichung des Alters—Chancen und Risiken für Arbeitnehmer aus innovativen Arbeitszeitmodellen. [Alternatives to withdrawal of employment due to age—Opportunities and risks for employees in innovative working-time models.] In: Behrens J, Morschhäuser M, Viehbrol H, Zimmermann E (eds) Länger erwerbstätig—aber wie? [Being longer employed—but how?]. Westdeutscher Verlag, Opladen/Wiesbaden, pp 116–172

Zimmermann E (2003) Chancen und Risiken innovativer Arbeitszeitmodelle für ältere Arbeitnehmer. [Opportunities and risks of innovative working-time models for older employees.] In: Bandura B, Schellenschmidt H, Vetter C (eds) Fehlzeiten-Report 2002. [Report of absenteeism in business—2002.] Springer, Berlin, Heidelberg, New York, Hongkong, London, Mailand, Paris, Tokio, pp. 167–183

Effects of an Ageing Workforce on the Performance of Assembly Systems

Gert Zülch, Martin Waldherr and Marcel Becker

Lessons Learned

In this chapter we present the results of a four-year research project structured in two phases. The topic is a simulation supported approach to analyse and re-structure existing assembly systems (phase 1) respectively to develop a new approach for planning pulsed assembly lines (phase 2).

It is undisputed that a person's working capacity and performance supply changes throughout the course of his or her life. As well as other authors we cannot confirm the sweeping judgement that older workers are generally more inefficient than younger ones, but with respect to biologic- and functional-oriented approaches, work performance is subject to a certain differentiated change. This leads to a reduced output from older workers in some situations, which, however, can be compensated, at least in part, through their accumulation of competencies. In certain activity areas, they might even perform better than their younger colleagues. In particular, a difference between the physical and mental performance can be observed during ageing. While the first aspect will be in focus of this chapter, we may refer to Schaie (2005) and Zülch and Schmidt (2011) with respect to the latter one.

The type of change and the expected performance changes are influenced strongly by earlier occupational activities as well as general life circumstances, and can thus only be estimated for certain personnel types. The maintenance of occupational efficiency is subject to a number of influencing factors, whose continuous activation

G. Zülch (✉) · M. Waldherr · M. Becker
ifab-Institute of Human and Industrial Engineering, Karlsruhe Institute of Technology, Karlsruhe, Germany
e-mail: gert.Zuelch@kit.edu

C. M. Schlick et al. (eds.), *Age-Differentiated Work Systems*,
DOI: 10.1007/978-3-642-35057-3_4,

throughout the entire occupational life play a significant role. Long-term examinations through targeted personnel development can help prevent future production bottlenecks. At this aim, enterprises must learn to appreciate the performance changes, both for individuals, for personnel types and for teams. Ergonomic working conditions as well as the training of physical and mental abilities will gain in importance in enterprises in order to influence the "ageing process" so that occupational efficiency can be maintained up until an old age.

Due to the possibilities of early retirement and the relocation of older workers to easier jobs, the emerging problem of the demographic development was not initially perceived in many production companies. Only the change in the socio-legal framework led to greater attention to this issue. Negative effects of ageing workers will increasingly be found in the extreme ranges of manual activities, in which, up until now, no older worker potential was expected.

Simulation-supported configuration of scenarios can cover part of the planning problem since each simulation scenario can merely represent one prognosis of the future reality. Due to the lack of models for a quantitative representation of the influences of age, the article at hand shows a coupling of some indices and effects, in particular with respect to the efficiency rate (see the definition below). Under that assumption of efficiency rate changes, the consequences of an ageing workforce for a specified assembly system were prognosticated in this pilot study.

With respect to our findings, the following recommendations due to effects of an ageing workforce on the performance of assembly systems can be derived:

- Simulation is an approach which helps to analyse effects of age-differentiated capability changes on system performances.
- Risk-related aspects should be discussed while planning new production systems.
- Characteristics like work flow principle and qualification has impact on the performance of ageing-robust production systems.
- Reducing the average metabolic energy consumption supports ageing-robustness of production systems.

Introduction

As already explained in the first chapter of this book, demographic development has an effect on employment, i.e. increased employment of older workers and fewer young employees joining the workforce. This is particularly the case for assembly systems that are strongly characterized by physical stress (Willnecker 2001, p. 34). In the future, it will be important to recognize the impact of an ageing workforce on the performance of assembly systems and to counteract it using appropriate measures. The original question behind our research project described here was therefore what effect a change in the performance of the workforce would have on a work system. This also raises the question of the extent to which innovative approaches such as simulation procedures can be used in the targeted design of an ageing-robust work system.

In the course of the research project, knowledge was gained on how assembly systems can be adjusted with regard to ageing-related effects. This knowledge led to the development of a planning process that takes these changes in a pulsed assembly line into account through balancing the stations' physical workloads as early as in the planning stage.

In order to investigate the effects of a workforce with age-related changing capabilities we need concrete information on the relationship between changes in the performance of individual employees and the effect on their productivity. It should also be possible to estimate the effect of ergonomic and organisational measures on the performance of the work system as a whole.

Challenges in the Design and Evaluation of Ageing-Robust Work Systems

The performance capability of a worker is inextricably intertwined with the development of the physical and mental skills, with the existing qualification and the state of health. Landau et al. (2009, p. 55) shows that the biological parameters of the physical performance capability will change with age, and that this change is mainly of a negative nature, e.g. changes in the muscular strength or in the maximum oxygen uptake of the blood. Tendencies towards a change of both the sensory skills and the motor skills have been confirmed in other research works (cf. e.g. Landau et al. 2007, p. 19; Wild-Wall and Falkenstein 2009; Hegele and Heuer 2010; Trautmann et al. 2011).

Rademacher et al. (2006, p. 232) highlights the increasing risk of the worker performance capabilities being reduced for work-related reasons resulting from highly repetitive tasks. This is particularly pronounced in manual assembly systems that are strongly based on the division of labour. This often leads to the "deficit hypothesis" which states that this development of reduced worker capabilities is accompanied by a performance drop at the assigned assembly system.

This statement is opposed by numerous studies which prove that workers compensate or even overcompensate for such a reduction in performance capabilities by substitution with other abilities (experience, methodical handling of tasks; Landau et al. 2009, p. 56). Furthermore, physical stress in a certain work situation can be reduced by interventions related to ergonomic workplace design or work organisation (cf. Enríques Diaz et al. 2009, p. 581; Landau et al. 2009, p. 60).

Figure 1 shows three studies conducted on the performance capability of ageing employees. Each study is scaled to 100 % with its highest performance related value. Two of the studies are based on the Work Ability Index (WAI; Ilmarinen and Tempel 2002, p. 339) with one study focusing on employees performing mainly manual tasks and the other study looking into employees performing mainly non-manual tasks. The Work Ability Index is based on the subjective assessment of a worker concerning his or her ability to work.

The third study by Baines et al. (2004, p. 524) evaluates the performance ability against the efficiency rate. The efficiency rate is the ratio of the standard time for a specific operation to the time required by the worker. Baines et al. (2004, p. 518) used this approach to model an assembly line in the test dress area of an automotive engine that is the last section prior shipping. The graph of performance efficiency has its seeds in two human performance models. The first one is referring to general—both cognitive and physical—performance decreases with increasing age (e.g. Snel and Cremer 1995). Baines et al. (2004, p. 519) assumed that the non-gender-specific performance decrease due to ageing increases linearly from 0 % at age 30 to 1 % at age 65 (see Fig. 1). The second model refers to human basal and higher-level systems which can be displayed in arousal and efficiency variations over time, daily, weekly, monthly and even over years (Monk and Leng 1982). Hence, this model focuses more on short term variations and less on long term changes of performance level.

Unfortunately, there are no findings concerning the correlation of the WAI and the efficiency rate. However all three studies postulate a decrease in performance ability with age. The two curves representing the physical performance capability show a stronger decrease in performance capability at higher age in comparison to mental work.

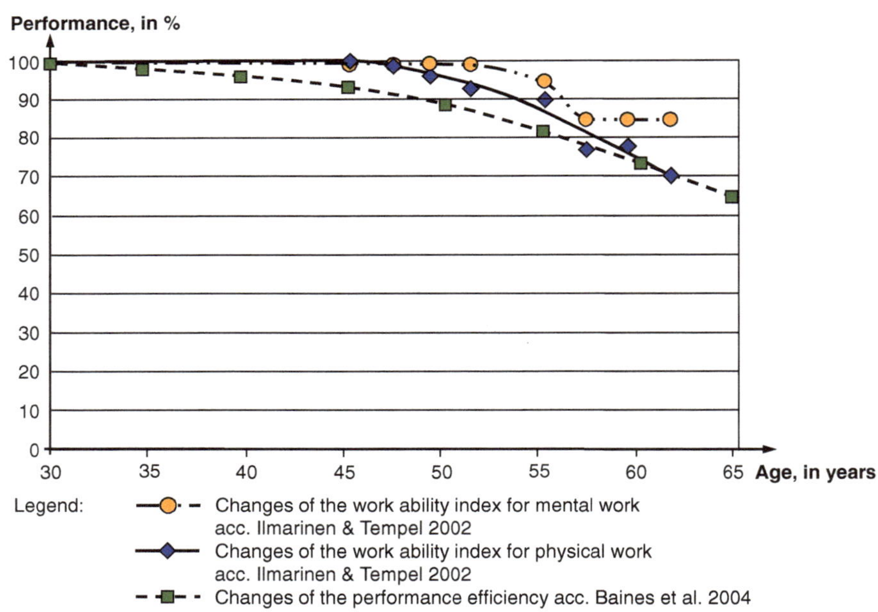

Fig. 1 Studies of performance changes due to ageing (acc. to Zülch and Waldherr 2011, p. 1–197)

However Landau et al. (2007, p. 32) state that the decrease in physical performance capability has been observed to be approximately 4 % over a period of 20 years. This study is referring to non-repetitive lifting and carrying tasks. Our following simulation study assumes a development of the performance efficiency according to this statement (also cf. Zülch and Waldherr 2010, p. 352).

Obviously, the statements of Ilmarinen and Tempel (2002, p. 339) as well as Baines et al. (2004, p. 524) conflict with the statement of Landau et al. (2009, p. 56). Findings of Rademacher et al. (2010, p. 193) support more the statement of Landau et al. (2007, p. 32). Due to this fact we modified our former assumptions upon the individual performance changes related to physical work (see Zülch and Waldherr 2010, p. 352). The decrease of the efficiency rate in our further work will likely be less abrupt as postulated by Baines et al. (2004, p. 524).

Simulation of Assembly Systems with an Ageing Workforce

Research Outline

In the first phase of our project, we developed a computer-based simulation procedure that allows human and technical resources to be modelled in detail with regard to an ageing workforce. For this purpose, the existing simulation procedure *ESPE* (German acronym for "Bottleneck-oriented simulation of personal structures") was enhanced by the module *AS* ("Module for ageing-robust structural planning"; see Zülch and Becker 2006, p. 152). The simulation procedure, which we have continuously developed since 1994 cannot only be used for assembly systems, but also for the simulation of parts manufacturing and service systems (for a more detailed description of this simulation procedure see Zülch et al. 2009).

In the framework of this project, the procedure was used for a pilot application in the automotive industry to forecast the performance of a component assembly line over a period of fifteen years using the same workforce. *ESPE-AS* allows criteria relating to production logistics (e.g. lead time, delivery reliability, degree of system output) to be taken into account as well as personnel-related criteria (e.g. workforce utilisation, age-related development of efficiency rates). This multi-criteria approach was used to calculate a total degree of goal achievement based on the individual criteria (see e.g. Zülch and Waldherr 2010). The individual degree of goal achievement standardizes a given key criteria to a value range between 0 and 100 %, where the former value is the pessimistic value and the latter the optimal one. The total goal achievement is then the weighted average of all individual degrees taken into account.

In the second phase, we combined this simulation procedure with a planning heuristic to balance the physical workloads of stations in pulsed assembly lines. A genetic algorithm that had been specifically adapted to this problem was developed for this purpose.

The pilot studies conducted as part of the research project revealed that not only manual but also hybrid assembly systems are affected by demographic development. Particularly in pulsed assembly lines, human performance variability is difficult to incorporate using existing planning methods. Furthermore, technical and organisational measures in this regard are often seen as economically unjustifiable from a business perspective. From an ergonomic perspective, this will lead to the challenge of making work systems ageing-robust in the long term.

Modelling an Ageing Workforce Using Simulation

In order to balance assembly lines, mainly technology-oriented approaches are known, which try to reduce line balancing losses, i.e. the residual time between the takt time and the operation times assigned to each assembly station. As an example, Schad (1986, p. 33) states: "The purpose is to allocate the tasks to the workers in an existing work system at minimum costs with the production programme typically being predefined" (transl. by the authors). This definition emphasises the economic goal for the planning process, which obviously is of high importance for any manual or hybrid assembly system. It does, however, not consider, whether a worker is capable of coping with the physical workload which derives from the assigned operations and can have short-term or long-term negative effects to the workers' health and efficiency, in particular when it comes to older workers.

There are specific planning procedures for assembly systems; the underlying methods, however, do not consider the skill development of ageing workers. When planning new work systems, it is frequently assumed that the skills to fulfil the tasks will actually exist. In some planning tools the skills of the workers are merely reflected by their wage group or their functional allocation to a certain work station or task. Moreover, it is often even assumed that all workers have the same level of qualification (cf. Müller 2002, p. 103). But these assumptions are by no means realistic.

None of the available planning processes allows for a specific ageing-robust balancing of assembly systems assuming inter- and intra-individual changes in skills over several years. Considering age-related differences in skills and changes thereof, however, is the key to designing ageing-robust assembly systems (Zülch and Becker 2008, p. 385). The idea is to apply new methods which strongly consider the ability changes of ageing workers. If this turns out to be successful, most cost-intensive improvements of existing assembly systems at a later stage of their operating life can be avoided.

The importance for new planning processes lies in taking the skill development of ageing workers into account both, for the technical equipment of an assembly system and for its organizational design, in particular with regards to work structuring and personnel assignment. By bearing in mind both these aspects, the workers should ideally be able to continue in their job until reaching the statutory retirement age. A method which considers these aspects in a new type of planning process has been developed in the second phase of the research project.

In an application study, Baines et al. (2004) have simulated the decreasing efficiency of assembly workers in an automotive company above the age of 30 and compared the data with existing real data from an assembly system. According to Baines et al. (2004, p. 524), the graph efficiency rate against age follows a declining curve (see Fig. 1), but this graph holds for all operations in the regarded assembly line. Yet, the extent to which the intensity of the assembly work influences the age-related efficiency rate of a worker is different for each assembly operation (cf. Zülch and Becker 2008, p. 379). Therefore, this refined assumption was incorporated into the simulation procedure *ESPE-AS*.

Furthermore, it is not sufficient to design the technical equipment of an assembly system in such a way that facilitates its productive operation during the entire operating life of the assembly system. Instead, the new challenge consists of adapting the assembly system to the changing individual performance capabilities of its core workforce, which is predicted to be ageing, with regards to their technical and organisational aspects early on during the planning phase (e.g. working aids or group work). Furthermore, it will be necessary to adapt the assembly system in response to quantity variability, workforce changes due to retirements, or any other types of changes.

Procedure for the Ageing-Robust Balancing of the Workloads of Assembly Systems

Both manual and hybrid assembly lines are characterized by the direct linkage of equipment according to a flow pattern. These assembly lines are most affected by an ageing workforce as a delay at one single work station due to an older worker requiring more time can bring about the stoppage of the entire assembly line (Willnecker 2001, p. 34).

Balancing the workloads of assembly stations is generally referred to in the literature as the Assembly Line Balancing Problem (ALBP; see e.g. Boysen et al. 2007, p. 674). Willnecker (2001, p. 122) defines assembly line balancing as "determining the takt time of an assembly system and… summarizing subactivities into workflows at individual work stations" (translated by the authors). In the case of a single-product assembly line with a fixed takt time, as in the case described below, the corresponding optimization problem is assigned to the problem category SALBP-1, or simple ALBP variant 1 (refer to Scholl 1999, p. 23; Boysen et al. 2007, p. 512, for the classification of ALBPs).

Procedures for solving SALBPs can be found many times in the literature (see the overviews given by Domschke et al. 1997, p. 179; Scholl and Becker 2006, p. 669). Because the complexity of the problem means that an exact solution cannot be found within reasonable time, heuristic methods are used in practice (see Gerdes et al. 2004,

p. 12). These methods use rules that make it possible to reduce the huge number of potential solutions, allowing a satisfactory solution to be found within relatively short time, e.g. in terms of the minimum number of work stations. However, such an algorithm runs the risk that it terminates in a local optimum and thus not finds the global optimum solution (see Gerdes et al. 2004, p. 15).

Genetic algorithms are a commonly used heuristic procedure for solving SALBPs (see Chong et al. 2008, p. 1273). A genetic algorithm is a computer-based optimization procedure based on evolutionary theory. In biological terms, evolution is an ongoing process in which a system adapts to a constantly changing environment (Schöneburg et al. 1994, p. 185). However, instead of being a linear improvement, changes to the planning solution occur randomly (see Schöneburg et al. 1994, p. 84): Solutions only survive if they have an advantage over the best previous solution ("survival of the fittest"; Schöneburg et al. 1994, p. 152). Genetic algorithms therefore mimic the process of evolution in order to improve a set of existing solutions. This makes the final solution-finding process iterative. However, previous algorithms developed to solve SALBPs, and genetic algorithms in particular, do not take the abilities of the staff into account, especially the effects of ageing that can arise within an assembly system over a long operating life.

Genetic Algorithm for the Planning of an Ageing-Robust Assembly System

In light of this, we used a genetic algorithm to develop a planning process for the ageing-robust balancing of assembly lines that also takes into account the physical workload of the assembly workers. The metabolic energy consumption required by the individual assembly operations were estimated using the method described by Spitzer et al. (1982, p. 21). The results of the calculation were included in the planning process. The process comprises the following stages (see also Fig. 2):

1. Creation of initial solutions

Because a genetic algorithm can only be used to improve existing solutions, a number of initial solutions are required which are then evaluated (stage 2) and modified (stage 3). When an existing assembly system is being redesigned, this system already serves as one initial solution. However, one single initial solution is insufficient for the algorithm, and additional feasible solutions must be created. When an assembly system is being planned for the first time, on the other hand, no initial solutions are available to start with and have therefore to be newly created.

2. Simulative evaluation of each solution

Each existing or newly created solution is evaluated by means of simulation on the basis of its total degree of goal achievement (see the definition above), relating to a given planning period spanning several years of operating life (the principle

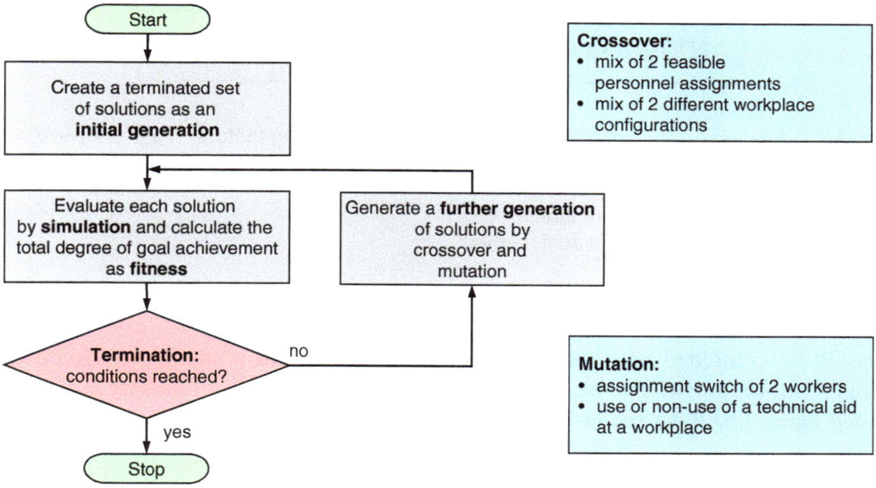

Fig. 2 Flow chart of the algorithm for ageing-robust assembly line balancing

for calculating a total degree of goal achievement is shown at the end of this paragraph). Each new solution must then be evaluated in terms of the total degree of goal achievement. This is done because only a set of the best solutions are used in subsequent iterations ("survival"). When planning an ageing-robust assembly system, it is not enough merely to examine the work situation in the early stages of the planned assembly system. Instead, a planning solution can only be described as ageing-robust if productivity declines only slightly or not at all for age-related reasons when the assembly system is in operation for many years (see Zülch et al. 2009, p. 2). This is why the personnel-oriented simulation procedure *ESPE-AS* is used to evaluate the solutions arrived at.

3. Creation of additional planning solutions

The reproduction and subsequent selection of the best planning solutions from the previous stage is used to generate the next planning solutions. New planning solutions can be created using crossover and mutation (Gerdes et al. 2004, p. 33). Mutation involves making random changes to individual elements of a planning solution, in the same way that random mutations occur in DNAs. This could mean swapping the assignments of two assembly workers in a particular year of operating life, for example. Recombination results in changes to a planning solution by taking parts of two existing solutions and forging them into a new solution. For example, the assignment of staff from planning solution 1 for years 1 through 7

$$TDG = a \bullet GAE + b \bullet GAM + c \bullet GAS + d \bullet GAW$$

GAE	Efficiency rate curve of staff	Weighting factors:	$a + b + c + d = 1$
GAM	Metabolic energy consumption		
GAS	System output		$0 \leq a, b, c, d \leq 1$
GAW	Average workload of staff		
TDG	Total degree of goal achievement		

Fig. 3 Calculation of the total degree of goal achievement (following Zülch et al. 2009, p. 5)

could be combined with the personnel assignment from planning solution 2 for years 8 through 14. The new planning solutions generated in this way are evaluated using *ESPE-AS* (as in stage 2).

4. Selecting the best planning solutions

A set of the best planning solutions are selected for the next iteration on the basis of this simulative evaluation. This selection is random, with the likelihood of selection increasing with the quality of the solution as measured by the total degree of goal achievement. This mirrors the evolutionary principle that only the "best survive" (Schöneburg et al. 1994, p. 152).

Stages 3 and 4 are repeated until at least one of the following two terminating conditions has been met: The first condition that brings about the termination of the algorithm is reaching a predetermined maximum number of iterations. The second condition may terminate the algorithm sooner if the assessment of the best planning solution changes only marginally in the course of several iterations. It is necessary to define two terminating conditions in order to avoid the endless repetition of the iteration steps (see Chong et al. 2008, p. 1275).

To evaluate each planning solution in stage 2, we used a specific total degree of goal achievement for this case, which consists of four individual degrees goal achievement (see Fig. 3). First of all there is the goal achievement system output that represents the average productivity of the assembly line. Second, the goal achievement average workload of staff stands for the average staff utilisation of the assembly line. Third, the goal achievement of efficiency rate development is defined as the average change in efficiency rate of all workers who are virtually allocated to this assembly system between the first and the last year of its operating life. The dedicated individual change in efficiency rate depends on the task and worker-related efficiency rate. Last not least the goal achievement metabolic energy consumption is defined as 1 minus the average needed metabolic energy consumption per shift divided by a maximum of energy consumption per shift. The latter is set to 10,000 kJ per shift in case of this study.

Then we calculated the total degree of goal achievement which in general can be defined as the weighted sum of all included individual degrees. In the given case we calculated equal parts of the four degrees of goal achievement.

In the genetic algorithm, this total degree of goal achievement of a planning solution equals the fitness of the solution. It is thus decisive for the solution being transferred to the next iteration of planning solutions.

Simulation Studies in the Automotive Industrie

In the following we describe the simulation studies in the automotive industry conducted as part of the research project. The study used in the first phase is a decoupled assembly system for passenger car components. The simulated period was 15 years. In the second phase we used an assembly system that was organised as pulsed line according to a flow pattern. All planning solutions of this second phase have their seeds in the precedence diagram of the related assembly operations. The study used here was further refined for a third simulation study. As common equivalent of two car generations we reduced the simulated period to 14 years in the second and third simulation study. We set the different simulation periods according to our partner's specifications. Due to vacancy of female workers in all three assembly systems we forbear from a gender-related analysis.

Pilot Study to Forecast the Productivity Development of a Decoupled Component Assembly

The functionalities of the *ESPE-AS* simulation procedure were used for a real-life example. The assembly system examined consisted of three, decoupled assembly sections supported by a rail-linked conveying system (Fig. 4). In all three sections manual assembly activities with different levels of severity were carried out. In order to reduce the physical workload, some workers were able to switch to a pressing station that was decoupled from the other assembly stations. One appropriately trained worker was responsible for monitoring the automated screw fixing stations following the manual work stations 1 and 3. The system manager and logistics worker completed the personnel structure in this example.

As things stood, it was not possible for workers to switch between the individual work stations as most of them involved in work station 1 of the assembly system were no longer qualified for the work stations 2 and 3 due to the fact that their physical abilities were already limited. These were older workers whose career paths had already led to them being transferred to these lighter jobs. New workers were initially assigned to work station 1 before they could qualify for other tasks.

As not all assembly operations contribute to age-related changes, the individual operations were subdivided using an age-differentiated degree of severity based on a scale of 0 to 10. This gave the highest time-averaged level of work severity for work

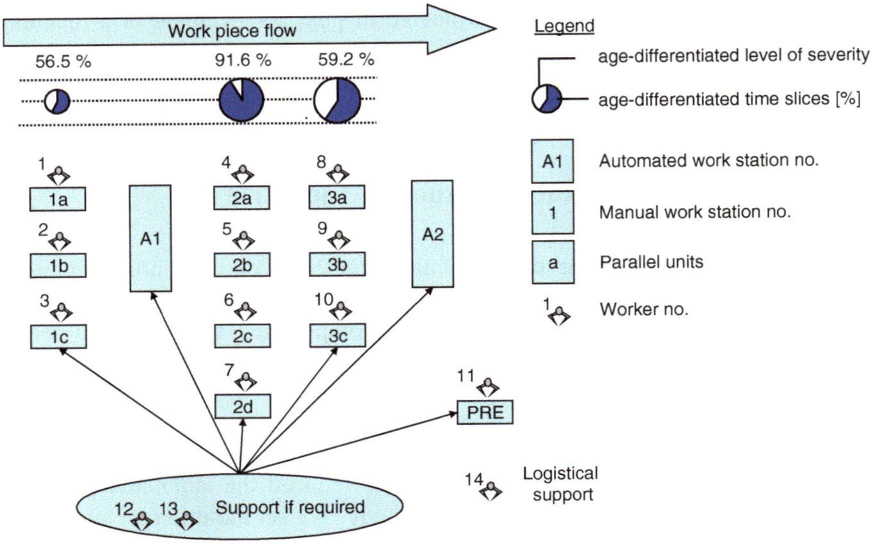

Fig. 4 Principle structure of the decoupled assembly system for the first simulation study (following Zülch and Becker 2008, p. 384)

station 3 (illustrated by the diameter of the circles in Fig. 4). However, the age-differentiated times accounted for only 59 % of the execution time for the section. This means that only about two thirds of the execution time was affected by the ageing of the workers, even though this included some activities with a high level of physical stress (including lifting and carrying heavy parts). Work station 2 consisted almost entirely of age-differentiated time elements, whose time-weighted levels of work severity were only marginally smaller than those in work station 3. Although the physical workload was less intense, the activities had to be performed continuously. In comparison to work station 2, work station 1 was less physically stressful for older workers, as a result of which older workers and those with reduced capacity in particular preferred this work station.

As evaluation criteria regarding production logistics the utilisation of the staff and the equipment respectively as well as the system output within the assessed period were used. In addition, the coefficient of variation for staff utilisation was taken into account in order to reflect the workload distribution among the assembly workers. The theoretical efficiency rate was also calculated using the ratio between the time-averaged efficiency rate and the period of time for which an individual or the entire staff was assigned to assembly operations. This ratio can be compared against the effective efficiency rate, which is calculated in proportion to the theoretical working hours. Zülch and Becker (2008, p. 383) illustrated this relation using individual efficiency rate development for the starting scenario and an alternative scenario in comparison to traditional results that do not take into account the changing efficiency rate patterns of the workers. The theoretical efficiency rate develops similarly in both—the starting and alternative—scenarios. The alternative scenario was able to improve on traditional planning for certain periods of the system's operation (years 0 through 2, 4, 7 and 11 through 15) by making use of waiting times previously imposed by the process or "takt compensation times".

The production logistics criterion of system output can also be translated into a monetary assessment criterion in the form of personnel and equipment costs per unit produced. The personnel costs are based on the hourly personnel rates, while the equipment costs for each work station are further subdivided into fixed and variable elements. The results clearly show the rising personnel costs in years 3 through 7, 11 and 13 through 15 in comparison to the relatively low level of personnel costs in period 12. On the whole, the alternative scenario is significantly more cost-effective than the starting scenario with regard to the output volume. For further details please refer to Zülch and Becker (2008, p. 383).

Pilot Study of Planning Process for an Ageing-Robust Assembly Line

In the second pilot study we looked at the hypothetical planning of a pulsed assembly line for another passenger car component (see Fig. 5). Because there was no difference between the characteristics of the (few) individual product variants, the problem was categorised as SALBP-1. This means that sub-tasks had to be allocated to assembly stations given a fixed takt time.

The data supplied for this example related to an assembly process consisting of 65 manual and 5 automated assembly operations, with the latter not being subdividable any further into individual activities. The metabolic energy consumption required by the individual assembly operations were estimated using the calculation method described by Spitzer et al. (1982, p. 21). The "posture" parameter was either "standing normally" or "walking on a level surface at 4 km/h", in all cases. There were no adverse postures. Following the mentioned calculation method, we gave 14 assembly operations the rating "severe". These included lifting and carrying an assembly component weighing 10 kg. This means that an efficiency rate development curve with a work severity factor of 1.0 was assumed for this operation for the worker assigned to it in the simulation (see Fig. 5; for further details see Zülch et al. 2009, p. 2). The work severity factor has got limits at 0.8 (easy) and 1.2 (very severe).

The staff of the assembly system consisted of 17 workers who were assumed to be equally qualified, meaning that they could be assigned to any operation. The average age was 41 years. Planning was based on a one-shift scheme over a period of 14 years. Within this period we assumed that the composition of the workforce would not change. For the purposes of planning, the age, qualifications (defined by assignment to assembly operations and work stations) and starting dates of new

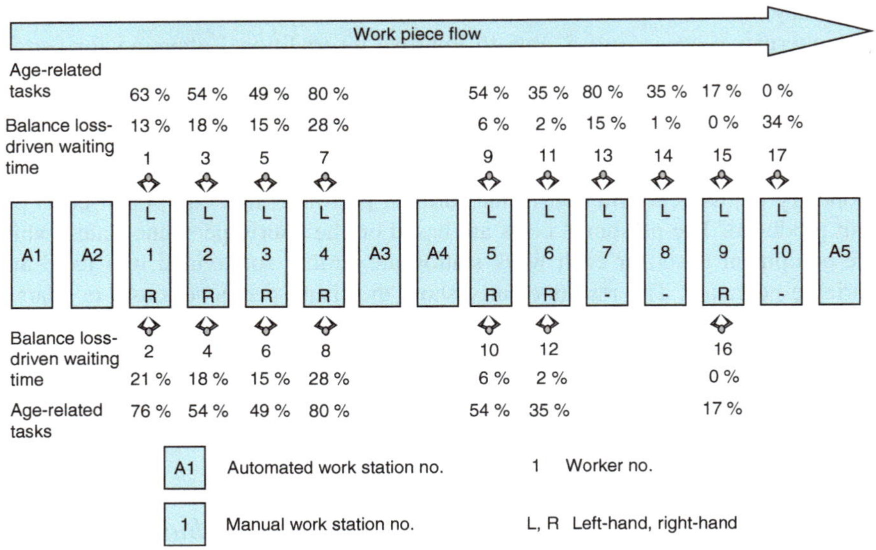

Fig. 5 Principle structure of an assembly line planned by the genetic algorithm

workers must be known. Replacing retired workers (reaching the retirement age of 67) was not needed in this example, but the planning process does allow for it in principle.

The various initial solutions were created using the Ranked Positional Weight Method by (Helgeson and Birnie 1961; see also Domschke et al. 1997, p. 181). The procedure used in this method was varied in order to produce several planning solutions. For example, activities were either assigned to assembly stations in parallel (e.g. station 1 left and station 1 right in Fig. 5) with or in opposition to the flow of materials. Further initial solutions were created by changing the allocation of workers to assembly stations, and varying the use of technical equipment.

Figure 6 shows the progress of the algorithm over 25 iterations (so called generations), with the total degree of goal achievement (TDG) of the best planning solution for each iteration shown as a column. The best TDG, which starts from 60.9 % in the first iteration, improves to 81.6 % by the 25th iteration.

During the study, the company had already developed its own solution. We also modelled and simulated this in *ESPE-AS* for the sake of comparison, here referred to as simulated realised solution. The TDG of this solution is 79.7 % (dotted line in Fig. 6) and thus lower than the planning solution arrived at using the algorithm.

This shows that a better TDG relative to the simulated realised solution can be achieved using a genetic algorithm. This is a result of the relatively high degree of goal achievement for system output (GAS). The degree of goal achievement for staff utilisation (GAW) is also improved. The degree of goal achievement for metabolic energy consumption (GAM) is slightly higher than in the simulated

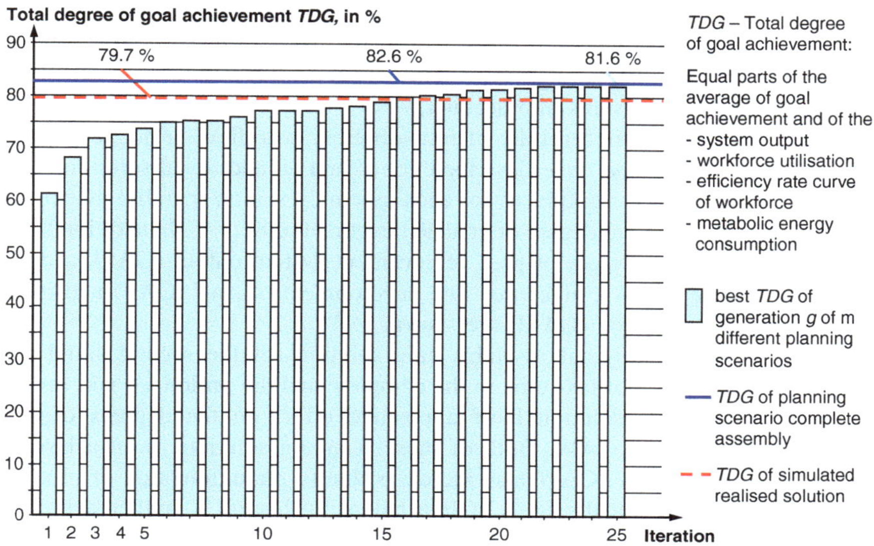

Fig. 6 Optimisation process in comparison with the realised solution and a further solution focusing on organisational measures

realised solution. This is essentially the same as a reduction in work severity, which is also a positive outcome.

The example also shows that the planning solutions of the last iteration mainly differ from those of the initial iteration in two ways: Firstly, the allocation of workers to assembly stations changes more frequently over time, while the allocation of the first-iteration planning solutions remains the same for the entire planning period of 14 years. The second difference is in the purposeful and well-timed use of assembly aids in the planning solutions of the last iteration. While technical equipment is in use for the entire period in the first-iteration planning solutions, it is only used from the middle of the operating life in the planning solutions of the last iteration.

Analysis of Different Workflow Principles

Figure 6 also shows a planning solution (out of the third simulation study, see section Analysis of the Sensitivity on Demanded Output Variation) that takes the question one step further with regard to issues of workflow principle. The approach had previously been limited to pulsed assembly lines. The new planning solution—which takes the form of complete assembly—incorporates aspects such as additional duties and process changes. In this case, one worker is responsible for the complete assembly of one unit. This means that there is no division of labour and no working according to takt times. Instead, there are 17 assembly stations working in parallel. It has been shown that with a total degree of goal achievement 82.6 %, this planning solution returns an even better result than the solution arrived at using the genetic algorithm. This is due to the fact that the algorithm does not take such measures into account.

Figure 7 shows detailed results of four different workflow principles for the component assembly: the three described above and an additional solution based on two complementary workplaces for each assembly station (parallel assembly). This solution gives a total degree of goal achievement of 82.5 %. In addition to the TDG, the individual degrees of goal achievement system output (GAS), average workload of staff (GAW), efficiency rate development of staff (GAE) and metabolic energy consumption (GAM) are shown.

Compared to the original assembly line solution (of the second simulation study), the elimination of takt time in the complete and the prolonged takt time in the parallel assembly solution have a positive effect on the degrees of goal achievement GAS and GAW. Both of these planning solutions even achieve a constant GAS of 100 %. GAW also increases due to the elimination of process-related waiting times in the form of line balancing losses. A higher degree of efficiency rate development (GAE) results as each worker benefits equally from ergonomic measures (e.g. assembly aids, job enlargement). Because the system output remains constant in both of these planning solutions, the metabolic energy consumption (GAM) is at the same level as for the assembly line planned with the help of the genetic algorithm. Due to well known findings in work structuring the result is not surprising. There is no balance loss driven waiting time in complete

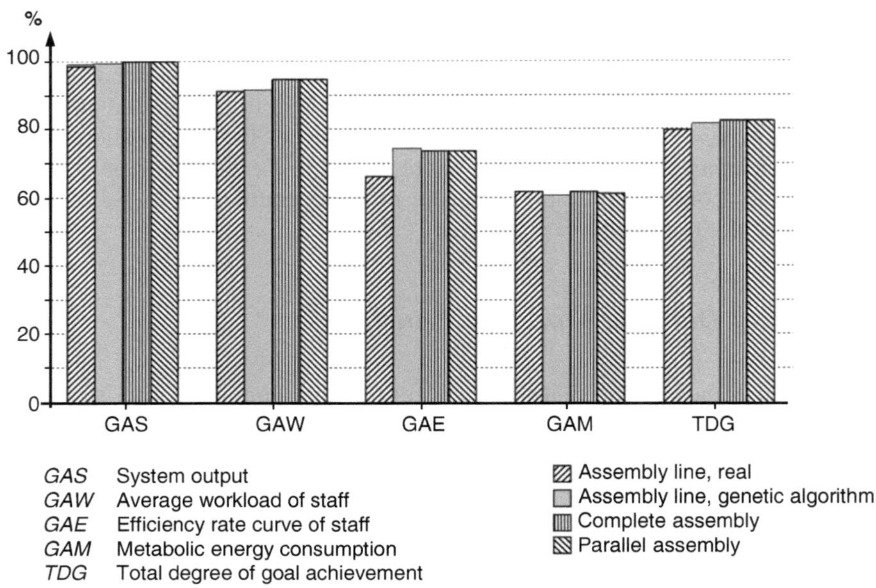

Fig. 7 Simulation results for different workflow principles

assembly compared to pulsed line. Though it was not part of our simulation study we expect a further improved result in a piece rate system compared to a time rate system.

Analysis of the Sensitivity Regarding the Efficiency Rate

In the above simulation studies we assumed a certain development of the workforce performance, modelled by efficiency rates depending upon assembly operation and age of the assembly worker. In the following sensitivity study we analysed the impact of a change of this assumption of ±10 % on the system parameters.

Figure 8 shows the results of the sensitivity analysis in the pulsed assembly line example with 17 workers we introduced in section Pilot Study of Planning Process for an Ageing-Robust Assembly Line. In case of the system output a +10 % deviation of the original efficiency rate tends to result in a path of a straight line at 100 %. On the other hand a –10 % deviation of the efficiency rate causes a loss of 1 % system output in the 1st year up to a loss of 2.5 % in the 14th year.

The development of the workforce utilisation curves is similar to the curves of the system output. In case a +10 % deviation of the original efficiency rate the utilisation is slightly increasing until the 13th year and decreases to the level of 92 % in the 14th year which is exactly the same level as in the 1st year. Due to the fact that the increase of the utilisation varied +0.2 % at a maximum to the initial value the curve is approximately a straight line at 92 %. Similar to the case of

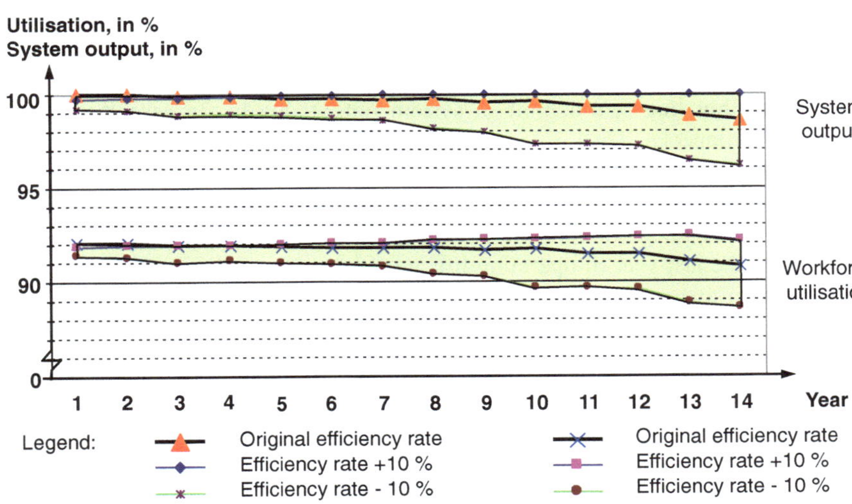

Fig. 8 Development of system output and workforce utilisation varying the efficiency rate

system output a –10 % deviation of the efficiency rate causes a loss of 0.8 % system output in the 1st year up to a loss of 2.1 % in the 14th year.

As a consequence, in the case of the pulsed assembly line an incorrect assumption of ±10 % of the efficiency rate causes a risk in system output between +1.5 and –2.5 %. The risk concerning the workforce utilisation is within a similar range.

Analysis of the Sensitivity on Demanded Output Variation

These findings were motivation of the following third simulation study. The study explores the effect of different output demands ranging between 70 and 120 % of the planned system output (Zülch and Waldherr 2011, p. I–198). By this, the assembly system in the case of both under-utilisation and excessive workload is reflected. The number of workers is the same for all simulation scenarios.

The results show that the assembly line is less robust in its response to quantity variability than the other two workflow principles. The results also show that the age-related drop in the system output of the assembly line remains at an average of 2 % during the 14 year period for all regarded quantities. The drop in the dynamic system output of the other workflow principles increases along with the demanded output quantities. For a quantity of 120 % of the target quantity, their system outputs decrease by 6 % in 14 years, which is three times the value of the assembly line. It is difficult to find the causality of this effect. We assume that this effect is linked to shifting bottlenecks in pulsed lines. In a complete assembly (without bottlenecks and balance loss) any worker's performance change causes a change in system output. Due to the possibility of shifting a bottleneck to another station (and consequently to another worker) in assembly lines a worker's performance change does not influence the system output as much as in a complete assembly. Furthermore, it is sure that (without additional shifts or overtime) a pulsed assembly line sets a limit in system output due to (in case fluctuating) emerging bottlenecks. Thus, it is obvious that in any case the system output of a complete assembly is much higher compared to the assembly line.

The results for the dynamic workforce utilisation are similar. Figure 9 shows the results of the related simulation runs. For reasons of clarity, only the result for the first and the last year of operating life are shown. Due to line balancing losses, the maximum average workforce utilisation is 90 % in the case of the assembly line. This value does not increase with the demanded output due to fluctuating bottlenecks. In comparison, the staff utilisation in the complete and in the parallel assembly system rises continuously to almost 100 %.

The age-related increase in workforce utilisation is smaller in a coupled assembly line than that in a non-coupled system. The under-utilized assembly line shows an average increase in the workforce utilisation of 1.5 % over 14 years

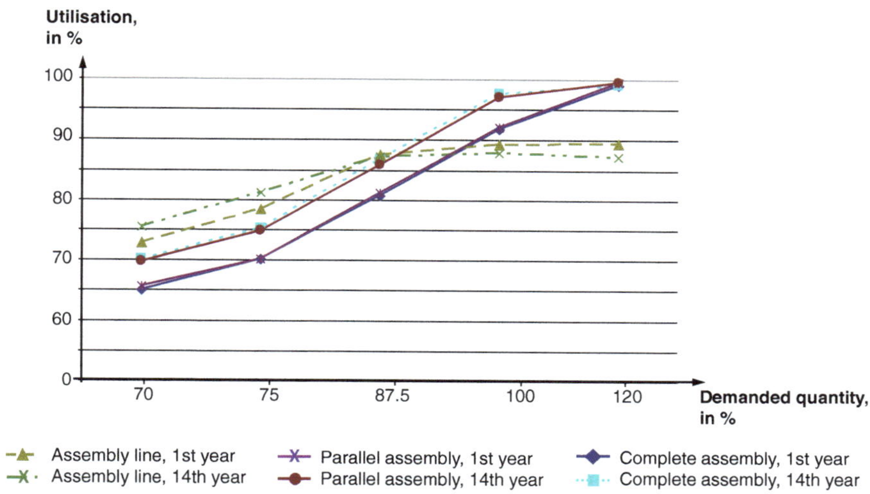

Fig. 9 Development of utilisation due to changing demanded output quantities

resulting from the fact that it takes older workers longer to perform each operation because they can used theoretical balances losses in their assembly station as buffer times for individually extended operation times.

An additional increase of the system output is not possible for the excessive workload. This means that the longer operation times needed by the older workers cannot be compensated for without any productivity losses. The age-related increase in staff utilisation in both, the complete assembly system and the parallel work system amounts to an average of 5.5 % over 14 years. Within the range of under-utilisation, there is still enough buffer time that can be used to achieve the demanded quantity. This holds true for both the 1st year and the 14th year.

Discussion

The ageing-robust design of assembly systems is set to become increasingly important in light of demographic development. The newly developed method to plan an ageing-robust pulsed assembly line uses the traditional Ranked Positional Weight Method for balancing assembly lines in order to develop a certain number of initial planning solutions together with the corresponding allocation of workers to assembly stations. These solutions are then evaluated using the novel *ESPE-AS* simulation procedure, and improved over the course of multiple iterations using a genetic algorithm. This algorithm takes the results of the simulation to determine the likelihood of reproduction, and therefore the survival of good planning solutions.

The aim of ageing-robust planning is to design an assembly system in such a way that it no longer needs to be adjusted in the course of its operating life, and instead maintains nearly the same output without the need for changes. This is actually achieved in the second study. It is therefore possible to forecast that the performance of the assembly system will remain constant up to the end of operations in 14 years. Provided the planning parameters are accurate, this can be described as an ageing-robust assembly system.

Through a series of simulation studies we observed the following trends: Firstly, the flexibility in terms of quantity variability is higher in decoupled assembly systems than in assembly lines. In the course of the operating life, the productivity of an assembly system with less division of labour changes along with an increase in demanded quantities compared to an assembly line. Secondly, the age-related decrease in the system output is not related to the demanded output quantity because the assembly line is a coupled line with inherent bottlenecks. Thirdly, the simulation studies show that age-related effects have a more pronounced impact on productivity and logistics-related parameters in assembly lines than in decoupled work systems.

However, any changes to the initial assumptions (unchanged workforce, no assembly of product variants) would require a review of these findings. This makes it necessary to further expand the planning algorithm. Another case would be the replacement of an assembly worker when he reaches the statutory retirement age. In such cases, the statement that the system is ageing-robust would have to be verified using the method described by Zülch and Becker (2008). It would also be necessary to integrate a method for evaluating more physical stressors beyond the metabolic energy consumption used here (see e.g. Schaub and Ghezel-Ahmadi 2007). It should also be noted that some of the parameters that we have assumed to be constant can change during operations in reality. The assumption that the composition of the staff will remain the same for the whole operating life of a work system will rarely be confirmed in the real world. This is another area where it would be useful to expand the planning method developed here.

However, events such as a change of jobs or the unforeseen absence of a worker before retiring cannot be foreseen during the planning phase of an assembly system, and are therefore difficult to take into account. Changes in the composition of the staff due to workers reaching the statutory retirement age can, however, already be taken into account with the developed planning procedure. In such cases, the workforce is adapted by replacing the departing assembly worker with a younger worker of a predetermined age. A similar approach can be taken with regard to changes in assembly activities (resulting from the continuous improvement process, use of new technologies etc.). However, the planning algorithm can only incorporate such changes if they are already foreseeable or apparent at the planning stage.

Further Research in the Field of Age-Adjustable Assembly Systems

As the changes described occur quite frequently in the real world, further research should not be limited to how an assembly system can be designed to be ageing-robust. Rather, investigations should be conducted into how flexible or adjustable an assembly system is with regard to an ageing workforce (see Zülch and Waldherr 2010).

The result of our research project gives cause to use the morphology of assembly systems as the basis for an assessment in terms of ageing (Zülch and Waldherr 2011). This would be helped by defining a numerical indicator for the degree to which a production system is ageing-robust. This could for example be calculated based on the average percentage change in system output per year of operation of the system.

In addition, the planning of new assembly systems should also take into account the risk-related aspect of whether and at what expense an assembly system can be adapted to unforeseen changes in the availability and capacity of workforce. In this case it would also be helpful to have a defined numerical indicator. This indicator should express the assembly system's ability to adapt, such as the opportunity to install equipment or automation for physically stressful work activities. To this end, the potential ergonomic characteristics of the assembly system should be assessed during planning, above all its fundamental logistical structure (workflow principle of assembly, division by product type), and the qualification of the workers. These indicators would allow us to compare different planning alternatives in terms of ageing-related effects.

The concept for assessing ageing-related robustness, flexibility and adaptability is also in need of further refinement. The so-called ageing-adaptability indicator takes the adaption options of the assembly system into account, e.g. the possibility to install assembly aids, or the automation replacing work operations that cause severe physical stress. For this to be possible, potential variations of an assembly system in terms of the organisation of the work must also be evaluated during planning. These evaluations can then be aggregated into an ageing-flexibility indicator as a weighted average. This indicator can then be used to compare different alternatives with regards to age-related effects and could be weighted depending on the level of physical stress (further details are shown in Zülch and Waldherr 2010, 2011). The balance of ergonomic factors with logistical factors would strengthen the ergonomic issues in comparison to traditional approaches only referring to logistical factors. The values influencing these numbers could again be evaluated using a personnel-oriented simulation.

Furthermore, reference should be made to the potentials and effects of the applied working time model (see e.g. Knauth 2007, p. 27). These aspects were not considered here but nevertheless important to cope with the demographic development on the industry level.

Acknowledgments From January 2006 until March 2010 this project with the title "Effects of an Aging Workforce on the Productivity of Manufacturing Systems" and the support code ZU 55/ 56-1 and ZU 55/56-2 was part of the DFG research program "Age-differentiated Work Systems". Members of the project research team were Prof. Dr.-Ing. Dipl.-Wirtsch.-Ing. Gert Zülch, Dipl.-Wirtsch.-Ing. Marcel Becker until 2008 and Dipl.-Wi.-Ing. Martin Waldherr since 2008.

References

Baines T, Manson S, Siebers PO, Ladbrook J (2004) Humans: the missing link in manufacturing simulation? Simul Model Pract Theor 12:515–526

Boysen N, Fliedner M, Scholl A (2007) A classification of assembly line balancing problems. Eur J Oper Res 183:674–693

Chong KE, Omar MK, Bakar NA (2008) Solving assembly line balancing problem using genetic algorithm with heuristics-treated initial population. In: Proceedings of the world congress on engineering 2008, vol II. International association of engineers, London, pp 1273–1277

Domschke W, Scholl A, Voß S (1997) Produktionsplanung. Springer, Berlin

Enríquez Diaz JA, Weichel J, Frieling E, Nöring R (2009) Arbeitsgestaltung in der deutschen Automobilindustrie: Ergonomische Intervention am Beispiel der Getriebemontage. In: Gesellschaft für Arbeitswissenschaft (ed) Arbeit, Beschäftigungsfähigkeit und Produktivität im 21. Jahrhundert, GfA Press, Dortmund, pp 579–582

Gerdes I, Klawonn F, Kruse R (2004) Evolutionäre Algorithmen. Vieweg, Wiesbaden

Hegele M, Heuer H (2010) Adaptation to a direction-dependent visuomotor gain in the young and elderly. Psychol Res 74:21–34

Helgeson WP, Birnie DP (1961) Assembly line balancing using the ranked positional weight technique. J Ind Eng 12(6):384–398

Ilmarinen J, Tempel J (2002) Arbeitsfähigkeit 2010. VSA, Hamburg

Knauth P (2007) Arbeitszeitgestaltung für die alternde Belegschaft. In: GfA: Gesellschaft für Arbeitswissenschaft (ed) Die Kunst des Alterns. GfA Press, Dortmund, pp 27–44

Landau K, Weißert-Horn M, Rademacher H (2007) Altersmanagement als betriebliche Herausforderung. Ergonomia, Stuttgart

Landau K, Ueberschär I, Weißert-Horn M (2009) Ältere Arbeitnehmer. In: Landau K, Pressel G (eds) Medizinisches Lexikon der beruflichen Belastungen und Gefährdungen. Gentner, Stuttgart, pp 54–61

Monk TH, Leng VC (1982) Time of day effects in simple repetitive tasks: some possible mechanisms. Acta Psychol 51:207–221

Müller R (2002) Planung hybrider Montagesysteme auf Basis mehrschichtiger Vorranggraphen. Shaker, Aachen

Rademacher H, Sinn-Behrendt A, Schaub K (2006) Ein Verfahren zur Ermittlung altersbedingter Engpässe hinsichtlich muskuloskelettaler Belastungen. Zeitschrift für Arbeitswissenschaft 60:230–244

Rademacher H, Sinn-Behrendt A, Schaub K (2010) Tätigkeitsbezogene Analyse körperlicher Fähigkeiten von jüngeren und älteren Beschäftigten in der Produktion. Zeitschrift für Arbeitswissenschaft 64:187–203

Schad G (1986) Entwicklung und Einsatz eines interaktiven Verfahrens zur Leistungsabstimmung von Montagesystemen. Springer, Berlin

Schaie W (2005) Developmental influences on adult intelligence. The Seattle longitudinal study. Oxford University Press, New York

Schaub K, Ghezel-Ahmadi K (2007) Vom AAWS zum EAWS—Ein erweitertes Screening-Verfahren für körperliche Belastungen. In: GfA: Gesellschaft für Arbeitswissenschaft (ed) Die Kunst des Alterns, GfA-Press, Dortmund, pp 27–44

Scholl A (1999) Balancing and sequencing of assembly lines. Physica, Heidelberg

Scholl A, Becker C (2006) State-of-the-art exact and heuristic solution procedures for simple assembly line balancing. Eur J Oper Res 168:666–693

Schöneburg E, Heizmann F, Feddersen S (1994) Genetische Algorithmen und Evolutionsstrategien: eine Einführung in Theorie und Praxis der simulierten Evolution. Addison, Bonn

Snel J, Cremer R (1995) Work and aging: a European perspective. Taylor and Francis, London

Spitzer H, Hettinger T, Kaminsky G (1982) Tafeln für den Energieumsatz bei körperlicher Arbeit. Beuth, Berlin

Trautmann M, Voelcker-Rehage C, Godde B (2011) Fit between workers' competencies and job demands as predictor for job performance over the work career. J Labour Mark Res 44:339–347

Wild-Wall N, Falkenstein M (2009) Age-dependent impairment of auditory processing under spatially focused and divided attention: an electrophysiological study. Biol Psychol 83:27–36

Willnecker U (2001) Gestaltung und Planung leistungsorientierter manueller Fließmontagen. Herbert Utz, München

Zülch G, Becker M (2006) Simulationsunterstützte Prognose der Leistungsfähigkeit von Fertigungssystemen bei alternder Belegschaft. Zeitschrift für Arbeitswissenschaft 60:151–159

Zülch G, Becker M (2008) Nachhaltigkeit bei der Planung von Fertigungssystemen: Simulation unter dem Aspekt der Leistungsentwicklung einer Belegschaft. In: Rabe M (ed) Advances in simulation for production and logistics applications. Fraunhofer IRB, Stuttgart, pp 377–386

Zülch G, Schmidt D (2011) Simulation von routinemäßigen Tätigkeiten mit vorwiegend mentalen Anforderungen. In: Spath D (ed) Wissensarbeit—Zwischen strengen Prozessen und kreativem Spielraum. GITO, Berlin, pp 303–324

Zülch G, Waldherr M (2010) Simulative Unterstützung der Planung alternsrobuster Fertigungssysteme. In: Nyhuis P (ed) Wandlungsfähige Produktionssysteme. GITO, Berlin, pp 349–364

Zülch G, Waldherr M (2011) Planning of adaptable assembly systems due to an ageing workforce In: Göbel M et al (ed) Human factors in organizational design and management—X, vol 1. IEA, Santa Monica, pp I-195–I-200

Zülch G, Waldherr M, Stock P (2009) Assembly line balancing in automotive industry referring to staff age-related performance changes. In: International Ergonomics Association (ed) 17th world congress on ergonomics. IEA, Bejing, 6 p

Age Diversity and Team Effectiveness

Birgit Claudia Ries, Stefan Diestel, Meir Shemla, Susanne Christina Liebermann, Franziska Jungmann, Jürgen Wegge and Klaus-Helmut Schmidt

Lessons Learned

Increasing age diversity in work groups due to demographic changes in recent decades raises the question of whether age diversity benefits or harms teamwork. According to the social categorization approach (Tajfel and Turner 1986), age diversity in teams is problematic because it is likely to lead to a formation of subgroups (young vs. old) within teams. This, in turn, could activate age-related stereotypes and emotional conflicts that might deteriorate group effectiveness. In contrast, models of information processing (Kerschreiter et al. 2003) posit that

B. C. Ries (✉) · S. Diestel · K.-H. Schmidt
Work and Organizational Psychology,
Leibniz-Research Centre for Working Environment and Human Factors,
Technical University of Dortmund, Dortmund, Germany
e-mail: BC.Ries@web.de

S. Diestel
e-mail: diestel@ifado.de

K.-H. Schmidt
e-mail: schmidtkh@ifado.de

M. Shemla · S. C. Liebermann · F. Jungmann · J. Wegge
Institute of Work-, Organizational- and Social Psychology,
Technical University of Dresden, Dresden, Germany
e-mail: shemla@psychologie.tu-dresden.de

S. C. Liebermann
e-mail: liebermann@psychologie.tu-dresden.de

F. Jungmann
e-mail: jungmann@psychologie.tu-dresden.de

J. Wegge
e-mail: wegge@psychologie.tu-dresden.de

C. M. Schlick et al. (eds.), *Age-Differentiated Work Systems*,
DOI: 10.1007/978-3-642-35057-3_5, © Springer-Verlag Berlin Heidelberg 2013

age diversity is beneficial for team effectiveness because age related differences in knowledge, motives, and work styles promote elaboration on task-related information that could lead to enhanced group effectiveness, particularly in tasks requiring innovation and complex decision-making. Recent research provides support for both perspectives, suggesting that these seemingly conflicting theoretical formulations address different aspects of the psychological processes through which diversity influences team outcomes. Building on this insight, we developed a new model of effectiveness in age diverse teams that integrates both theoretical approaches. Aiming to shed light on the conditions and processes that determine and account for the relationship between age diversity and team effectiveness, our model describes several mediator (e.g. salience of age differences) and moderator variables (e.g. task complexity). We tested the model using data from more than 745 natural teams with 8,848 employees in three occupational sectors (car production, administrative work, financial services). In addition, central propositions of this model were examined with a representative survey of the German workforce (N = 2,000). Based on our findings, the following seven recommendations for effective use of age diverse teams were derived:

- Provide complex team tasks without high time pressure
- Reduce age diversity salience in teams
- Establish a positive team climate
- Promote high appreciation of age diversity in teams
- Reduce age stereotypes and age-discrimination at work often enacted by supervisors
- Promote the use of age-differentiated leadership
- Improve the ergonomic design of work places within teams

In order to support the application of these recommendations in organizational practice, we designed and evaluated a new training for supervisors using a sample of 47 supervisors (221 employees) working in tax offices. This training program aims to reduce age stereotypes, to enhance appreciation of age diversity, and to develop strategies for effective use of age-related differences in experience and knowledge. It was found that the training program successfully reduced age stereotypes and team conflicts and enhanced innovation and health. In addition, we developed a new leadership questionnaire for assessing age-differentiated leadership. Based on the consistent results of our studies we conclude that effective interventions for the integration of elderly employees in work groups are available and that combinations of measures that address the best team composition, leadership processes and ergonomic design issues in teams are strongly recommended.

Introduction

The composition of a team is an important determinant of team effectiveness (Wegge 2003). Indeed, given current demographic changes, team age composition and its effects on group functioning is of growing interest to researchers. The main reason for the special interest in age diversity is the expected increase in age diversity in the coming years due to demographic changes that include a trend towards an earlier entrance into the job market as well as a later retirement. As a consequence, much younger employees will have to work with much older colleagues.

In recent years, a growing number of studies reported a negative relationship between age diversity in teams and various indicators of team effectiveness (see Wegge and Schmidt 2009, for a review). For example, in their meta-analysis, Joshi and Roh (2009) found a negative correlation between age diversity and team innovation, job satisfaction, and performance. In contrast to these studies, some empirical findings indicate a positive relationship between age diversity and performance or innovation (Wegge et al. 2008). However, there are only a few studies that investigated potential mediators and moderators of this relationship (e.g. Jehn 1995; van Dick et al. 2008). Whereas the positive effects of age diversity fall in line with expectations drawn from models of information processing in groups (Kerschreiter et al. 2003), the negative effects of age diversity are in accordance with the social categorization theory (Tajfel and Turner 1986). In an attempt to bridge between these two general perspectives, van Knippenberg et al. (2004) integrated these two theoretical approaches into a single model that emphasizes the role of social categorization processes as well as information elaboration in determining and driving the impact of team diversity on team outcomes. Consistent with this integrative perspective, we developed a new model that describes mediating and moderating processes *specific for age diversity in teams*. In addition, based on the insights gathered from testing this model, we designed and evaluated a new training program for supervisors.

State of the Art

According to the concept of diversity as "separation" Harrison and Klein (2007) define age diversity as the degree to which the age of all group members differs within the group. In that sense, age diversity is at its minimum if all group members are at the same age, as maximum separation of diversity occurs when the group is split into two subgroups at both ends of the age continuum (i.e. the older and the younger). To measure age diversity as separation, the standard deviation of age of all group members would be the most appropriate statistical measure (Harrison and Klein 2007). We use this conceptualization of age diversity throughout this chapter.

Both the social categorization theory (Tajfel and Turner 1986) and models of information processing (Kerschreiter et al. 2003) predict age diversity to have a significant impact on several team processes and outcomes. However, the two theories differ in the underlying mechanisms they ascribe to these effects. In particular, whereas the social categorization perspective posits that age diversity exerts its effects via the salience of age differences in teams (e.g. the activation of age categories as the basis of self-definition), models based on the information processing perspective suggest that age diversity influences team functioning due to the inherent differences between older and younger persons with regards to knowledge, skills, and experience.

According to the social categorization theory, when age differences among group members become salient, age as a demographic category is likely to be used to describe one's group members (Bell et al. 2011). The probability that age diversity becomes salient is supposed to be higher the greater the age differences among group members are. Further, the salience of age diversity promotes the emergence of two age homogeneous subgroups (e.g. the older and the younger). In its extreme manifestation, this process of sub-categorization may result in in-group favoritism and out-group discrimination and exclusion. Moreover, these processes are known to be strengthened by age stereotypes. Age stereotypes are attributes assigned to people within a specific age category. On the one hand, they are supposed to guide people in social interactions. For example, having the stereotype that older people's hearing is poor might result in automatically adapting one's volume of speech. On the other hand, stereotypes are often based on partial knowledge about changes in skills, competences and motives with increasing age (Hassell and Perrewe 1995). In fact, many age stereotypes are colored negatively. For example, people expect older employees to be inflexible, to be reluctant to change, and to underperform (Kite et al. 2005). Despite the predominance of negative age stereotypes against older people, there are also some negative prejudices against younger employees, including presuppositions about younger employees' lack of experience, patience, or social competences (Hummert 1999). As mentioned above, it is likely that age stereotypes will be activated when age diversity becomes salient within work groups. Because of the negative nature of most stereotypes, devaluation of the out-group will likely result. Such evaluations and cognitive biases are expected to be reflected in emotional conflicts among group members. Emotional conflicts in age diverse teams refer to tensions between older and younger group members on a relational level, which are distinct from conflicts with respect to the task. Emotional conflicts within a work group consume important resources like attention and time to deal with the group task. Furthermore, they are likely to impair interactions among group members. As a consequence, team effectiveness will suffer (Dewitt et al. 2012; see chapter Age-Related Differences in the Emotion Regulation of Teachers in the Classroom, Philipp and Schüpbach).

Models of information processing in groups lay out different propositions about the mechanisms through which age diversity influences team functioning. Whereas social categorization theory refers to negative and incorrect age stereotypes, information processing models relate to factual differences in task experience as well as task and methodological knowledge. In many studies, such different task perspectives or work styles are described as cognitive conflicts (Kerschreiter et al. 2003). Models of information processing in groups posit that cognitive conflicts mediate the relationship between age diversity and positive effects for team performance. However, we expect this positive relationship between cognitive conflicts and team effectiveness to be dominated by social categorization processes quite often since negative age stereotypes are very prominent (Kite et al. 2005). Thus in general, we expect not only a negative correlation between emotional conflicts and team performance but also a negative correlation between cognitive conflicts and team effectiveness. The positive impact of cognitive conflicts is supposed to be restricted to specific team conditions described in the following sections.

Hypothesis 1: The relationship between age diversity in teams and team effectiveness will be mediated by salience of age diversity and emotional/cognitive conflicts. Specifically, higher age diversity will enhance salience; salience, in turn, will result in higher emotional and cognitive conflicts. Both types of conflicts are supposed to decrease team effectiveness on various indicators.

Hypothesis 2: The relationship between the salience of age diversity and emotional conflicts will be mediated by age stereotypes. Specifically, the higher age salience is the more age stereotypes will come to mind; the increase in age stereotypes, in turn, will yield higher emotional conflicts.

Based on the above mentioned theories and additional insights from team composition research (see Wegge 2003; Wegge and Schmidt 2009, for a more comprehensive discussion), we have considered four moderator variables in our model that influence interaction between team members and are supposed to have an impact on both social categorization as well as information processing in teams. These are team climate, appreciation of age diversity, task complexity, and age discrimination.

A positive team climate is characterized by (1) a strong task orientation (e.g. team members pursue high quality task performance), (2) a safe and trustful environment (e.g. team members are motivated to bring in new ideas), and (3) the invitation of all team members to develop and promote new and innovative ideas (Brodbeck et al. 2000). These characteristics of a team are supposed to enhance information exchange and information-elaboration among team members. In contrast, a lack of task orientation and trustful environment is expected to promote detrimental social categorization processes. In other words, negative team climate provides a fruitful platform for conflicts and inhibits a productive exchange of age related experiences and knowledge. In line with these arguments, we posit that:

Hypothesis 3: The relationship between age diversity and team effectiveness will be moderated by team climate. Specifically, the relationship is expected to be positive in teams with a positive team climate and negative in teams with a negative team climate.

Appreciation of age diversity "reflects the extent to which individuals believe there is value in age diversity" (van Dick et al. 2008, p. 1464). Such beliefs promote the exchange of information and knowledge among members of different age groups. In that way, appreciation of age diversity prevents the building of age based subgroups and thus compromises the development of emotional conflicts. Furthermore, appreciation of age diversity is supposed to lead team members to solve existing conflicts successfully within their group. In contrast, such efforts to elaborate on age-related differences in work perspectives will hardly be found in work teams with a low appreciation of age diversity. Instead, the emergence of subgroups will be facilitated, and conflicts will flourish. As a consequence, team effectiveness is expected to decrease with respect to behavioral and emotional indicators. Thus, the following hypothesis was derived:

Hypothesis 4: The relationship between age salience and team effectiveness will be mediated by conflicts such that higher age salience yields higher conflicts and is detrimental for team effectiveness. Moreover, this mediation should be strengthened by a low appreciation of age diversity (mediated moderation effect).

Moreover, the exchange of age related differences in knowledge and expertise should be especially beneficial if the group task requires *complex decision making*. Otherwise, the existence of conflicting task perspectives of older and younger team members should be detrimental for team effectiveness (see chapters Age-Differentiated Work Systems Enhance Productivity and Retention of Old Employees, Zwick et al.; Successful Aging Strategies in Nursing: The Example of Selective Optimization with Compensation, Müller et al.). In line with these arguments, we suggest that:

Hypothesis 5: The relationship between age diversity and team effectiveness will be moderated by task complexity. Specifically, this relationship will be positive under conditions of high task complexity and negative under low task complexity.

Age discrimination occurs when "individuals are refused employment, dismissed from jobs, paid less, or denied promotions, trainings, or other benefits because of their age" (Warr 1994). This, in turn, can lead to a low self-esteem of the discriminated subgroup (Hassell and Perrewe 1995) and has been shown to be detrimental for well-being. In diverse teams, discriminated subgroups perceive a threat to the social identity that increases intergroup biases and, in turn, the negative impact of diversity. In our project we examine whether age discrimination enhances negative effects of age salience on health and we propose that:

Hypothesis 6: The relationship between the salience of age diversity and team effectiveness will be moderated by age discrimination, such that age diversity will be negatively associated with team effectiveness, especially well-being, under conditions of high age discrimination.

In sum, our basic research model is depicted in Fig. 1.

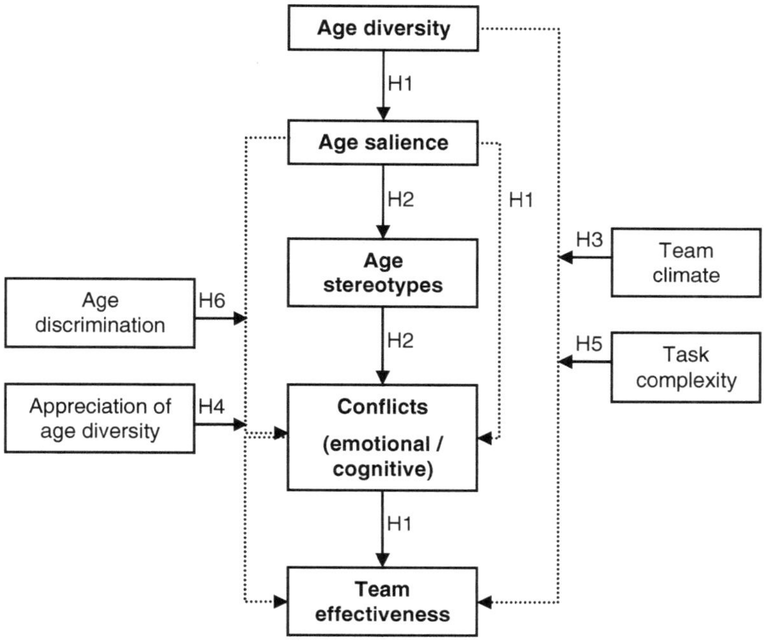

Fig. 1 Main propositions (H1–H6) derived from the ADIGU model. *Spotted arrows* visualize the specific relations investigated in the single studies. *Stressed arrows* mark the supposed relations within the whole ADIGU model

Samples/Methods/Instruments/Procedure

Samples

Several (longitudinal) studies were conducted to examine the validity of the integrative model (see Fig. 1). Due to restrictions in data collection in some fields, it was not always possible to assess the complete set of variables at all points of measurement. Three different types of teams were analyzed in this project (see Table 1).

First, we studied administrative teams working in pension- and tax-offices in North-Rhine Westphalia in Germany. The main tasks for employees in the pension offices include processing requests concerning child-rearing allowances as well as allowances for severe disabled persons. The main tasks for employees in the tax offices include processing tax computations. In both pension- and tax-offices, work is organized in teams and all employees work in front offices. Every team has to deal with similar tasks, but there are slight differences in the number of cases

Table 1 Overview of all samples

	2005	2006	2007	2008	2010
Administrative teams					
Pension offices[a]		$N_{team} = 66$	$N_{team} = 51$		
		$N_{employee} = 410$	$N_{employee} = 281$		
Tax offices[a]		+ archival data (2003)	$N_{team} = 157$	$N_{team} = 80$	$N_{team} = 47$
		$N_{team} = 222$	$N_{employee} = 722$	$N_{employee} = 397$	$N_{employee} = 221$
		$N_{employee} = 4{,}538$	$N_{supervisor} = 139$	$N_{supervisor} = 82$	$N_{supervisor} = 28$
					Extended questionnaire[b]
Financial service teams[c]	$N_{team} = 259$	$N_{team} = 246$	$N_{team} = 247$	$N_{team} = 236$	
	$N_{employee} = 2{,}337$	$N_{employee} = 2{,}468$	$N_{employee} = 2{,}514$	$N_{employee} = 2{,}384$	
Car production teams[c]					$N_{team} = 56$
					$N_{employee} = 623$
German workforce[d]					$N_{employee} = 2{,}000$
	2010 Pre	**2011a** Post 4 months	**2011b** Post 12 months		
Training with tax offices					
Training group	$N_{supervisor} = 16$	$N_{employee} = 109$	$N_{team} = 19$	The same for all three measure points	
Waiting group	$N_{supervisor} = 8$	$N_{employee} = 112$	$N_{team} = 17$	The same for all three measure points	

[a] Standard questionnaire: age diversity, age salience, emotional and cognitive conflicts, burnout, innovation, team identification, job satisfaction, team climate, task complexity, appreciation of age diversity
[b] Extended questionnaire: additional data of age stereotypes, age discrimination
[c] Data from company's personnel records on demographics, performance and turnover or absenteeism (car production)
[d] Phone interviews: age diversity, age salience, age stereotypes, age discrimination, appreciation of age diversity, health status

depending on team size. Besides predetermined rules to process single cases, teams have to organize the assignment of cases on their own. Furthermore, single cases were usually handled by several team members together.

Sample statistics of pension-offices. Data from pension-offices were collected at two different times. Because of the structural reorganization of all pension-offices, it was not possible to collect additional data. In 2006, 66 teams (n = 410 employees) participated in the investigation. Fifty-one teams participated in the survey one year later, in 2007 (n = 281 employees).

Sample statistics of tax-offices. Data from tax-offices were collected at several times. In this sample, additionally to team members' ratings of team processes, supervisor ratings were made available. In 2007, 722 employees and 139 supervisors from 157 teams completed the questionnaire. This questionnaire was also answered by 397 employees and 82 supervisors in 2008. In 2010, we administered an extended version of the questionnaire used previously. At this third point of measurement, 221 employees and 28 supervisors of 47 teams filled out the questionnaire. Additional data from a prior study (222 teams with n = 4,538 tax officers) were also re-analyzed (Wegge et al. 2008) to test basic assumptions of our model.

Second, we also gathered data from a large financial services consulting company in Germany over four years (see for example Shemla et al. submitted). Employees in this company work as individual consultants selling insurances and other financial products directly to their private and small enterprise customers. Consultants are organized within more than 230 locally separated teams across Germany. Within these teams, team members rely on and interact with each other such that they deal with the same product information and software and share a general secretary and a branch leader. Thus, these teams resemble a pooled type of teams. Team members also got together once a week for formal as well as informal meetings and exchange of information. Although they interacted with each other, team members made independent contributions to the team and their contributions were measured separately for each individual member.

Third, data from car production teams were available. The sample consisted of employees working on the final assembly line (Fritzsche 2010; Fritzsche et al. submitted). Each of the work teams are responsible for a certain section of the production line and thus spread across different workstations. Usually, team members rotate between workstations within their team. Demographic, innovation, and performance data as well as data on absenteeism were collected in 56 teams (n = 623).

Finally, in order to test the generalizability of the ADIGU model, a representative survey of the German workforce was conducted (n = 2,000, see Wegge et al. 2011a).

Measures

The questionnaire used in the project consisted of all variables of the ADIGU model depicted in Fig. 1. Whereas some of the scales were drawn from valid German measures, others were adapted from measures in English. In the following, a brief overview of all scales is given (see single publications for detailed information).

Age diversity. Following Harrison and Klein (2007) and their notion of age diversity as "separation", this variable was typically measured using the standard deviation of age within the teams. In the car production sample we measured age diversity also with the Blau-index and as part of a general faultline measure that included differences in gender and organizational tenure (Fritzsche 2010; Breu et al. 2010). The concept of diversity faultlines is based on the idea that alignment of multiple demographic differences between team members may cause a team to split into subgroups. Faultline strength was assessed using a statistical algorithm developed by Thatcher et al. (2003).

Salience of age diversity. We measured age salience using Wegge and Schmidt's (2009) 6-item scale that assesses the extent to which team members focus on age differences within the team (e.g. "Sometimes, I think about age differences in our team"). The response format of this scale ranges from 1 (*does not match*) to 5 (*matches totally*). As this scale is developed within the ADIGU project, all items are illustrated in Table 2. Alpha reliability of the salience scale ranged between Cronbach's $\alpha = 0.74$ (Liebermann et al. submitted) and Cronbach's $\alpha = 0.82$ (Ries et al. 2010a).

Age stereotypes. This variable was measured via six items assessing negative biased attitudes towards older employees (e.g. "Older employees of our team are less cooperative on the job than younger employees"). The items were adopted from Hassell and Perrewe (1995) and Kluge (2006). The response format ranges from 1 (*do not agree at all*) to 5 (*agree totally*). Alpha reliability of age stereotypes was $\alpha = 0.77$.

Conflicts. Two types of conflicts were measured using a scale by Jehn (1995). Cognitive conflicts (seven items) refer to the disagreement among team members with regard to task-related perspectives and attitudes (e.g. "Sometimes, members of our team disagree about opinions regarding the work being done"). Emotional

Table 2 Item description of salience of age diversity

No.	Item
1	If asked for a description of our team, age composition comes in my mind (e.g. three younger and two older colleagues)
2	Age differences between my colleagues are very present for me
3	Sometimes, I think about age differences in our team
4	Different age of team members are considered by team decisions (e.g. at task assignment)
5	If problems within our team arise, this is due to age differences in our team
6	We talk about the differences in age of our team members

conflicts refer to interpersonal clashes and hostile interactions between team members (e.g. "There are tensions among members of our team"). The response format ranged from 1 (*does not match*) to 5 (*matches totally*). Alpha reliability ranged between $\alpha = 0.88$ for cognitive conflicts and $\alpha = 0.92$ for emotional conflicts (e.g. Ries et al. 2010a).

Innovation. We assessed innovation using a scale by Janssen (2001). The seven items address the development, promotion and implementation of new ideas to deal with the group task (e.g. "Our team generates original solutions to problems"). Innovation was rated on a 7-point rating format (1 = *never*, 7 = *very often*). In the car production sample, innovation was measured using objective data. Specifically, based on the procedure for converting improvement ideas to production enhancement described by West (1990), two variables were measured in this study: the number of proposed ideas and the number of successfully implemented ideas. Alpha reliability of the Janssen-scale was $\alpha = 0.92$ (e.g. Ries et al. 2010a).

Burnout. We measured *emotional exhaustion* as a core dimension of burnout. The measurement comprises five items of an adopted version of the Maslach Burnout Inventory (Maslach and Jackson 1986). The items refer to physical and psychological reactions to work overload that are manifested in a negative attitude against work (e.g. "At the end of a working day, I'm feeling exhausted"). The response format of this scale ranges from 1 (*not at all*) to 5 (*very often*). Alpha reliability ranged between $\alpha = 0.89$ (Ries et al. 2012) to $\alpha = 0.93$ (Ries et al. 2010b).

Identification with the team. We assessed identification with the team using the scale from Mael and Ashforth (1992). The five items refer to group cohesion (e.g. "When I talk about this team, I usually say 'we' rather than 'they'") and response format ranges from 1 (*do not agree at all*) to 5 (*agree at all*). Alpha reliability was $\alpha = 0.75$ (e.g. Ries et al. 2010a).

Job satisfaction. We measured job satisfaction via three items from Neuberger and Allerbeck (1978) that cover satisfaction with global work, specific group task, and work environment. The extent to which group members are satisfied with those aspects was assessed on a 7-point rating scale (1 = *not right at all*, 7 = *absolutely right*). Alpha reliability was $\alpha = 0.85$ (e.g. Ries et al. 2010a).

Team climate. We used a scale from Moltzen and van Dick (2002) to assess team climate. The nine items refer to high task orientation and the degree of a trustful work environment (e.g. "In our team, everyone feels accepted and well-understood"). Response format is ranging on a 5-point rating format (1 = *does not match*, 5 = *matches totally*). Alpha reliability was $\alpha = 0.92$ (e.g. Ries et al. 2010b).

Appreciation of age diversity. We assessed appreciation of age diversity with a scale developed by Wegge et al. (2011b). The six items refer to the individual belief that a team may benefit from age diversity (e.g. "A team is more effective if its members have different age"). Intensity of appreciation of age diversity is rated on a 5-point rating format (1 = *does not match*, 5 = *matches totally*). As this scale was developed within the ADIGU project, all items are illustrated in Table 3.

Table 3 Item description of appreciation of age diversity

No.	Item
1	Our team profits from contributions from older as well as younger team members
2	In our team, one can learn new things from different perspectives of older and younger team members
3	In our team, we deal constructive with proposals from team members with different age
4	A team is more effective if its members have different age
5	A team is more functional if its members have different age
6	Team climate is better if team members have different age

Alpha reliability of the appreciation of age diversity scale was high ($\alpha = 0.84$ in Wegge et al. 2011b, and $\alpha = 0.84$ in Ries et al. 2012) suggesting that the scale is internally consistent.

Objective performance. Data gathered from the financial services company included three objective performance measures: (1) *commission target*, (2) *family target*, and (3) *new customers target*. Commission target indicates the level of commission that a consultant achieved based on their total volume of sales in a certain year. Family target refers to the acquisition of new customers and was assessed by the number of new acquired customers through existing ones during the same year. New customers target assessed the acquisition of new customers. For each consultant, the performance indicators were measured in percentage, such that 100 % was defined as the average of all consultants in the previous year. The company designed these measures deliberately in order to set challenging goals for all branches of the company and to make the performance of all consultants comparable. In the car production teams, performance was measured in terms of the number of assembly errors identified and registered by the quality management system.

In the representative survey of the German work population the following scales were used:

Age diversity was measured using a subjective scale assessing how participants rate the age-composition in their work teams. The four-point-rating scale ranged from 1 (*there are no age differences in my team (meaning the team members all are about the same age)*) to 4 (*the team is very age-diverse, meaning there are young as well as old colleagues*).

Salience of age diversity and *age stereotypes* were measured using the same scales as in the administrative sample. Alpha reliabilities were $\alpha = 0.74$ and $\alpha = 0.79$.

Age discrimination in the workplace was measured using the Nordic Age Discrimination Scale (NADS; Furunes and Mykletun 2010). On six five-point rating-scales the subjects indicated the degree of age discrimination in their workplace with regard to promotion, training, development, appraisals, wage increases, and change processes. Alpha reliability was $\alpha = 0.79$.

Current health status was assessed with four items. First, participants had to rate their general health status on a five-point rating scale ranging from 1 (*very*

bad) to 5 (*very good*). Second, they were asked to indicate the number of days, within the last four weeks, in which they experienced physical and psychological impairments (items 2 and 3). Finally, they were requested to give the number of days within the last month when they were not able to accomplish their everyday activities (item 4). The health indicator was calculated using the mean of the four z-standardized items. Alpha reliability was $\alpha = 0.75$.

Procedure

Data collection. In the pension- and tax-offices samples data were collected with the questionnaire described above. In order to get these data, we visited all offices and instructed the employees in groups with 15–20 members about the purpose of the study and how to complete the questionnaire. The longitudinal data gathered at the financial services company were drawn from the company's personnel records and thus includes only objective measures.

Statistical analysis. Questionnaire data were collected on the individual level. First, they were aggregated to group level by averaging (except data from supervisors). To justify this procedure, we computed ICC(1) and ICC(2); all variables met the criteria of >0.12 for ICC(1) and >0.60 for ICC(2).

Second, we computed explorative and confirmatory factor analysis with SPSS and AMOS for every variable at each survey (pension- and tax-offices). We used Comparative Fit Index (CFI, >0.97), Global Fit Index (GFI, >0.95), Root Mean Square Error of Approximation (RMSEA, <0.08), Standardized Root Mean Residual (SRMR, <0.09) and Cronbach's Alpha (>0.70) for quality criteria. All scales were found to be sufficiently distinct with respect these indices.

Third, we checked correlations among all variables. They were all found to be in the proposed directions.

Finally, we carried out statistical analyses to test Hypotheses 1–6. To avoid multicollinearity we used standardized variables. Several statistical procedures were applied. One of these procedures was Structured Equation Modeling (SEM), estimated with Mplus. This method estimates the supposed indirect effects of the independent variables on the dependent variables through one or more mediators applying bias-corrected and accelerated bootstrap. Furthermore, we used hierarchical regression analysis to test mediation and moderation effects.

Results

The data analysis of this project is not completely finished yet. Thus, in the following, we can just present selected results to illustrate our basic findings.

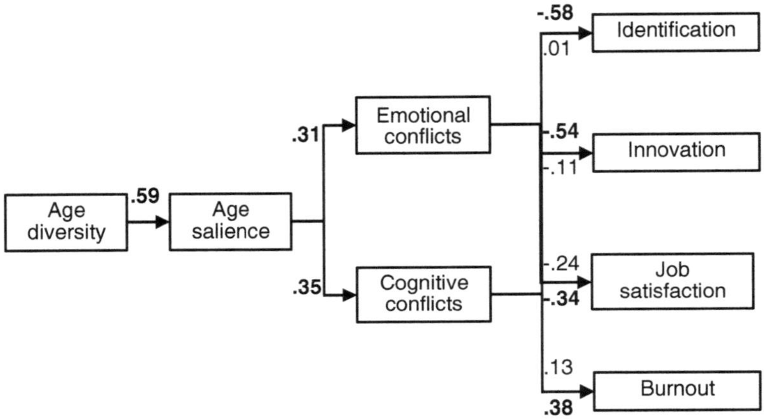

Fig. 2 Mediating effects of salience of age diversity and conflicts. The β above the arrows refer to emotional conflicts; the β below the arrows refer to cognitive conflicts. Significant ($p < 0.05$) parameters are stressed

Salience of Age Diversity and Conflicts

We analyzed the mediating role of age salience and conflicts in the relationship between age diversity and team effectiveness with SEM in Ries et al. (2010a). As illustrated in Fig. 2, an increment of age diversity leads to an increase in age salience. Age salience, in turn, goes along with higher emotional conflicts and cognitive conflicts. Interestingly, emotional conflicts have a negative influence on identification with the team (but not on burnout and job satisfaction). In contrast, cognitive conflicts exert a significant main effect on job satisfaction and burnout (but not on identification with the team and innovation). In sum, the model yielded a good fit, all expected paths reached significance and all relationships turned out to be in the proposed directions. Thus, our findings provide support for the mediating role of age salience and both types of conflicts.

Age Stereotypes

The mediation effect of age stereotypes within the relationship between age salience and both types of conflicts was tested with data from tax offices collected 2010. Results of regression analysis are summarized in Table 4. As expected, step one of the analyses revealed no significant influence of age diversity and gender on emotional and cognitive conflicts. Step two in the regression analysis yielded an

Table 4 Mediator effects of age stereotypes

	Conflicts (emotional) β	Conflicts (cognitive) β
Step 1		
Age diversity	0.19	0.05
Gender	−0.00	−0.04
Step 2		
Age salience	0.41*	0.50*
Step 3		
Age salience	0.17	0.20
Age stereotypes	0.40*	0.50*

* $p < 0.05$

important impact of age salience on emotional conflicts ($\beta = 0.41, p < 0.05$) and cognitive conflicts ($\beta = 0.50$, $p < 0.05$). Step three demonstrates that the significant relationship between age salience and conflicts disappears when age stereotypes are included in the model (emotional conflicts: $\beta = 0.17$, n. s.; cognitive conflicts: $\beta = 0.20$, n. s.). Therefore, age stereotypes were found to have significant influences on both types of conflicts (emotional conflicts: $\beta = 0.40, p < 0.05$; cognitive conflicts: $\beta = 0.50$, $p < 0.05$). Thus, the results reveal that age stereotypes mediate the relationship between age salience and emotional and cognitive conflicts, such that age salience increases age stereotypes, which, in turn, enhances emotional and cognitive conflicts. A very similar result was also found in Wegge et al. (2011a) for the German workforce. Here it was found that the negative impact of age salience on age discrimination at work was mediated by age stereotypes.

Age Discrimination

The postulated moderating effect of age discrimination on the relationship between age salience and health was tested using two steps (Liebermann et al. submitted). In the first step, age salience was found to affect health ($r = -0.16$, $p < 0.01$). This effect is even stronger when age discrimination in the team is high. The interaction between age salience and age discrimination also influences health ($r = -0.09, p < 0.01$), while at the same time age discrimination has an impact on health ($r = -0.08$, $p < 0.01$). When testing the moderation effect for different age-groups it is found that age discrimination only moderates the relationship between age salience and health for employees older than 45 years of age. Thus, our hypothesis is supported only for older employees (see also chapter Age Differences in Motivation and Stress at Work, Hertel et al. for similar age differences). Figure 3 depicts the moderating effect of age discrimination on the relation between salience and health for the age-group over 44 years.

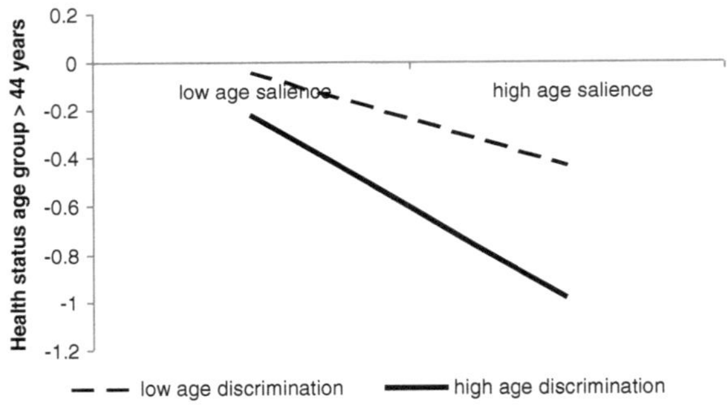

Fig. 3 Moderating effect of age discrimination (of employees ≥ 44 years)

Team Climate

The hypothesized moderating effect of team climate on the relationship between age diversity and team effectiveness was tested using regression analysis. In Ries et al. (2010b) team effectiveness was operationalized through burnout and innovation. With respect to innovation, ratings from both team members as well as supervisors were considered. As expected, team climate was found to have a significant main effect on team effectiveness. Thus, as team climate becomes more positive, burnout decreases and innovation increases, respectively. The interaction between age diversity and team climate also exerts an important effect on all three indicators of team effectiveness. To facilitate the interpretation of these findings, interaction plots were generated (Fig. 4). In support of Hypothesis 3, the relationship between age diversity and burnout was positive when team climate was bad and negative when team climate was good. The opposite interaction pattern was found for innovation ratings. The relationship between age diversity and innovation was negative when team climate was bad and positive when team climate was good. In sum, Fig. 4 portrays different regression slopes depending on the level of team climate. Thus, our results provide support for the moderating role of team climate (see also Roth 2008).

Appreciation of Age Diversity

We used regression analyses to test for the moderating role of appreciation of age diversity and the mediating impact of conflicts (e.g. Ries et al. 2012). To simplify our analyses, emotional and cognitive conflicts were combined into a single scale; this procedure was justified by factor analysis.

Conflict as a mediator. Analyses reveal a significant relationship between age salience and burnout ($r = 0.27$, $p < 0.01$) and innovation (employee rating: $r = -0.17$, $p < 0.05$; supervisor rating: $r = -0.21$, $p < 0.05$), respectively. Furthermore, age salience exerts an influence on conflicts ($b = 0.26$). Conflicts, in

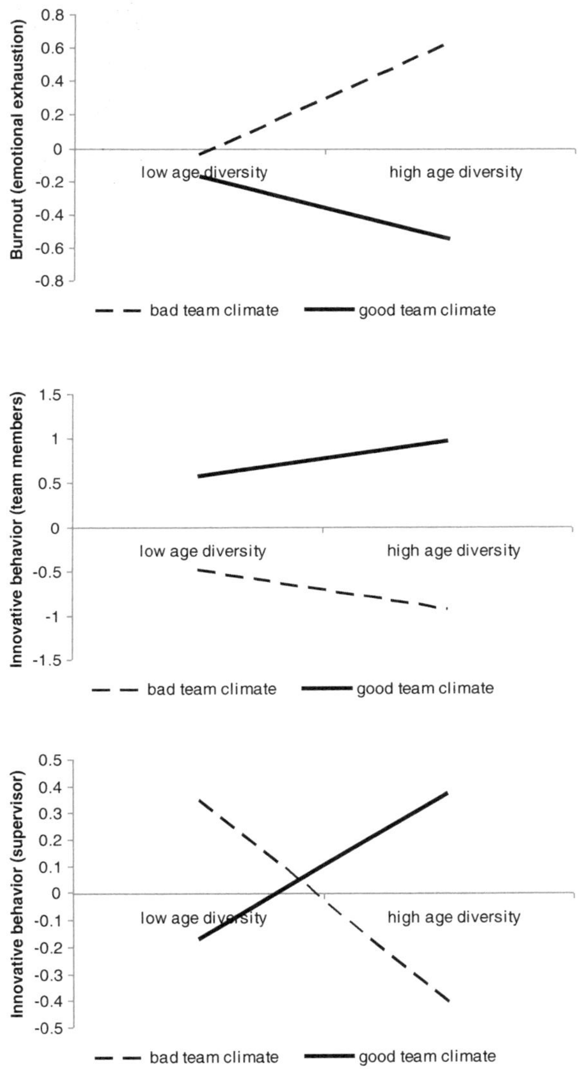

Fig. 4 Moderating effects of team climate

turn, are associated significantly with innovation (team members ratings) ($b = -0.23$) and with burnout ($b = 0.20$). Indirect effects of age salience on innovation ratings and burnout through conflicts demonstrate full mediation. In contrast, mediation of conflicts between age salience and supervisor ratings of innovation was not supported.

Appreciation of age diversity as a moderator. Correlations show a significant relationship between appreciation of age diversity and conflicts ($r = -0.51$, $p < 0.01$), innovation rated by team members ($r = 0.54$, $p < 0.01$), and burnout ($r = -0.35$, $p < 0.01$). Furthermore, the interaction between age salience and

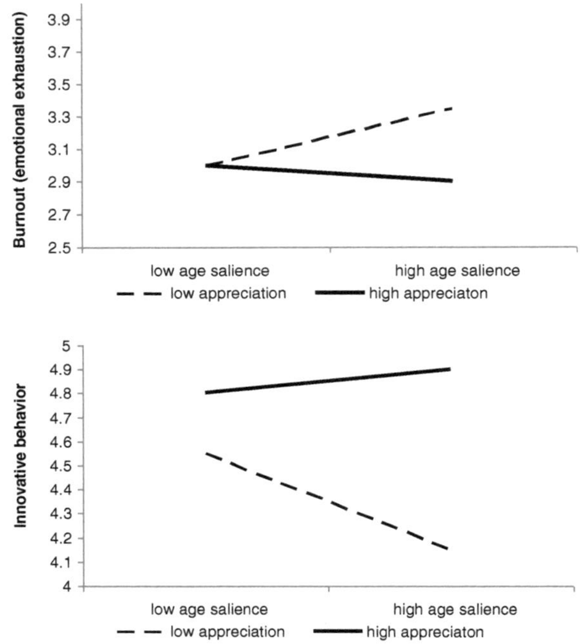

Fig. 5 Moderated mediation of appreciation of age diversity and conflicts

appreciation of age diversity has a significant impact on conflicts ($b = -0.17$). Similarly, the interaction between conflicts and appreciation of age diversity has a significant influence on innovation rated by team members ($b = 0.16$) and on burnout ($b = -0.13$). Thus, appreciation of age diversity moderates the relationship between age salience and conflicts as well as the relationship between conflicts and team effectiveness.

Specifically, the higher appreciation of age diversity is the lower are conflicts and burnout, and the higher is innovation (Fig. 5). In contrast to our expectations, the data do not support the moderating effect on supervisor ratings of innovation. The importance of appreciation of age diversity as a moderator variable was also supported in a cross-lagged panel analysis (Wegge et al. 2011b) and the dissertation of Roth (2008). Thus, this variable was included in the training that was developed in the ADIGU-project.

Task Complexity

The moderating effect of task complexity was examined using data from tax-offices as well as from financial service teams (Wegge et al. 2008). Examination of the tax-office data reveals positive correlations between age diversity and performance when teams engaged in complex decision making tasks ($r = 0.22$, $p < 0.05$).

In contrast, a negative correlation between age diversity and performance was found when teams engaged in routine tasks ($r = -0.19$, $p < 0.05$). This moderator effect of task complexity could be replicated by performance data collected one year later, demonstrating a long-range effect of task complexity. To the same token, only in teams working on routine tasks were health disorders positively linked to age diversity ($r = 0.21$, $p < 0.05$). In teams with complex tasks no relationship between age diversity and health disorders were observable ($r = -0.07$, n. s.). These results are in line with postulations drawn from information processing models, suggesting that diversity can positively impact performance outcomes when teams have to fulfil complex tasks. These positive effects of age diversity could also be confirmed when looking at the average team performance in the financial service teams. Our findings reveal, for example, a significant positive correlation ($r = 0.13$, $p < 0.05$) between age diversity and performance in the year 2005 (Shemla et al. submitted). This effect was hypothesized because selling a large number of different financial products to private and small enterprise customers requires complex and creative decision making.

Ergonomic Design in Team Work and Age Diversity

Prior research suggests that appropriate ergonomic workplace design may reduce the decline of productivity in aging employees working at paced assembly lines (Fritzsche 2010; see chapter Assembly Tasks in the Automotive Industry: A Challenge for Older Employees, Frieling et al.), We also had the opportunity to investigate the simultaneous effects of both team level factors on individual absenteeism (time lost and frequency) and team performance (22,821 errors) over one year in a sample of 56 natural car-manufacturing teams (n = 623). Results show that age was positively associated with absenteeism and mistakes in work planning. In contrast, controlling for physical workload, it was found that age diverse teams were more effective than age homogenous teams, but only if diversity was measured as a balanced mix across age categories (Blau-index) rather than as separation of old and young (standard deviation, SD). Hierarchical linear modeling (HLM) analyses further demonstrated that productivity was most strongly affected by workplace ergonomics because high physical workload amplified age-related increases in absenteeism and was associated with more assembly errors (Fritzsche et al. submitted). These results indicate that both team diversity and ergonomic workplace design may reduce age-related productivity risks in manufacturing by maintaining the work ability of older employees and improving production quality.

Training for Supervisors

In the following section we present the development and evaluation of a new training for supervisors, which aims to improve attitudes towards older employees, reduce salience of age differences among team members, improve team performance, and provide practical advice to companies for a successful management of the aging workforce. Section Conceptualization of a Supervisors Training addresses the development of the training. The second part (section Evaluation of the Training) describes the way in which the training was evaluated and provides first results of that evaluation.

Conceptualization of a Supervisors Training

Based on the aforementioned results, we developed a new modular training for supervisors. The training was developed by an expert team, consisting of scientists and future instructors, designated leaders of the potential audience and several members of an in-house HR development team. Potential topics and the design of the training were intensively discussed, as well as some components of the training (such as case studies) were adapted to typical situations in everyday work of the supervisors to ensure understanding and prepare for transfer of topics into work-life. The training consists of two modules that build upon each other and is planned for two days (see Table 5 for an overview of topics). It primarily addresses supervisors who already have to manage age-diverse teams; on the other hand, supervisors of age-homogeneous teams can use the training to prepare for future tasks with age-diverse teams. Both modules are presented in the following (see Table 5).

First module: Recognizing age diversity as a resource. The first module aims at sensitizing supervisors for the topic and providing them with the required knowledge. In particular, the following aspects were addressed: First of all, participants of the training developed a definition of diversity and received information on the relationship between age and health, as well as on age-related changes in learning abilities, motivation and social skills (referring to the positive, correct age stereotypes). Second, they learned about age diversity as a resource that may exert a positive impact on team outcomes by eliciting cognitive conflicts, as often described in models of information processing. To enhance the elaboration of the presented input, participants applied the information within their teams.

Second module: Practical implications of age diversity as a resource. Building on the background laid out in the first module, in the second module the supervisors were instructed to discuss strategies for dealing with age diversity and draw practical implications for their everyday work life. The second training module consists of three subjects: promotion of age-differentiated leadership, reduction of age stereotypes and enhancement of appreciation of age diversity. In all three

Table 5 Overview of training topics

	Module	Topics	Methods
Training (2 days)	*Module I*		
	Recognizing age diversity as a resource	• What is diversity?	
	Focus:	• How can a team benefit from diversity?	• Presentation of information
	Definition and consequences of age diversity	• What factors influence the relationship between diversity and team effectiveness?	• Group discussion
	Module II		
	Practical implications of age diversity as a resource	• What changes in performance/motivation/ competences are linked to age and how can they be used for age-differentiated leadership?	• Presentation of information
	Focus:	• What are age stereotypes and how can they be diminished?	• Exercises
	Acquisition of practical guidelines	• What is appreciation of age diversity and how can it be enhanced?	• Case study
Booster session (1 day)	*Module III*	• Recapitulation and reflection of main topics from the training	• Group discussion
	Recapitulation and reflection	• What problems evolved in applying practical guidelines from the training session?	• Coaching/ intervision

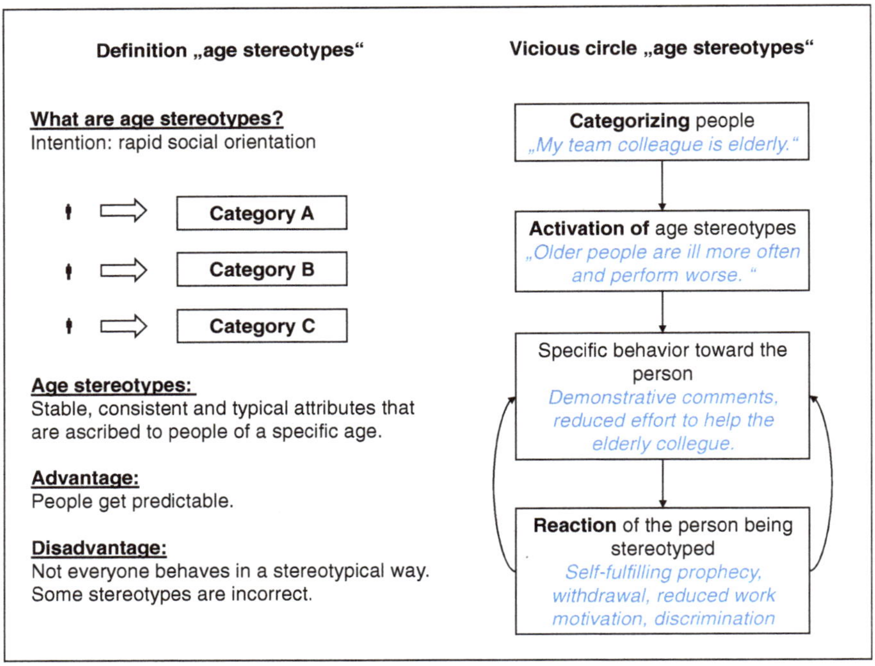

Fig. 6 Two examples of information provided in the module "age stereotypes"

subjects, the course of instruction was similar. At the beginning, supervisors were informed about facts concerning age-related changes (with regard to health, learning abilities, motivation and social skills, but in more detail than in the first module), the definition and vicious circle of age stereotypes, and appreciation of age diversity (see Fig. 6 for an example). Thereafter, supervisors discussed communication styles to promote age-differentiated leadership, age stereotypes, and appreciation of age diversity. Finally, they were instructed to develop practical guidelines based on the information provided. Again, participants applied this information in their own teams. To gain a deeper impact on behavioral changes, group discussions, group work, and case studies dominated this second training module.

Booster session. Four months after the training, a single day booster session was carried out. After a brief recapitulation of the main training topics, problems that the supervisors had experienced in applying the training topics to their leading routine were discussed. At the end of the session, strategies that supervisors found most promising for dealing with age diversity were summarized.

Evaluation of the Training

Method

The training evaluation was conducted with 47 supervisors (221 employees) working in tax-offices in North Rhine Westphalia, Germany (Jungmann et al. 2011). For evaluation purposes, participating supervisors were split into a training group (23 supervisors of 109 team members) and a waiting control group (24 supervisors of 112 team members). It was ensured that participants from the training and waiting groups were not drawn from the same tax-office to avoid a spill over effect. We excluded teams that consisted of less than three team members and in cases where data were not available for pre- and post-measures (see below). Thus, we obtained data from 36 teams (19 training groups and 17 waiting groups) and 24 supervisors (16 in training group and 8 in control group). For the training group, the average age of the team members was $M = 41.1$ years ($SD = 11.0$) with an average tenure of $M = 21.9$ years ($SD = 11.6$). The percentage of women was 69.9 %. Team size ranged from 3 to 7 members with an average team size of $M = 4.8$ ($SD = 1.5$). The average age of team members in the control group was $M = 41.7$ years ($SD = 9.8$) with an average tenure of $M = 22.1$ years ($SD = 10.9$). The percentage of women was 61.3 %. Team size ranged from 3 to 10 members with an average team size of $M = 6.1$ ($SD = 1.9$). There was no significant difference between both kinds of groups in regard to these characteristics. Two trained psychologists conducted the training. Both had been working with the subjects of this study in previous data collections and were involved in the development of the training modules. Data were collected before the training and four months after. An additional follow up measure was conducted 12 months after the training.

Results

The training had an impact on focal variables four months later, indicating decreased salience of age differences, increased appreciation of age diversity and reduced age stereotypes in supervisors. Supervisors' age stereotypes were slightly reduced in the training group between pre and post measure as compared with the waiting group. The changes are not statistically significant, but can be seen on a descriptive level, with the mean ratings falling from 1.85 at pre measure to 1.46 at post measure for the trained leaders. The waiting group did not show significant changes. The supervisors' ratings on appreciation of age diversity were enhanced in the training group compared with the waiting group. Again, changes were found on a descriptive level, but failed to reach statistical significance. Supervisors in the waiting group showed no significant changes. No significant changes revealed for the degree of age stereotypes and the appreciation of age diversity on the level of

Fig. 7 Training effects on supervisors' rating of appreciation of age diversity

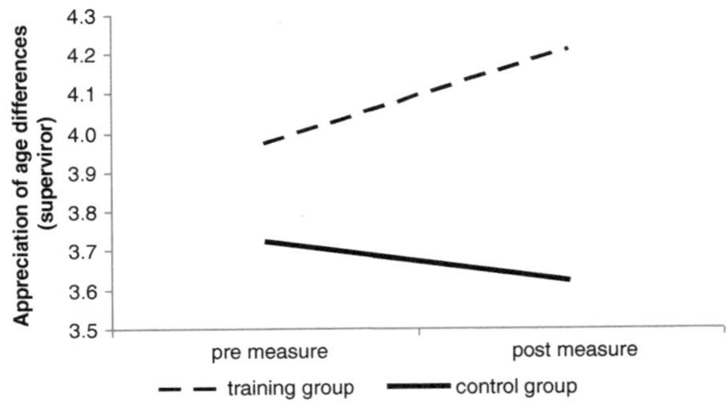

team members. However, there was a tendency of reduction in age stereotypes in groups of trained supervisors between pre- to post-measure.

Ratings of conflicts in teams and innovative behaviors as measures of team performance in groups of trained supervisors were also positively influenced. There was a significant decrease in the trained supervisors' ratings of conflicts in the team, in such a way that trained supervisors reported fewer conflicts within their teams after the training ($p < 0.01$). No significant changes in ratings of conflicts were found for team members. Supervisors who participated in the training also reported higher ratings of the innovative behavior of their teams after the training ($p < 0.05$). The same pattern was found at the team member level ($p < 0.05$). Results are illustrated in Figs. 7 and 8.

As expected, the training had an impact on both trained variables and measures of team performance. The effects were small. Nevertheless, these effects are relevant, especially as attitudinal constructs such as stereotypes are often seen as stable and trait-like in social psychological research. Influences on team members can be seen, however, mostly in small degrees. We assume that four months are a short time for changes of attitudes and stereotypes. Changes in supervisors' attitudes have to manifest in their mindset before they subsequently imply changes in behavior and communication of supervisors. Therefore, the according actions of supervisors might be delayed and not be that predominant. Thus, even small effects

Fig. 8 Training effects on supervisors' rating of age stereotypes

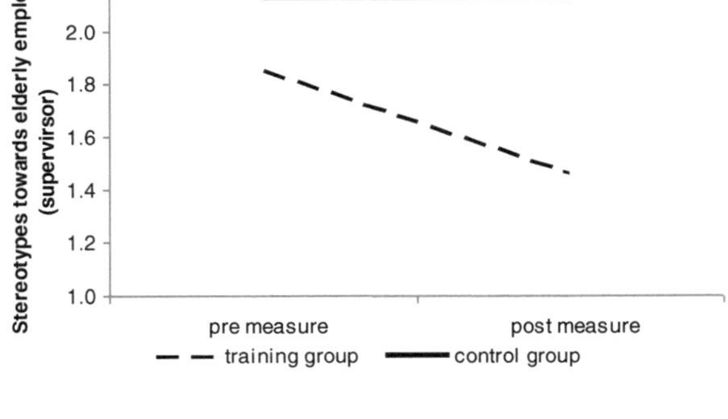

should be seen as important. The data of the follow-up measures are not analyzed yet, but we expect changes on team members' level to be found at the follow-up measure.

Discussion

The main goal of the ADIGU project was to investigate the consequences of age diversity in teams. In particular, the relationship between age diverse teams and team effectiveness was studied, shedding light on several important mediators and moderators. These findings lend substantial support to the ADIGU model, which integrates the propositions drawn from both the social categorization and the information processing perspectives. In the following, the key conclusions of our empirical findings are summarized.

Mediators of the Relationship Between Age Diversity and Team Effectiveness

Analyses show full mediation effects of age salience (Ries et al. 2010a, Liebermann et al. submitted), age stereotypes (Wegge et al. 2012), and emotional as well as cognitive conflicts (Ries et al. 2010a, 2012) of the relationship between age diversity and team effectiveness. Specifically, increasing age diversity in teams enhances salience of age differences in the team. In turn, increased salience of age activates age stereotypes, namely negative attitudes towards older employees. Such negative attitudes towards older team members then manifest in emotional conflicts as well as cognitive conflicts within the team, which may decrease several indicators of team effectiveness and well-being.

These findings are consistent with the principles of the social categorization theory. In line with this theory, age salience activates age stereotypes, which in turn give rise to emotional or relational conflicts, but also give rise to cognitive conflicts deteriorating team effectiveness. Maybe negative attitudes towards older team members stress invincible differences between work styles of older and younger employees. The measure of cognitive conflicts by Jehn (1995) focuses on such differences in work styles. Thus, it is not surprising that cognitive conflicts are associated with less identification with the team, less innovation, and higher burnout. This negative relationship between cognitive conflicts and team effectiveness has been established elsewhere (Dewitt et al. 2012; van Knippenberg et al. 2004) using similar measures. In contrast, models of information processing in groups suppose positive effects of cognitive conflicts. The lack of empirical support for that assumption may partly arise also from these methodological aspects, suggesting a need to construct a new scale to measure cognitive conflicts that better fits the potential benefits of age related differences in work styles and perspectives on the task. Furthermore, inappropriate mind-sets may be involved in preventing teams from gaining the benefits of cognitive conflicts. The ADIGU project identified several factors that constitute beneficial and unfavorable mind-sets for age diverse teams. These moderators are discussed in the next section.

Moderators of the Relationship Between Age Diversity and Team Effectiveness

Age discrimination. Findings reveal a significant negative relationship between age salience and health as an indicator of team effectiveness (Liebermann et al. submitted). These negative influences of age diversity are enhanced by age discrimination. Although age discriminating behavior bears a risk for all age-groups, the moderation effect of age discrimination was demonstrated only for employees over 44 years of age. Age discrimination enhances the negative effect of subgroup building on the health of the older team members. When older workers are aware of age differences in a team, it depends on their perception of age discrimination in their workplace whether or not this salience would affect their health negatively. As proposed by the social-categorization theory, discrimination as a perceived threat to the in-group enhances the negative effects of diversity.

Team climate. Studies show a positive relationship between age diversity and innovation under the condition of a positive team climate (e.g. Ries et al. 2010b). The direction of that relationship is reversed under the condition of a negative team climate. Similarly, a positive team climate promotes a negative relationship between age diversity and burnout; otherwise, the relationship turns positive if team climate is negative. Thus, a positive team climate constitutes an important framework for teams to benefit from age diversity. Moreover, this study demonstrates the necessity of a model that integrates both perspectives—social categorization and information processing—as was done by the ADIGU model.

Appreciation of age diversity. In addition to the mediation of conflicts in the relationship between age salience and team effectiveness, appreciation of age diversity was found to moderate this relationship (Ries et al. 2012; Roth 2008; Wegge et al. 2012). In particular, when appreciation of age diversity is low, the relationship between age salience and innovation is negative and mediated by conflicts. Similarly, the relationship between age salience and burnout is positive under the condition of a low appreciation of age diversity. In contrast, the mediation effect of conflict diminishes if appreciation of age diversity is high. Thus, a low appreciation constitutes a risk factor in age diverse teams while a high appreciation of age diversity marks a protective resource.

Task complexity and ergonomic design of team work. Our findings show that there is a positive relationship between age diversity and team performance in teams engaged in complex tasks (Roth et al. 2006; Wegge et al. 2008). This relationship turns negative in teams engaged in routine tasks. Thus, the potential benefits are dependent on the group task. This may explain why cognitive conflicts do not promote team effectiveness in general, but only if the group task requires different cognitive perspectives. Moreover, it was also found that ergonomic workplace design may reduce age-related productivity risks in manufacturing by maintaining the work ability of older employees and improving production quality in addition to a balanced team composition regarding age of team members.

General Reflections

Findings from the ADIGU project replicate and extent single results from prior research, fitting these results into an integrative theory. In addition to the identification of several mediators and moderators, the present findings generalize prior results from teams in the management area to administrative teams and teams within the production sector.

In the course of our research we have come to the conclusion that two constructs have an especially important role in driving the effects of age diversity on team functioning. First, appreciation of age diversity was found to moderate the relationship between age diversity and several indicators of team functioning. Therefore, it is important to identify in future studies the various antecedents of this critical attitude, e.g. inter-individual differences, work design in teams and leadership behavior. Second, whereas our findings suggest that cognitive conflicts result in deteriorated team effectiveness, there is room to further investigate this relationship and the contingencies that determine it. For example, it might be that the elaboration on task-relevant information among team members would be more likely to foster team performance when the emotional significance members assign to interacting with other team members and to their membership in the team is high. This assumption was recently supported in a meta-analysis (Dewitt et al. 2012). Therefore, we propose that any effort directed at improving teamwork through discussion, elaboration, and interaction among members should be accompanied by efforts to enhance the bond among members.

Outlook

As shown in this chapter, age diversity is an important factor for predicting group effectiveness, especially team innovation and burnout. This impact is characterized by a heightened age salience as well as the activation of age stereotypes and conflicts (mediation effects). As social categorization theory mainly refers to negative and incorrect age stereotypes, it remains to be seen how positive stereotypes regarding older and younger employees influence team functioning. For example, older group members are expected to have greater experience and to have better social skills (Hummert 1999). In contrast, younger group members are expected to have up to date technological knowledge. To the best of our knowledge, we do not know of any study that has analysed potential benefits of the activation of positive stereotypes. Future studies, therefore, should seek to differentiate between the activation of negative and positive age stereotypes.

Our project has revealed several moderating factors of the relationship between age diversity and team effectiveness, especially team climate, appreciation of age diversity, task complexity, age discrimination and age-differentiated leadership. These moderating factors constitute the starting points to successfully manage age-diverse teams (see recommendations in section Lessons Learned). Thus, demographic change requires managers to change their mind-sets and to learn new leadership skills. Consistent with our findings, other projects in the SPP 1184 found that older people have to be motivated differently than younger people (see chapter Age Differences in Motivation and Stress at Work, Hertel et al.) and that leaders must carefully consider how to react to age-related performance changes (see the editorial, Schlick et al.), promoting the continuous use of age-differentiated ergonomic design in work systems and available trainings (see chapters Successful Aging Strategies in Nursing: The Example of Selective Optimization with Compensation, Müller et al.; Integrating Training, Instruction and Design into Universal User Interfaces, Sengpiel et al.; Ergonomic Design of Human-Computer Interfaces for Aging Users, Schlick et al.). In this chapter, we have outlined in some detail a new management training, designed specifically for managing mixed-age teams. The usefulness of training for supervisors has been tested empirically. Results reveal several positive effects of the training four months after. However, to support this learning process, supervisors also need feedback regarding the quality of their age-differentiated leadership behavior. With this goal in mind, we have recently constructed a new questionnaire with 16 items measuring age-differentiated leadership with regards to (a) general principles for leading age-mixed teams, (b) specific leadership behaviors directed to older employees and (c) specific leadership behaviors directed to younger employees (see FAF 16, Wegge et al. 2012a). The first empirical findings regarding the validity from two samples (192 nurses and 106 production workers) where this new instrument was used are also promising. We found, for example, that young and old employees who perceive their supervisors to lead in an age-differentiated manner are much more satisfied with their work, have less burnout, emotional conflicts, turnover intentions and higher self-efficacy. We therefore suggest that future training studies should include also measures of age-differentiated leadership quality of supervisors in order to support the long-term success of such interventions.

Acknowledgments This research was supported by multiple grants from the German Research Foundation (WE 1504/8-1; WE 1504/8-2; WE 1504/8-3; SCHM 851/1-1; SCHM 851/1-2; SCHM 851/1-3).

References

Bell ST, Villado AJ, Lukasik MA, Belau L, Briggs AL (2011) Getting specific about demographic diversity variable and team performance relationships: a meta-analysis. J Manage 37:709–743

Brodbeck F, Anderson N, West M (2000) TKI—Teamklima-Inventar. Hogrefe, Göttingen

Breu C, Wegge J, Schmidt K-H (2010) Alters-, Geschlechts- und "Tenure"-diversität in Verwaltungsteams—Erklären Faultlines mehr Varianz bei Teamkonflikten und Burnout als traditionelle Diversitätsindikatoren? [Age-, gender- and tenure-diversity in administrative teams: do faultlines explain more variance in team conflicts and burnout compared to traditional diversity indicators?] Zeitschrift für Arbeitswissenschaft 64:147–159

de Witt FRC, Greer LL, Jehn KA (2012) The paradox of intragroup conflict: a meta-analysis. J Appl Psychol 97:360–390

Fritzsche L (2010) Work group diversity and digital ergonomics as new approaches for compensating the aging workforce in automotive production. Dissertation, TU Dresden

Fritzsche L, Wegge J, Schmauder M, Kliegel M, Schmidt KH (submitted) Team diversity, workplace ergonomics and age-related declines in health and performance of car manufacturing teams

Furunes T, Mykletun RJ (2010) Age discrimination in the workplace: validation of the Nordic age discrimination scale (NADS). Scand J Psychol 51:23–30

Harrison DA, Klein KJ (2007) What's the difference? Diversity constructs as separation, variety, and disparity in organizations. Acad Manag Rev 32:1199–1228

Hassell BL, Perrewe PL (1995) An examination of belief's about older workers: do stereotypes still exist? J Organ Behav 16:457–468

Hummert ML (1999) A social cognitive perspective on age stereotypes. In: Hess TM, Blanchard-Fields F (eds) Social cognition and aging. Academic, San Diego, pp 175–196

Janssen O (2001) Fairness perceptions as a moderator in the curvilinear relationships between job demands, and job performance and job satisfaction. Acad Manag J 44:1039–1050

Jehn KA (1995) A multi method examination of the benefits and detriments of intragroup conflict. Adm Sci Q 40:256–282

Joshi A, Roh H (2009) The role of context in work team diversity research: a meta-analytic review. Acad Manag J 52:599–627

Jungmann F, Ries BC, Liebermann S, Wegge J, Schmidt K-H (2011) Konzeption und Evaluation eines Führungskräftetrainings zu "Altersheterogenität als Ressource erkennen und nutzen", 7. Kongress Fachgruppe der Arbeits-, Organisations- und Wirtschaftspsychologie der DGPS, Rostock, pp 7–9

Kerschreiter R, Mojzisch A, Schulz-Hardt S, Brodbeck FC, Frey D (2003) Informationsaustausch bei Entscheidungsprozessen in Gruppen: Theorie, Empirie und Implikationen für die Praxis. In: Stumpf S, Thomas A (eds) Teamarbeit und Teamentwicklung. Hogrefe, Göttingen, pp 85–118

Kite ME, Stockdale GD, Whitley BE, Johnson BT (2005) Attitudes toward younger and older adults: an updated meta-analytic review. J Soc Issues 61:241–266

Kluge A (2006) Die Einstellungen zur Leistungs-, Lern- und Anpassungsfähigkeit älterer Arbeitnehmer/-innen und die subjektiv erlebte Diskriminierung—eine Untersuchung in Schweizer Unternehmen. Arbeit, Heft 1(15):3–17

Liebermann S, Wegge J, Jungmann F, Schmidt K-H (submitted) Age diversity as a health hazard for specific age groups. Evidence from a representative sample of the German workforce

Mael F, Ashforth BE (1992) Alumni and their almamater: a partial test of the reformulated model of organizational identification. J Organ Behav 13:103–123

Maslach C, Jackson SE (1986) Maslach burnout inventory, 2nd edn. Consulting Psychologist Press, Palo Alto C.A

Moltzen K, van Dick R (2002) Arbeitsrelevante Einstellungen bei Call Center-Agenten: Ein Vergleich unterschiedlicher Call Center-Typen. Zeitschrift für Personalpsychologie 1:161–170

Neuberger O, Allerbeck M (1978) Messung und Analyse von Arbeitszufriedenheit: Erfahrungen mit dem Arbeitsbeschreibungsbogen (ABB). Huber, Bern

Roth C, Wegge J, Schmidt K-H, Neubach (2006) Altersheterogenität als Determinante von Leistung in Arbeitsgruppen der öffentlichen Verwaltung. [Age heterogeneity as determinant of performance in teams of a public administration]. Zeitschrift für Arbeitswissenschaft 60:266–273

Ries BC, Diestel S, Wegge J, Schmidt K-H (2010a) Die Rolle von Alterssalienz und Konflikten in Teams als Mediatoren der Beziehung zwischen Altersheterogenität und Gruppeneffektivität. Zeitschrift für Arbeits- und Organisationspsychologie 54:117–130

Ries BC, Diestel S, Wegge J, Schmidt K-H (2010b) Altersheterogenität und Gruppeneffektivität—Die moderierende Rolle des Teamklimas. Zeitschrift für Arbeitswissenschaft 64:137–146

Ries BC, Diestel S, Wegge J, Schmidt K-H (2012) Altersheterogenität und Gruppeneffektivität: Der Einfluss von Konflikten und Wertschätzung für Altersheterogenität. Zeitschrift für Arbeitswissenschaft 66:58–71

Roth C (2008) Age diversity in teams. LMU, Dissertation

Shemla M, Thies M, Wegge J, Schmidt KH (submitted) Effects of diversity on performance in financial service teams: The impact of age-, tenure- and gender diversity

Tajfel H, Turner JC (1986) The social identity theory of intergroup behavior. In: Austin S, Austin WG (eds) Psychology of intergroup relations. Nelson Hall, Chicago, pp 7–24

Thatcher SMB, Jehn KA, Zanutto E (2003) Cracks in diversity research: the effects of diversity faultlines on conflict and performance. Group Decis Negot 12:217–241

van Dick R, van Knippenberg D, Hägele S, Guillaume YRF, Brodbeck FC (2008) Group diversity and group identification: the moderating role of diversity beliefs. Human Relations 61:1463–1492

van Knippenberg D, de Dreu CKW, Homan AC (2004) Work group diversity and group performance: an integrative model and research agenda. J Appl Psychol 89:1008–1022

Warr P (1994) Age and employment. In: Triandis HC, Dunette MD, Hough LM, and Leaetta M (Eds.) Handbook of industrial and organizational psychology, vol 2, 2nd edn. Consulting Psychologists Press, Palo Alto, CA, pp 485–550

Wegge J (2003) Heterogenität und Homogenität in Gruppen als Chance und Risiko für die Gruppeneffektivität. In: Stumpf S, Thomas A (eds) Teamarbeit und Teamentwicklung. Hogrefe, Göttingen, pp 119–141

Wegge J, Jungmann F, Schmidt KH, Liebermann S (2011a) Das Miteinander der Generationen am Arbeitsplatz [The cooperation of generations at work]. Iga Report 21:64–97

Wegge J, Jungmann F, Liebermann S, Shemla M, Ries BC, Diestel S, Schmidt K-H (2012) What makes age diverse teams effective? Results from a six-year research program. Work: J Prev Assess Rehabil 41:5145–5151

Wegge J, Roth C, Neubach B, Schmidt K-H, Kanfer R (2008) Age and gender diversity as determinants of performance and health. The role of task complexity and group size. J Appl Psychol 93:1301–1313

Wegge J, Schmidt K-H (2009) The impact of age diversity in teams on group performance, innovation and health. In: Antoniou A-S, Cooper CL, Chrousus GP, Spielberger CD, Eysenck MW (eds) Handbook of managerial behavior and occupational health. Northhamptom, Elgar, pp 79–94

Wegge J, Schmidt K-H, Liebermann S, van Knippenberg D (2011b) Jung und alt in einem Team? Altersgemischte Teamarbeit erfordert Wertschätzung von Altersdiversität. In: Gellèri P, Winter C (eds) Potenziale der Personalpsychologie. Hogrefe, Göttingen, pp 35–46

Wegge J, Schmidt KH, Piecha A, Ellwart T, Jungmann F, Liebermann S (2012a) Führung im demografischen Wandel [Leadership during demographic change]. Report Psychol 37(9):344–354

West MA (1990) The social psychology of innovation in groups. In: Farr M (ed) Innovation and creativity at work: psychological and organizational strategies. Wiley, Chichester, pp 309–333

Age Differences in Motivation and Stress at Work

Guido Hertel, Markus Thielgen, Cornelia Rauschenbach, Anna Grube, Christian Stamov-Roßnagel and Stefan Krumm

Lessons Learned

Age differences at work not only go along with differences in job-related abilities and expertise, but also with diverging attitudes and experiences at work. Indeed, there seem to be systematic age differences both in work motivation and stress which might provide important guidelines for age-differentiated human resource management. Among these systematic age differences, the following seem to be quite robust across samples and studies:

- Learning goals are more attractive for younger as compared to older workers.
- Goals directed towards emotion regulation and positive experiences at work are more important for older as compared to younger workers.

G. Hertel (✉) · M. Thielgen · C. Rauschenbach · A. Grube · S. Krumm
Department of Organizational and Business Psychology,
University of Münster, Münster, Germany
e-mail: ghertel@uni-muenster.de

M. Thielgen
e-mail: markus.thielgen@uni-muenster.de

C. Rauschenbach
e-mail: rauschenbach@uni-muenster.de

A. Grube
e-mail: anna.grube@uni-muenster.de

S. Krumm
e-mail: stefankrumm@uni-muenster.de

C. Stamov-Roßnagel
Organisational Psychology, Jacobs University Bremen, Bremen, Germany
e-mail: c.stamovrossnagel@jacobs-university.de

C. M. Schlick et al. (eds.), *Age-Differentiated Work Systems*,
DOI: 10.1007/978-3-642-35057-3_6,

- Need for autonomy is more pronounced in older as compared to younger workers.
- Generativity motives (helping others, legacy, etc.) are more prioritized by older as compared to younger workers.
- Older workers have developed higher congruencies between their implicit and their explicit motive systems, reducing the need for self-regulation.
- Older as compared to younger workers suffer more when implicit and explicit motives are incongruent, as well as when individual work values are in conflict with given task characteristics.
- Older workers experience lower stress levels as middle-age workers, which is not only a consequence of differences in external stressors.
- Even though older workers seem to have a lower need for self-regulation and often have more opportunities for active control at work, they also possess higher self-regulation skills as compared to younger workers.

These empirical research findings are in line with the general theoretical framework of our research project, connecting work psychology models of motivation and stress with life-span models of aging. Notably, both the theoretical predictions as well as the empirical findings clearly speak against a mere deficit perspective of aging at work. Older workers do have various strengths compared to their younger colleagues both in motivational and in stress-related work competencies (see also Hertel et al. 2013a). Understanding these potential strengths and their dynamics is central for a successful leverage and development of these potentials in modern work environments.

Introduction

Demographic changes and the increasing lack of qualified workers in many branches [e.g. Ilmarinen (2005), see also the editorial to this volume, Schlick et al.] foster the interest in age-related differences at work as a starting point for a more differentiated human resource management and workplace design. Initially, however, researchers merely focused on age differences in *work capabilities,* predicting the *maximum* performance potential a worker might have. Today, age differences in fluid intelligence, job knowledge, or physical strength and flexibility are considerably understood though further research is still needed (e.g. Salthouse 2012). However, in addition to work capabilities, job performance (defined broadly as behavior that supports organizational goals) is also determined by worker's *attitudes* and *experiences* at work that particularly predict their *average performance levels*. For instance, age-related declines in work capacities might be compensated by additional effort or training if workers are highly motivated. Thus, in order to understand job performance as well as voluntary work behavior (e.g. Hertel et al. 2000; Organ 1990), motivational and stress-related changes are central for a complete picture of an increasingly age-diverse workforce.

Between 2005 and 2012, the research project reported in this chapter was devoted to systematically investigating age-related differences in motivation and stress experience at work. Based on more than 25 primary studies conducted in various organizational settings and one meta-analysis, data from more than 40,000 workers of different ages were considered. The obtained results provide fruitful insights both for scientific and applied perspectives. Interestingly, while age differences in work motivation and stress were largely ignored by the time the first project proposal was written (2004), in the last years the number of relevant empirical studies has significantly increased (e.g. Kooij et al. 2009; Zacher and Frese 2009), indicating that these issues are now well received both in the academic community as well as among practitioners and work organizations facing the current demographic changes.

Below, we describe the main insights from this research project. We structure our findings according to genuine phases in an action regulation process (e.g. Heckhausen and Gollwitzer 1987; Hertel and Wittchen 2008), integrating rather stable person-immanent factors (implicit motives and explicit motives/work values) with context-contingent experiences at work (stress and job satisfaction), the latter considered as consequences of both congruency between implicit and explicit motives as well as the relative fit between work values and task characteristics. Moreover, control strategies as workers' active attempts to influence and change their current work situation are considered as well. As this book chapter intends to provide a comprehensive overview that might also be useful for practitioners, the presentation will be rather cursory and large-scale instead of research oriented (also to avoid redundancies with the empirical and theoretical research papers coming out of this project). A more detailed description of the various studies can be found in the single publications of the primary studies (see reference list). Moreover, a detailed presentation of the underlying theoretical model is provided in Hertel et al. (2013).

Basic Theoretical Assumptions

Apart from a few exceptions (e.g. Super 1970), theoretical work on age-related differences in work motivation is relatively new (e.g. Kanfer and Ackermann 2004; Truxillo et al. 2012; Warr 2001). Indeed, most research on *work motivation* has been conducted with young adults (e.g. student participants), so that age biases in the psychological insights on motivational dynamics are possible. In a similar way, age differences in *stress at work* have only recently been addressed (see Rauschenbach et al. 2013, for an initial review). And even age contingencies with *job satisfaction* are not completely understood apart from repeated findings of older workers reporting higher job satisfaction on average than their younger and middle-aged colleagues (e.g. Hochwarter et al. 2001; Ng and Feldman 2010).

The current research project started with the general assumption that work motivation and stress experience at work are systematically moderated by the chronological age of workers. Moreover, we believe that these changes are not simply directed towards "more" or "less" motivation or stress, but imply qualitative shifts (cf. Kanfer and Ackermann 2004; Stamov-Rossnagel and Hertel 2010). These assumptions are built on a theoretical framework that connects established motivation and stress theories (e.g. Heckhausen 2001; Lazarus 2007; McClelland et al. 1989) with approaches on aging processes from developmental and life-span psychology (e.g. Baltes and Baltes 1990; Carstensen 2006, 2009). The integration of these two rather independent research lines enables both innovative assumptions of age effects on motivation and stress at work, as well as the integration of seemingly contradicting findings. Below, we briefly describe the main assumptions of the central theories as part of our theoretical framework. For a more detailed discussion see Hertel et al. (2013).

Socio-emotional Selectivity Theory

Socio-emotional Selectivity Theory (SST) (e.g. Carstensen 2006, 2009) predicts that changes in people's future time perspective (e.g. from time since birth to remaining time until death) results in changes of individuals' priorities of goals and motives. More precisely, SST maintains that under an open-ended future time perspective, persons prioritize goals related to learning, individual growth, and building expertise. However, under a limited future time perspective, persons rather prioritize goals with emotional contents and are more interested in optimizing their psychological well-being in the "here and now" (Carstensen 2006). Hence, SST predicts that older individuals de-prioritize goals related to growth and advancement and instead emphasize social and affective values.

Model of Selection, Optimization, and Compensation

The (normative) Model of Selection, Optimization, and Compensation (SOC) (e.g. Baltes and Baltes 1990) suggests three major strategies of successful development over the lifespan. Selection refers to focusing resources on those goals that are regarded as most valuable (e.g. family-related goals as opposed to work-related goals). Optimization denotes striving for higher levels of performance or functioning (e.g. by training). Finally, compensation strategies are directed towards counteracting age-related losses, e.g. by re-arranging task processes to compensate declines in specific task related skills. In general, the SOC model predicts that older person's functioning and well-being should positively correlate with the use of these strategies. Applied to the work context, we would expect that the relative degree of older worker's performance and well-being is also affected by the level of SOC strategies used at work.

Work Values

Values can be defined as "concepts or beliefs about desirable end states or behaviors that transcend specific situations, guide selection or evaluation of behavior and events, and are ordered by relative importance" (Schwartz and Bilsky 1987, p. 551). Work values denote "specific expressions of general values in the work setting" (Ros et al. 1999, p. 54; see also Super 1970). Building on the taxonomy of ten motivationally distinct types of values (Schwartz 1994), translations to the workplace may include the value domains intrinsic, extrinsic, social/affective, and power/growth (cf. Krumm et al. 2013). Age-related changes in work values might provide important guidelines for age-differentiated human resource management (e.g. Grube and Hertel 2012; Kooij et al. 2011).

Theory of Psychosocial Development

Erikson's (1963) Theory of Psychosocial Development stressed generativity as an important value domain that has been neglected in the context of work motivation for a long time. The theory of psychosocial development highlights generativity as a central motive located in the seventh of eight stages of development, reflecting "the communal need to be nurturant and the agentic desire to do something or be something that transcends death" (McAdams et al. 1993, p. 222). Generativity values are assumed to be increasingly prioritized through adulthood (Erikson 1963), a prediction that is in line with the assumed increase in social value salience made by the Socio-emotional Selectivity Theory (Carstensen 2006). Older workers' priorities for generativity motives might be a particular fruitful starting point for age-differentiated human resource management if this assumption can be confirmed empirically (Grube and Hertel 2012).

Theory of Dual Motive Systems

The Theory of Dual Motive Systems (McClelland et al. 1989) posits that self-attributed motives that are openly acknowledged by a person (e.g. self-reported goals, motives, and values) differ fundamentally from implicit motives that affect behavior without conscious awareness of a person. Specifically, McClelland and colleagues state that the "big three" motives (i.e. achievement, affiliation, and power) divide into these functionally independent but interacting motive systems. Implicit and explicit motive systems have different developmental origins (pre-linguistic versus post-linguistic developmental state), are aroused by different classes of incentives (task versus social incentives), and predict different classes of behavior (operant versus respondent behaviors). Typically, explicit motives are measured by self-report questionnaires (e.g. Havercamp and Reiss 2003), whereas implicit motives are measured with projective tests (e.g. the Thematic Apperception Test; Murray 1943) or semi-projective measures (e.g. the Multi-Motive-Grid; Sokolowski et al. 2000). McClelland and colleagues as well as other authors repeatedly showed that a combination of implicit and explicit motives leads to a better prediction and better understanding of motivation, well-being, and behavior (e.g. Schultheiss and Brunstein 2010). In the current project, we extend this theory by postulating (and demonstrating) that the interplay between these two motive systems is systematically moderated by individual's age, particularly at the workplace (Hertel et al. 2013; Thielgen et al. 2012).

Transactional Stress Model

The Transactional Stress Model (Lazarus 2007; Lazarus and Folkman 1987) maintains that individuals differ in their reaction to stress factors as a function of two interdependent processes: appraisal and coping. Appraisal refers to person's evaluation whether external stressors might exceed their personal resources or threaten their well-being. If this is the case, the stress reaction is qualified by the availability of adequate coping strategies for a person. Coping strategies can broadly be distinguished into strategies that actively change external stressors (e.g. divide an overwhelming task into subtasks that can be delegated) and strategies that regulate one's own emotions (e.g. de-prioritize current goals). In sum, persons experience stress when they evaluate characteristics of the environment as harmful and/or threatening to their well-being, and feel unable to apply adequate coping strategies. In order to understand age differences in stress experience more thoroughly, we consider aging effects at different stages of the stressor-stress chain as outlined in the Transactional Stress Model (e.g. Rauschenbach et al. 2013).

Life-Span Theory of Control

The Life-span Theory of Control (Heckhausen 1997) posits specific age trajectories for primary and secondary control striving. Primary control striving denotes attempts to actively change the external world (problem-focused coping). Secondary control striving summarizes attempts to regulate emotions (emotion-focused coping, see also Lazarus and Folkman 1987). The Life-span Theory of Control maintains that the capacity for applying primary control strategies is impacted by resource losses over the life span as well as other age-related constraints, leading to a decline of primary control strategies in older age groups. In contrast, secondary control striving is assumed to steadily increase across the life span, presumably reflecting constant gains in emotion regulation skills (Heckhausen et al. 2010). It should be noted, however, that these assumptions have been developed with reference to the general life-span and not to the time frame of the usual occupational life. Applying and adapting this theory to the work context provides both additional insights as well as more differentiated strategies to increase well-being at work (cf. Rauschenbach and Hertel 2011).

Samples, Methods, Instruments and Procedure

Samples

In order to test the different assumptions derived from our theoretical framework, we conducted more than 25 field studies with more than 12,000 working people from a wide area of occupations and branches (e.g. consumer goods, logistics, retail and wholesale, and social services). Sample sizes in these studies varied between 40 and 5,132 workers. Moreover, we conducted a meta-analysis on age and stress reactions comprising more than 29,000 workers overall. In addition to participants recruited from work organizations, we used online panels (i.e. databases consisting of registered members who are willing to take part in online surveys and receive money or prizes as compensation; e.g. Krumm et al. 2012) that facilitate repeated measurement of persons across time, thereby enabling us to distinguish between genuine age effects and cohort/generation effects. Some of these panel studies are conceptualized for long time intervals (more than 20 years) and are still ongoing. Given the rather short time scope of the current project cycle, however, we concentrate on cross-sectional results in this chapter. While cross-sectional results can not sufficiently distinguish between genuine age effects and genuine cohort effects, they nevertheless provide important information for applied issues because organizations have to handle age diversity regardless of the exact origin of differences between age groups.

Measures

While different measures of age and aging are discussed in the literature (including biological age, social age, etc.; e.g. Sterns and Doverspike 1989), we focused on chronological age as central age measure in our studies. Chronological age can be easily and reliably assessed, and it is the dominant indicator used in HR policies in the field. Given that chronological age is closely related to the other age concepts (Kooij et al. 2008), we assume that focusing on this age concept will not lead to severe misconceptions. However, in order to better understand the psychological mechanisms underlying age effects, we also included potential mediator variables that are highly contingent with age in our studies, for example, tenure, job experience, and future time perspective at work (cf. Carstensen 2006, see also Zacher and Frese 2009).

The other constructs in this research project have been mostly assessed with validated and established scales from the literature.

- *Implicit motives* were either assessed with projective tests (Picture Story Exercise, PSE; Pang and Schultheiss 2005) or semi-projective measures (Multi-Motive-Grid, MMG; Sokolowski et al. 2000). The PSE (Pang and Schultheiss 2005) is the current research version of Murray's (1943) Thematic Apperception Test. During the PSE, participants are asked to write imaginative stories in response to six ambiguous pictures. Strength of implicit motives (i.e. achievement, affiliation, and power) is assessed by coding the content of the written stories for motive imagery. For instance, the affiliation motive is scored if "establishing, maintaining or restoring friendship or friendly relations among persons…" (Winter 1994, p. 12) is indicated in the text. Standardized manuals are provided in order to facilitate objectivity of coding (e.g. Winter 1994). The MMG is a measure of implicit motives combining features of classical projective measures (i.e. pictures) with features of questionnaires (i.e. items with standardized responses), thereby avoiding difficulties of manual coding procedures. During a MMG session, 14 ambiguous pictures are presented and participants are asked for each picture to rate their agreement on different statements referring to that picture (e.g. "Hoping to get in touch with other people" as indicator of implicit affiliation motives). Today, the MMG is a widely used alternative to projective measures, and has been validated in several studies (for details, see Schüler et al. 2010).
- *Explicit motives* were measured with the German version of the Personality Research Form (PRF; Jackson 1984; Stumpf et al. 1985). The PRF is a widely used and validated measure of personality traits. Building on Henry Murray's theory of personality, the measure comprises 14 scales, including achievement, affiliation and dominance motives. Each scale consists of 16 self-report statements (e.g. "I seldom put out extra effort to make friends", reverse coded, as a measure of explicit affiliation motives).
- *Work-related future time perspective* as potential mediator of age effects in the context of the Socio-emotional Selectivity Theory (Carstensen 2006) was measured with 10 items from Carstensen and Lang (1996) and adapted by Zacher (2009). The scale is divided into two subscales that focus on remaining opportunities (e.g. "My occupational future is filled with possibilities") and on remaining time ("Most of my occupational life lies ahead of me"). Both subscales show good reliability estimates as indicated by good fit indices of the two-factorial model (Zacher and Frese 2009).
- *Work-related demands and resources* such as social stressors (e.g. "I get told off for every little thing") or job control (e.g. "Looking at your work overall, how many opportunities do you have to make your own decisions?") were measured using scales from the Instrument for Stress Oriented Task Analysis (ISTA;

Semmer et al. 1999). The ISTA is widely used in Germany as measure of demands and resources at work.

- *Goal orientation* of individuals refers to different general strategies when pursuing values at work. Based on the SOC-Model (Baltes and Baltes 1990), we differentiate between goal striving, goal maintenance, and prevention of losses. These three orientations were operationalized for each of the different work values considered. For instance, the work value "Interesting Work" was differentiated in "I like to have more interesting work activities" (goal striving), "I like to maintain the level of interesting work I have" (goal maintenance) as well as "I try to prevent losing the interesting parts of my work"). Participants rated their agreement with these statements on 5-point Likert scales.
- *Self-regulation strategies* were assessed with three subscales derived from the Coping Orientations to Problems Experienced inventory (COPE; Carver et al. 1989), each of these subscales consisting of four items. We used the subscales active coping (e.g. "I concentrate my efforts on doing something about it"), positive reinterpretation (e.g. "I try to grow as a person as a result of the experience"), and disengagement (e.g. "I admit to myself that I can't deal with it, and quit trying"). Participants were asked to indicate what they usually do when under considerable stress at work.
- *Work Motivation* was measured with three items from Hertel (2002) (e.g. "During the last three months, I have been highly motivated at work").
- *Job satisfaction* was assessed with eight items from the Work Description Inventory (Neuberger and Allerbeck 1978), considering satisfaction with colleagues, direct supervisor, task, pay, work conditions, developmental opportunities, organization and management as well as general job satisfaction. This scale is among the best validated measures of general job satisfaction with German samples.
- *Irritation* as measure of proximate stress reactions was assessed with six items from the Irritation scale (Mohr et al. 2005). This scale includes both perceived emotional stress (e.g. "I get irritated easily, although I don't want this to happen") and perceived cognitive stress (e.g. "Even at home I often think of my problems at work"). Both the overall scale and both subscales have been successfully validated in several work-related settings (Mohr et al. 2005).

In addition, we developed various new measures when existing measure were not sufficiently specific to capture age differences:

- The *Muenster Work Value Measure* (MWVM; Krumm et al. 2013) was developed to particularly include work values of importance for older workers (e.g. generativity values). The MWVM is based on Schwartz's (1994) taxonomy of general values and includes the following value domains: Intrinsic Growth, Extrinsic Growth, Affective, Generativity, and Context-related Values. Intrinsic Growth Values denote the acquisition and extension of resources related to the work itself, e.g. developing new skills. Extrinsic Growth Values cover the

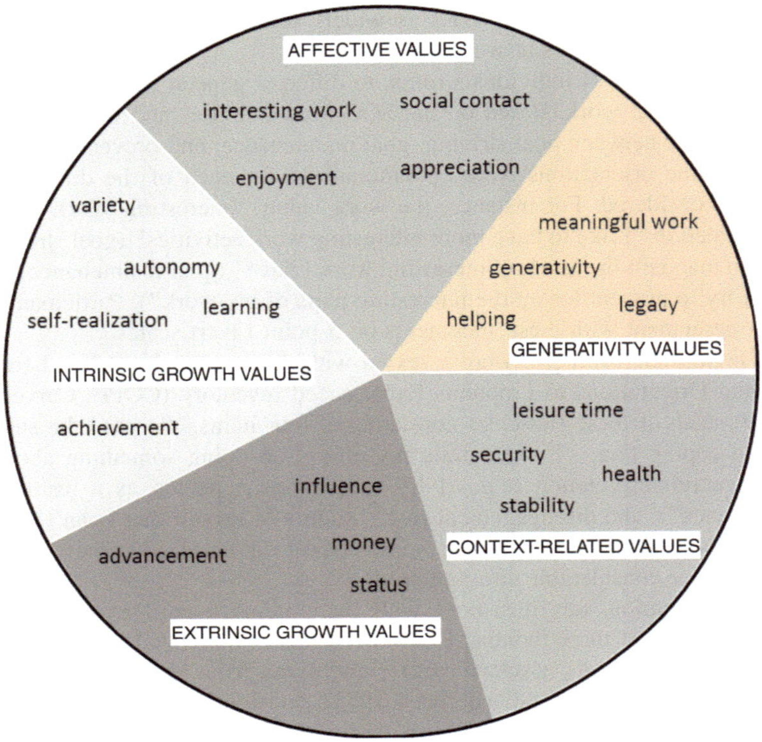

Fig. 1 Mapping of the work values of the MWVM in a circumplex model (for details, see Krumm et al. 2013b)

acquisition of resources contingent on outcomes of successful work, e.g. advancement. Affective Values describe positive affective experiences and emotion regulation, such as appreciation. Generativity Values can be defined as "the concern in establishing and guiding the next generation" (Erikson 1963, p. 267), e.g. passing on knowledge. Finally, Context-related Values focus on desirable states outside of the work context that may be negatively affected by work (e.g. leisure time). These values are also referred to as deficit values because they are characterized by individual's striving to prevent potential losses. The value domains and subordinated values are depicted in Fig. 1. The structure of the MWVM has been confirmed in initial studies, supporting the construct validity of the MWVM both in its ranking and its rating version (Krumm et al. 2013).

- In order to address the assumption that older workers might show a higher congruency between global (attitudinal) and experience-based experiences at work (i.e. stress and job satisfaction), we developed a new procedure to measure experience-based constructs at work. Building on a more general reconstruction approach (Kahneman et al. 2004), the underlying idea of the Event Reconstruction Method (ERM; Grube et al. 2008) is to instruct workers to reconstruct specific job events vividly in their memory. By tapping on episodic memory traces, both affective and cognitive details of these experiences should be accessible without being as invasive as experience sampling methods (e.g. Csikszentmihalyi and Lefevre 1989). At the same time, the explicit re-experience of the job events might be even more reliable for measuring affective aspects of job events than diary methods (for a more detailed discussion, see Hertel and Stamov-Rossnagel 2012). In one initial validation study, the ERM provided nearly identical data patterns for experience based job satisfaction as found in a study using classic experience sampling (Grube et al. 2008). Thus, the reconstruction approach has been initially validated at least for short time intervals (reconstructed job events were only a few days ago; for initial data on stress experience see Rauschenbach and Hertel 2011).

- Finally, in an initial attempt to address cohort effects in a cross-sectional design, we extended the reconstruction idea to longer time intervals and developed the "Mental Time Journey" approach in which participants are instructed to re-experience their life situation 15 and 30 years ago. Again, specific memory cues were given in order to facilitate the re-experience of former life stages (where did you live, which job did you have, who were the people you worked and lived with, etc.). Initial data using this approach showed indeed systematic differences between cohorts in work values, for instance between young workers (20–35 years) today and middle-aged workers (35–50 years) re-experiencing their young adultness (Hertel et al. 2007). Although these differences are similar to differences described to hold between the related generations (e.g. Twenge et al. 2010), further validation in longitudinal studies are desirable in order to estimate how reliable the reconstruction approach might be for long time intervals.

Procedure

Data collection. Data were either collected with paper-and-pencil or with online questionnaires. Organizational samples were recruited in cooperation with organizations who were interested in the topic of age differences in work motivation and stress. In these cases, data were conceded in exchange for status-quo analyses and feedback at the organizational level. In each case, confidentiality of individual answers and compliance to ethical research guidelines were guaranteed in a written agreement. Online panel data were recruited either by service providers specialized in online surveys or by cooperation partners. Surveys were handed to the providers, who processed recruiting, reminders and rewards of the survey.

Statistical analyses. The various statistical analyses were selected depending on the specific research questions of the respective studies. In addition to standard analyses (e.g. correlation and regression analyses), we used more specific methods in order to examine the structure of our Work Value Measure (MWVM), the relationship between two or more independent and one dependent variable, to test assumed moderating and mediating effects, as well as to aggregate effects in a meta-analysis.

The structure of the *rating version* of the MWVM was tested using confirmatory factor analyses, examining whether a given data set fits a predefined structure—in our case the five-factorial structure of the MWVM. Model fit was evaluated on the basis of established combinations of fit indices (e.g. Beauducel and Wittmann 2005). In contrast, the structure of the *ranking version* of the MWVM was assessed with multidimensional unfolding as a special case of multidimensional scaling. This approach enabled us to graphically locate items and individuals in the same space in such a way that distances between items and individuals optimally represent individuals' preferences.

The congruency or fit between two (or more) independent variable and one dependent variable (e.g. intrinsic needs, intrinsic supplies, age, and job satisfaction) was tested using polynomial regression analyses and subsequent response surface methods (cf. Edwards and Perry 1993). Polynomial regression analysis provides tests of the predictive power of any linear and non-linear combination of two or more variables on one dependent variable. In case of significant two- or three-way interactions between the independent variables, the nature of this interaction is further examined using response surface methods which provide graphical depictions of the interaction. The shape of the surface can be further scrutinized with regard to its slope and curvature (for simplified examples, see Figs. 2 and 4). Moderator variables were identified by conducting response surface methods separately for different age groups, creating different surface charts for different partitions of the moderator.

Finally, meta-analyses were conducted following Hedges and Olkin (1985). We used a random-effects model since we expected that the true effects could vary from study to study. All effect sizes were converted into correlation coefficients and were weighted by the respective sample size. Correlations were considered as

Fig. 2 Two-dimensional surface charts relating need-supply-misfit of the extrinsic growth values cluster to job satisfaction at two age levels (Krumm et al. 2012)

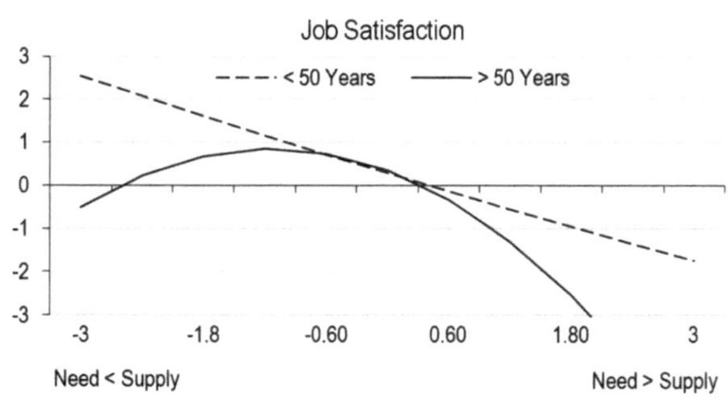

statistically significant when the 95 % confidence interval did not include zero. Moderation analyses were performed either by regression analysis techniques on meta-analytically obtained data for continuous moderators or subgroup analyses for categorical moderators (Borenstein et al. 2009). Mediators were identified testing direct and indirect paths from independent to dependent variables (cf. Baron and Kenny 1986). Complete mediation was assumed when the indirect paths (from independent variables to mediators and from mediators to dependent variables) showed significant coefficients whereas the simultaneously estimated direct paths revealed no significant coefficients.

Results

Age and Explicit Work Values

In the initial studies of this research project, we addressed age differences in the perceived importance of different work values as the first step in a motivational action control cycle (e.g. Heckhausen and Gollwitzer 1987; Hertel and Wittchen 2008). Work values are understood as outcomes that workers consciously desire or feel they should attain through their work (e.g. Dose 1997; Twenge et al. 2010), and thus directly affect the way workers perceive jobs and workplace conditions. Therefore, knowing individuals' work values and related needs and interests is a central cornerstone for effective recruiting and retention plans, attractive incentives and career systems, as well as differentiated leadership strategies. As work values are usually consciously represented, they are relatively easily assessed using self-report questionnaires.

While work values have been shown to vary only slightly between early adolescence and early adulthood (e.g. Low et al. 2005), we assume that they systematically vary and change later in a worker's career once persons have entered their professional life. Indeed, work values are not isolated but embedded in a broader system of values important for an individual, and these values are continuously affected by different phases and steps in individuals' development. For instance, Erikson's Theory of Psychosocial Development describes various stages and developmental "tasks" during adulthood (Erikson 1963) that should influence the priority of values persons pursue when working. In addition to taking over responsibilities for others (e.g. marriage and raising children), Erikson stressed generativity as a central theme of the middle adulthood that has often been neglected in research on work motivation. However, if such age-related changes in work values are neglected, human resource management strategies might only address the needs and interests of a selected part of workers, and thus may lead to inefficient results.

In order to derive specific and testable assumptions on age differences in work values, we connected existing research on work values with theoretical approaches from life-span research, such as Socio-emotional Selectivity Theory (Carstensen 2006, 2009) and the Selection, Optimization, and Compensation Model (Baltes and Baltes 1990). Moreover, given that existing measures of work values are rather focused on specific value dimensions and often neglect other relevant values, we developed the Muenster Work Value Measure (MWVM; Krumm et al. 2013) which provides an efficient assessment of the subjective importance of 21 work values distributed across five general value domains (see above).

Having used the MWVM in many studies with currently more than 4,600 participants from different companies and branches (e.g. logistics, utilities, retail and wholesale, social service), our data indicate that age groups indeed do differ systematically in their work value profiles. For instance, the results of two questionnaire studies (with 303 and 386 age-diverse workers from different branches, respectively) consistently showed that older workers place more importance on Affective Values and Generativity Values (Grube and Hertel 2012). This finding is in line with the idea that emotion-related goals increase in importance (Carstensen 2006, 2009). Moreover, in order to pursue goals related to generativity, workers first need to have acquired certain experiences and skills. A second robust pattern in these studies is that older as compared to younger workers prioritize autonomy more strongly (Grube and Hertel 2012). This pattern reflects the fact that a minimum of expertise and occupational skills are necessary if autonomy is to be experienced as a benefit instead of a burden. However, autonomy also enables older workers to cope with age-related deficits in physical and cognitive capacity (e.g. Salthouse 2012). Last but not least, autonomy might signal appreciation and occupational success that are part of the normative expectations towards persons at the end of their professional life. In other words, still having low levels of autonomy at the end of a career might signal that a worker has not been very successful at work.

On the other hand, age was negatively related to Extrinsic Growth Values (e.g. advancing ones career, gaining influence), which is in line with our expectation based on Socio-emotional Selectivity Theory that these goals are more relevant for workers with a longer occupational time perspective. Moreover, Context-related Values (e.g. health, job-security) showed a curvilinear relationship with age, confirming our assumption that middle-aged workers instead of older workers place the highest importance to security issues due to high responsibilities for others (young children etc.) and the sandwich position between multiple role expectations at work and in their private life (Heckhausen 2001), which is also reflected in particularly high stress ratings of this age group (Rauschenbach and Hertel 2011). Finally, and perhaps most importantly, we found no significant age differences in the subjective importance of high performance, interesting work, learning and social contacts at work, speaking clearly against negative age stereotypes in these domains.

The reported age differences are similar in ranking and rating versions of the MWVM, and mostly consistent with recent data from other research groups (e.g. Kooij et al. 2011) suggesting reliability and generalizability of our findings. Moreover, in addition to age differences in subjective importance of work values, we could also show that at least some of these age differences are mediated by participant's future work-time perspective as a central process derived from the Socio emotional Selectivity Theory (Carstensen 2006, 2009; see also Zacher and Frese 2009). For instance, age differences in the subjective importance of career advancement and gaining influence were mediated by future work time perspectives, such that the effect of younger as compared to older workers reporting higher striving for career advancement and influence was mediated by longer future work time perspectives of younger workers (Grube and Hertel 2012). In a similar way, the results documented stronger orientations towards compensation and loss prevention for older workers (Grube and Hertel 2012), which is in line with assumptions of the Selection, Optimization, and Compensation Model (Baltes and Baltes 1990). However, one important finding was that older workers not always reported higher maintenance or prevention orientations. In case of Generativity Values, older workers reported even higher optimization orientation than younger workers (Grube and Hertel 2012). In general, these findings demonstrate that instead of effects of chronological age per se, many of the age differences in work motivation seem to be mediated by age-contingent constructs such as future time perspective or the existence of sufficient job experiences.

In sum, our research reveals robust age differences in various work values and goal orientations consistent with assumptions derived from life-span theories. In particular, emotion-oriented values, other-oriented values (generativity) and autonomy seem to be particularly relevant for older workers, while career-related goals and incentives are particular interesting for younger workers. Security-related values were most relevant for middle-aged workers. Together with a better understanding of the underlying psychological mechanisms, these findings might provide interesting guidelines for age-differentiated human resource management (Table 1).

Table 1 Summarized trends of age differences in work values (adopted from Grube and Hertel 2012)

Work values	Age differences in relative importance
Affective values Generativity Autonomy	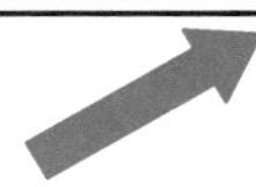
Career advancement Gaining influence	
Achievement Interesting work Learning Social contact at work	
Health Security at work	

Age and Person-Organization (Need-Supply) Fit

While it at least might appear plausible that considering workers' values in human resource management and job design might lead to better outcomes, a next step in our research project was to provide empirical data for this assumption. Thus, extending the work on age differences in absolute rankings (or ratings) of work values, we contextualized these absolute preferences within the different work environments. In addition to the well-established finding that perceived fit between work values and job characteristics (so-called "need-supply fit") leads to higher job satisfaction (e.g. Kristof-Brown et al. 2005), we again assumed that these effects are additionally moderated by workers' age. More specifically, based on the Socio-emotional Selectivity Theory (Carstensen 2006, 2009), we assumed that affective consequences of the perceived fit between worker's values and task characteristics should be more important for older as compared to younger workers.

Data from 471 age-diverse workers supported this presumption showing that job satisfaction of older as compared to younger workers was more strongly affected by the perceived need-supply fit. Polynomial regression analyses yielded significant three-way interactions between needs, supplies, and age in predicting job satisfaction. This was true for needs and supplies related to four out of five value domains included in the MWVM: Extrinsic Growth Values, Generativity Values, Affective Values, and Context-related Values (for details see Krumm et al. 2012).

The nature of the interaction was further examined by scrutinizing response surface charts. For each value domain, separate charts were created for young to middle-aged workers (age ≤ 50) and for older workers (age > 50). These surface charts unanimously indicated that if needs exceeded supplies (cf. right hand side in Fig. 2), the relationship between need-supply fit and job satisfaction was more pronounced for older workers. In other words, job satisfaction of older as compared to younger workers was more negatively affected when job characteristics fell behind their needs. However, age did not moderate the relationship between need-supply fit and job satisfaction in the Intrinsic Growth Value domain. This result might be seen as consistent with the Socio-emotional Selectivity Theory, assuming lower general importance of Intrinsic Growth Values for older workers.

Together, the results of our studies complement findings from earlier work reporting a positive relationship between age and job satisfaction (e.g. Hochwarter et al. 2001; Ng and Feldman 2010). Our results revealed that a steady increase of job satisfaction (on average) across the life span may be corroborated by misfits between needs and supplies in older age groups. Hence, human resource management might be well-advised to avoid need-supply misfits particularly by constant adaptations of job characteristics to workers' needs. Moreover, misfits between the subjective importance of values and job characteristics were detrimental across almost all value domains. We assume that perceived need-supply misfits threaten emotional well-being at work. In line with predictions made by Socio-emotional Selectivity Theory (Carstensen 2006), older workers react more strongly to such threats even when tenure or status were controlled as potential covariates.

Age and Congruency Between Implicit and Explicit Motives

In addition to age differences in need-supply fit as described above, we also assumed that chronological age might be an important moderator of the congruency between worker's explicit and implicit motives ("i-e congruency"). High i-e congruency is indicated by a positive correlation between implicit and explicit measures of motives. In contrast, low i-e congruency is indicated by low or negative correlations between implicit and explicit motive measures, either because the implicit motive exceeds the strength of the explicit motive, or the explicit motive exceeds the strength of the implicit motive (Thrash et al. 2010). According to our general framework (Hertel et al. 2013), we expected that older as compared to younger workers achieve a better i-e congruency due to learning and self-reflection processes. However, similar as to reactions of need-supply misfit, we also assume that older workers react more negatively to low i-e congruencies that might provide an additional stressor and impede self-regulation and well-being at work (e.g. Brunstein 2010; Brunstein et al. 1998; Kehr 2004). Whereas high i-e congruency should facilitate satisfaction of implicit motives and promote well-being, low i-e congruency should cause motivational conflicts resulting in frustration of implicit motives and decreased well-being. Although this basic idea has received empirical support (e.g. Brunstein 2010), this relationship has neither been examined in an occupational context, nor in the context of aging.

In two initial empirical studies (Thielgen et al. 2012), we examined the interaction between implicit and explicit affiliation motives and the moderating function of age. The results provided support for our assumptions in the affiliation motive domain. First of all, polynomial regression analyses revealed that, overall, i-e congruency of affiliation motives was positively correlated with job satisfaction. More importantly, this pattern was moderated by chronological age of workers in two ways. First, higher levels of i-e congruency were found for older as compared to younger workers (cf. Fig. 3). Second, older workers reacted more negatively to low i-e congruency as indicated by lower levels of job satisfaction (cf. Fig. 4). Noteworthy, this effect was generally stronger when the strength of the implicit motive exceeded the strength of the explicit motive, providing additional

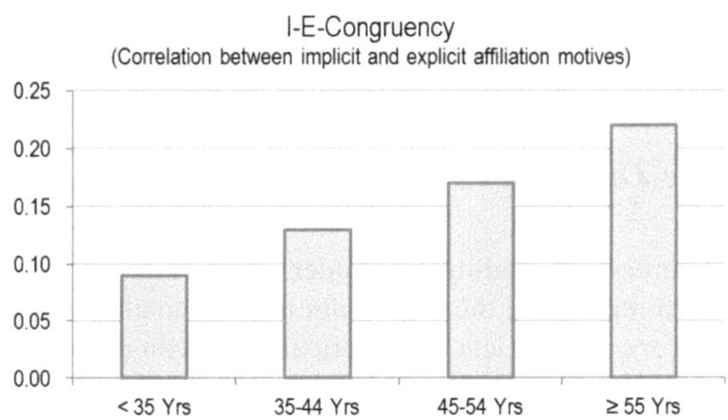

Fig. 3 Bar chart showing bivariate correlations between implicit and explicit affiliation motives in four age groups (based on data from Thielgen et al. 2012). Higher correlations indicate a higher level of i-e congruency

Fig. 4 Two-dimensional surface chart relating implicit-explicit motive congruency of affiliation motives to job satisfaction at two age levels (based on data from Thielgen et al. 2012). Please note that, other than in Fig. 2, the lines depicted in this chart are based on a continuous regression model showing predicted values of 30 vs. 60 year old workers

support for the assumption that satisfaction of implicit motives is a prerequisite of well-being at work, particularly for older workers.

In sum, the results of this initial study support our idea that older workers not only develop a better congruency between their implicit (and thus more generic) motives and what they consciously consider to be their motives (i.e. their explicit motives), but that they also react more negatively in case such congruency is not achieved, causing motivational conflicts and related frustration of implicit motives. These results might be of particular interest for personnel development and coaching processes. Indeed, while explicit motives are usually considered in these programs, implicit motives are often neglected because they are more difficult to assess.

However, it should be noted that we found the described pattern most clearly for affiliation motives but not for achievement or power motives. A plausible explanation in line with Socio-emotional Selectivity Theory (Carstensen 2006) is that affiliation and social relationships become particularly important with restricted future time perspective (i.e. higher age). Therefore, frustration of affiliation motives might be more severe for older workers as compared to frustration of power or achievement motives. However, these initial ideas have to be further examined in future research.

Age Differences in Stress Experience

Whereas the finding that older workers report higher job satisfaction has been often replicated, the interrelation between chronological age and stress experience at work is less clear. Theoretically, we can derive different patterns of covariation between age and stress experience at work (cf. Fig. 5).

Fig. 5 Illustration of potential covariations between age and stress experience (cf. Rauschenbach and Hertel 2011)

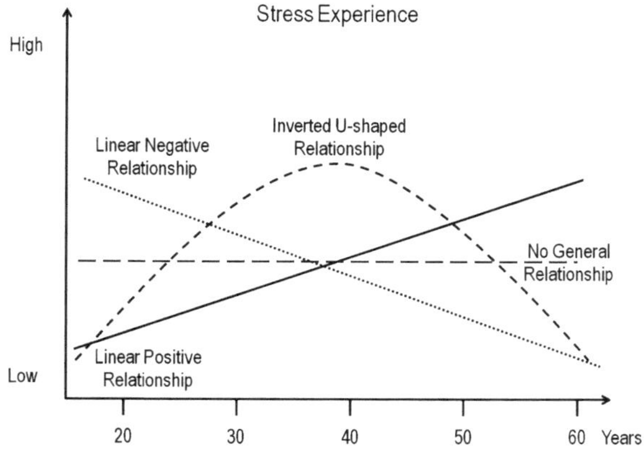

A first potential pattern is based on potential losses of physical capabilities (e.g. Charles 2010) or mental resources due to aging (e.g. Salthouse 2012), that might increase the vulnerability of older workers. As a consequence, one would expect that older workers report higher levels of stress experience at work than their younger colleagues, resulting in a positive relationship between age and stress experience. A second potential pattern, however, can be derived from theories of emotional development (e.g. Labouvie-Vief et al. 1989), predicting increasing coping skills when persons mature. As a consequence, older persons not only should know more about their own emotional reactions than younger persons, but older persons should also have acquired higher skills to manage their emotional experiences, for instance, by applying secondary control strategies. Moreover, older persons should also have higher skills to influence emotions of interaction partners. As a consequence, one would expect that older workers report lower levels of stress experience at work than their younger colleagues, resulting in a negative relationship between age and stress reactions. Indeed, Ng and Feldman (2010) provide meta-analytical evidence for such a relation between age and rather distal consequences of stress, such as emotional exhaustion and depersonalization.

Finally, individuals pass through different stages not only during their professional career but also during their private life, providing various demands that might interfere with their jobs. In particular, middle-aged workers often have to handle both high demands at work and in their families (e.g. young children, elderly care, etc.), creating a so-called sandwich position (e.g. Heckhausen 2001) with particular high demands on coping skills and stress resilience. As a consequence, the relationship between age and stress might also follow a curvilinear pattern with highest stress reactions for middle-aged workers.

As an initial empirical test of these different theoretical predictions, a cross-sectional study with 274 workers between 18 and 65 years provided most support for the third option (Rauschenbach and Hertel 2011). Middle-aged workers between 36 and 50 years of age reported higher levels of stress than both younger and older workers. Interestingly, these results were not only found with global measure of stress reactions (Irritation scale; Mohr et al. 2005) but also with the Event Reconstruction Method (Hertel and Stamov-Rossnagel 2012) as experience-based measure of stress "in situ". As the interrelation between the two stress measures was rather low, the two measures captured different parts of stress variance. Thus, middle-aged workers not only reported higher stress levels in general, but also experienced higher stress levels (on average) in specific situations during their daily work. Notably, the specific job events reported as stressful during the Event Reconstruction Method were rather similar across the different age groups, suggesting that the observed age differences were not simply a consequence of different task demands as might have been expected by a job change hypothesis (e.g. Wright and Hamilton 1978), assuming that, on average, older workers have jobs with fewer stressors. Instead, age differences in individual factors (coping skills, level of additional demands outside of work, etc.) seem to be more plausible.

Although instructive, the results of this initial study need to be replicated in more and varying job settings. Therefore, we conducted a meta-analysis (Rauschenbach et al. 2013) addressing the robustness of the observed findings on age and stress reactions as well as on potential moderating conditions. In contrast to an existing meta-analysis (Ng and Feldman 2010) considering rather distal stress reactions such as depersonalization, we focused on proximal stress reactions using an established measure of irritation (Mohr et al. 2005). Overall, the literature search gained 66 primary studies which assessed irritation and reported correlations with age (usually as control variable), representing 29,806 workers (Rauschenbach et al. 2013). The results of this meta-analysis showed a covariation between chronological age and proximal stress reactions close to zero; older workers reported neither more nor less stress at work than their younger colleagues. However, this overall pattern was slightly moderated by the type of occupation. In blue-collar jobs with higher physical demands (e.g. heavy lifting, heat) stress reactions were higher for older than for younger workers. In jobs with higher demands on social and emotional skills (e.g. teachers, nurses), the relationship between age and stress reactions followed an inverted U-shape which is in line with the earlier described primary study (Rauschenbach and Hertel 2011).

Together, the results of these studies suggest that age affects stress experience at work. However, the results do not support a general deficit perspective on older workers as the overall relationship between age and stress experience was close to zero. Instead, the data support a more differentiated view, underlining the importance of the specific job demands in relation to age-related skills and challenges. From this perspective, older workers seem to be particularly apt when social and emotional job demands are high, whereas high physical demands seem to require rather younger workers (see also chapter Assembly Tasks in the Automotive Industry a Challenge for Older Employees, Frieling et al.). In general, for a more thorough understanding of age effects on stress experience at work it will be fruitful to further differentiate between age effects on different steps in the stressor-stress chain (Rauschenbach et al. 2013) as these can partly contradict and/or neutralize each other.

Age Differences in Coping Strategies and Control

As a final example of age effects on motivation at work, we examined age differences in coping and control strategies as reactions to stress at work, and with consequences on further motivational steps in an action regulation approach (Hertel et al. 2013). In general, the Life Span Theory of Control (Heckhausen and Schulz 1995; Heckhausen et al. 2010) predicts that primary control strategies and problem focused coping (e.g. direct steps to eliminate stressors) are increasingly used up to middle age, and later decrease again as persons realize that their personal resources and opportunities to change the environment diminish with higher age. On the other hand, the use of secondary control strategies and emotion focused coping (e.g. positive reinterpretation of stressors) is assumed to constantly increase across age. However, when applying these general dynamics to the work context, it has to be noted that Heckhausen and her colleagues refer to the complete life cycle of individuals. As a consequence, declines in primary control strategies are assumed particularly for persons in their 70s and 80s. Given that most persons are 65 and younger when retiring from work, they should have still high levels of primary control according to the Life Span Theory of Control. This assumption is additionally supported by the fact that aging at work is indeed often connected with an increase in power and autonomy, providing increasing opportunities for primary control strategies. Therefore, we expect that primary control strategies at work increase with age within the usual timeframe of a career (usually up to 65 years of age).

In addition to different foci (external vs. internal), coping strategies can also be distinguished according to their effectiveness in reducing stress, discerning between adaptive versus maladaptive strategies (Carver et al. 1989). This holds particularly for secondary control and emotional coping strategies. For instance, positive re-interpretation of stressors at work should reduce subsequent stress experiences. In contrast, disengagement (i.e. reduction of effort and abandoning of goals; cf. Carver et al. 1989) is rather ineffective in the work context (e.g. Torkelson and Muhonen 2004). Based on the idea that age is positively related with emotional competencies (see above), we expect that *adaptive* secondary coping at work is *positively* related with workers' age, whereas *maladaptive* secondary coping at work is *negatively* related with workers' age.

These assumptions were tested in an online survey with 700 age-diverse workers from various branches (Rauschenbach et al. 2012b). The results revealed that older as compared to younger workers reported more frequent use of primary coping (e.g. direct steps to eliminate stressors at work) as well as adaptive secondary coping (i.e. positive reinterpretation of stressors), but less frequent use of maladaptive secondary control strategies (i.e. disengagement). Interestingly, among these coping strategies, disengagement occurred to be the strongest mediator of age differences in stress reactions. This was particularly the case in jobs with low to moderate autonomy at work. In these jobs, secondary control strategies might be particularly crucial. The fact that younger workers reported disengagement more often than older works seems to indicate that younger workers have not yet developed positive reinterpretation as more effective secondary coping strategy. Whether this lack is due to insufficient time (not yet learned) or insufficient motivation (younger workers might prefer and/or have more opportunities to change jobs or employer) needs to be answered in future research.

Together, the present results support our assumption that age differences are relevant at different steps in the stressor-stress chain, moderating not only the perception of given stressors but also coping and self-regulation strategies (see also chapter Age-related Differences in the Emotion Regulation of Teachers in the Classroom, Philipp and Schüpbach, on age-related differences in the emotion regulation of teachers; and chapter Successful Aging Strategies in Nursing: The Example of Selective Optimization with Compensation, Müller et al. on the use of self-regulation strategies among nurses). In general, these data add to the positive perspective of older workers in terms of stress resilience, documenting maladaptive coping strategies to be more likely for younger than for older workers.

Discussion

This research project started with the general assumption that work motivation and stress experience are systematically affected by worker's age. After six years now, we have indeed documented numerous examples where younger and older workers show considerable differences. For instance, older workers seem to prioritize work values differently than their younger colleagues, placing higher importance on values related to their emotional well-being and experience of meaningfulness in the "here and now". Developmental issues such as career and gaining influence, however, seem to loose importance with higher age. In a similar way, older workers seem to focus less strongly on striving for new gains but rather on maintenance of current gains and on loss prevention. These effects are well in line with general predictions from life-span theories of aging (e.g. Baltes and Baltes 1990; Heckhausen 2001) when applied to the work context. Moreover, we could demonstrate that the degree to which worker's work values fit to the specific job opportunities (need-supply fit) or to their own dispositional profiles (implicit-explicit motive congruency) is related to satisfaction and well-being. This result underlines once again the importance of an age-differentiated approach both in job design as well as in individual coaching activities. Interestingly, our data also reveal that the impact of high need-supply fit or of high congruency between implicit and explicit motives increases with worker's age. This result is also consistent with life-span theories of aging, predicting that emotional well-being becomes more important when persons grow older and future time perspective decreases (Carstensen 2006, 2009). Thus, a differentiated or individualized approach of human resource management and job design becomes even more important for older than for younger workers.

In addition to age differences in work values and goal orientations, systematic age differences occurred also for later phases in the action regulation process at work. Older as compared to younger workers showed significant differences both in the reaction to stressors as well as in self-regulation strategies. Notably, our data disconfirm the pessimistic stereotype of rather weak and passive older workers (e.g. Rauschenbach et al. 2012a, b). Instead, the average of older workers in our studies reported lower stress and more active control strategies than middle-aged or younger workers, respectively. These results are encouraging in light of an increasingly aging workforce.

In addition to revealing systematic age differences in work motivation and stress, we also sought to further understand the psychological mechanisms and dynamics of age differences at work. Again in line with life-span theories, the results demonstrate that observed age differences are often not contingent to chronological age per se, but are caused by associated psychological processes such as future time perspectives at work. Despite of value for scientific progress, these insights are also fruitful for practitioners because they enable the development of general strategies to decrease or even disable aging effects. For instance, if age differences are caused by contingent differences in future work time perspective, extending the future work time perspective for older workers—for instance by offering employees to continue some work on a self-employed basis after retirement—might reduce or even annihilate age differences. Systematic research on these ideas is pending.

Last but not least, some of the scrutinized constructs seem to be rather unaffected by worker's age in general, for instance, performance orientation, valuing interesting tasks and social contacts at work, and first and foremost, job performance (see also Ng and Feldman 2008). Again, these are positive news in light of an aging workforce and disconfirms pessimistic age stereotypes. However, it should be noted that a zero-relationship between age and an outcome variable does not "proof" the absence of any age effects. In fact, age might have partly contradicting effects on different mediating mechanisms related to the outcome. A more differentiated analysis of age differences in stress experience at work, for instance, has revealed that chronological age can have different and partly reverse effects on different stages of the stressor-stress chain (Rauschenbach et al. 2013). A thorough understanding of the complex dynamics of age differences is an important precondition for building on the strengths of older workers without neglecting potential weaknesses.

Future Prospects

Limitations and Future Research

While the reported research is on age differences, perhaps the biggest constrain is time because most of the empirical research so far has been cross-sectional and therefore falls short to distinguish genuine age effects from cohort or generation effects (e.g. Twenge et al. 2010). Indeed, the rapid development of technologies in the last decades together with impressive changes of learning opportunities, work safety, food, and medicine have dramatically changed the context condition of workers today. The work environment of, for instance, a 30 year old factory employee, office clerk, or farmer in 2012 is quite different from the conditions s/he would have 25 or 50 years ago. Disentangling these different origins of age differences is important for both theory development and applied issues, and requires longitudinal approaches. Moreover, given that these changes are relatively slow, long time intervals are necessary.

In light of these concerns, we have started longitudinal studies on work values in our group using a large and diverse sample and expect interesting insights in the next 10–20 years. Moreover, such longitudinal studies are also needed to validate creative ideas to collect longitudinal data in a cross-sectional design (e.g. using a reconstructing approach as described above). In the meantime, however, cross-sectional research on age differences does provide helpful guidelines for applied issues given that organizations have to deal with age diversity now, regardless of the origins of these differences.

In addition to the timeframe of studies, further research on the specific mechanics of age effects at work is needed. Initial work on future work time perspectives as mediator of age effects (e.g. Carstensen 2006; Grube and Hertel 2012; Zacher and Frese 2009) illustrate the strengths of such an approach that not only deepens the understanding of age effects but also helps us to develop strategies to avoid or compensate for unwanted age effects. In a similar way, the current initial findings have to be further explored in different work contexts in order to reveal important moderating conditions, such as job content, tenure, job experience, employment status, gender, cultural background, etc.

Implications for Practitioners

The findings reported in this chapter have various implications for practitioners. First, managers might be well advised to grant more autonomy to older workers. Our results show that older workers value autonomy more highly than their younger colleagues. Moreover, we revealed that older workers may be able to use their autonomy more effectively than their younger colleagues when dealing with work-related stressors, e.g. by exerting more primary control.

Second, organizations might want to adjust the work environment of older workers to satisfy affective and generativity needs. This can be, for example, achieved by mentoring programs or age-diverse teams giving older workers the opportunity to train their younger colleagues. The findings reported in this chapter as well as established theories suggest that older workers emphasize affective and generativity values. Importantly, we also revealed that a misfit between needs and supplies is more detrimental for older than for younger workers.

Third, organizations can benefit from tools that help to monitor the needs of older workers. Employee surveys conducted on a regular basis or yearly performance appraisals may be used to gain insights into workers' current priorities of needs and values. Subsequently, work conditions may be adjusted accordingly and individually (if possible) to ensure job satisfaction across the working life-span. This example also illustrates that the implications of our research are not only relevant for older workers. Indeed, the current demographic changes and related reflections on work processes might lead to a more differentiated and individualized approach of human resource management that is beneficial for all workers.

Fourth, in order to retain their workforce motivated and satisfied, HR management should provide sufficient opportunities for self-selection in order to enable a good fit between implicit and explicit affiliation motives. We revealed that age is not only associated with higher congruencies between implicit and explicit motives but also with more negative reactions to incongruencies in this respect. Thus, older workers seem to be more sensitive to subtle signals of misfit and to self-select more often into different jobs if possible.

Fifth, organizations should particularly support middle-aged workers to deal with stress, even if some of the stressors are unrelated to their work. Opportunities such as flexible work schedules or temporary reduction of responsibilities might facilitate a better work-life balance and more "balanced" and motivated workers in the long run. Given that the average life expectation has significantly increased in the last decades, a more flexible change between working times, family times, education and training, as well as community-related activities is highly recommended (see Carstensen 2009, for intriguing ideas on such general changes).

Together, the current research demonstrates the strengths of a theory driven approach. Connecting traditional work psychology models of motivation and stress with life-span models of aging, the current project provides fruitful predictions and ideas not only to reduce or compensate aging effects at work but to benefit from the various strengths of older and younger workers that might create additional gains instead of losses in an increasingly aging workforce. One precondition for this, however, will be that managers and workers critically reflect their pre-assumptions and stereotypes about older (and younger) workers in order to not fall short of self-fulfilling prophecies (Hertel et al. 2013a, b). As has been shown in research on age diverse teamwork, the attitudes team members have towards age diversity are important moderators of positive diversity effects (Wegge et al. 2012; see also chapter Age Diversity and Team Effectiveness, Ries et al.). One of the best ways to reflect and—if necessary—correct socially transmitted stereotypes is systematic empirical research.

Acknowledgments This research was supported by three grants from the German Research Foundation to the first author (HE 2745/11-1; HE 2745/11-3 and HE 2745/11-4).

References

Baltes PB, Baltes MM (1990) Psychological perspectives on successful aging: the model of selective optimization with compensation. In: Baltes PB, Baltes MM (eds) Successful aging: perspectives from the behavioral sciences. Cambridge University Press, New York, pp 1–34

Baron RM, Kenny DA (1986) The moderator-mediator variable distinction in social psychological research: conceptual, strategic and statistical considerations. J Pers Soc Psychol 51:1173–1182

Beauducel A, Wittmann WW (2005) Simulation study on fit indexes in CFA based on data with slightly distorted simple structure. Struct Equ Model: A Multidisc J 12:41–75

Borenstein M, Hedges LV, Higgins JPT, Rothstein HR (2009) Introduction to meta-analysis. Wiley, Chichester

Brunstein JC (2010) Implicit motives and explicit goals: the role of motivational congruence in emotional well-being. In: Schultheiss OC, Brunstein JC (eds) Implicit motives. Oxford University Press, New York, pp 347–374

Brunstein JC, Schultheiss OC, Grässmann R (1998) Personal goals and emotional well-being: the moderating role of motive dispositions. J Pers Soc Psychol 75:494–508

Carstensen LL (2006) The influence of a sense of time on human development. Science 312:1913–1915

Carstensen LL (2009) A long bright future: an action plan for a lifetime of happiness, health, and financial security. Broadway Books, New York

Carstensen LL, Lang FR (1996) Future time perspective scale. Stanford University, Stanford

Carver CS, Scheier MF, Weintraub JK (1989) Assessing coping strategies: a theoretically based approach. J Pers Soc Psychol 56:267–283

Charles ST (2010) Strength and vulnerability integration: a model of emotional well-being across adulthood. Psychol Bull 136:1068–1091

Csikszentmihalyi M, LeFevre J (1989) Optimal experience in work and leisure. J Pers Soc Psychol 56:815–822

Dose JJ (1997) Work values: an integrative framework and illustrative application to organizational socialization. J Occup Organ Psychol 70:219–240

Edwards JR, Perry ME (1993) On the use of polynomial regression equations as an alternative to difference scores in organizational research. Acad Manag J 36:1577–1613

Erikson EH (1963) Childhood and society. Norton, New York

Grube A, Hertel G (2012) Age differences in work motives. Manuscript submitted for publication

Grube A, Schroer J, Hentzschel C, Hertel G (2008) The event reconstruction method: an efficient measure of experience-based job satisfaction. J Occup Organ Psychol 81:669–689

Havercamp S, Reiss S (2003) A comprehensive assessment of human strivings: test-retest reliability and validity of the Reiss profile. J Pers Assess 81:123–132

Heckhausen J (1997) Developmental regulation across adulthood: primary and secondary control of age-related challenges. Dev Psychol 33:176–187

Heckhausen J (2001) Adaptation and resilience in midlife. In: Lachman ME (ed) Handbook of midlife development. Wiley, Hoboken, pp 345–391

Heckhausen H, Gollwitzer PM (1987) Thought contents and cognitive functioning in motivational versus volitional states of mind. Motiv Emot 11:101–120

Heckhausen J, Schulz R (1995) A life-span theory of control. Psychol Rev 102:284–304

Heckhausen J, Wrosch C, Schulz R (2010) A motivational theory of life-span development. Psychol Rev 117:32–60

Hedges LV, Olkin I (1985) Statistical methods for meta-analysis. Academic, Orlando

Hertel G (2002) Management virtueller teams auf der Basis sozialpsychologischer Theorien: das VIST Modell [Management of virtual teams on the basis of social psychological theories: the VIST model]. In: Witte EH (ed) Sozialpsychologie wirtschaftlicher Prozesse. Pabst, Lengerich, pp 172–202

Hertel G, Bretz E, Moser K (2000) Freiwilliges Arbeitsengagement: Begriffsklärung und Forschungsstand. Gruppendynamik und Organisationsberatung [Voluntary work engagement: Conceptual clarification and current state of research] 31:121–140
Hertel G, Grube A, Hilsebein U (2007) Mental time journey: a method to estimate cohort effects based on guided reconstructions. University of Würzburg, Würzburg
Hertel G, Stamov-Rossnagel C (2012) Reconstruction methods: Using episodic memory traces to capture experiences at work efficiently. In: Bakker AB, Daniels K (eds) A day in the life of the happy worker. Psychology Press, London, pp 100–113
Hertel G, Thielgen M, Krumm S (2013) Aging and motivation at work: a conceptual model. Manuscript in preparation
Hertel G, van der Heijden BIJM, De Lange AH, Deller J (Guest Editors) (2013a). Facilitating age diversity in organizations—Part I: challenging popular misbeliefs. Spec issue J Manag Psychol
Hertel G, van der Heijden BIJM, De Lange AH, Deller J (Guest Editors) (2013b). Facilitating age diversity in organizations—Part II: managing perceptions and interactions. Spec issue J Manag Psychol
Hertel G, Wittchen M (2008) Work motivation. In: Chmiel N (ed) An introduction to work and organizational psychology. A European Perspective. Blackwell, Oxford, 29–55
Hochwarter W, Ferris G, Perrewé P, Witt L, Kiewitz C (2001) A note on the nonlinearity of the age-job-satisfaction relationship. J Appl Soc Psychol 31:1223–1237
Ilmarinen J (2005) Towards a longer worklife! Ageing and the quality of worklife in the European Union, Finnish Institute of Occupational Health, Helsinki
Jackson DN (1984) Personality research form manual. Research Psychologists Press, Port Huro
Kahneman D, Krueger AB, Schkade DA, Schwarz N, Stone AA (2004) A survey method for characterizing daily life experience: the day reconstruction method. Science 306:1776–1780
Kanfer R, Ackerman PL (2004) Aging, adult development, and work motivation. Acad Manag Rev 29:440–458
Kehr HM (2004) Implicit/explicit motive discrepancies and volitional depletion among managers. Pers Soc Psychol Bull 30:315–327
Kooij D, Jansen P, Dikkers J, de Lange A (2009) The influence of age on the associations between HR practices and both affective commitment and job satisfaction: a meta-analysis. J Organ Behav 31:1111–1136
Kooij D, de Lange A, Jansen P, Dikkers J (2008) Older worker's motivation to continue to work: five meanings of age: a conceptual review. J Manag Psychol 23:364–394
Kooij D, de Lange A, Jansen P, Kanfer R, Dikkers J (2011) Age and work-related motives: results of a meta-analysis. J Organ Behav 32:197–225
Kristof-Brown AL, Zimmerman RD, Johnson EC (2005) Consequences of individual's fit at work: a meta-analysis of person-job, person-organization, person-group, and person-supervisor fit. Pers Psychol 58:281–342
Krumm S, Grube A, Hertel G (2012) No time for compromises: age as a moderator of the relation between needs-supply fit and job satisfaction. Eur J Work Organ Psychol
Krumm S, Grube A, Hertel G (2013) The Munster work value measure. J Manag Psychol
Labouvie-Vief G, Hakim-Larson J, DeVoe M, Schoeberlein S (1989) Emotions and self-regulation: a life span view. Hum Dev 32:279–299
Lazarus RS (2007) Stress and emotion: a new synthesis. In: Monat A, Lazarus RS, Reevy G (eds) The Praeger handbook on stress and coping. Praeger Publishers/Greenwood Publishing Group, Westport, pp 33–51
Lazarus RS, Folkman S (1987) Transactional theory and research on emotions and coping. Eur J Pers, 1(3, Special Issue):141–169
Low KSD, Yoon M, Roberts BW, Rounds J (2005) The stability of vocational interests from early adolescence to middle adulthood: a quantitative review of longitudinal studies. Psychol Bull 131:713–737
McAdams DP, de St Aubin E, Logan RL (1993) Generativity among young, midlife, and older adults. Psychol Aging 8:221–230

McClelland DC, Koestner R, Weinberger J (1989) How do self-attributed and implicit motives differ? Psychol Rev 96:690–702
Mohr G, Rigotti T, Müller A (2005) Irritation: an instrument assessing mental strain in working contexts. Scale and item parameters from 15 studies. Zeitschrift für Arbeits- und Organisationspsychologie 49:44–48
Murray HA (1943) Thematic apperception test. Harvard University Press, Cambridge
Neuberger O, Allerbeck M (1978) Messung und Analyse von Arbeitszufriedenheit. Erfahrungen mit dem "Arbeitsbeschreibungsbogen (ABB) [Measurement and analysis of job satisfaction. Experiences with the 'Job Description Questionnaire'], Huber, Bern
Ng T, Feldman D (2008) The relationships of age to ten dimensions of job performance. J Appl Psychol 93:392–423
Ng T, Feldman D (2010) The relationships of age with job attitudes: a meta-analysis. Pers Psychol 63:677–718
Organ DW (1990) The motivational basis of organizational citizenship behavior. In: Staw BM, Cummings LL (eds) Research in organizational behavior. JAI Press, Greenwich, pp 43–72
Pang JS, Schultheiss OC (2005) Assessing implicit motives in U. S. college students: effects of picture type and position, gender and ethnicity, and cross-cultural comparisons. J Pers Assess 85:280–294
Rauschenbach C, Hertel G (2011) Age differences in strain and emotional reactivity to stressors in professional careers. Stress Health 27:e48–e60
Rauschenbach C, Göritz A, Hertel G (2012a) Age stereotypes about emotional resilience at work. Educ Gerontol 38:511–519
Rauschenbach C, Thielgen M, Krumm S, Hertel G (2012b) Why do older workers experience lower strain? Effects of coping strategies and job control. Manuscript submitted for publication
Rauschenbach C, Krumm S, Thielgen M, Hertel G (2013) Age and work stress: a review and meta-analysis. J Manag Psychol
Ros M, Schwartz SH, Surkiss S (1999) Basic individual values, work values, and the meaning of work. Appl Psychol Int Rev 48:49–71
Salthouse TA (2012) Consequences of age-related cognitive declines. Annu Rev Psychol 63:201–226
Schultheiss OC, Brunstein JC (2010) Implicit motives. Oxford University Press, New York
Schwartz SH (1994) Are there universal aspects in the structure and contents of human values? J Social Issues 50:19–45
Schwartz SH, Bilsky W (1987) Toward a universal psychological structure of human values. J Pers Soc Psychol 53:550–562
Semmer N, Zapf D, Dunckel H (1999) Instrument zur Stressbezogenen Tätigkeitsanalyse (ISTA) [Instrument for stress-related job analysis (ISTA)]. In: Dunckel H (ed) Handbuch psychologischer Arbeitsanalyseverfahren [Handbook of psychological job analysis measures]. vdf, Zürich, 179–204
Sokolowski K, Schmalt HD, Langens TA, Puca RM (2000) Assessing achievement, affiliation, and power motives all at once: the Multi-Motiv-Grid (MMG). J Pers Assess 74:126–145
Schüler J, Sheldon KM, Fröhlich SM (2010) Implicit need for achievement moderates the relationship between competence need satisfaction and subsequent motivation. J Res Pers 44:1–12
Stamov-Rossnagel C, Hertel G (2010) Older worker's motivation: against the myth of general decline. Manag Decis 48:894–906
Sterns HL, Doverspike D (1989) Aging and the training and learning process. In: Goldstein IL (ed) Training and development in organizations. Jossey-Bass, San Francisco, pp 99–332
Stumpf H, Angleitner A, Wieck T, Jackson DN, Beloch-Till H (1985) Deutsche Personality Research Form (PRF) [German Personality Research Form (PRF)]. Hogrefe, Göttingen
Super DE (1970) Work values inventory manual. Houghton Mifflin, Boston

Thielgen M, Krumm S, Rauschenbach C, Hertel G (2012) Older but wiser: Age moderates congruency effects between implicit and explicit motives on job satisfaction. Manuscript submitted for publication

Thrash TM, Cassidy SE, Maruskin LA, Elliot AJ (2010) Factors that influence the relation between implicit and explicit motives: a general implicit-explicit congruence framework. In: Schultheiss OC, Brunstein JC (eds) Implicit motives. Oxford University Press, New York, pp 308–346

Torkelson E, Muhonen T (2004) The role of gender and job level in coping with occupational stress. Work Stress 18:267–274

Truxillo DM, Cadiz DM, Rineer JR, Zaniboni S, Fraccaroli F (2012) A lifespan perspective on job design: Fitting the job and the worker to promote job satisfaction, engagement, and performance. Organizational Psychology Review 2: 340-360

Twenge JM, Campbell SM, Hoffman BJ, Lance CE (2010) Generational differences in work values: leisure and extrinsic values increasing, social and intrinsic values decreasing. J Manage 36:1117–1142

Warr P (2001) Age and work behaviour: physical attributes, cognitive abilities, knowledge, personality traits and motives. In: Cooper CL, Robertson IT (eds) International review of industrial and organizational psychology. Wiley, New York, 16:1–36

Wegge J, Jungmann F, Liebermann S, Shemla M, Ries BC, Diestel S, Schmidt K-H (2012) What makes age diverse teams effective? Results from a six-year research program. Work 41:5145–5151

Winter DG (1994) Manual for scoring motive imagery in running text. University of Michigan, Ann Arbor

Wright JD, Hamilton RF (1978) Work satisfaction and age: some evidence for the job change hypothesis. Soc Forces 56:1140–1158

Zacher H (2009) Maintaining a focus on work-related opportunities at higher ages. Justus-Liebig-University, Gießen

Zacher H, Frese M (2009) Remaining time and opportunities at work: relationships between age, work characteristics, and occupational future time perspective. Psychol Aging 24:487–493

Age-Related Differences in the Emotion Regulation of Teachers in the Classroom

Anja Philipp and Heinz Schüpbach

Lessons Learned

In a project on "Age-related differences in the emotion regulation of teachers facing demanding classroom situations" we investigated how teachers can be supported in regulating their emotions during demanding situations in their lessons so that they remain well, healthy and able to work until regular retirement age. Results of this project demonstrated that emotion regulation is indeed relevant to maintain emotional well-being and work ability of teachers. Being able to regulate one's emotions successfully can, thus, be considered a resource of teachers which contributes to their well-being and health. On the other hand, a loss spiral (conservation of resources theory, Hobfoll et al. 1996) might develop: once depleted of resources and impaired in their work ability, teachers are more likely to express an emotion not actually felt, which in turn, may result in even lower work ability. Such a spiral should be interrupted in teachers' careers as early as possible.

Second, teachers frequently report that the interaction with their students can be quite demanding (Kyriacou 2001). Results from an observational instrument, which allows for a more objective assessment of lessons than teacher self-report, supports this by showing that an average lesson is disturbed 20 times and teachers spent approximately 6 min on solving these demanding situations. Many of the situations only lasted a few seconds, other situations, however, could not be solved quite as quickly. In line with Gross (1998), we examined emotion regulation over the course of such episodes.

A. Philipp
Goethe-University Frankfurt, Frankfurt am Main, Germany
e-mail: philipp@paed.psych.uni-frankfurt.de

H. Schüpbach (✉)
University of Freiburg, Freiburg im Breisgau, Germany
e-mail: heinz.schuepbach@fhnw.ch

C. M. Schlick et al. (eds.), *Age-Differentiated Work Systems*,
DOI: 10.1007/978-3-642-35057-3_7, © Springer-Verlag Berlin Heidelberg 2013

Third, we examined whether teachers of different age groups differ in their emotion regulation over the course of such emotional episodes. Results show that especially teachers in the last phase of their career seem to struggle with the emotional demands of teaching. Older, more experienced teachers report higher levels of dissonance between the emotions they spontaneously feel and the emotions they consider appropriate during a demanding situation than their mid-career colleagues. Experienced teachers also try to regulate their emotions more intensely, either by using a strategy aimed at reconciling the spontaneously felt emotion and the emotion considered appropriate (so-called deep acting) or by regulating their emotional expression because they acknowledge that students have a right to be dealt with in a calm way even if this is not how the teacher feels (so-called faking in good faith). Teachers in the first career phase also tend to experience higher levels of emotional dissonance than their mid-career colleagues. Thus, they also seem to struggle with the emotional demands of teaching—but to a lesser extent than their older, more experienced colleagues. We assume that younger, less experienced teachers might still have more resources available to counterbalance emotional demands.

Major findings are:

- For teachers, the ability to regulate their emotions successfully can be considered a resource which contributes to health and well-being.
- Especially older teachers in the last phase of their career—and to a lesser extent their young colleagues—seem to struggle with the emotional demands of teaching.
- Thus, interventions should address young teachers as well as their older colleagues to support them in their emotion regulation.
- Teachers with impaired work ability who are depleted of resources are more likely to regulate their emotions dysfunctionally.
- Thus, a loss spiral might develop which should be interrupted as early as possible.

Based on our results, we recommend interventions on the basis of peer consulting, containing modules on health-beneficial emotion regulation. Such interventions should especially include teachers at the beginning and end of their careers—but also their mid-career colleagues who could (a) establish functional emotion regulation to prevent potential loss-cycles and (b) support their younger and older colleagues in their emotion regulation.

In an exploratory field study, we singled out potentially advantageous emotion regulation techniques which could be implemented into such an intervention in the future.

Demographic Change in the Teaching Profession

At present, the demographic change which is expected to hit other occupations in the next decades has already arrived in the teaching profession. On average across the countries within the Organisation for Economic Co-operation and Development

(OECD), about one-third of teachers are 50 years or older (OECD 2010). The percentage of teachers over 50 years is even higher in Germany (50 % of the primary, 52 % of lower secondary teachers), Sweden (49 % of primary teachers), or Italy (42 % of primary teachers, 60 % of lower secondary education teachers; OECD 2010).

Many teachers feel not fit to teach any longer which makes the situation even more precarious. It is estimated that approximately 30 % of the teacher workforce is severely emotionally exhausted (Farber 1991) which is a key component of a developing burnout process (Maslach 1998). If possible, these teachers tend to leave the profession prematurely. In 2000, 64 % of the German teachers leaving the profession early did this because they were unable for service (Federal Statistical Office 2007). A representative study of teachers from Bavaria (a federal state of Germany) showed that more than half of the teachers retiring early do so with burnout-related diagnoses (Weber et al. 2003). Thus, German teachers are an example for (a) a profession with a high proportion of older employees indicating that the demographic change has already arrived and (b) a profession with severe emotional demands which a considerable number of teachers feel not fit to cope with any longer and leave the profession early because of burnout-related diagnoses.

However, in 2001 the regulatory conditions for early retirement of German teachers had been changed making it more difficult to retire early. Since 2001 the number of teachers retiring early because they are unfit for service has decreased from 64 % in 2000 to 28 % in 2004 (Federal Statistical Office 2007). This may help to buffer the immense costs associated with early retirement (according to estimations by the Bavarian teachers union from 2003 approximately 250 million Euro per year were spent in the federal state of Bavaria only). Yet, these decreasing numbers also imply that many teachers who do not feel fit to teach any more remain in the profession. Other factors which may be associated with this trend can only be estimated (e.g., costs for sick leave of teachers). Moreover, research has provided some indication that burned out teachers are also less likely to provide high quality teaching (Klusmann et al. 2008). Thus, it can be assumed that keeping burned out teachers in the profession might have an impact on an economic as well as an educational level.

If teachers are supposed to remain longer in the workforce, it is essential to address the question which factors help them to remain well and healthy until regular retirement age of 65 years (many federal states of Germany are gradually increasing the retirement age to 67 years). The considerable number of burned out teachers indicates that teaching is associated with high emotional demands. It is, therefore, crucial to examine how teachers can cope with the emotional demands of their profession more effectively over their entire career cycle.

Emotion Regulation of Teachers

Since the seminal book of Hochschild (1983) emotion regulation in service interactions (also called emotional labor, Hochschild (1983); or emotion work, Zapf (2002)) is thriving and also the emotional aspects in the teacher-student interaction have become focus of research.

Antecedents of Teacher Emotion Regulation

Frequency and duration of service interactions have been discussed as one possible antecedent of emotion regulation (e.g., Morris and Feldman 1996). Many teachers experience the interaction with their students as an especially demanding aspect of their profession (Kyriacou 2001) and some of these situations may cause intense emotions. However, not all emotions are appropriate in service interactions. Service professions often require a specific emotional expression from their employees and in some organizations implicit norms on an appropriate emotional expression exist, whereas other organizations even establish explicit rules (display rules, Zapf 2002). In schools, however, it is more likely that implicit norms about appropriate emotional expressions in the classroom exist which teachers have internalized during their career: "During face-to-face or voice-to-voice interactions, many employees are required to express appropriate emotions as a job requirement. Examples are [...] nurses or teachers who have to show empathy toward patients or children, and bank employees who have to signal trustworthiness by putting on a friendly but solemn face" (Zapf 2002, p. 238).

Fischbach (2003) described specific emotion regulation requirement profiles for different service professions: teachers, policemen, and travel agents. According to her, teaching is associated with a high level of emotion regulation job requirements. Teachers need to be sensitive to students' emotions, express sympathy, and show positive emotions—like travel agents in the interaction with their customers. However, teachers—like policemen—can also express negative emotions to a certain extent. Despite the potential to also be able to express negative emotions, teachers experience states of discrepancy between the emotion spontaneously felt and the emotion expressed (Fischbach 2003). Such a state of discrepancy (so-called emotional dissonance, Zapf 2002) has been shown to be associated with increased emotional exhaustion and depersonalization and increased psychosomatic complaints of service professional (e.g., Hülsheger and Schewe 2011).

Emotion Regulation Strategies and Their Influence on Teacher Well-Being and Health

An "emotional dissonance is likely to create psychological discomfort that individuals are motivated to reduce" (Rubin et al. 2005, p. 194). To do so, emotion regulation is required. Morris and Feldman (1996) define emotion regulation as the "effort, planning, and control needed to express organizationally desired emotion during interpersonal transactions" (p. 987). Two core emotion regulation strategies are discussed: deep versus surface acting. Deep acting refers to a change of the emotion itself (Grandey 2000; Zapf 2002). It is characterized as an antecedent-focused form of emotion regulation which means that the regulation takes place before an emotion is generated (Gross 1998). Surface acting, on the other hand, refers to a change in the emotional expression (Grandey 2000) which takes place after the emotion has already developed and it can, thus, be considered a response-focused form of emotion regulation (Gross 1998). Whereas deep acting helps to resolve a state of emotional dissonance, surface acting does not (Zapf 2002). This is one of the reasons why surface acting is considered less health-beneficial. Furthermore, a superficial expression of an emotion not felt may be associated with reduced levels of rewarding relationships and, thus, results in feeling inauthentic and alienated from oneself (Grandey 2000; Zapf 2002).

Influencing one's inner feelings by deep acting, on the contrary, results in being more authentic. Grandey (2000) summarized: "Although this process is still effortful, it may lead to an expression that is perceived as more genuine than when an employee surface acts" (p. 105).

In their meta-analysis, Hülsheger and Schewe (2011) confirm that surface acting is indeed associated with indicators of impaired well-being and job attitudes. Deep acting, however, is weakly associated with increased emotional exhaustion and negatively associated with psychological strain indicators (Hülsheger and Schewe 2011). This indicates that influencing one's emotions by deep acting may be effortful at first and an expected positive effect may only unfold over time. Surface and deep acting may also have a more enduring effect on health indices, i.e., the work ability of employees as indicator for perceived health and work capacity. High work ability is associated with a longer working life, whereas low work ability is accompanied by a higher risk of premature retirement. The work ability of teachers is especially precarious and factors preventing low work ability should be investigated (Seibt et al. 2007).

Emotion regulation should not only be investigated with regard to impaired well-being and health but also with regard to positive outcomes. Despite the high number of burned out teachers (approx. 30 %; Farber 1991), many teachers do not suffer from burnout and may be engaged. Engagement is characterized as "a positive, fulfilling, work-related state of mind that is characterized by vigor, dedication, and absorption" (Schaufeli and Bakker 2003, p. 4) with dedication as a key component.

To date the majority of studies are cross-sectional and hardly any results on the effects of emotion regulation over longer periods of time exist. However, as Côté and Morgan (2002) point out, effects of emotion regulation may only unfold over time. Teachers who regulate their emotions in a dysfunctional way over years might burn out eventually. On the contrary, reverse effects of well-being and health on the regulation of emotions are equally plausible. Once substantially exhausted, depleted of resources (conservation of resources theory, Hobfoll et al. 1996) and impaired in their work ability, teachers might not be able to invest effort into being authentic by applying deep acting anymore and turn to surface acting instead. On the other hand, dedicated teachers may have more resources available and should be able to influence their feelings by deep acting and are thus less likely to use surface acting. Therefore, we investigated in a study with two measurement points separated by one year whether deep and surface acting have an effect on emotional exhaustion, dedication as well as work ability of teachers and if potential reverse effects occurred.

Age-Related Difference in Teacher Emotion Regulation

Dahling and Perez (2010) found that with increasing age of employees, the use of deep acting increased and, thus, the application of surface acting decreased. These results can be explained by drawing on the Socioemotional Selectivity Theory (Carstensen et al. 2003) which postulates that as individuals age, positive experiences become more and more important. To achieve such positive emotional states, it becomes necessary to regulate ones emotions successfully.

Yet, there are some contradictory results from research on emotion regulation of service professionals. Lee and Brotheridge (2011) reported that young, less experienced child care workers deep act more frequently than their older, more experienced colleagues. This indicates that no general optimization of the emotion regulation over the career of a service professional can be expected. Results from a large sample of German teachers support this assumption by showing that many older teachers have inefficient recovery strategies (Schaarschmidt and Kieschke 2007). The authors report that young teachers face high emotional demands and already establish unfavorable regulation strategies during the first five years after graduation. Thus, emotion regulation might be especially relevant in certain phases of teachers' professional careers.

Huberman et al. (1993) describe different stages in the professional life-cycle of teachers: many teachers experience a phase of "survival" at the beginning of their career in which they become acquainted with their job and students. After this they stabilize and many teachers go on into a phase of experimenting in which they are confident about teaching, test new teaching strategies and become serene teachers. Thus, in the mid-career phase, the many serene teachers seem to be confident about teaching and less prone to be struggling with the emotional demands. Some, however, start to reassess themselves after the stabilization phase. These teachers develop self-doubts and feel exhausted from the teaching routine and may enter a phase of conservatism. At the end of their careers, retirement is imminent and all teachers enter a phase of disengagement in which they reduce their involvement in school activities. Some do this with a feeling of serenity, others with bitterness (Huberman et al. 1993).

Drawing on Huberman et al. (1993) and the results from Lee and Brotheridge (2011) or Schaarschmidt and Kieschke (2007), especially young teachers at the beginning of their professional career as well as older teachers seem to struggle with the emotional demands of their profession which might bring about increased states of emotional dissonance. Thus, we expect teachers at the beginning and end of their careers to regulate their emotions more intensely than their mid-career colleagues. Therefore, we investigated in a quasi-experimental study if teachers would use different emotion regulation strategies during emotionally demanding episodes depending on their career phase.

Techniques of Emotion Regulation of Teachers in Demanding Situations in Class

If teachers' strategies of emotion regulation prove to be important for their long-term well-being and health it is crucial to translate these global strategies into different emotion regulation techniques which can eventually be implemented into teacher trainings.

Gross (1998) describes in his process model how emotion regulation develops over the course of demanding situations and which strategies and techniques of emotion regulation are associated with each stage of an emotional episode. According to him, an individual could first of all *select situations* which are likely to be associated with positive emotions. A teacher might, for example, decide to reduce the number of teaching hours. Gross (1998) further suggests that an individual can also use strategies to change an emotionally demanding situation itself in order to modify its emotional impact (*situation modification*). A teacher might, for example, change the seating plan in the classroom in order to separate problematic students. Some situations, however, leave little room for modification. Individuals can, in this case, also direct their attention to positive aspects of the situation (*attentional deployment*) by focusing the attention to pleasant things or thoughts (Grandey 2000). Teachers could focus on disciplined and cooperative students instead of worrying about unmotivated students. Even after selecting or modifying a situation and attending to it selectively, some potential for changing its emotional impact remains. A required emotional state could also be achieved by reappraising the situation (*cognitive change*). According to Grandey (2000), these techniques can be characterized as antecedent-focused emotion regulation altering the onset of an emotion and are, thus, techniques of deep acting. Gross (1998) also describes emotion modulation in response to a situation by changing the emotional response tendencies with effects on physiological, behavioral, and experiential level. Grandey (2000) characterizes such emotional suppression techniques as surface acting.

Another conceptualization has been introduced by Zapf (2002), who describes four different strategies of emotion regulation. First, an emotion is felt and then the expression occurs automatically without conscious effort. Such an *automatic regulation* might result in being perceived as authentic. The author also describes *surface acting* as one potential strategy which is associated with an ongoing state of emotional dissonance between inner feelings and the emotional expression. Third, he defines *deep acting* as a strategy which refers to an effortful regulation of the inner feelings. Zapf (2002) summarizes that deep acting involves techniques like actively invoking thoughts, memories, and images by thinking of one's role or using metaphors. Fourth, if individuals disobey to the display rules of their organization and show inacceptable emotions, they act *emotionally deviant*.

Such an emotional deviance as well as authentic displays of emotions were also described by Rafaeli and Sutton (1987). The authors refer to the latter as emotional harmony. It is, however, not clear if effort has to be invested to achieve this harmony and, thus, this category might also include states after successful deep acting. Furthermore, they provide a differentiated description of surface acting as: (a) an emotional expression shown because an employee acknowledges that this is part of the job and the interaction partner deserves to be treated in a friendly way—even if the employee does not automatically feel happy (*faking in good faith*) or (b) an employee does not acknowledge that such an emotional expression should be part of the job (*faking in bad faith*). Rafaeli and Sutton (1987) assume that a regulation of an emotional expression by faking in good faith might be associated with financial benefits or effective employee performance and is, therefore, not necessarily health-detrimental. They argue, for example, that physicians are expected to appear concerned but should not be too concerned to be able to work effectively. The authors expect that faking in bad faith believing that "pasting on a smile should not be part of the job" (Rafaeli and Sutton 1987, p. 32) is detrimental to well-being because of a clash between personal values and role requirements.

In study 3 we assessed which of these emotion regulation techniques teachers use in their lessons, how they are they related to teacher well-being, and if teachers of different age groups differ in the use of these techniques.

Research Model

The research model of the project on "Age-related differences in the emotion regulation of teachers facing demanding classroom situations" was developed on the basis of Rubin et al. (2005) and supplemented (Fig. 1).

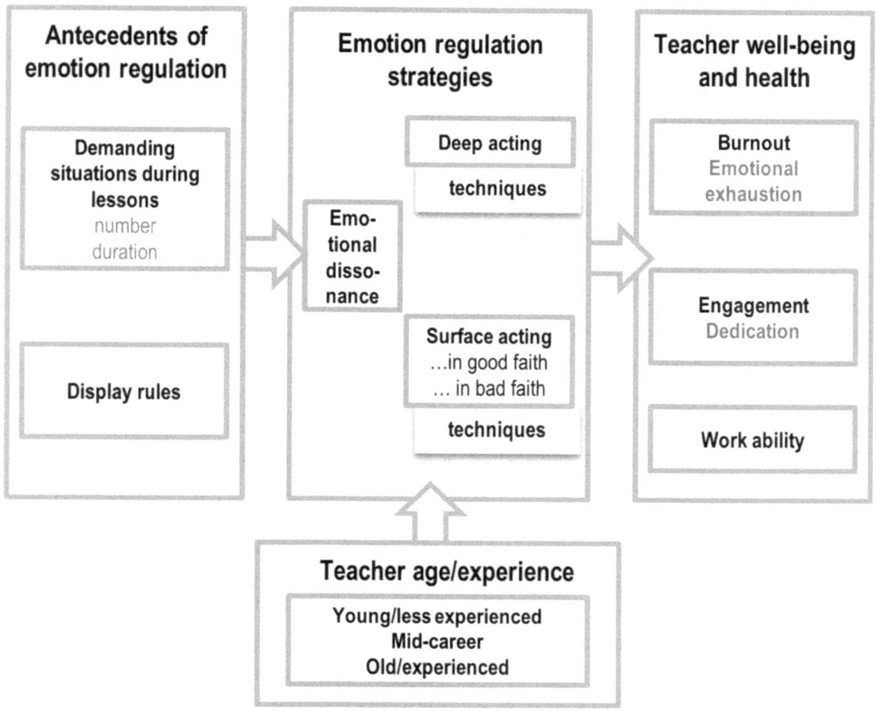

Fig. 1 Research model of the project "Age-related differences in the emotion regulation of teachers facing demanding classroom situations"

Methods, Samples, and Measures

Three different studies were conducted to (a) examine the effect of emotion regulation on teacher well-being and work ability over the period of one year (study 1; realized in 2006 and 2007), (b) assess age-related differences in emotion regulation (study 2; in 2007), and (c) explore which emotion regulation techniques teachers use in demanding situations in their lessons (study 3; in 2009).

Method, Sample, and Measures of Study 1

A full two-wave panel design was realized in order to address the question whether different emotion regulation strategies have an effect on the well-being and health of teachers and if reverse causation occurred. Measurement point 1 and 2 were separated by one year. 102 teachers provided complete information at both points in time and their data were, thus, included in the analyses. At time 2, teachers were on average 44 years old ($SD = 12.35$; age range 24–63 years); 80 % of them were female (for details on method and sample, see Philipp and Schüpbach 2010a).

At both measurement points, identical instruments were used to assess emotion regulation strategies as well as well-being and work ability. Emotion regulation: surface acting and deep acting were measured with three items each. Two deep acting items were taken from the Emotional Labour Scale (ELS, Brotheridge and Lee 2003). They were translated into German and supplemented by one self-developed item. Well-being and health: Emotional exhaustion was assessed by seven items of the Maslach Burnout Inventory (MBI, German version: Büssing and Perrar 1992). Dedication was measured with six items of the Utrecht Work Engagement Scales (UWES, German version: Schaufeli and Bakker 2003). Participants were asked to rate their agreement on a 5-point Likert-type response scale (1 = strongly disagree to 5 = strongly agree). Work ability was assessed with the short version of the Work Ability Index (WAI) questionnaire (Tuomi et al. 1998). The WAI is a widely used individual screening and health promotion method in the occupational health sector. Work ability was assessed by teacher self-reports on the seven subscales of the WAI. Most subscales consist of several questions (Tuomi et al. 1998): (1) current work ability in comparison with lifetime best (1 item, 0 = completely unable to work to 10 = work ability at its best); (2) work ability in relation to physical and mental work demands (2 items, 1 = very poor to 5 = very good); (3) number of current diseases diagnosed by a physician (14 items, 0 = no, 1 = yes); (4) work impairment because of diseases (1 item, 1 = entirely unable to work to 6 = no hindrance/no disease); (5) sickness absence in the past year (1 = 100–365 days, 2 = 25–99 days, 3 = 10–24 days, 4 = at the most 9 days, 5 = none at all); (6) prognosis of work ability two years from now (1 = unlikely able to work, 4 = not certain, 7 = relatively certainly able to work); and (7) mental resources (3 items: recently been able to enjoy daily activities, recently been active and alert, recently felt filled with hope for the future; 0 = never, 2–4 = often/always/continuously, respectively). WAI scores range between 7 and 49: scores between 2 and 29 indicate "poor", scores between 28 and 36 indicate "medium", scores between 37 and 43 indicate "good" and scores between 43 and 49 indicate "very good" work ability.

Due to the limited number of respondents, we chose to calculate two separate models. Model A examines the cross-lagged, reverse, and simultaneous effects of emotion regulation on well-being (i.e., emotional exhaustion and dedication) on the basis of latent constructs (for details on statistical analyses of model A, see Philipp and Schüpbach 2010a).

Cross-lagged effects of emotion regulation on work ability, reverse causation and simultaneous causation were tested in model B. Work ability was measured with the WAI short version which results in an index and, therefore, it was included as a manifest variable in the model (Fig. 2).

In two different structural equation models (SEM), we examined in a stepwise procedure if cross-lagged, reverse causation or simultaneous effects are predominant: first a null model (all variables are mutually independent) was specified, followed by a stability model (containing all autoregressive paths). Third, a

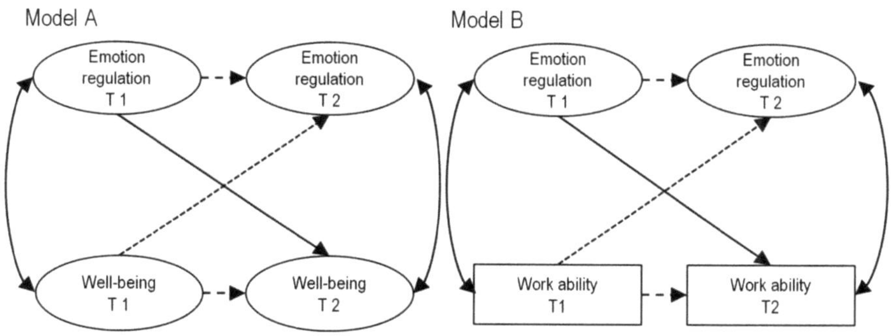

Fig. 2 Cross-lagged structural equation models of emotion regulation and teacher well-being (*model A*) and work ability (*model B*), *dashed lines* = stabilities, *solid lines* = cross-lagged effects, *dotted lines* = reverse causation

cross-lagged model (containing autoregressive paths and paths from deep/surface acting on well-being or work ability) as well as, fourth, a reverse causation model (containing autoregressive paths and paths from emotional well-being or work ability on deep/surface acting) were tested. Fifth, a full model (containing all paths) was tested.

For model A (for details, see Philipp and Schüpbach 2010a) as well as model B, the full model containing all paths fitted the data best and, thus, simultaneous effects are predominant.

Method, Sample and Measures of Study 2

This quasi-experimental study focused on the question whether teachers in different phases of their career differ in the use of emotion regulation strategies. For this, two typical demanding situations had been identified in a pre-study (episode A: "A girl plays with her MP3-player and talks to her neighbor. The teacher sends her out of the classroom. But she does not calm down and disturbs the class from outside"; episode B: "A boy arrives much too late for class and instead of sitting down quietly he starts to chat with his neighbors. He increasingly disturbs the entire class"). These were then re-enacted with 8th grade students of a secondary school and recorded on video. In order to examine the emotion regulation over the course of these two episodes, each episode was then cut into three scenes. These standardized video vignettes were then used as stimulus material in the following steps of the study.

This study was conducted with 205 German teachers from 12 secondary schools (German: Realschulen) in Baden-Württemberg (a federal state of Germany). The teachers were on average 44 years old (SD = 11.30; age range 24–64 years) and very experienced (M = 16 years teaching experience, range 1–41 years). The majority (64 %) was female.

Groups of teachers watched the standardized video vignettes of the emotionally demanding episodes scene by scene and were asked to put themselves in the position as teachers of these students. After each scene, teachers filled in a short questionnaire on how they would regulate their emotions at this stage of the situation. For a time-economic assessment of the emotion regulation strategies, we used two-item short versions. Emotional dissonance was assessed with two items from the Frankfurt Emotion Work Scales (FEWS, Zapf et al. 2005). Deep acting was assessed with two items from the Emotional Labour Scale (ELS, Brotheridge and Lee 2003). Surface acting was assessed as faking in good faith (acknowledging emotion regulation as part of the job) or faking in bad faith (without this acknowledgement) with two items each from a German questionnaire developed by Nerdinger and Röper (1999).

At the end of each episode, teachers were asked if the situations are realistic and emotionally demanding: Both episodes were perceived as realistic (episode A: $M = 3.09$, $SD = 1.01$; episode B: $M = 2.64$, $SD = 1.03$) as well as moderately emotionally demanding (episode A: $M = 3.15$, $SD = 1.02$; episode B: $M = 3.59$; $SD = 1.04$). All items were assessed on a 5-point Likert-type response format from 1 = strongly disagree to 5 = strongly agree.

In order to assess differences in emotion regulation between less experienced, mid-career and more experienced teachers the sample was divided into three equally large subgroups according to their experience. Previous results had shown that the first (approx. 5) years of work experience as well as the last years (approx. from 20 years onwards) are especially associated with high emotional demands. The subgroups were equally large: younger teachers with less than 8 years of work experience ($N = 75$), mid-career teachers with 8–19 years of work experience ($N = 62$) and, older teachers with more than 19 years of work experience ($N = 68$).

Each episode consisted of three scenes. Therefore, repeated-measure analyses of variance were conducted to assess changes over the course of the episodes (scene as within-subject factor) as well as age-related differences (experience as between-subject factor). (For details on sample and statistical analyses, see Philipp and Schüpbach 2010b).

Method, Sample and Measures of Study 3

In this exploratory field study, we examined which techniques teachers use to regulate their emotions in real situations in class. In total, 13 teachers from five secondary schools (German: Realschulen) in the federal district of Baden-Württemberg took part. The schools had not taken part in previous studies.

First, two typical lessons of each teacher were recorded on video (except for one teacher who faced organizational problems and, thus, only one lesson could be filmed). Prior to the recording of the lessons, the parents of all students filmed had to give their written consent. Overall, 25 lessons (45 min each) were filmed; nine of them were double periods. Seven lessons were recorded in a 5th grade, one in a 6th grade, twelve in an 8th grade, five in a 9th grade; different subjects were taught. Average class size was 23 students (range 12–28 students).

Second, more than 1.100 min of video were screened, inventoried and analyzed. The analysis was conducted with the RHIA-Unterricht instrument (Meder et al. 2008). RHIA-Unterricht assumes that stress occurs when the goal-oriented process of working is obstructed, goals and working conditions conflict with each other (so-called regulation hindrances occur) and, hence, extra work is required (so-called additional effort). RHIA-Unterricht is an observational instrument which allows for an objective assessment of periods of lessons (e.g., teaching the subject or assessing performance of students) as well as regulation hindrances for teachers in class. Such hindrances can either be capacity-overtaxing factors (e.g., high noise-level in class) or regulation barriers (e.g., diverging goals of teacher and students or lack of students' skills). Additionally, the duration of each regulation barrier was quantified as an indicator for the additional effort the observed teacher had to invest (Meder et al. 2008).

In this study, RHIA-Unterricht analyses were conducted by five trained raters. They quantified how many regulation barriers and which type of regulation barriers occurred and how long it took to solve each situation (additional effort). The quality of the ratings was supervised by a rater with high experience in RHIA-Unterricht.

Third, after the lessons were filmed, the teachers were handed out a short questionnaire to fill out at home. In this questionnaire, emotional exhaustion and dedication were measured with the same scales as in study 1 (see section Method, Sample, and Measures of Study 1). Again, both scales showed acceptable to good internal consistencies. Years of teaching experience and gender were assessed with a single item each. The 13 teachers had on average 13 years of teaching experience ($M = 13.04$, $SD = 11.47$, range 1.5–36 years). They were moderately emotionally exhausted ($M = 2.34$, $SD = 0.72$) and dedicated to teaching ($M = 3.71$, $SD = 0.45$).

Fourth, in a video-stimulated recall situation, each of the 13 teachers was separately shown five to eight of the demanding situations (regulation barriers) from his/her lessons which had been identified with RHIA-Unterricht. For time-economic reasons, only the most time consuming and intense demanding situations were included in the video-stimulated recall instead of the entire lesson. In total, 79 demanding situations were included in the video-stimulated recall. Each teacher was interviewed separately regarding the emotion regulation techniques used in the demanding situations in his/her lesson(s). The standardized interviews lasted one hour on average and were conducted by five trained interviewers. After presenting a demanding situation in the video-stimulated recall, teachers were asked to comment on a number of questions regarding their states of emotional dissonance ("Did you find it appropriate in this situation to show your spontaneously felt emotion to your students?"; developed on the basis of the Frankfurt Emotion Work Scales, FEWS, Zapf et al. 2005) or their techniques of emotion regulation ("Can you please describe how you experienced this situation?; How did you cope with this situation emotionally?"). At the end of each interview, teachers were asked to rate if the situations are typical for their lessons. Of the 13 teachers 9 agreed that the filmed lessons were typical, 1 teacher usually experiences fewer and 2 teachers a higher number demanding situations in their regular lessons.

Fifth, the interviews were recorded on tape and transcribed for the following qualitative content analysis. This study was of exploratory nature and the aim was to inductively develop a coding scheme following a procedure described by Mayring (2000) which finally resulted in eight categories of emotion regulation techniques.

Due to the exploratory nature of the study, we mainly conducted descriptive and inferential statistical analyses. Codes were used to assign the data from video analysis, interview and questionnaire to each teacher. For health-related analyses the sample was divided into two subsamples according to emotional exhaustion by median split (median = 2.14; $N_{healthy} = 6$, $N_{less\text{-}healthy} = 7$). To assess mean differences between the two groups, we conducted independent *t*-test. Albeit the small sample size, we divided the sample into three groups according to experience because study 2 had demonstrated that mid-career teachers differ in their emotion regulation from their less or more experienced colleagues and we intended to examine such differences in this exploratory study. Thus, similar cutoff-points as in study 2 were used to divide the sample into three subgroups: young, less experienced teachers (<10 years of work experience; N = 7), mid-career teachers (10–19 years of work experience; N = 3), older, experienced teachers (>19 years of work experience; N = 3). To assess mean differences between the three subgroups, we conducted univariate analyses of variance. Due to the small sample size, results have merely exploratory character and should, therefore, be interpreted cautiously.

Results

Effects of Emotion Regulation on Teacher Well-Being and Work Ability Over the Period of One Year (Study 1)

In a full two-wave longitudinal study over the period of one year, we found that emotion regulation has an effect on teacher emotional exhaustion (Philipp and Schüpbach 2010a) as an indicator for burnout (Maslach 1998). Teachers who were able to regulate their emotions successfully by influencing their emotions to feel the emotion appropriately in a situation (deep acting) were significantly less emotionally exhausted after one year. Deep acting had, however, no significant effect on dedication as an indicator for teacher engagement (Philipp and Schüpbach 2010a).

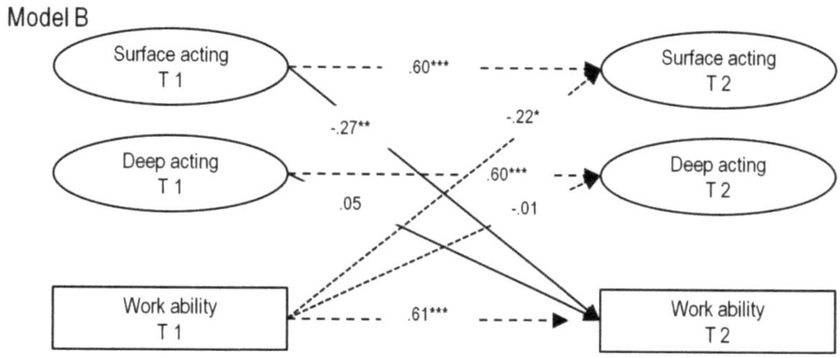

Fig. 3 Results of a cross-lagged structural model on the simultaneous impact of deep and surface acting on teacher work ability and reverse causation (*model B*; all values are standardized weights), *dashed lines* = stabilities, *solid lines* = cross-lagged effects, *dotted lines* = reverse causation, $^{***}p < 0.001$, $^{**}p < 0.01$, $^{*}p < 0.05$

Simultaneously, well-being of teachers influenced if they are able to regulate their emotions effectively after one year: once exhausted and emotionally drained, teachers rather regulated their emotional expression by surface acting than regulating the emotion itself by engaging in deep acting. More dedicated teachers, however, did neither regulate their emotions nor their emotional display and can, thus, be assumed to be less likely to act (Philipp and Schüpbach 2010a).

Analyses on the effect of emotion regulation on the work ability of teachers as indicator for an advanced health impairment revealed that surface acting had a significant cross-lagged effect on the work ability of teachers after one year (Fig. 3).

Teachers who used more surface acting at time 1 reported lower work ability scores one year later. Low work ability, on the contrary, led to an increased use of surface acting after one year. In order to assess if either cross-lagged or reverse effects are predominant or if both effects occur simultaneously, we examined different versions of model B (for details, see section Method, Sample, and Measures of Study 1). We found that the full model has acceptable to good model fit indices and provided the best fit (model fit of the full model: $X^2 = 41.89$; df = 38, $p = 0.31$, Comparative Fit Index (CFI) = 0.99, Tucker-Lewis Index (TLI) = 0.99, Root Mean Square Error of Approximation (RMSEA) = 0.03). Thus, the model fit parameters indicate that cross-lagged and reverse effects occur simultaneously.

Age-Related Differences in Emotion Regulation of Teachers During Emotionally Demanding Episodes (Study 2)

The 207 teachers involved in this study experienced the two episodes as emotionally demanding (see section Method, Sample and, Measures of Study 2). Each episode had been divided into three scenes and after each scene teachers were asked to fill out a short questionnaire on their emotion regulation. Over the course of both episodes, states of emotional dissonance between spontaneously felt emotions and the emotions considered appropriate (emotional dissonance) increased slightly, indicating that over the duration of the episodes, the emotion regulation requirement tended to increase. However, this result is only significant in episode B (Philipp and Schüpbach 2010b).

Over the course of episode A as well as after scenes two and three of episode B, teachers with at least 20 years of teaching experience felt a significantly stronger emotional dissonance than their mid-career colleagues. Also, younger teachers with up to seven years of work experience reported slightly higher levels of emotional dissonance than their mid-career colleagues, albeit this result is only a non-significant tendency.

In a second step we tested whether the three groups differed in their use of emotion regulation strategies. Results showed that teachers with at least 20 years of work experience deep acted more intensely over the course of both episodes than their mid-career colleagues (Philipp and Schüpbach 2010b). Less experienced teachers showed a similar tendency which was, however, only significant at the end of episode B and marginally significant after scene two and three of episode A.

Faking in good faith—a strategy of expressing an emotion not actually felt because it is acknowledged that this is a part of the professional role of teachers—decreased over the course of both episodes. Over the course of episode B, experienced teachers engaged significantly more in faking in good faith than their mid-career colleagues. Younger, less experienced teachers made more use of this strategy at the end of this episode. In episode A, however, no such difference occurred. Regulation of an emotional expression not acknowledging it as part of the job (faking in bad faith) decreased over the course of episode B. Teachers of all three groups used this strategy alike and no significant difference occurred.

Overall, emotion regulation changed over the course of the episodes with small changes between the episodes. Differences in emotional dissonance and the use of emotion regulation between the three groups occurred. The episodes were considered emotionally demanding and especially older, more experienced teachers and to some extent also their young, less experienced colleagues regulated their emotions (deep acting) and their emotional expression (faking in good faith) more intensely than mid-career teachers (for details on results, see Philipp and Schüpbach 2010b).

Techniques of Emotion Regulation in "Real" Lessons (Study 3)

In a field study, we explored which techniques teachers use to regulate their emotions in "real" classroom situations. Lessons of 13 teachers were filmed and analyzed with RHIA-Unterricht (Meder et al. 2008). The RHIA-Unterricht instrument provided a more objective measure of the demands of teachers during lessons than teacher self-reports. In the 25 lessons filmed, 504 demanding situations (so-called regulation barriers) occurred.

On average, the 25 lessons were disturbed 20 times ($M = 19.84$, $SD = 13.07$). Lesson 1 was disturbed 18 times on average and lesson 221 times. The number of disturbances per lesson ranged severely from 4 to 41 in lesson 1 and from 2 to 61 in lesson 2. On average, the disturbances lasted 19 s ($M = 18.83$, $SD = 13.11$) ranging from 6 to 83 s. In an average 45 min lesson, teachers spent 6 min on dealing with demanding situations ($M = 6.3$, $SD = 4.82$; range 2–14 min). Thus, the additional effort teachers have to invest in order to be able to continue teaching can be quite high with up to 14 min in a 45 min lesson. An exploratory analysis showed that less experienced, mid-career and experienced teachers differ in the number of demanding situations in their lessons ($F(2,12) = 5.18$, $p < 0.05$). The lessons of the younger teachers are significantly less often disturbed by demanding situations than the lessons of their experienced colleagues. Yet, no significant difference in the length of the demanding situations occurred ($F(2,12) = 1.09$, $p > 0.05$).

The 79 most time consuming and complex situations were identified for a video-stimulated recall situation in which the 13 teachers were interviewed about their emotion regulation techniques in these situations. Answers to the questions were classified into eight categories (see section Method, Sample and, Measures of Study 3). The teachers used a broad range of different emotion regulation techniques (Fig. 4).

The teachers most frequently reported that they tried to change perspective in order to understand their students' needs and motives, followed by seeking a dialogue with the students to solve the problem so that it will not come up again in the future or handling the situation with humor by making a funny or ironic remark. Some also tried to ignore the situation and focused on cooperative students

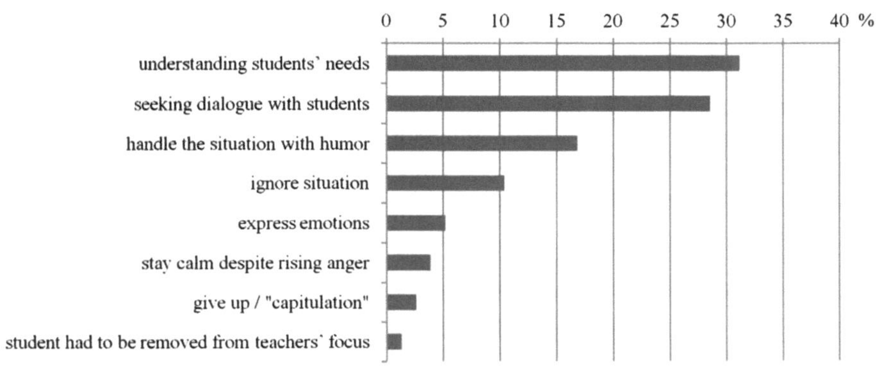

Fig. 4 Emotion regulation techniques used in the 79 demanding situations presented in a video-stimulated recall

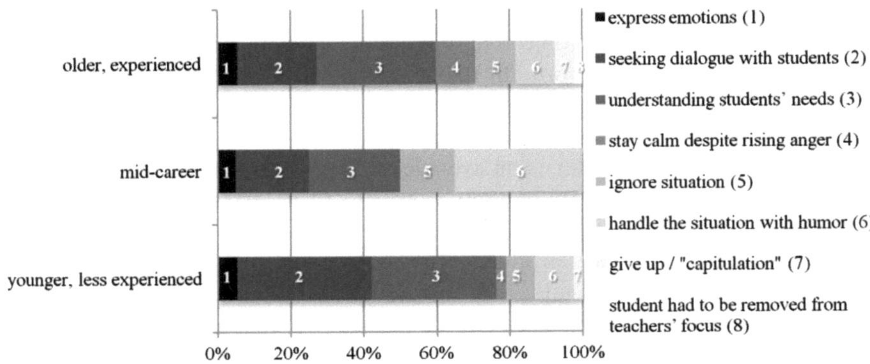

Fig. 5 Techniques of emotion regulation according to level of experience

instead. Least mentioned were techniques such as being authentic and expressing upcoming emotions (e.g., anger), staying neutral and calm despite rising anger or giving up ("capitulation").

Some teachers even described that they gave up when faced with an especially difficult situation; in one situation a teacher became so angry that the difficult student had to be removed from the class so that the teacher was able to go on with teaching.

Furthermore, descriptive analyses show that less experienced teachers (with less than 10 years of work experience) and their more experienced colleagues (with more than 10 years of work experience) used a broader range of techniques than their mid-career colleagues. Furthermore, young, mid-career, and older teachers differed in their pattern of techniques (Fig. 5).

Less experienced teachers predominantly sought a dialogue with their students and tried to understand their motives and needs. Their mid-career colleagues also reported that they used these two techniques, however, to a (slightly) lesser extent. Mid-career teachers rather tried to solve the situation with humor. In some cases they decided it is best to ignore the demanding situation and focus on cooperative students instead. Older teachers predominantly tried to seek a dialogue with their students or understand their motives and needs, however, to a lesser extent than their young colleagues. In some of the situations they tried to handle the problem with humor and in some they gave up addressing the problem but still continued teaching because they thought that the other students in their class deserved this. One older teacher could not continue the lesson without sending the student out of the room.

In a next step, these techniques were assigned to the different emotion regulation strategies based on the theoretical categories provided by Zapf (2002), Gross (1998), Grandey (2000) and Rafaeli and Sutton (1987; see section Techniques of Emotion Regulation of Teachers in Demanding Situations in Class). Table 1 provides an overview over the allocation of techniques to strategies of emotion regulation.

Table 1 Emotion regulation (ER) techniques reported by the 13 teachers in the video-stimulated recall and their classification into strategies

ER-technique	Gross (1998)	Grandey (2000)	Zapf (2002)	Rafaeli and Sutton (1987)
Expressing an emotion	Express emotional response tendency	–	Automatic regulation	Emotional harmony
Seeking dialogue	Situation modification/ antecedent-focused ER	Deep acting	Deep acting	Emotional harmony
Understanding students' needs	Cognitive change/ antecedent-focused ER	Deep acting	Deep acting	Emotional harmony
Ignoring the situation	Attentional deployment/ antecedent-focused ER	Deep acting	Deep acting	Emotional harmony
Humor	Attentional deployment/ antecedent-focused ER	Deep acting	Deep acting	Emotional harmony
Staying calm despite anger	Response-focused ER	Surface acting	Surface acting	Faking in good faith
Giving up	Response-focused ER	Surface acting	Surface acting	Faking in bad faith
Remove student from focus	Response-focused ER	Surface acting	Surface acting	Faking in bad faith

We then explored if teachers use different emotion regulation techniques depending on their level of emotional exhaustion (Fig. 6). Less exhausted teachers preferably acted automatically and expressed their emotions on most of the situations. They talked to the difficult students so that the situation was solved and did not come up again (situation modification/deep acting). Thus, they may have been more likely to reach a state of emotional harmony. Like their more exhausted colleagues they also used humor as technique to highlight potential positive aspects of the situation (attentional deployment/deep acting) or changed perspective by trying to understand their students' needs (deep acting).

Emotionally exhausted teachers showed a somewhat different pattern of techniques. They did not report that they automatically express their emotions. Instead, they preferably tried to change perspective (deep acting) by understanding their students' needs, sought a dialogue with their students to solve the situation (situation modification, deep acting) or tried to solve the situation with humor in most of their demanding situations. They ignored some of their problematic situations and focused on cooperative students instead (attentional deployment/deep acting) or tried to stay calm despite rising anger. In a few of the situations they had to give up. In one situation the teacher became so angry and overwhelmed that faking in bad faith by sending the student out of the room was the only possibility to be able to go on with teaching.

Trying to understanding students' needs aims at cognitive change (Gross 1998) and according to Grandey (2000) it can be categorized as deep acting technique, which should be a strategy associated with increased well-being. However, teachers with higher levels of emotional exhaustion used this technique in more of the situations than their less exhausted colleagues. One explanation might be that their aim was to understand their students better—but maybe the teachers were not successful in gaining that understanding and turned to faking in good faith instead.

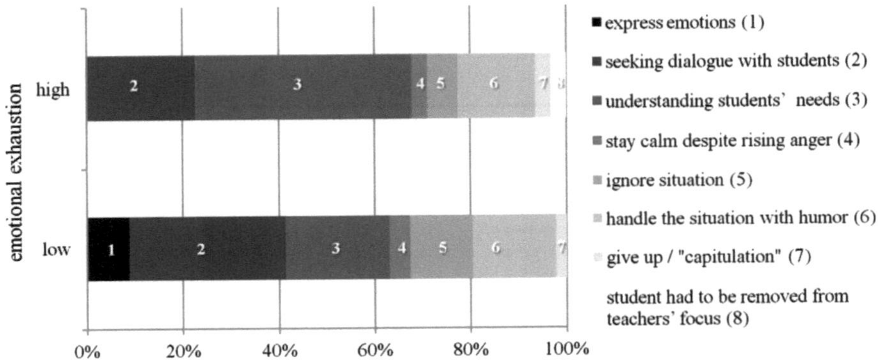

Fig. 6 Techniques of emotion regulation according to teacher emotional exhaustion

It is also possible that the answers are not entirely free of social desirability processes and teachers thought that this was the appropriate answer to our question. Future studies are necessary to investigate if deep acting was indeed aimed at and also successful so that teachers could fully benefit from this potentially health-promoting technique.

Discussion

The three mayor research questions of the project on "Age-related differences in emotion regulation of teachers facing demanding classroom situations" were (1) does emotion regulation have an effect on teacher well-being over the period of one year, (2) are there age-related differences in the use of strategies of emotion regulation and, (3) which techniques do teachers use to regulate their emotions during demanding situations in their lessons?

In order to best address these research questions, we realized a combination of quantitative (study 1: cross-lagged panel design; study 2: quasi-experimental design) as well as qualitative (study 3: video-stimulated recall and interviews of teachers) elements in a multimethod design.

In the study 1 we assessed cross-lagged effects of emotion regulation on well-being and work ability as well as reverse causation over the period of one year. Results show that teachers who regulated their emotions successfully (by deep acting) were one year later less emotionally exhausted (as an indicator of a burnout process under way, Maslach 1998). Others, who preferably regulated their emotional expression by surface acting, showed significantly lower work ability scores (as an indicator for impaired health, Seibt et al. 2007) one year later. A study with 155 German trainee teachers (Hülsheger et al. 2010) also provides evidence for cross-lagged effects over two months. In their study, surface acting led to increased irritation and deep acting to increased job performance. This indicates that emotion regulation does influence well-being, job performance and work ability of teachers not only over short periods of time, but our results also showed that the effects may unfold over one year. Thus, the ability to deep act should be fostered in order to positively influence teacher well-being. Established surface acting strategies should be changed towards less dysfunctional ways of regulating emotions in class. Bearing in mind that a considerable number of teachers feel emotionally exhausted (Farber 1991) and half of those teachers retiring early leave the profession with burnout-related diagnoses (Weber et al. 2003) it seems especially relevant to support teachers in their emotion regulation.

Moreover, once teachers are drained of resources and impaired in their work ability, they rather regulated their emotional display by using a dysfunctional emotion regulation strategy (surface acting). In line with Hobfoll et al. (1996) we argue that individuals whose resources are depleted (because of reduced work ability) are less likely to invest in deep acting and turn to surface acting instead. Engaging in surface acting, in turn, leads to lower work ability over time. Thus, over time, a so-called loss spiral according to the conservation of resources theory (Hobfoll et al. 1996) might unfold. Contrary to our results, Hülsheger et al. (2010) found no reverse causation of irritation or job performance on the use of emotion regulation over two months. In contrast to our study with teachers from a broader age range, it is possible that the young and moderately exhausted trainee teachers in the study by Hülsheger et al. (2010) still had enough resources available and a loss spiral was not (yet) likely to develop.

Hobfoll et al. (1996) also postulate a so-called gain spiral in which individuals with ample resources (e.g., dedicated teachers) are more likely to further invest in new resources (e.g., by deep acting). However, we found no reverse effects of dedication on emotion regulation. Drawing on Grandey's (2003) conclusion, we assume that engaged individuals are less likely to act and, thus, less likely to regulate their emotions at all.

Results of study 2 show that especially more experienced but also young, less experienced teachers seem to struggle with the emotional demands of their profession. This is in line with results on teachers' career cycles from Huberman et al. (1993) who also described a "survival" phase of young teachers as well as their last career phase as potentially difficult stages. In our study these career phases (adaptation to a new field or having been exposed to the emotional demands for a long time) were associated with increased feelings of emotional dissonance that especially the more experienced teachers tried to down-regulate by using different strategies of emotion regulation (i.e., deep acting and faking in good faith) more intensely than their mid-career colleagues. The results from Hertel et al. (see chapter Age Differences in Motivation and Stress at Work) support our finding. The authors report that older employees from different professions make use of coping strategies aimed at the positive reinterpretation of stressors more frequently than their colleagues. In our study, young, less experienced teachers preferably engaged in deep acting. We conclude that teachers at the beginning and at the end of their careers might especially profit from support on beneficial emotion regulation in order to (a) foster functional emotion regulation, (b) not establish dysfunctional emotion regulation strategies or (c) change dysfunctional strategies towards functional deep acting.

Gross (1998) describes in his process model that emotion regulation develops over the course of episodes. In study 2, we found changes in the use of emotion regulation strategies over the course of two emotionally demanding episodes. Apart from the episodic character of emotion regulation, it also seems fruitful to investigate deep and surface acting on a more detailed level by describing specific emotion regulation techniques. Such techniques could then be implemented in interventions which help teachers to successfully cope with the emotional demands in class.

In the third study, we interviewed teachers in a video-stimulated recall situation which emotion regulation techniques they used in the demanding situations in their lessons. Results show that a number of different techniques were used in these situations. In line with results from study 2, we found that older and also younger teachers use a broader variety of techniques than their mid-career colleagues. These techniques were assigned to different emotion regulation strategies according to categorizations by Grandey (2000), Gross (1998), Rafaeli and Sutton (1987) and Zapf (2002). In a next exploratory step, we found different (and somewhat unexpected) patterns of emotion regulation techniques for teachers with lower or higher levels of emotional exhaustion. These techniques will have to be tested further regarding their effect on well-being and health in larger teacher samples and could then, be implemented into teacher interventions.

On the basis of our results it seems necessary to develop interventions on health-beneficial emotion regulation to help teachers remain well and healthy until regular retirement age. How such an intervention could be designed will be discussed in section Theoretical and Practical Implications.

Theoretical and Practical Implications

An important strength of our studies is the multimethod design realizing a combination of quantitative and qualitative methods. However, they also have some limitations which should be addressed in future studies. We cannot rule out that a healthy-worker-effect occurred. Such an effect occurs if teachers with lower well-being or work ability have already left the profession so that they could not take part in a study any more or not feel fit to take part in such a study.

The samples sizes of our studies were relatively small. Especially study 3 is of exploratory nature. Future studies should draw on larger (and if possible) representative samples. To evaluate the impact of emotion regulation over time, studies with different time lags should be realized. Longitudinal studies with shorter or longer time lags and over the course of months or even years would have the advantage that not only potential loss or gain spirals but also age-related changes in the emotion regulation of teachers could be investigated. Study 2 only provides indications for differences between age groups and possible changes over the career cycle were thus interpreted cautiously. Age-related results of study 2 and 3 might also be explained by generation differences in dealing with emotional demands in the classroom. Longitudinal studies over many years would also allow for ruling out such a cohort effect. Also, future studies should test the influence of the identified techniques on teacher well-being in larger samples so that they can be implemented into teacher interventions. We conclude that emotion regulation can be considered a health-beneficial strategy of teachers. Other strategies aiming at effective resource regulation such as selection, optimisation and compensation which have been shown to be associated with higher levels of work ability in a sample of nurses (see chapter Successful Aging Strategies in Nursing: The Example of Selective Optimization with Compensation, Müller et al.) should also be considered in future studies.

Interventions for teachers on health-beneficial emotion regulation should be developed and we conclude from our studies that in such an intervention dysfunctional surface acting should be reflected and gradually replaced by health-beneficial deep acting. Study 3 provided some indication as to which techniques are associated with deep acting and should, thus, be implemented in an intervention as well as which techniques of surface acting should be reflected critically. We also showed that especially teachers in the last career phases but also those at the beginning of their career need support in their emotion regulation. Their mid-career colleagues who seem more confident in their emotion regulation might be a good source for support. Thus, we suggest interventions on the basis of peer consulting containing modules on health-beneficial emotion regulation which include younger and older teachers as well as their mid-career colleagues.

Acknowledgments We gratefully acknowledge funding for the project on "Age-related differences in the emotion regulation of teachers facing demanding classroom situations" by the German Research Foundation (DFG) within the priority program 1184 "Age-differentiated Work Systems". The project was conducted at the University of Freiburg (Germany); in the first project phase from 2006 to 2007 under the scientific supervision of Prof. Dr. Andreas Krause and Prof. Dr. Heinz Schüpbach (KR 3304 1-1) and from 2008 to 2009 under the scientific supervision of Prof. Dr. Heinz Schüpbach in co-operation with Prof. Dr. Andreas Krause (SCHU 2090/1-2 and SCHU 2090/1-3). We gratefully acknowledge the contribution of our co-operation partners, especially Prof. Dr. Andreas Krause. We would also like to thank the Regierungspräsidium Freiburg for their support in recruiting the schools for our studies as well as the teachers who took part in our studies.

References

Brotheridge CM, Lee RT (2003) Development and validation of the emotional labour scale. J Occup Organ Psychol 76:365–379

Büssing A, Perrar KM (1992) Die Messung von Burnout. Untersuchung einer Deutschen Fassung des Maslach Burnout Inventory (MBI-D) [Measurement of burnout. Investigation of a German version of the Maslach Burnout Inventory (MBI-D)]. Diagnostica 38:328–353

Carstensen LL, Fung HH, Charles ST (2003) Socioemotional selectivity theory and the regulation of emotion in the second half of life. Motiv Emot 27:103–123

Côté S, Morgan LM (2002) A longitudinal analysis of the association between emotion regulation, job satisfaction, and intentions to quit. J Organ Behav 23:947–962
Dahling JJ, Perez LA (2010) Older worker, different actor? linking age and emotional labor strategies. Pers Individ Differ 48:574–578
Farber BA (1991) Crisis in education: stress and burnout in the American teacher. Jossey-Bass, San Francisco
Federal Statistical Office (2007) Lehrerinnen und Lehrer gehen im Schnitt mit 62 Jahren in Pension [Teachers retiring at an average age of 62 years]. Pressemitteilung Nr.104 vom 13.03.2007 [Press Release No.104/2007-03–13]
Fischbach A (2003) Determinants of emotion work. Dissertation, Georg-August University of Goettingen, Goettingen. http://webdoc.sub.gwdg.de/diss/2003/fischbach/fischbach.pdf
Grandey AA (2000) Emotion regulation in the workplace: a new way to conceptualize emotional labor. J Occup Health Psychol 5:95–110
Grandey AA (2003) When the show must go on: surface acting and deep acting as determinants of emotional exhaustion and peer-rated service delivery. Acad Manag J 46:86–96
Gross JJ (1998) The emerging field of emotion regulation: an integrative review. Rev Gen Psychol 2:271–299
Hobfoll SE, Freedy JR, Green BL, Solomon SD (1996) Coping in reaction to extreme stress: the roles of resource loss and resource availability. In: Zeidner M, Endler NS (eds) Handbook of coping: theory, research, applications. Wiley, Oxford, pp 322–349
Hochschild AR (1983) The managed heart: the commercialization of human feeling. UCP, Berkeley
Huberman M, Gronauer MM, Marti J (1993) The lives of teachers. Cassell, London
Hülsheger UR, Lang JW, Maier GW (2010) Emotional labor, strain, and performance: testing reciprocal relationships in a longitudinal panel study. J Occup Health Psychol 15:505–521
Hülsheger UR, Schewe AF (2011) On the costs and benefits of emotional labor: a meta-analysis of three decades of research. J Occup Health Psychol 16:361–389
Klusmann U, Kunter M, Trautwein U, Lüdtke O, Baumert J (2008) Teachers' occupational well-being and quality of instruction: the important role of self-regulatory patterns. J Educ Psychol 100:702–715
Kyriacou C (2001) Teacher stress: directions for future research. Educ Rev 53:27–35
Lee RT, Brotheridge CM (2011) Words from the heart speak to the heart: a study of deep acting, faking, and hiding among child care workers. Career Dev Int 16:401–420
Maslach C (1998) A multidimensional theory of burnout. In: Cooper CL (ed) Theories of organizational stress. Oxford University Press, Oxford, pp 68–85
Mayring P (2000) Qualitative content analysis. Forum: qualitative social research, 1: http://www.qualitative-research.net/fqs
Meder L, Dorsemagen C, Krause A (2008) Observational stress analysis at school: classroom teaching as an example of interaction work. Psychol Everyday Act 1:23–32
Morris JA, Feldman DC (1996) The dimensions, antecendents, and consequences of emotional labor. Acad Manag Rev 21:986–1010
Nerdinger FW, Röper M (1999) Emotionale Dissonanz und Burnout. Eine empirische Untersuchung im Pflegebereich eines Universitätskrankenhauses [Emotional dissonance and burnout: an analysis in a university hospital nursing sector]. Zeitschrift für Arbeitswissenschaft, 53:187–193
OECD (2010) Education at a glance 2010. OECD indicators. OECD Publishing. http://www.oecd.org/dataoecd/45/39/45926093.pdf
Philipp A, Schüpbach H (2010a) Longitudinal effects of emotional labour on emotional exhaustion and dedication of teachers. J Occup Health Psychol 15:494–504
Philipp A, Schüpbach H (2010b). Berufsaltersbedingte Veränderungen im Umgang mit belastenden Unterrichtsepisoden [Age-related changes in dealing with emotional episodes in the classroom]. Zeitschrift für Arbeitswissenschaft 3:161–171
Rafaeli A, Sutton RI (1987) Expression of emotion as part of the work role. Acad Manag Rev 12:23–37

Rubin RS, Staebler Tardino VM, Daus CS, Munz DC (2005) A reconceptualization of the emotional labor construct: on the development of an integrated theory of perceived emotional dissonance and emotional labor. In: Haertel CEJ, Zerbe WJ, Ashkanasy NM (eds) Emotions in organizational behavior. Erlbaum, Mahwah, pp 189–211

Schaarschmidt U, Kieschke U (2007) Lehrer-Burnout [Teacher Burnout]. In: Zumbach J, Mandl H (eds.) Pädagogische Psychologie in Theorie und Praxis. Ein fallbasiertes Lehrbuch [Educational psychology in theory and practice. A case-based textbook]. Hogrefe, Goettingen pp 245–254

Schaufeli WB, Bakker AB (2003) The Utrecht work engagement scale. Preliminary manual. Occupational Health Psychology Unit, Utrecht University

Seibt R, Luetzkendorf L, Thinschmidt M (2007) Predictors of work ability in teachers and office workers. In: Richter P, Peiró JM, Schaufeli WB (eds) Psychosocial resources in health care systems. Rainer Hampp Verlag, München, pp 67–80

Tuomi K, Ilmarinen J, Jahkola A, Katajarinne L, Tulkki A (1998) Work ability index (2nd ed). Finnish institute of occupational health, Helsinki

Weber A, Weltle D, Lederer P (2003) Fruehpension statt Praevention? Zur Problematik der Frühinvalidität im Schuldienst [Early retirement instead of prevention? The problem of early invalidity in the education services]. Arbeitsmedizin Sozialmedizin Umweltmedizin 38:376–384

Zapf D (2002) Emotion work and psychological well-being: a review of the literature and some conceptual considerations. Hum Resour Manag Rev 12:237–268

Zapf D, Mertini H, Seifert C, Vogt C, Isic A, Fischbach A (2005) Frankfurt emotion work scales—FEWS (Version 4.2.), Goethe Universität Frankfurt am Main

Successful Aging Strategies in Nursing: The Example of Selective Optimization with Compensation

Andreas Müller, Barbara Heiden, Matthias Weigl, Jürgen Glaser and Peter Angerer

Lessons Learned

Future care can only be provided when employed nurses age in a healthy manner and are capable to perform their jobs. Life-span research shows that action strategies in terms of selective optimization with compensation (SOC) contributes to healthy and capable aging (Baltes and Baltes 1990). SOC therefore ought to be considered as a strategy to help maintain the work ability of nurses. Research is needed in particular to understand the job-specific manifestations of SOC in nursing, as well as its effects on work ability, and to identify working conditions that may support the application and effectiveness of SOC in nursing.

Our findings show that nurses apply job-specific SOC to cope with demands in their daily care work. We further demonstrate that SOC in nursing significantly contributes to work ability of nurses and that the positive relationship between

A. Müller (✉) · P. Angerer
Institute of Occupational and Social Medicine, Heinrich Heine University Düsseldorf, Düsseldorf, Germany
e-mail: andreas.mueller@uni-duesseldorf.de

P. Angerer
e-mail: peter.angerer@uni-duesseldorf.de

B. Heiden · M. Weigl
Institute and Outpatient Clinic for Occupational, Social and Environmental Medicine, Ludwig-Maximilians-University, Munich, Germany
e-mail: barbara.heiden@med.lmu.de

M. Weigl
e-mail: Matthias.Weigl@med.uni-muenchen.de

J. Glaser
Institute of Psychology, University of Innsbruck, Innsbruck, Austria
e-mail: juergen.glaser@uibk.ac.at

C. M. Schlick et al. (eds.), *Age-Differentiated Work Systems*,
DOI: 10.1007/978-3-642-35057-3_8, © Springer-Verlag Berlin Heidelberg 2013

SOC in nursing and work ability is stronger for older nurses compared to younger nurses. Moreover SOC in nursing seems to assist cope with detrimental effects of musculoskeletal impairments on work ability of nurses. Finally our results indicate that nurses' latitude to schedule work, make own decisions, or choose methods—i.e. job control—substantially enhances the effectiveness of resourceful SOC in aging nurses and nurses with musculoskeletal impairments.

In line with these findings, the following recommendations for health care organizations may support the "successful aging" of employed nurses:

- Enable nurses to apply individualized action strategies in terms of SOC.
- Enhance the job control of nurses in terms of autonomy and decision latitude.
- Provide trainings to acquaint nurses with the principles of the SOC model.

Introduction

Due to demographic changes health care organizations in Western countries are facing enormous challenges: An aging nursing workforce that has to deal with job demands that are known to be particularly stressful for older employees, like intense physical lifting or bending, quick information processing, high work speed, and/or shift work. Moreover a further considerable increase of job demands can be expected because also the number of patients increases. Thus, in the future most likely less and older nurses will have to provide care to more patients. As an example, the German Federal Statistical Office forecasts a lack of more than 112.000 nurses in 2025 in Germany (Afentakis and Maier 2010).

In order to cope with these challenges, it becomes essential for health care organizations to support nurses in remaining healthy, capable, and motivated in their job until retirement age. In other words, future health care delivery can be only secured if nurses are aging successfully at work (Camerino et al. 2006; Simoens et al. 2005).

One key to successful aging is the adaptation of individuals to loss of resources (Riediger et al. 2006). One of the most prominent psychological concepts that specifies important adaptive processes of successful aging is the model of selective optimization with compensation (SOC) developed by Baltes and Baltes (1990). Basically the SOC model assumes that by applying the three action strategies of SOC, a loss of resources should be minimized (Baltes and Baltes 1990). Thus, especially in face of an age-related decline of individual resources (e.g. the reviews in chapters Capability Related Stress Analysis to Support Design of Work Systems, Rademacher et al.; Ergonomic Design of Human–Computer Interfaces for Aging Users, Schlick et al.), SOC may allow for an improved adaptation, enhanced mastery, and functioning. To put it in a nutshell: Successful aging is possible when a person is able to focus more on less but important goals, pursuit these goals in an optimized way, and in doing so applying more compensatory means (Baltes 1999).

This chapter reports several studies that attempt for the first time apply the SOC model to nursing work. Our paramount research questions were:

- Do nurses apply job-specific SOC?
- Does SOC in nursing help maintain work ability of older nurses?
- Does the degree to which the nursing profession allows discretion to schedule work, make decisions, and choose methods—i.e. job control—contribute to the application and effectiveness of SOC in nursing?
- Does SOC in nursing contribute to work ability of nurses with musculoskeletal impairments?
- Does SOC in nursing predict well-being or does well-being of nurses predict the application of SOC in nursing?

From a practitioners' perspective, our study would like to provide a basis for the successful development of new measures that promote the health of nurses and their capabilities to perform their daily care work.

Successful Aging at Work

The Model of Selective Optimization with Compensation

In the course of aging, resource losses (e.g. physical fitness, health, sensory abilities, and basic cognitive functions) successively outweigh resource gains (e.g. knowledge, experience, and social status). However, this continuously impaired ratio of losses and gains does not necessarily correspond with diminished functioning or capability of older employees. Comprehensive meta-analyses on the relationship between age and performance indicate for instance that older employees are as capable as their younger colleagues (Ng and Feldman 2008; see also chapter Age-Differentiated Work Systems Enhance Productivity and Retention of Old Employees, Zwick et al.). Thus, one of the core research questions on aging at work is how employees can maintain functioning, personal development and well-being over the whole life span despite a relative loss of resources (Baltes and Dickson 2001).

Baltes and Baltes (1990) proposed a theoretical model to stipulate research on this question: The proposed model suggests that every developmental process encompasses a combination of three kinds of adaptive behaviors: Selection, optimization, and compensation (SOC). It is assumed that successful personal development and life management requires a focus on specific goals—in contrast to dividing energy among multiple goals. *Selection* refers to the setting and prioritization of goals, based on personal motives and preferences (elective selection) or due to perceived loss of internal or contextual resources (loss-based selection). Thus selection guides and organizes individual behavior, directs personal development, and creates a feeling of purpose and meaning in one's life

(Freund and Baltes 2002). *Optimization* refers to useful means of goal achievement. It is crucial for personal development to constantly adopt and refine personal means that help achieve a desired goal. Thus optimization involves the obtainment, improved and coordinated use of individual means to pursue important goals. *Compensation*, like optimization, also refers to means. It specifically addresses the question of how people facing permanent or temporary loss of resources are still able to maintain a desired level of functioning. Therefore compensation involves the acquisition and application of alternative individual means or the utilization of external or technological aid to substitute lost means and to maintain a desirable level of performance.

SOC at Work: Literature Review

The SOC model has stimulated various research and has received numerous empirical support (for an overview see Riediger et al. 2006). Within the last decade a growing number of studies on SOC at the work place also demonstrate that the SOC model can provide a valuable framework for explaining organizational behavior and coping with age related changes in occupational contexts (Baltes and Dickson 2001).

SOC at work is related to subjective well-being, job satisfaction, and positive expectations about future opportunities at work: In a survey by Wiese et al. (2000), young white-collar professionals that reported the utilization of more SOC strategies also reported higher levels of well-being both in the work and in the family domain. Two surveys by Baltes and Heydens-Gahir (2003) and Young et al. (2007) replicated this supporting effect of SOC on work-family balance. In line with these findings, a survey by Zacher and Frese (2011) with employees of a manufacturing company indicate that employees can maintain positive expectations about future opportunities at work when they use SOC. Results of a recent diary study among administrative employees by Schmitt et al. (in press) indicated that SOC is significantly related to job satisfaction. Finally, in an extension of these cross-sectional findings, Wiese et al. (2002) were also able to demonstrate the longitudinal predictive power of SOC on subjective well-being and job satisfaction.

There is also some evidence that SOC at work is positively associated with career success: In a survey with young professionals, Abele and Wiese (2008) observed that generalized optimization strategies are indirectly—i.e. through specific strategies of career planning—linked with various indicators of career success such as better pay, higher position or career satisfaction. However, in a longitudinal study by Wiese et al. (2002), the initial use of SOC by young white-collar professionals did not predict subjective career success.

Studies also suggest that the use of SOC at work contributes to competency maintenance and job performance: Bajor and Baltes (2003) observed in a survey with bank employees that the positive effects of conscientiousness on job performance are partly operating through the use of SOC. In a survey by Abraham and Hansson (1995), SOC interacted significantly with the age of white-collar professionals in predicting competency maintenance, thus, indicating that older employees can maintain professional competencies when they apply SOC. This result was confirmed by a diary study by Yeung and Fung (2009) among Chinese sales workers. The two studies by Abraham and Hansson (1995), Yeung and Fung (2009) are especially noteworthy because these studies provide evidence that SOC at work is especially helpful for older employees in their attempt to maintain capability and functioning at work.

Few of the aforementioned studies indicate that SOC at work is not independent from the work environment: In the Abraham and Hansson (1995) study stressors at work in terms of role conflicts, and role ambiguity were positively related to selection and compensation. In Yeung and Fungs (2009) study a stronger positive relation between SOC and job performance was observed for older workers who were handling more difficult tasks. Zacher and Frese (2011) found that only in low-complexity jobs SOC was related to positive expectations about future opportunities at work. Finally, in a study by Schmitt et al. (in press), high use of SOC significantly buffered the relationship between problem solving demands and fatigue. These studies consistently indicate that SOC is especially effective when only sparse contextual resources at work are available, i.e. when working conditions do not adequately support the achievement of work goals, job demands and associated individual costs are high, or personal development is not stimulated (Bakker and Demerouti 2006).

Research Questions

Notwithstanding these encouraging findings on the supportive effects of SOC at work some important questions remain unanswered:

First, research so far has neglected that the type of job may be of utter importance for specific manifestations of SOC (Bajor and Baltes 2003). For example, if one considers in terms of action theory SOC as goal-related behavior (Freund and Baltes 2002), the possibility to execute selection at the workplace—i.e. setting goals—should be influenced by employees' latitude to decide about work goals (Abraham and Hansson 1995; Bajor and Baltes 2003). Accordingly, specific manifestations of optimization or compensation at the workplace should be dependent on which particular means or aids are relevant or available in a specific job. Thus, the first aim of our study was to explore job-specific manifestations of SOC in nursing.

Research Question 1: Do Nurses Apply Job-Specific SOC?

Second, the majority of research on SOC at work examined white-collar workers in managerial, highly educated professions (Abraham and Hansson 1995; Bajor and Baltes 2003; Baltes and Heydens-Gahir 2003; Schmitt et al. in press;

Wiese et al. 2002; Young et al. 2007). Compared to employees in most other jobs, white-collar workers usually have more latitude and opportunities to apply self-directed behaviors like SOC (Bajor and Baltes 2003). Moreover, in white-collar jobs, age related losses usually do not predominate sufficiently to make SOC a highly important strategy of older employees (Abraham and Hansson 1995). Thus we do not know whether most of the available results on the effects of SOC at work are representative to other kinds of jobs. We argue that SOC might be more important in jobs with age-sensitive demands which lay a substantial burden on individual reserve capacity of older employees (Abraham and Hansson 1995). Nursing contains many age-sensitive job demands such as intense physical lifting or bending, or high work speed, resulting in a lower work ability of older nurses—i.e. lower perceived physical and mental capability to perform their work (Ilmarinen 2007, 2009; Tuomi et al. 1997)—compared to their junior colleagues (Camerino et al. 2006). In health care, work ability is shown to be closely related to mental and physical well-being, general health perception, emotional exhaustion, intention to leave the nursing profession, and disability (Ahlstrom et al. 2010; Camerino et al. 2006). Thus, due to its age-sensitive nature and its predictive validity work ability, SOC may serve as an important criterion for successful aging at work (van den Berg et al. 2009).

We assume that the negative relationship between age and work ability is less pronounced for nurses with high use of SOC than for nurses with low use of SOC (Abraham and Hansson 1995; Yeung and Fung 2009). According to Ilmarinen (2009), work ability is based on the balance between the resources of an employee and his/her work demands: If age-sensitive job demands permanently exceed available recourses, work ability may decrease. Thus, an efficient use of individual resources by SOC may contribute to work ability especially of older nurses in terms of a better balance between individual resources and job demands.

Research Question 2: Does SOC in Nursing Help to Maintain Work Ability of Elder Nurses?

Third, so far studies on the interplay between working conditions and SOC mainly have focused on how *limited* contextual resources can be addressed by SOC (e.g. Abraham and Hansson 1995; Yeung and Fung 2009; Zacher and Frese 2011). As an extension to that perspective, we argue that SOC should also benefit from *enhanced* contextual resources at work (Freund 2006). This assumption draws on action theory (Frese and Zapf 1994; Hacker 2003) and COR-theory (Hobfoll and Wells 1998).

In terms of action theoretical interpretations of the SOC model (Freund and Baltes 2002; Zacher and Frese 2011) selection can be considered as selecting or prioritizing goals whereas optimization and compensation can be considered in terms of optimizing means or developing alternative means to achieve selected goals. Job control can be considered as a contextual resource that provides employees with the opportunity to apply individual action strategies such as selecting goals or applying alternate means or methods of job task execution (Humphrey et al. 2007). Conversely, jobs with less control might provide employees limited opportunities to apply SOC. In terms of conservation of resources theories' notion on the interrelation between individual and contextual resources (Hobfoll and Wells 1998), enhanced job control should provide additional support for employees' attempts to cope effectively with age-related changes and losses—such as applying SOC. Thus, we argue that enhanced job control should facilitate the implementation and effectiveness of resourceful SOC at work.

Research Question 3: Does Job Control Contribute to the Application and Effectiveness of SOC in Nursing?

Fourth, recent publications suggest that functional age, i.e. the state of capacity of employees, is considered to be more relevant for being considered as an old employee, than chronological age, i.e. the number of lived years (Schalk et al. 2010). Following this argumentation, we aimed to look at functional age in terms of musculoskeletal impairments (MSI) of nurses. Compared to most other professions; nurses have an increased risk of MSI (e.g. Leino-Arjas et al. 2002). Nursing incorporates also a multiplicity of physical demands such as frequent lifting of heavy loads, patient transfers, extreme flexion, bending, or twisting (e.g. Engels et al. 1996; Trinkoff et al. 2003). Thus musculoskeletal capacity can be considered as an important resource for nurses to perform their work, and MSI of nurses such as limited mobility or pain may constitute a relevant loss of resources. We examined whether SOC can help to balance out assumed negative effects of MSI on work ability of nurses.

Research Question 4: Does SOC in Nursing Contribute to Work Ability in Nurses with Musculoskeletal Impairments?

Fifth, our final research question refers again to the notion of resources: On the one hand, the application of SOC should help to maximize the efficient use of available resources (Baltes and Baltes 1990); while on the other hand, individuals can better apply SOC if they have many resources (Baltes and Lang 1997). Thus, beyond cross-sectional associations we aimed to investigate whether SOC in nursing may longitudinally predict better well-being or whether only nurses with better well-being are capable to apply resourceful job-specific action strategies like SOC in nursing. Moreover, we attempted to replicate longitudinal findings by Wiese et al. (2002) that SOC predicts wellbeing 3 years later, by showing that this finding holds also in the context of nursing and for a shorter time lag.

Fig. 1 Research model

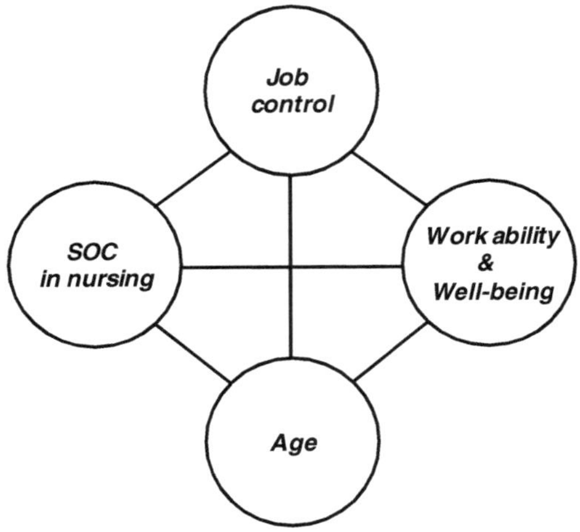

Research Question 5: Does SOC in Nursing Predict Well-Being or Does Well-Being of Nurses Predict the Application of SOC in Nursing?

Figure 1 provides an overview on our research model.

Empirical Studies

We conducted a qualitative study to explore job-specific SOC in nursing (research question 1). Subsequent to this study we performed a survey study to deal with the questions whether the explored job-specific SOC in nursing helps maintain work ability of older nurses and whether job control contributes to the application and effectiveness of SOC in nursing (research questions 2 and 3). In a next step we complemented the data of the survey study by data of a physical examination to investigate whether SOC in nursing especially contributes to work ability of nurses with musculoskeletal impairments (research question 4). Finally, we did a panel study by repeating our survey 6 months later to address the question whether SOC in nursing predicts well-being or whether well-being of nurses predicts the application of SOC in nursing (research question 5).

Qualitative Study

Method

Participants

To explore job-specific SOC in nursing, we asked 20 nurses who were at least 45 years old for an interview. The age cutoff is consistent with studies that point to an impaired self-reported work ability of older nurses (e.g. Camerino et al. 2006). Finally, 17 nurses (14 women, 3 men) agreed to participate (participation rate 85 %). The mean age of the interviewees was 57.0 years (standard deviation, SD = 3.29, Range: 51–62 years). They worked on average for 35.6 years in nursing (SD = 4.9, Range: 25–43 years).

Procedure and Measures

The semi-structured interviews had the following scheme: First, we asked for "age-sensitive" physical and psychical job demands in nursing (e.g., "What are in your experience physical job demands in nursing that are more difficult to accomplish with increasing age?"; "From your personal experience, which are the physical job demands in nursing that are easier to accomplish with increasing age?"). Following each question on job demands, we asked an open-ended question about strategies to cope with these job demands (e.g. "Which strategies do you apply to cope with these physical job demands? Please, provide specific examples.").

We performed a qualitative content analyses (Miles and Huberman 1984) of the transcribed interviews: First, we determined the units to code in the transcripts. A unit was every meaningful entity (either a word, or a group of words, or a sentence or a section) that described a strategy to cope with a job demand in nursing. Second, we assigned these units either to the three pre-defined SOC dimensions or to the category "other behaviors". The coding was based on predefined inclusion and exclusion criteria for every SOC behavior. The coding criteria were in accordance with action theoretical definitions of SOC by Freund and Baltes (2002) (for a more detailed description of the coding rules see Müller et al. in press). Additionally, to test for inter-rater agreement, the two raters independently coded a random sample of four interviews to test the reliability of our coding system. Cohen's Kappa = 0.75 (83 %, 149 coded units) indicates a good reliability of the coding scheme for SOC in nursing.

Results on Research Question 1: Do Nurses Apply Job-Specific SOC?

Results of the interview study indicate that older nurses apply a high proportion of SOC to cope with job demands (Müller et al. in press): On average 77 % (60–83 %) of the reported behaviors could be labeled as SOC. 88 % of the nurses reported the selection strategy "concentrating on the most important job tasks or setting priorities". Second and third ranked optimization behaviors such as "making suggestions for improvement" (76 %), "doing exercises", and "informing oneself about the current state of professional knowledge" (respectively 70 %). The most frequently reported compensation strategy was "asking for help" (52 %).

Whereas the content of the reported selection strategies are quite similar to general SOC measured by the established SOC questionnaire by Baltes et al. (1999), the most specific SOC behavior in nursing seems to be compensation: The reported compensation strategies were mainly related to the physical demands in nursing. This finding is consistent with studies on the increasing health risks of older nurses because of physical demands (Müller et al. 2010). Accordingly, 14 (82 %) of the interviewed nurses said that their physical job demands would be more difficult to accomplish with increasing age. In summary, this qualitative study indicates that older nurses apply to a great extent job-specific SOC in their daily care work.

Survey Study

Method

Participants

The study population was composed of nurses from a University hospital in Southern Germany (n $\sim$ 3,000). Our study focused on nurses in all intensive care units, all operation units, all anesthesia units and three general wards. According to the data of the hospitals human resources department, 953 nurses of these units (about 30 % of the population) were available for participation. These nurses formed the convenience sample of our study.

438 out of 953 nurses (participation rate 46 %) participated in the cross-sectional survey. 366 (84 %) nurses were female and 71 male (16.0 %; missing = 1). Age ranged from 21 to 63 years with an average of 38.5 years (standard deviation, SD = 11.25). 190 nurses were <35 years (43.7 %), 111 between 35 and 44 years (25.5 %) and 134 $\geq$ 45 years (30.8 %, missing n = 3). The average professional tenure was 18.1 years (SD = 10.9). 238 nurses (54.5 %) came from intensive care units, 85 (19.5 %) from the operating room, 68 (15.5 %) from anesthesia units, 39 (8.9 %) from general wards, and 7 (1.6 %) male nurses were exclusively responsible for patient transfer to the operating theatres (missing n = 1).

Measures

Work ability: Work ability was assessed with the German version of the Work Ability Index Scale (WAI; Hasselhorn and Freude 2007). The original scale was developed by Tuomi et al. (1997). The applied item measures the current work ability in relation to life-time best work ability (WAI dimension 1). We chose this item for the following reasons: (a) The overall WAI is often criticized because it incorporates disparate constructs such as chronic illness, which impairs the construct validity of the index (Ahlstrom et al. 2010). (b) The overall WAI contains questions on neck or back pain. Thus measures of musculoskeletal impairments (MSI) and the overall WAI are confounded and associations between MSI and overall WAI are trivial (see research question 4). (c) The applied WAI-dimension

1 is strongly predictive to well-being outcomes whereas other WAI-dimensions have less predictive value (Ahlstrom et al. 2010). The item used a scale ranging from 0 = "completely unable to work" to 10 = "work ability at its best".

Age: We assessed chronological age in years via questionnaire. To perform analyses on age-group differences, we categorized age into three groups: <35, 35–44 and ≥ 45 years. The group <35 years should be considered separately because nurses have on average a rather short tenure in their profession. The age-group ≥ 45 years is worth considering separately because there is evidence that nurses at that age report meaningfully reduced work ability (Camerino et al. 2006).

Selection, optimization, and compensation (SOC) in nursing-scale: Based on the interview results in study 1, we converted nine of the extracted behaviors into survey questions. The item selection procedure was based on consensus by involved nursing experts and work psychology professionals. Finally, three items for each of the SOC behaviors were selected that constituted a SOC in nursing-scale. Item examples are: "I concentrate on the most important tasks to do my job well" (Selection); "I keep myself constantly informed about the current professional knowledge in nursing" (Optimization); "I organize my work in a way that allows me to counterbalance one-sited physical stressors" (Compensation). Items used a 5-point Likert scale ranging from 1 = "not at all" to 5 = "to a very great extent". The overall scale as well as the subscales show good internal consistencies (Cronbach's Alpha): $\alpha = 0.76$ for the overall SOC-scale; $\alpha = 0.72$ for the subscale 'selection'; $\alpha = 0.63$ for 'optimization'; $\alpha = 0.71$ for 'compensation'. SOC in nursing and its subscales show substantial positive correlations with the established general SOC-short scale by Baltes et al. (1999). This indicates the validity of the new scale and its subscales. For more details on our scale construction see Müller et al. (in press).

Job control: Job control refers to the degree in which work organization allows autonomous decision-making and personal discretion (e.g., "This work allows for decisions on which work methods one pursues"). It was assessed with a German self-report instrument for work analysis in hospitals (Büssing and Glaser 2002). The instrument is well-validated, repeatedly used in German speaking countries, and adapted to health-care professions (Weigl et al. 2010). All items used a 5-point Likert scale from 1 = "not at all" to 5 = "to a very great extent". Job control was assessed with nine items. Cronbach's alpha was 0.94.

Results

Descriptive Results on the Main Study Variables

Table 1 displays the descriptive results and intercorrelations of the main study variables. According to the classification of the WAI dimension 1 proposed by Ahlstrom et al. (2010), an average work ability score of nearly eight scale-points indicates a "good" work ability of our sample. As expected, age of participants was negatively related to work ability, whereas job control and SOC in nursing and

Table 1 Means (M), standard deviations (SD), and intercorrelations of work ability, age, job control, and SOC in nursing

	M (SD)	01	02	03	04	04a	04b
01 Work ability	7.72 (1.60)						
02 Age	38.53 (11.25)	−0.13**					
03 Job control	3.19 (0.82)	0.15**	−0.07				
04 SOC in nursing	3.62 (0.54)	0.21**	0.22**	0.19**			
04a Selection	4.00 (0.59)	0.15**	0.12*	0.13**	0.64**		
04b Optimization	3.39 (0.73)	0.19**	0.19**	0.10*	0.82**	0.34**	
04c Compensation	3.45 (0.82)	0.13**	0.18**	0.20**	0.80**	0.24**	0.49**

Note **$p \leq 0.01$, *$p \leq 0.05$, two-tailed

its sub-dimensions were positively related to work ability. Interestingly, the age of participants correlated positively with SOC in nursing, indicating a higher use of job-specific SOC by older than younger nurses.

Results on Research Question 2: Does SOC in Nursing Help to Maintain Work Ability of Older Nurses?

In line with the majority of studies on age and work ability (see van den Berg et al. 2009), results of hierarchical moderated regression analyses (Müller et al. in press) suggest that higher chronological age of nurses goes along with diminished work ability ($\beta = -0.32$; $p < 0.01$). In addition, our results indicate that enhanced use of SOC in nursing is related with better work ability of nurses ($\beta = 0.12$; $p < 0.05$).

Beyond that, there is a significant two-way interaction between chronological age and SOC in nursing on work ability ($\beta = 0.18$; $p < 0.01$), such that for older nurses there is a positive relation between SOC in nursing and work ability and for younger nurses there is almost no relation between SOC in nursing and work ability (Müller et al. in press; see Fig. 2). In other words, older nurses who report a high use of SOC in nursing report approximately the same work ability level than younger nurses. Older nurses who report a low use of SOC in nursing report a considerably lower level of work ability than younger nurses. This interaction effect exists over and above the effects of lifestyle factors, working conditions, as well as the effects of "general SOC" as measured by the SOC short scale of Baltes et al. (1999). The interaction explains an additional 3 % of work ability variance ($R^2 = 0.03$, $p < 0.01$).

We further analyzed the main effects of the three subscales of SOC in nursing on work ability as well as their interaction effects with age: Of all three subscales only optimization in nursing is significantly positively related with work ability ($\beta = 0.14$; $p < 0.05$). The interaction effects between age and all three sub-dimensions of SOC in nursing on work ability are the same as those of the total scale displayed in Fig. 2: For younger nurses there is almost no relation between

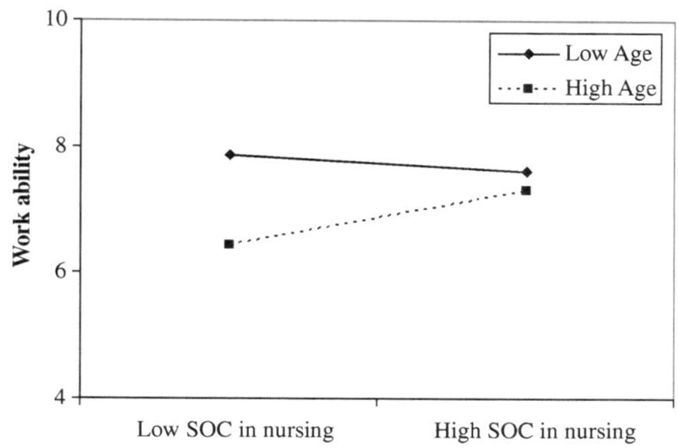

Fig. 2 Two-way interaction of age, selection, optimization and compensation in nursing (*SOC*) on work ability (Müller et al. in press)

selection/optimization/compensation and work ability whereas for older nurses there is a positive relation between SOC and work ability.

Overall, our results indicate that older nurses can maintain work ability when they apply SOC in nursing.

Results on Research Question 3: Does Job Control Contribute to the Application and Effectiveness of SOC in Nursing?

We expanded our models by testing a model that also incorporates the assumed supportive effects of job control. More specifically, we hypothesized that the independent variable 'job control' determines the dependent variable 'work ability' through the mediator variable 'SOC in nursing' (mediation hypothesis; path c* in Fig. 3).

In accordance with the reported age effects, we additionally assumed that the strengths of the indirect effect of job control through SOC in nursing on work ability is stronger for older nurses (moderation hypothesis; path d in Fig. 3).

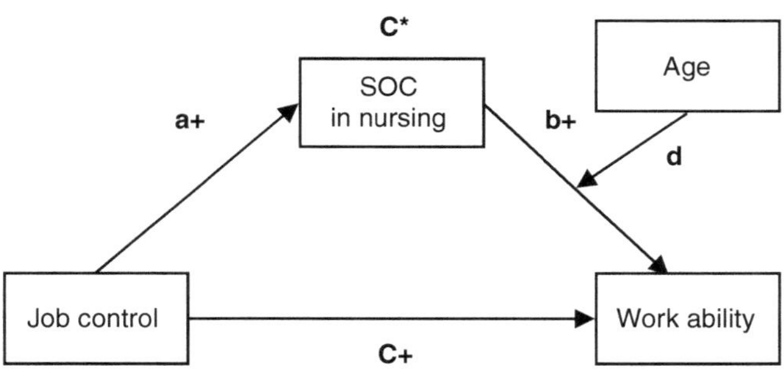

Fig. 3 The moderated mediation model of work ability, job control, age, and strategies of selective optimization with compensation (*SOC*) in nursing

Results indicate that job control is significantly positively related with SOC in nursing ($\beta = 0.18$; $p < 0.00$), thus indicating that job control may promote the application of SOC. Moreover, SOC in nursing significantly mediated the effect of job control on work ability ($\beta = 0.05$; 95 % $CI = 0.01, 0.11$). This mediation effect suggests that the well-known supporting effects of job control on work ability (van den Berg et al. 2009) potentially operate partially through SOC. This interpretation is in line with the assumption from action theory and job design that working conditions are "redefined" by employees in terms of self-set goals or choices of task accomplishment (Hacker 2003; Hackman and Oldham 1976). In other words, different employees may act differently under the same degree of job control. Stronger supporting effects of job control on work ability can be expected for those employees that use the opportunities of job control more efficiently by applying SOC. Further analyses revealed that the indirect effects of job control on work ability through SOC become stronger with age (youngest age group: $\beta = -0.01$ n.s.; middle-age group: $\beta = 0.06$; 95 % $CI = 0.02$, 0.13; oldest age group: $\beta = 0.13$; 95 % $CI = 0.05$, 0.24). This effect suggests that job control can help older nurses maintain important capacities and competencies at work despite experiencing age-related losses through the use of SOC (see also Müller et al. 2012 for similar analysis).

The previous model assumed a "functional chain", i.e. job control effects work ability through SOC in nursing. We tested another model that does not include direct effects of job control on SOC in nursing but rather an interactional effect between job control and SOC in nursing on the work ability. Results of hierarchical moderated regression models indicated main effects of job control and SOC in nursing on work ability, but *no* interaction effect between job control and SOC in nursing on work ability. This means that contrarily to our assumptions high job control seems not to directly strengthen the positive relation between SOC in nursing and work ability.

However, we observed in another sample of health care personnel working in a psychiatric in-patient clinic and a neurological rehabilitation clinic a three-way-interaction between SOC, job control, and work ability ($\beta = 0.19$; $p < 0.05$) (Weigl et al. in press). The beneficial effect of high job control on maintaining work ability among older health care personnel was accentuated when they used SOC. This finding is especially noteworthy because we applied supervisor ratings of work ability as an independent outcome measure. Supervisor-based reports of employee's work ability provide an alternative approach to obtain valid information as well as to diminish common method bias (Podsakoff et al. 2003). This study demonstrates that our previous findings cannot be referred to single source or common method biases that could potentially increase the risk of spurious results.

In summary, the results indicate that SOC in nursing as well as job control should each have a unique positive effect on work ability. However, the findings also point to synergistic effects: On the one hand, job control might be an important precondition for the application of SOC in nursing. On the other hand, positive effects of job control seem to be partly effective through the use of SOC in nursing. Thus SOC in nursing can also supplement the effectiveness of job control. Moreover, although job control in general seems not to strengthen effects of SOC in nursing on work ability, it may help especially older nurses to utilize SOC in nursing in order to cope with age-related changes or impairments (cf., Kanfer and Ackerman 2004).

Physical Examination

Method

Participants

273 of the nurses who participated in the survey study reported above took also part in a physical examination. 232 (85.0 %) nurses were female and 41 male (15.0 %). Participants' age ranged from 21 to 63 years with an average age of 39.4 years (standard deviation, SD = 11.6). 110 nurses were <35 years (40.3 %), 64 between 35 and 44 years (23.4 %) and 99 ≥45 years (36.3 %). Nurses did not differ from the total sample of participating nurses regarding gender and work ability, but were older (physical examination: 39.3 years, total sample: 37.2 years; $T = 1.94$, $p < 0.05$).

Measures

The examination was divided into a detailed structured anamnesis and a systematic orthopedic screening-examination of all body sites. Musculoskeletal impairments (MSI) were assessed with a screening-examination suggested by Spallek et al. (2007) which is based on widely accepted clinical tests. A trained physician judged functional impairments, restrictions of mobility, and pain in six body regions (neck, shoulder, arms, lower back, knees, ankles) in regard to predefined criteria (e.g. a left or right rotation of the cervical spine of less than 70° from the neutral starting position is considered as pathological). The variable MSI was the sum of body sides with either functional impairment, or restrictions of mobility, or pain (Range 0–6). A more detailed description of the examination is reported in Heiden et al. (2011).

Results

Descriptive Results on Musculoskeletal Impairments

Musculoskeletal Impairments (MSI) of at least one body site was observed in 134 participants (49.1 %). Regarding the location of MSI, impairments of the lower back (24.9 %) were most prevalent, followed by neck (22.0 %), shoulders (16.5 %), and knees (10.3 %). Elbows/hands and feet were rarely affected (3.3; 2.9 %).

The frequency of having at least one MSI significantly increased with age, ($X^2(2) = 74.98$, $p < 0.01$). There was no significant increase of MSI from the young (<35 years) to the middle age group (35–44 years), but to the old age-group (≥45 years), ($X^2(2) = 67.98$, $p < 0.01$) and also from the middle age to the old age-group ($X^2(2) = 40.17$, $p < 0.01$).

Results on Research Question 4: Does SOC in Nursing Help to Maintain Work Ability of Nurses with Musculoskeletal Impairments?

In line with publications on functional age (Schalk et al. 2010), we examined whether SOC in nursing helps to offset assumed negative effects of musculoskeletal impairments (MSI) on work ability of nurses (Müller et al. 2011).

The results confirm our assumptions: The more MSI of a nurse were diagnosed by the physician, the lower the work ability the nurse reported ($\beta = -0.15$; $p < 0.05$). There is a significant two-way interaction between MSI and SOC in nursing on work ability ($\beta = 0.13$; $p < 0.05$), such that for nurses with low use of SOC in nursing the negative relation between MSI and work ability remains and for nurses with high use of SOC in nursing there is no relation between MSI and work ability (Müller et al. 2011). These interaction effects explain 1 % of the variance in work ability ($R^2 = 0.01$, $p < 0.05$).

In line with the results reported in the previous section, we further observed a three-way-interaction between SOC in nursing, job control, and MSI ($\beta = 0.16$; $p < 0.01$), indicating that the detrimental effects of MSI on work ability can be best equilibrated when nurses use to a great extent SOC in nursing and have high job control (see Fig. 4).

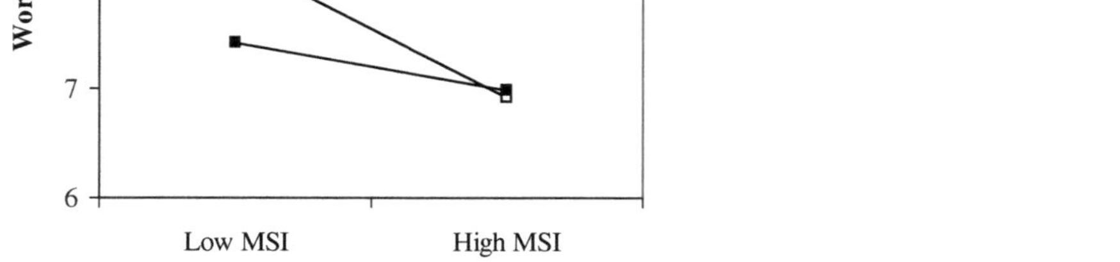

Fig. 4 Three-way interaction of musculoskeletal impairments (*MSI*), job control, and selection, optimization and compensation in nursing (*SOC*) on work ability

For a better interpretation of this effect, we provide a short case report of an interviewed nurse: A female nurse who is working for over 30 years in the operation theatre suffers from backache, which highly impaired her capability to fulfill her job demands. She developed small but effective strategies to compensate this constraint by reorganizing her surgical work. She moved for instance the waste bin for dabbers away from the operating table to avoid long standing and to be able to switch repeatedly to favorable body postures. The precondition to efficiently apply this strategy was the opportunity to make own decisions, employing individual work approaches, and using personal discretion, i.e. job control.

Panel Study

Methods

Participants

346 of the nurses who participated in the first survey study took also in a second survey 6 month later. 289 (83.5 %) nurses were female and 57 male (16.5 %). Participants' age ranged from 21 to 63 years with an average age of 39.1 years (standard deviation, SD = 11.4). 149 nurses were <35 years (43.1 %), 81 between 35 and 44 years (23.4 %) and 116 ≥45 years (33.5 %). Nurses did not differ from the sample of the first survey regarding gender, work ability and age.

Measures

Well-being: With the Irritation-Scale (Mohr et al. 2006) we additionally applied a well-validated and widely used measure for impaired well-being in occupational contexts. Irritation is defined as a state of mental impairment that goes along with ruminative thoughts about problems at work (e.g. "Even at home I often think of my problems at work") and irritability (e.g. "I get grumpy when others approach me"). Tests of re-test reliability indicate that the Irritation-Scale is especially useful for the assessment of short-term changes of mental states related to stress (Mohr et al. 2007). Thus, the scale ideally serves our purpose to detect short-term effects of SOC in nursing on well-being. Irritation was assessed with six items. All items used a 7-point Likert scale from 1 = "strongly disagree" to 7 = "strongly agree". Cronbach's alpha was 0.86 (time 1) and 0.87 (time 2).

Results

Results on the Main Study Variables

Table 2 summarizes the descriptive statistics, mean-differences and correlations of variables over the 6 months period. Mean differences indicate that the well-being level of nurses did not change over time. However, nurses report a significant increase of SOC in nursing which can be attributed to changes of the two sub-dimensions optimization and compensation. Correlations indicate a throughout high stability of impaired well-being as well as of SOC in nursing.

Results on Research Question 5: Does SOC in Nursing Predict Well-Being or Does Well-Being of Nurses Predict the Application of SOC in Nursing?

Lagged correlations indicate a significant negative effect of SOC in nursing at time 1 on impaired well-being in terms of irritation at time 2 ($r = -0.23$, $p < 0.01$). This means that nurses who reported a higher use of initial SOC in nursing reported better well-being 6 months later. Looking at the sub-dimensions, optimization at time 1 ($r = -0.22$, $p < 0.01$) had the strongest effect on well-being at time 2, followed by compensation ($r = -0.16$, $p < 0.01$), and selection ($r = -0.15$, $p < 0.01$). Altogether the three SOC sub-dimensions predicted 6 % of well-being variance at time 2 ($R = 0.24$, $p < 0.01$). In line with our previous findings on the interplay between SOC in nursing and chronological age, there is a stronger lagged association between SOC in nursing and well-being in the two older age groups than in the youngest age group (< 35 years: $r = -0.11$, n.s.; 35–44 years: $r = -0.32$, $p < 0.01$; ≥45 years: $r = -0.33$, $p < 0.01$), indicating again that older nurses especially benefit from the application of SOC in nursing. Although, these lagged correlations do not allow drawing conclusions about causal relations between the variables, they do demonstrate that our cross-sectional findings on SOC in nursing are not artificial correlations due to unmeasured occasional factors like mood of the participants (Zapf et al. 1996).

In a further analysis we applied cross-lagged panel models that simultaneously estimate the cross-sectional (path a and b in Fig. 5), as well as reciprocal relationships between SOC in nursing and impaired well-being (path c and d in Fig. 5)

Table 2 Means (M), standard deviations (SD), mean-differences and correlations of variables over 6 months

	M (SD) $_{T1}$	M (SD) $_{T2}$	t test	$r_{T1/T2}$
Impaired well-being (irritation)	3.03 (1.22)	2.92 (1.19)	t (341) = 1.25 (n.s.)	0.62**
SOC in nursing	3.62 (0.54)	3.73 (0.51)	t (340) = 5.43 ($p < 0.01$)	0.74**
Selection	4.00 (0.59)	3.94 (0.56)	t (340) = 1.36 (n.s.)	0.61**
Optimization	3.39 (0.73)	3.69 (0.59)	t (340) = 10.11 ($p < 0.01$)	0.70**
Compensation	3.45 (0.82)	3.56 (0.82)	t (338) = 2.63 ($p < 0.01$)	0.64**

Note **$p \leq 0.01$, two-tailed

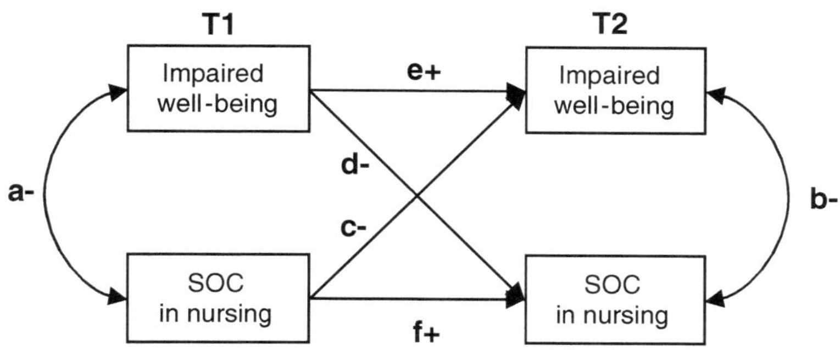

Fig. 5 Proposed cross-lagged associations between selective optimization with compensation (*SOC*) in nursing and impaired well-being

and simultaneously controlling for the stabilities (autocorrelations) of the variables over time (path e and f in Fig. 5). Such models have the advantage that the causal dominance of either of the variables can be tested (Zapf et al. 1996). On the overall level of SOC in nursing, the statistical models neither provide evidence of lagged effects of initial SOC in nursing on well-being 6 months later nor demonstrate reverse effects. However, looking at the level of SOC sub-dimensions, optimization behavior appears to have a significant negative lagged effect on posterior impaired well-being (path c: $r = -0.09$; $p < 0.05$). There were no reverse effects of well-being on neither of the single SOC sub-dimensions.

A separate analysis over three age groups (<35 years, 35–44 years, ≥45 years) revealed that SOC in nursing predicts well-being especially in the middle-age group of 35–44 years old nurses ($r = -0.14$; $p < 0.10$), but not in the other two age groups. Analyses on the level of SOC sub-dimensions in the age group of 35–44 years again indicated that the initial use of optimization in nursing may help to prevent impaired well-being 6 months later ($r = -0.18$; $p < 0.05$). However, there are also significant reverse effects ($r = 0.16$; $p < 0.05$). Contrarily to our assumption that impaired well-being might hamper the use of SOC in nursing, the observed reverse effect indicates that optimization behavior might be applied as a response to impaired well-being.

In brief, our longitudinal data corroborate the assumption that SOC in nursing—especially optimization—may predict well-being of employees 6 months later although to a small extent.

Discussion

We demonstrated for the first time that older nurses apply job-specific SOC behaviors in their daily care work. Thus, our results contribute to the growing evidence on the positive association of SOC at work and favorable individual outcomes. We further showed that SOC in nursing explain substantial variance of desired outcomes over and above the effects of "general SOC" behavior. With that our study is the first in the occupational context that demonstrates the benefits of the measurement of "contextualized" SOC behaviors (Riediger et al. 2006).

We further showed that SOC in nursing significantly contributes to the work ability, particularly for older nurses and—in terms of functional age—for nurses with musculoskeletal impairments (MSI). These findings support one of the central hypotheses of the SOC model that people who are applying SOC can use available resources more efficiently and may therefore cope better with detrimental conditions or limited resources (Baltes and Lang 1997).

Moreover, although SOC in nursing as well as job control respectively seem to have a unique positive effect on work ability, we provide first direct empirical evidence that for older employees high job control seems to substantially enhance the effectiveness of resourceful SOC behaviors. Recent research mainly concentrated on the assumption that SOC is especially effective when contextual resources are *limited* (Yeung and Fung 2009; Zacher and Frese 2011). Our findings broaden this perspective by showing that the use of SOC at work can also benefit from *enhanced* environmental resources. Drawing on the perspective of optimal resource allocation (Baltes and Lang 1997), our findings show that employees who face dwindling capacities may especially benefit from the efficient allocation of limited individual resources by using SOC and high job control that support the efficient use of SOC (Abraham and Hansson 1995). Thus, high job control seems to allow older workers to make use of individual action strategies in order to cope with age-related changes or impairments (Kanfer and Ackerman 2004; see also chapter Age Differences in Motivation and Stress at Work, Hertel et al.). From a practical point of view, we believe that our focus on *enhanced* contextual resources at work offers new avenues for occupational life-span models of work design by the generation of contextual resources that support SOC at work instead of the individual attribution of successful aging at work.

Finally, we observed longitudinal effects that SOC in nursing—especially optimization—may predict the well-being of employees 6 months later. Nevertheless, the observed longitudinal effects are weaker than those reported from Wiese et al. (2002), who so far conducted the only longitudinal study on SOC at work we are aware of. There are several explanations for this difference: Our study used a shorter time period than the Wiese et al. (2002) study (6 months vs. 3 years). Thus, our results indicate that SOC may require a longer time period than 6 months to unfold its effects. Another reason might be that the study of Wiese and colleagues used a much younger sample: Our results point to dynamic effects of SOC in nursing on well-being especially for middle-aged employees but not for older employees.

The study's results should also be interpreted in the light of several limitations: First, the data are based on a convenience sample, i.e. there was no random selection of participants but the sample was drawn from nurses that were available for this study. This sampling procedure limits the generalizability of results. However, the sample equals the typical German workforce in nursing regarding age, gender structure and job-tenure (Isfort et al. 2010). Thus, it seems plausible that the findings will be replicable in other nursing settings.

Second, our design does not allow inferences about underlying causes of the age differences. Consequently, the validity of the results may be limited due to cohort or generation effects (Smola and Sutton 2002). Potential selection bias may also have been occurred, such as that nurses with impaired work ability retiring prematurely and therefore only a healthier population took part in the study ("healthy worker effect"). Thus, future studies should apply longitudinal designs with longer time lags and multiple measurements to overcome these problems.

Third, our cross-sectional findings do not allow conclusions about causal determination between the variables under study. Effects might also be valid in the opposite direction. In surveys investigating two- or three-way interactions of SOC, age, job factors, and individual outcomes it is however very unlikely that opposite determination patterns occur (Zacher and Frese 2011).

Fourth, even though parts of this study aimed at overcoming the common single-source methods in SOC research, most of the data are based on self-report. Common method bias might lead to spurious results due to inflated correlations between study variables (Podsakoff et al. 2003). However, altogether there seems only limited empirical evidence that a common method itself produces systematic variance that inflates associations to a significant degree (Spector 2006). Due to reliance on self-report in the description of job control also the objectivity of the information might be questioned. However, a validation study on the instrument used for describing the work situation in the hospital environment has corroborated that work characteristics can be effectively measured by self-reports as there is high convergence between job incumbents' self-ratings and external observations (Büssing and Glaser 2002). Nevertheless, it is desirable to replicate the present results by applying further multi-method designs, such as using for instance job descriptions to assess job control, observational data to assess SOC in nursing or conducting standardized medical examinations of work ability.

Outlook

We think that our findings can substantially contribute to the successful development of measures that promote the health of nurses and their capabilities to perform daily care work. Generally, we suggest that nurses should be acquainted and trained with the SOC model. Potential trainings should for instance clarify important job related goals of nurses (selection), develop strategies and discuss supportive contextual resources to optimize goal striving. Moreover, internal and contextual factors that hamper goal striving should be clarified and compensation strategies should be developed. Referring to our example of an older female nurse who is working in the operation theatre suffering from backache: the SOC training would focus on the development of an individual strategy that helps to counterbalance the physical demand of "standing in the operating room", like organizing the work in such a way that the nurse can change his or her work position during an operation. The SOC training would further enhance the communication of the developed compensation strategy to the nurses' supervisor and accompany the implementation over a certain time period. In addition, it can be suggested to supplement such behavior-oriented SOC interventions with interventions aimed at working conditions. On the one hand, our results on research question 3 suggest that working conditions—here job control—as well as SOC each may have a unique effect on work ability. Consequently, SOC-oriented interventions and interventions targeted at working conditions should each have a substantial positive effect on work ability. To stay with the example above, reducing the physical demand "standing in the operating room" should have a singular positive effect on the work ability besides the effect of better coping with these physical demands by SOC. On the other hand, we also found evidence for synergistic effects of SOC and working conditions especially for older nurses. Thus SOC trainings can also supplement the effectiveness of job design interventions and vice versa.

Finally, future research should address the relation of SOC in nursing to performance measures, like quality of care or patient safety: Research in other professions indicates that SOC in general is associated with better performance (Bajor and Baltes 2003; Yeung and Fung 2009). However, also potential negative side effects of SOC in nursing on performance are thinkable. For example, nurses who apply selection behavior may intensify certain nursing activities, while at the same time disregarding other nursing activities. The latter might impair quality of care or patient safety. Moreover, we assume that SOC makes high demands on teamwork or team climate. For instance, selection behavior like "delegating tasks to colleagues" or compensation behavior like "asking for help" might be associated with a higher workload of colleagues. It is plausible that the latter might impair the collaboration and climate on wards. Therefore, negative consequences of SOC in nursing for teamwork cannot be excluded. Future research on age diversity in teams and group effectiveness (see the chapter Age Diversity and Team Effectiveness, Ries et al.) may further develop the effective organization of SOC on the team level.

Acknowledgments This research was supported by the German Research Foundation (Deutsche Forschungsgemeinschaft, DFG): MU 3079/1-1. The project is part of the DFG priority program 1184 "Age-differentiated Work Systems". The authors would like to thank Helga Asenbauer, Martina Bürger, Ines Englmann, Dana Lippert, and Birgit Müller for helping in the conduction of this study and Kristel Vanessa Barbosa for her advice on the English language.

References

Abele AE, Wiese BS (2008) The nomological network of self-management strategies and career success. J Occup Organ Psychol 81:733–749

Abraham JD, Hansson RO (1995) Successful aging at work: an applied study of selection, optimization, and compensation through impression management. J Gerontol B Psychol Sci Soc Sci 50:94–103

Afentakis A, Maier T (2010) Projektionen des Personalbedarfs und -angebots in Pflegeberufen bis 2025, Statistisches Bundesamt Wirtschaft und Statistik

Ahlstrom L, Grimby-Ekman A, Hagberg M, Dellve L (2010) The work ability index and single-item question: associations with sick leave, symptoms, and health—a prospective study of women on long-term sick leave. Scand J Work Environ Health 36:404–412
Bajor JK, Baltes BB (2003) The relationship between selection optimization with compensation, conscientiousness, motivation, and performance. J Vocat Behav 63:347–367
Bakker AB, Demerouti E (2006) The job demands-resources model: state of the art. J Manag Psychol 22:309–328
Baltes PB (1999) Alter und Altern als unvollendete Architektur der Humanontogenese. Z. Gerontol Geriatr 32:433–448
Baltes PB, Baltes MM (1990) Psychological perspectives on successful aging: the model of selective optimization with compensation. In: Baltes PB, Baltes MM (eds) Successful aging: perspectives from the behavioral sciences. Cambridge University Press, New York, pp 1–34
Baltes PB, Baltes MM, Freund AM, Lang FR (1999) The measure of selection, optimization, and compensation (SOC) by self-report. Max Planck Institute for Human Development, Berlin
Baltes BB, Dickson MW (2001) Using life-span models in industrial-organizational psychology: the theory of selective optimization with compensation. Appl Dev Sci 5:51–62
Baltes BB, Heydens-Gahir HA (2003) Reduction of work-family conflict through the use of selection, optimization, and compensation behaviors. J Appl Psychol 88:1005–1018
Baltes MM, Lang FR (1997) Everyday functioning and successful aging: the impact of resources. Psychol Aging 12:433–443
Büssing A, Glaser J (2002) Das Tätigkeits- und Arbeitsanalyseverfahren für das Krankenhaus—Selbstbeobachtungsversion (TAA-KH-S) [Work analysis instrument for hospitals—Self report version (TAA-KH-S)], Hogrefe, Göttingen
Camerino D, Conway PM, Van der Heijden BIJM, Estryn-Behar M, Consonni D, Gould D, Hasselhorn HM, Grp NS (2006) Low-perceived work ability, ageing and intention to leave nursing: a comparison among 10 European countries. J Adv Nurs 56:542–552
Engels JA, van der Gulden JW, Senden TF, van't Hof B (1996) Work related risk factors for musculoskeletal complaints in the nursing profession: results of a questionnaire survey. Occup Environ Med 53:636–641
Frese M, Zapf D (1994) Action as the core of work psychology: a German approach. In: Dunnette MD et al (eds) Handbook of industrial and organizational psychology. Consulting Psychologists Press, Palo Alto, pp 271–340
Freund AM (2006) Age-differential motivational consequences of optimization versus compensation focus in younger and older adults. Psychol Aging 21:240–252
Freund AM, Baltes PB (2002) Life-management strategies of selection, optimization and compensation: measurement by self-report and construct validity. J Pers Soc Psychol 82:642–662
Hacker W (2003) Action regulation theory—a practical tool for the design of modern work processes? Eur J Work Organ Psychol 12:105–130
Hackman JR, Oldham GR (1976) Motivation through the design of work: test of a theory. Organ Behav Hum Perform 16:250–279
Hasselhorn H-M, Freude G (2007) Der work ability index (The work ability index). Wirtschaftsverlag, NW, Bremerhaven
Heiden B, Weigl M, Angerer P, Müller A (2011) Association of age and physical work stressors with musculoskeletal disorders in nurses. Manuscript submitted for publication
Hobfoll SE, Wells JD (1998) Conservation of resources, stress, and aging: why do some slide and some spring? In: Lomranz J (ed) Handbook of aging and mental health: an integrative approach. Plenum Press, New York, pp 121–134
Humphrey SE, Nahrgang JD, Morgeson FP (2007) Integrating motivational, social, and contextual work design features: a meta-analytic of the summary and theoretical extension work design literature. J Appl Psychol 92:1332–1356
Ilmarinen J (2007) The work ability index (WAI). Occup Med-Oxford 57:160
Ilmarinen J (2009) Work ability-a comprehensive concept for occupational health research and prevention. Scand J Work Environ Health 35:1–5

Isfort M, Weidner F, et al (2010) Pflege-Thermometer 2009. Eine bundesweite Befragung von Pflegekräften zur Situation der Pflege und Patientenversorgung im Krankenhaus. (Nursing-Thermometer 2009. A nationwide survey of nuses to to the situation of care and patient care in hospitals), German Institute of Applied Nursing Research, Köln
Kanfer R, Ackerman PL (2004) Aging, adult development, and work motivation. Acad Manag Rev 29:440–458
Leino-Arjas P, Kaila-Kangas L, Notkola V, Keskimäki I, Mutanen P (2002) Inpatient hospital care for back disorders in relation to industry and occupation in Finland. Scand J Work Environ Health 28:304–313
Miles MM, Huberman AM (1984) Qualitative data analysis: a sourcebook of new methods. Sage, Newbury Park
Mohr G, Müller A, Rigotti T, Aycan Z, Tschan F (2006) The assessment of psychological strain in work contexts: concerning the structural equivalency of nine language adaptations of the irritation-scale. Eur J Psychol Assess 22:198–206
Mohr G, Rigotti T, Müller A (2007) Irritations-Skala zur Erfassung arbeitsbezogener Beanspruchungsfolgen. Hogrefe, Göttigen
Müller A, Weigl M, Heiden B, Herbig B, Glaser J, Angerer P (in press) Selection, optimization, and compensation in nursing: exploration of job-specific strategies, scale development, and age-specific associations to work ability. J Adv Nurs
Müller A, Weigl M, Lampert B, Glaser J, Angerer P (2010) Alterns- und altersspezifische Effekte von körperlichen Belastungen und Tätigkeitsspielraum auf die Gesundheit von Altenpflegekräften (Age-specific health-effects of physical stressors and autonomy in elderly care). Z. Arbeitswissenschaft 64:229–238
Müller A, Heiden B, Weigl M. (2011) Maintaining the work ability in nurses with musculoskeletal impairments: the moderating role of selection, optimization, compensation. EAWOP Small Group Meeting Age Cohorts in the Workplace, Rovereto, Italy
Müller A, Weigl M, Heiden B, Glaser J, Angerer P (2012) Promoting work ability and well-being in hospital nursing: the interplay of age, job control, and successful ageing strategies. WORK: J Prev Assess Rehabil 41:5137–5144
Ng TWH, Feldman DC (2008) The relationship of age to ten dimensions of job performance. J Appl Psychol 93:392–423
Podsakoff PM, MacKenzie SB, Lee JY, Podsakoff NP (2003) Common method biases in behavioral research: a critical review of the literature and recommended remedies. J Appl Psychol 88:879–903
Riediger M, Li S-C, Lindenberger U (2006) Selection, optimization, and compensation as developmental mechanisms of adaptive resource allocation: review and preview. In: Birren JE, Schaie KW (eds) Handbook of the psychology of aging, 6th edn. Elsevier, Amsterdam, pp 289–314
Schalk R, van Veldhoven M, de Lange AH et al (2010) Moving European research on work and ageing forward: overview and agenda. Eur J Work Organ Psychol 19:76–101
Schmitt A, Zacher H, Frese M The buffering effect of selection, optimization, and compensation strategy use on the relationship between problem solving demands and occupational well-being: a daily diary study. J Occup Health Psychol (in press)
Simoens S, Villeneuve M, Hurst J (2005) Tackling nurse shortages in OECD Countries. OECD, Directorate for Employment, Labour and Social Affairs
Smola KW, Sutton CD (2002) Generational differences: revisiting generational work values for the new millennium. J Organ Behav 23:363–382
Spallek M, Kuhn W, Schwarze S, Hartmann B (2007) Occupational medical prophylaxis for the musculoskeletal system: a function-oriented system for physical examination of the locomotor system in occupational medicine. J Occup Med Toxicol 2:12
Spector E (2006) Method variance in organizational research. Truth or urban legend? Organ Res Methods 9:221–232

Trinkoff AM, Lipscomb JA, Geiger-Brown J, Storr CL, Brady BA (2003) Perceived physical demands and reported musculoskeletal problems in registered nurses. Am J Prev Med 24:270–275

Tuomi K, Ilmarinen J, Seitsamo J, Huuhtanen P, Martikainen R, Nygard CH, Klockars M (1997) Summary of the Finnish research project (1981–1992) to promote the health and work ability of aging workers. Scand J Work Environ Health 23(Suppl 1):66–71

van den Berg TIJ, Elders LAM, de Zwart BCH, Burdorf A (2009) The effects of work-related and individual factors on the work ability index: a systematic review. Occup Environ Med 66:211–220

Weigl M, Müller A, Hornung S, Zacher H, Angerer P The moderating effects of job control and selection, optimization, and compensation strategies on the age–work ability relationship. J Organ Behav (in press)

Weigl M, Hornung S, Parker SK, Petru R, Glaser J, Angerer P (2010) Work engagement and accumulation of task, social, personal resources: a three-wave structural equation model. J Vocat Behav 77:140–153

Wiese BS, Freund AM, Baltes PB (2000) Selection, optimization, and compensation: an action-related approach to work and partnership. J Vocat Behav 57:273

Wiese BS, Freund AM, Baltes PB (2002) Subjective career success and emotional well-being: longitudinal predictive power of selection, optimization and compensation. J Vocat Behav 60:321–335

Yeung DY, Fung HH (2009) Aging and work: how do SOC strategies contribute to job performance across adulthood? Psychol Aging 24:927–940

Young LM, Baltes BB, Pratt AK (2007) Using selection, optimization, and compensation to reduce job/family stressors: effective when it matters. J Bus Psychol 21:511–539

Zacher H, Frese M (2011) Maintaining a focus on opportunities at work: the interplay between age, job complexity, and the use of selection, optimization, and compensation strategies. J Organ Behav 32:291–318

Zapf D, Dormann C, Frese M (1996) Longitudinal studies in organizational stress research: a review of the literature with reference to methodological issues. J Occup Health Psychol 1:145–169

Assembly Tasks in the Automotive Industry: A Challenge for Older Employees

Ekkehart Frieling, Daniel Kotzab, José Alonso Enríquez Díaz and Alina Sytch

Lessons Learned

It appears that the demographic problems reported in the Daimler-Motorengesellschaft in Stuttgart-Untertürkheim 100 years ago (Schumann 1911/2011) can still not be considered solved by empirical social research today. Schumann writes,

> Peak workability is reached at the age of 35. By the age of 45, workers are already having trouble finding employment. Short sickness leaves occur frequently. The influence of piece-work tasks on mental and physical strain during work is distinctive (p. 146f).

Growing physical complaints among older employees, the considerable increase in standardised processes, the partial decrease in task diversity and autonomy, and short cycle times all make it difficult for employees in assembly lines to remain healthy and active until they reach the legal retirement age. If this socio-politically meaningful goal is to be pursued in terms of age-based working conditions in automotive assembly, the following measures, which are based on the findings of the given project, should be taken into account when planning and implementing interventions:

- Ergonomic design of work equipment (standing/sitting workplaces, height-adjustable assembling tables and platforms, assisting systems for lifting and

E. Frieling (✉) · D. Kotzab · J. A. Enríquez Díaz · A. Sytch
Kassel University, Kassel, Germany
e-mail: Ekkehart.frieling@t-online.de

D. Kotzab
e-mail: kotzab@uni-kassel.de

J. A. Enríquez Díaz
e-mail: alonso.enriquez@volkswagen.de

A. Sytch
e-mail: a.sytch@uni-kassel.de

C. M. Schlick et al. (eds.), *Age-Differentiated Work Systems*,
DOI: 10.1007/978-3-642-35057-3_9,

carrying weights, etc.) and surrounding conditions (noise, lighting, temperature, etc).

- Systematic alternation of physical strain through the combination of value-adding tasks (for example, assembling by hand) and non-value-adding tasks (material provision, maintenance, core figures, etc).
- Consideration of each individual's abilities during competence development, work design and work organisation.
- Longer, non-cycled assembly processes and individually adaptable work speed and work rhythm.
- Participation-oriented design of work conditions and work processes.
- Development of a human resource management strategy that is person-oriented while still taking into account standardised processes.

By incorporating these criteria during the development of sustainable assembly concepts and systems and adopting a person-oriented view of the manufacturing processes, we can preclude the simple continuance of integrated production systems and assure employment of older persons.

Older Employees in the Automotive Industry: Conceptual Framework and Derivation of Central Research Questions

The automotive industry is a key industry in Germany and one that faces high competitive pressure at the international level. It employs approximately 400,000 people, with another 300,000 persons working in the automotive supply industry (see VDA 2011). A substantial percentage of these persons are involved in assembly tasks.

Between 5.4 and 6 million cars were assembled in Germany between 2006 and 2011. The entire world market produced approximately 64 million cars in 2010. The greatest growth in the automotive industry can be noted in Brazil, Russia, India and China (BRIC states). The production capacities of all automotive-producing countries combined are estimated at roughly 100 million vehicles per year (Wimmer et al. 2010).

The number of employed persons in the German automotive industry has only shown slight growth. Take VW as an example: Between the years 2006 and 2010, the number of VW employees in Germany rose from 174,000 to 178,000, while employee numbers at foreign production sites increased from 155,000 to 210,000. This trend developed further in 2011, providing evidence that the plants abroad employ a great deal more people than the German plants.

During the peak of the economic crisis in 2008 and 2009, the German automotive industry attempted to prevent lay-offs by granting older employees (55 years and up) tax concessions in conjunction with partial retirement models. For instance, 5.9 % (2008) and 4.5 % (2009) of BMW's German workforce opted for partial retirement in 2008 and 2009 (see BMW Annual Report 2011).

The average age of permanent employees in the German automotive industry varies between 40 and 48 years. Most of them are male (women make up only about 10–14 % of the workforce). The number of temporary staff and contract workers is rising, as is the number of employees who are retained by means of work contracts. In some assembling domains as much as 30 % of the workers are on a temporary contract, and in exceptional cases the percentage is even higher. In company agreements (between the management and the work council), quotations can be found in which temporary and contract workers make up 3–8 % of the overall workforce.

In addition to these economic conditions, the German and the international automotive and components supply industry has been copying and adapting the Toyota production system over the past 10 years. According to Jürgens (2007), most of the basic principles, instruments and practical routines of this integrated production system have already been introduced into German automotive plants. As far as Jürgens is concerned, Kaizen, Poka-yoke, Andon and many other Japanese terms have found their way into the operative business. However, such profound changes to the customary work structure can cause irritation among employees. Moreover, Jürgens (2007) sees a risk of succumbing to "management by stress" (2007, p. 45 and Conti et al. 2006).

The avoidance of waste [over production, buffering, needless movements, unnecessary walking, searching for parts, etc., see Ohno (2009)] is the paramount goal of value-oriented production systems. It is best achieved using the Toyota production system and can be found at nearly all German automotive manufacturers (cf. Neuhaus 2010, p. 80 ff). Some of the tools in this system are 5S orderliness and cleanliness, just-in-time logistics, total quality control, visual management, continuous improvement process, standardisation, total productivity maintenance (TPM), teamwork, process management, auditing, key figures, etc.

The implementation of these new production structures creates numerous risks in the assembly and fabrication work of the automotive industry and its components suppliers [see a more detailed report by Landau (2011)]. These include:

- Repetitive tasks with high manual operation frequencies, such as cycle times of 1 min or less.
- Enforced body postures, such as assembling inside of the vehicle interior.
- Work with increased energy/strength expenditure, such as when padding car seats.
- Utilisation of hands and arms as a tool (knocking, hammering, turning, pushing, such as when aligning parts or during clipping).
- Energy-impact/force-impact while attending to work appliances/equipment (such as screwdriver, riveting machine).
- Manual load handling (such as lifting and carrying pieces exceeding 10 kg).

These risks cause greater physical strain for older employees. That said, such working conditions ought to receive critical attention even among the younger workforce.

Employee surveys are a common phenomenon in the German automotive industry. However, their evaluation is treated as an internal matter. Systematic comparisons between individual automotive manufacturers concerning the effects of working conditions on employees engaged in assembling and fabrication are lacking, despite the fact that such data is of great importance when dealing with the impact of demographic change. The automotive industry exerts considerable influence on the work organisation of its components suppliers, so its work structures often serve as a standard for the components suppliers depending on this industry.

The present comparative study looks at two automotive manufacturers and their working conditions for components assembly (plant A) and final assembly (plant B). The object of the longitudinal study is to identify how employee attitudes toward work (tasks) differs in terms of company and point of assessment over a period of 6 years. Furthermore, their psychological and physical wellbeing is assessed in conjunction with the individual tasks carried out. How do the various work structures in the two companies affect employees during the project period? To what extent do workability and the selected psycho-physical performance parameters of employees change over these 6 years and what impact do the ergonomic measures adopted have on the employees?

Description of the Area of Research and Data Collection

Data was collected at two German automotive manufacturers (plant A and B) between 2005 and 2011 using a longitudinal study design (see Fig. 1). The study focused on aggregate assembly in plant A and final (vehicle) assembly in plant B. In plant A, various equipment assembling is considered: A1 gearbox assembly of a specific automatic gearbox, A2 handbrake lever assembly and A3 exhaust system assembly. Meanwhile, in plant B the assembly of the wiring harness is examined (see Table 1).

Simplified, the component assembly (in A1) can be described as a semi-automated assembly task (degree of automation approximately 50 %) on an assembly line. This means the employees take components (from a container), place them on

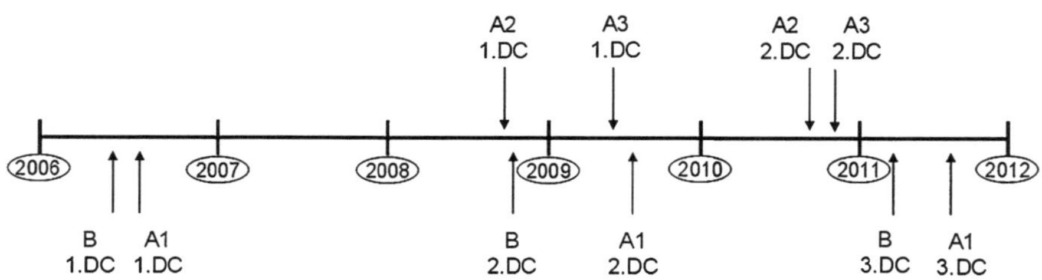

Fig. 1 Time schedule of the data collection (DC)

Table 1 Average age of persons in sample groups (in years); sample size (N)

	Cross sectional data			Longitudinal data
Data collection	1.DC	2.DC	3.DC	1.DC
Plant A1	38.2	38.8	41.3	37.5
Gearbox assembly	(N = 185)	(N = 207)	(N = 65)	(N = 19)
Plant B	38.1	40.4	40.7	36.5
Wiring harness assembly	(N = 242)	(N = 155)	(N = 110)	(N = 80)
Plant A2	32.1	36.7	–	34.4
Handbrake lever assembly	(N = 17)	(N = 28)		(N = 9)
Plant A3	41.9	42.6	–	40.3
Exhaust system assembly	(N = 22)	(N = 35)		(N = 12)

Note In the longitudinal design, the average age of the persons in sample groups increases by number of years passed between the points of 1.DC, 2.DC and 3.DC

the assembly rack or corresponding appliance and activate the process with an automatic assembling machine. The entire assembling process consists of approximately 50 cycles. The cycle time varied between 85 and 72 s from the first to the third point of data collection.

In A2, handbrake levers are installed by means of an automatic assembling machine. The entire assembling process consists of two cycles, each with a duration of roughly 25 s. The assembling process is based on the Chaku–Chaku–principle (see Kono 2004; Yagyu 2007): the employee sequentially operates two or three workstations by placing parts, activating the machining process, removing the component and placing it in another machine. In A3 exhaust systems are welded together, again according to the Chaku-Chaku principle. The cycle time here is around 50 s.

Table 2 displays the essential data characterising the three samples. While A1 features three points of data collection, the Chaku-Chaku assembling units (A2 and A3) report only two points of data collection since this work-system was only introduced between 2008 and 2009. In the following presentation of results, the focus will lie primarily on assembly A1, as this unit will be compared to plant B. A2 and A3 will only be considered in brief examples here [for a complete record of findings concerning the comparison of A1, A2 and A3 please refer to Enríquez Diaz(2012)].

The participants in plant B are active at the first 80 workstations in final vehicle assembly where they fit customer-specific wiring harnesses and install some

Table 2 Average cycle times [s] during the period of data collection in each sample group

	1.DC	2.DC	3.DC
Plant A1	85	74	74
Plant A2	25*	25	
Plant A3	50	49	
Plant B	72	69	64

* Indicates a theoretical numerical value

additionally components for the vehicle interior. The work setting is an assembly line, more precisely a walk-on conveyor belt on which the employees ride alongside the vehicle during assembling, with height-adjustable assembling platforms. The cycle time varied between 72 and 64 s over the data collection period. It should be noted here that cycle times can be marginally adapted to meet specified quantities.

The employees have been engaged in the four different work systems for an average of 4–18 years (see Table 3). The relatively long periods of employment within one work system derogates the flexible placement in other work systems but ensures a rather high degree of routine when accomplishing given tasks.

A substantial amount of organisational effort was required to follow up on the participating employees during the longitudinal survey period. In this process, it must be guaranteed that participants are distinctly yet anonymously assigned to their individual sets of previously collected data. The chosen procedure was approved by the relevant departments and stakeholders at the two respective companies including the commissioner for data protection, the workers council, the occupational health and safety division and the human resources department.

The employees, who participated in the survey voluntarily,answered the questionnaires during the regular working time. During this time, their tasks were taken over by a so-called swing-man (German: Springer), a stand-by man. The accrued costs (the result of inactive working time) were covered by the companies. The costs incurred during the second data collection in plant B were covered by the project sponsor, the German Research Foundation (DFG), since the financial situation of the company was very tense at the time due to the market downturn in 2008.

As Table 1 shows, the sample size diminishes in both plants over the course of the longitudinal study: In plant A1 a drastic decline of participants can be observed between the data collections, from 185 at 1.DC to 19 at 3.DC, while plant B shows a decline from 242 at 1.DC to 80 participants at 3.DC. The dwindling of participants can be attributed to in-plant fluctuation and the changing business situation of the individual manufacturing components.

To assess the economic effect and the impact of the respective work organisation on the employees in terms of an inter-company comparison, the surveys of the second and third data collection were extended to include a larger area of focus. This means that the cross-sectional samples are considerably larger in number (compare Table 1), enabling a comparison between the two plants (A1 and B).

Table 3 Average period of employment (in years) in the analysed work system at the time of data collection

	Cross section			Longitudinal section
Data collection	1.DC	2.DC	3.DC	1.DC
Plant A1	10.8	10.6	8.2	9.8
Plant B	11.9	14.4	15.9	11.9
Plant A2	4.4	4.1	–	4.8
Plant A3	14.8	8.3	–	18.4

Methods and Procedures

Standardised observational methods and interview methods were used to allow for comparative recording of employee strain and stress throughout the study period of 6 years. The two observational methods used (NIOSH and OWAS) require video recording for data analysis. The remaining data-collecting techniques are survey-based. Conversations with the management, work councils and employees are also considered. Using group discussions, interim results from surveys and workplace, observations were reviewed with employees in plant A. An appropriate presentation of collected data could not be arranged with plant B.

Assessment Criteria of Group Work

Through expert interviews (direct supervisors, group spokesman and manager for group activities) and the use of a checklist concerning group work (Freiboth 1998; Frieling and Sonntag 1999), the working structures are categorised according to 45 items into the following six dimensions: organisational framework, group activities, participation, group discussions/round table, qualification, and continuous improvement process (CIP).

NIOSH Method

The NIOSH method assesses load-handling by utilising various assessment criteria (such as energetic, bio-mechanical, psycho-physical and epidemiological criteria). Like the OCRA method, this technique takes criteria into account that are relevant to the specific task (e.g. load distance and lift height) (Bongwald et al. 1995). The current study uses the extended version of the NIOSH method (see Waters et al. 1993).

OWAS Method

Developed in the mid 1970s in a Finnish steel mill, the OWAS method (Ovako Working Posture Analysing System) is an approved technique in the occupational sciences for classifying and assessing/evaluating body postures, particularly in conjunction with load-handling (Ellegast 2005).

Questionnaire: Healthy Aging Through the Use of Work Design

To depict the subjective state of health of the employees, a questionnaire was compiled on the basis of existing tests. This comprehensive tool consists of more than one hundred items. On average, it takes an employee 50–60 min to fill out the entire questionnaire. A member of staff from the Institute of Industrial Psychology was present at all times during data collection to answer any questions that arose.

The following list displays the selected questionnaires that were chosen for the assessment (for further details please refer to the references):

- Questions regarding complaints of the musculoskeletal system: Nordic Questionnaire–NQ by Kuorinka et al. (1987).
- Short version of the Work Ability Index (WAI) by Nübling et al. (2004).

- Job satisfaction by Neuberger and Allerbeck (1978).
- Salutogenetic Subjective Work Analysis (Salutogenetische Subjektive Arbeitsanalyse)—SALSA by Rimann and Udris (1997).
- German short version of the Big Five Inventory—BFI-K by Rammstedt and John (2005).

Results

Based on comprehensive description of the work organisation, within the analyzed work systems we presented data regarding strain and attitude towards work.

The significance level applied to all analyses is $\alpha = 0.05$. A repeated measures analysis of variance (ANOVA) is used for the longitudinal analysis. Comparisons of groups are done using an ANOVA and post-hoc tests.

Work Organisation

Teamwork

Teamwork plays a major role in work organisation. The essential circumstances for this are formally upheld in both plants through collective agreement (plant A since 2007, plant B since 1995). A standardised checklist (Freiboth 1998; Frieling and Sonntag 1999) is used to ensure an objective evaluation of teamwork in the plants (Fig. 2).

The significant differences in the categories "organisational framework" and "expanded team activities" are particularly distinct between plant A2 and the other plants, where plant A2 achieves the higher percentile score. Of note are the low scores in plant B for the dimensions "participation", "qualification", "team discussions" and "CIP activities". These poor grades in plant B are ascribed to infrequent discussions within the group. Whilst in plant A group discussions take place once a week, in plant B this occurs only every 2 months, or when the conveyor belt stands still for longer periods of time, for example due to a technical malfunction (cf. Sytch et al. 2011). The number of participants attending such group discussions in plant B usually exceeds 50 persons (two groups). A comparison of the span of control at team level in plant A and B demonstrates that teamwork is barely practical in plant B due to the working structure adopted there.

Span of Control

The size of the span of control determines how present and approachable the direct supervisor is for his employees. Personal and close contact with the employees within an integral production system is necessary for ongoing troubleshooting, process optimisation and competency development. Table 4 shows a current distribution of leaders, starting from the team spokesperson to the foreman level.

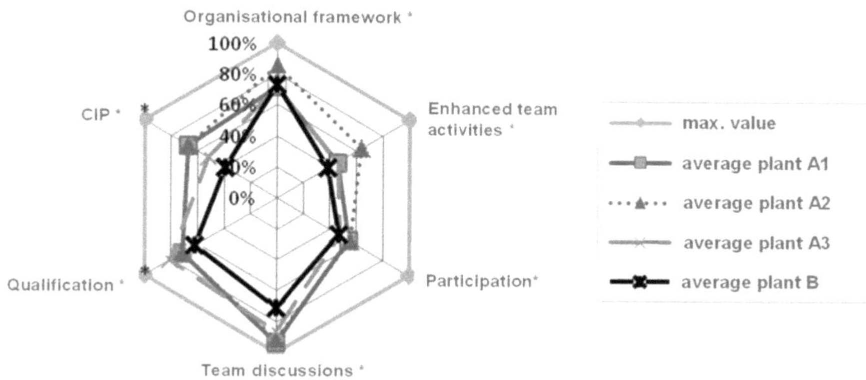

Fig. 2 Graphical comparison of the categories in the examined assembly units (mean normalized values) *The difference of the mean scores is significant at level 0.05

Table 4 Span of control in the four sample groups

	Plant A1	Plant A2	Plant A3	Plant B
Foremen total	3	1	3	6
Employees per foreman	About 30	About 18	About 25	About 50
Team spokesperson total	9	3	9	6
Employee per team spokesperson	About 10	About 6	About 7–8	About 25

It is striking that in plant B approximately 50 employees are assigned to one foreman per shift and 25 employees to one team spokesperson. During the study period (2006–2011) the foremen's areas were downsized from five to three divisions. Depending on the shift, two foremen are available for one area. The management in plant B very consciously weakens the role of the participatory team spokesperson in their newly implemented work structure (2011/12). Instead, the team spokesperson is replaced by a "deputy-foreman", who is responsible for approximately ten employees.

Shift Systems and Cycle Times

During the period of the three data collections, the employees in plant B worked in a two-shift system (early and late shift alternating on a weekly basis). The early shift begins at 5 a.m. The late shift begins at 1:30 p.m. and ends at 10:00 p.m.

The employees engaged in production in plant A mostly work in a backward-rotating shift system of early shifts, late shifts and night shifts. The early shift starts at 6:30 a.m., the late shift starts at 2:30 p.m., and the night shift starts at 10:30 p.m. However, this is only one of many different working time models found in plant A.

From an industrial psychology perspective, the shift system in plant B is particularly problematic, as the employees need to get up between 3 and 4 a.m. to get to work on time (see also Chap. Development and Evaluation of Working-Time

Models for the Ageing Workforce. Lessons Learned from the KRONOS Research Project, Knauth et al.).

Among the assessed assembling tasks, cycle times play a crucial role as they determine workflow, high repetition rates and standardised operations. When examining the four different assembling systems, one sees that in three of the four assembling units a reduction of the cycle time occurred during the study period, leading to a substantial productivity increase (see Table 2). One exception is the hand brake lever assembly. In this working system no exact cycle time could be determined, since the assembling system was in its start-up phase at the point of the first data collection and the average cycle times were still subject to great variation. Upon the implementation of the system, the cycle time played a subordinate role. Due to frequently occurring faults in the new system, troubleshooting, maintenance and repairs had to be optimised first. Therefore, only a theoretical value of 25 s, when everything runs smoothly, can be given as cycle time (Table 5).

The cycle time reduction brings about optimised performance and higher productivity. This is intended to avoid non-value adding tasks, such as walking, searching for material or preparation of components. This leads to a one-sided workload and the avoidance of hidden breaks (Kotzab et al. 2011).

Ergonomic Evaluation of the Work Systems

The following section focuses on body postures and handling of loads, as these are associated with the physical complaints we examined. For detailed data concerning the ergonomic analysis, see Enríquez Díaz (2012).

Loads (weights) have a greater impact on physical strain for equipment assembly than for wiring harness assembly. Consequently, the results of the gearbox assembly (A1) are displayed below. Overstepping of the critical value of 1.0 is interpreted in terms of a strain risk according to the NIOSH method. During the course of the study a slight risk decrease was observed, which can be attributed to the increased use of lifting aids and the introduction of ergonomic improvement measures (e.g. the reduction of awkward body postures, as well as a modification of the start and end position of the manipulated load weight) (Fig. 3).

The contributing effect of the lifting aids is revealed in the categorisation of the three workstations in plant B (Fig. 4) at which the work is performed with and without lifting aids. The analysis shows that the utilisation of ergonomically designed lifting aids is a prerequisite for their acceptance (among the employees) and that their implementation leads to a considerable lessening of physical strain risks.

The longitudinal study conducted in plant B using the OWAS method (Fig. 5) reveals that the strains related to working posture show a downward trend (for example, in the back, leg and head region) while a rise in strain can only be observed for the arm region. The reasons for this increase are the larger number of

Table 5 Shift times and break times of plant B and plant A1, A2 and A3

Shift	Begin	Break 1	Break 2	Break 3	End
Plant B					
Early	5:00 a.m.	7:55 a.m. (15 min)	10:45 a.m. (30 min)		1:30 p.m.
Late	1:30 p.m.	5:30 p.m. (30 min)	8:15 p.m. (15 min)		10:00 p.m.
Plant A1, A2 and A3					
Early	6:30 a.m.	8:30 a.m. (16 min)	10:30 a.m. (30 min)	12:30 p.m. (16 min)	2:30 p.m.
Late	2:30 p.m.	4:30 p.m. (16 min)	6:30 p.m. (30 min)	8:30 p.m. (16 min)	10:30 p.m.
Night	10:30 p.m.	12:30 a.m. (16 min)	2:30 a.m. (30 min)	4:30 a.m. (16 min)	6:30 a.m.

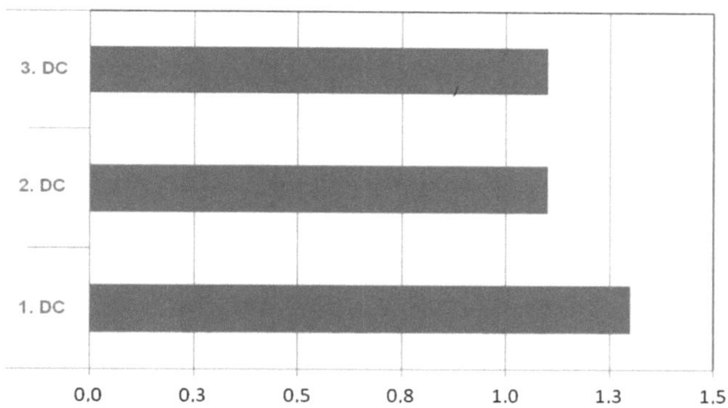

Fig. 3 Presentation of the combined lifting index (CLI) according to the NIOSH method for the analysed work places in plant A1 (N = 5 work stations)

assembling tasks that need to be done in the vehicle interior (utilisation of riveting machines) and the accessibility of the mounting site in the rear lid assembly.

The OWAS-rated working postures in workstation (8) vehicle interior assembly top, (7) vehicle interior assembly bottom and (6) exterior assembly (boot, engine compartment) are particularly problematic, as shown in Fig. 6. Due to the increasing complexity of the wiring harness and the interior design, new tasks emerge in the vehicle interior that must be performed on top of cycle time reduction and without compensatory movements.

As the evaluation of the survey data (see section Physical Complaints) demonstrates, the implemented ergonomic aids cannot significantly contribute to the reduction of physical complaints caused by awkward working postures; on the contrary, the physical complaints increase. This is primarily caused by cycle time reduction and the loss of compensatory movements and hidden micro-breaks. One of the production planners' targets is to raise the employees' degree of utilization

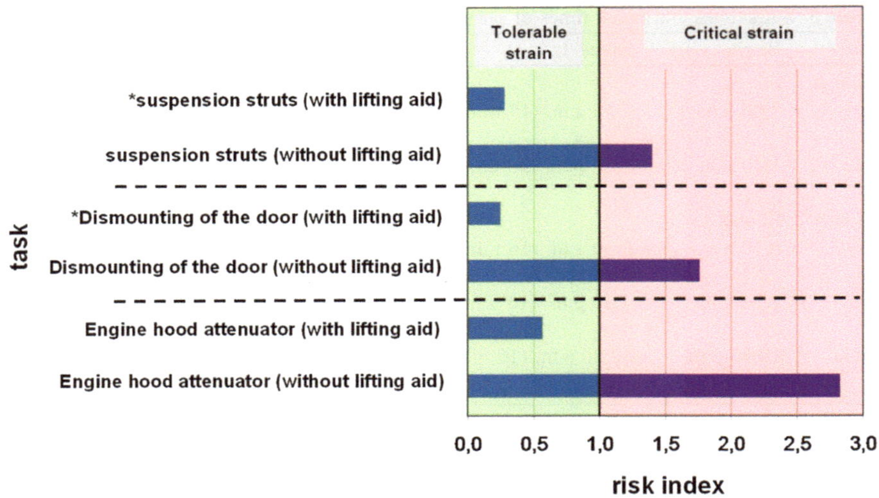

Fig. 4 NIOSH rating of assembly tasks with and without lifting aids at three selected work stations (plant B). *Note* The use of lifting aids reduces the risk index (NIOSH 91) by roughly 80 % and moves it from a red risk classification (critical) to a green risk classification

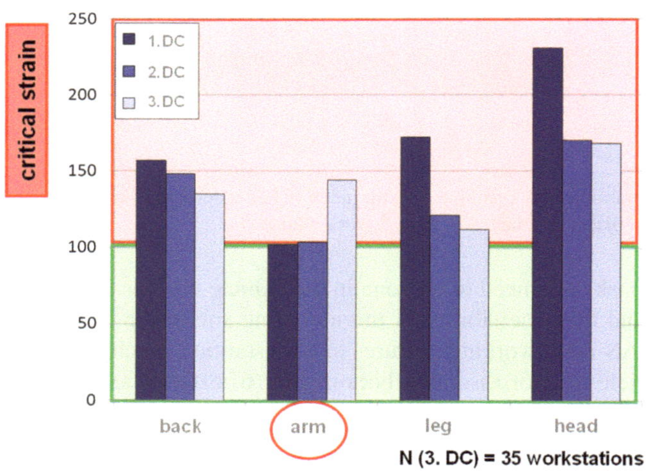

Fig. 5 OWAS rating of selected work places in plant B. *Note* Range 100–400 is (physically) stressful

to 95 % or higher by avoiding wastage. This goal increases physical strain, particularly among older employees.

In the Chaku–Chaku work systems (plant A2 and A3), the working postures and the required lifting and carrying of loads are inconspicuous (see also Enríquez

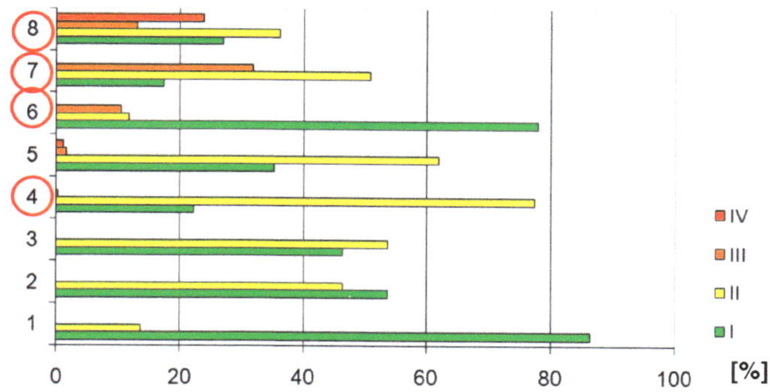

Assembling points on the vehicle	
1. Assembling on the outside, while the employee is outside the vehicle	5. Assembling on the interior (lower section), while the employee is outside the vehicle
2. Assembling under the bonnet, while the employee is outside the vehicle	6. Assembling on the interior (lower section), while the employee is outside the vehicle
3. Assembling under the tailgate, while the employee is outside the vehicle	7. Assembling on the interior (on the undertray and the lower section), while the employee is inside the vehicle
4. Assembling on the interior (on the undertray), while the employee is outside the vehicle	8. Assembling on the interior (upper section), while the employee is inside the vehicle

Fig. 6 OWAS categorisation of measures for assembling positions on the vehicle in plant B

Diaz 2012). The short cycle times (25 and 50 s respectively) and the avoidance of compensatory movements (wastage) cause the distinct physical complaints found in these systems (see below).

Physical Complaints

Below is a list of six body regions, recorded using the Nordic Questionnaire, and related physical complaints, both cross-sectional and longitudinal. Where available, the comparative data of the representative BiBB/BAuA study (Bundesanstalt 2005/2006) conducted among 20,000 employees in 2005/2006 is provided. The percentage of persons with the respective physical complaints is also given.

Neck

The scores relating to the neck region deteriorate significantly in plant B across the longitudinal data evaluation. This can be attributed to a large extent to the twisting of the head during the assembly in the interior. During the second data collection, 100 % of the employees of the Chaku–Chaku system in plant A2 complain about neck ache. Plant A1 on the other hand, shows a significant change that manifests itself only in the cross-section and that can be explained by the varying structure of the group (Table 6).

Shoulder

The extent of complaints concerning the shoulders is largely comparable to that of the neck. It shows a significant increase of complaints in plant B. More still, all employees engaged in the Chaku–Chaku assembly system in plant A2 report complaints in this body region. A likely explanation for this condition can be found in the short-cycled, monotonous assembling movement and the assembling platforms that are not individually adjustable in height. Plant A3 shows a relatively constant record of complaints over time (Table 7).

Wrist

The significant increase of complaints in the wrist in plant B (Table 8) makes it quite clear that the ergonomic conditions in this division have not improved. The utilisation of heavy riveting tools and their handling in narrow spaces (e.g. vehicle interior) is very strenuous in the hand and arm region; the same applies for clipping tasks. Here, thumb pressure is needed, which requires an expenditure of effort up to 130 N. A significant deterioration is observed in plant A2 and A3. These severe findings are presumably a result of the high repetition rates of the individual operations and the lessening of micro breaks.

Table 6 Longitudinal (L) and cross-sectional (CS) view of physical complaints (%)—neck pain

		1.DC	2.DC	3.DC
Plant A1	L	44.4 (N = 17)	55.6 (N = 17)	41.2 (N = 17)
	CS	38.3 (N = 175)	55.3* (N = 179)	53.2* (N = 62)
Plant B	L	53.7 (N = 67)	50.8 (N = 67)	68.7*/** (N = 67)
	CS	50.9 (N = 216)	49.3 (N = 144)	66.0 (N = 100)
Plant A2	L	88.9 (N = 9)	100.0 (N = 9)	–
	CS	62.5 (N = 16)	84.6 (N = 29)	–
Plant A3	L	45.5 (N = 11)	45.6 (N = 11)	–
	CS	61.9 (N = 21)	36.7 (N = 30)	–

Note
*Means a significant diff. to the first DC
**Means a significant diff. to the second DC; (BiBB/BAuA comparative value for persons below the age of 45, 42 %, and above the age of 45, 45 %)

Table 7 Longitudinal (L) and cross-sectional (CS) view of physical complaints (%)—Shoulder

		1.DC	2.DC	3.DC
Plant A1	L	35.3 (N = 17)	35.3 (N = 17)	41.2 (N = 17)
	CS	25.3 (N = 170)	44.6 (N = 177)	47.5 (N = 59)
Plant B	L	40.3 (N = 62)	54.8* (N = 62)	58.1* (N = 62)
	CS	44.4 (N = 216)	46.9 (N = 144)	55.7*/** (N = 100)
Plant A2	L	77.8 (N = 9)	100 (N = 9)	–
	CS	60.0 (N = 16)	89.3 (N = 26)	–
Plant A3	L	50.0 (N = 8)	50.0 (N = 8)	–
	CS	60.0 (N = 20)	50.0 (N = 28)	–

Note
*Means a significant diff. to the first DC
**Means a significant diff. to the second DC; (BiBB/BAuA comparative value for persons below the age of 45, 42 %, and above the age of 45, 45 %)

Table 8 Longitudinal (L) and cross-sectional (CS) view of physical complaints (%)—Wrist

		1.DC	2.DC	3.DC
Plant A1	L	23.5 (N = 17)	23.5 (N = 17)	17.7 (N = 17)
	CS	18.6 (N = 167)	29.5* (N = 173)	27.4 (N = 62)
Plant B	L	37.1 (N = 62)	48.4 (N = 62)	46.8 (N = 62)
	CS	38.7 (N = 204)	44.2 (N = 138)	52.4* (N = 98)
Plant A2	L	25.0 (N = 8)	50.0 (N = 8)	–
	CS	40.0 (N = 15)	55.6 (N = 27)	–
Plant A3	L	55.6 (N = 9)	55.6 (N = 9)	–
	CS	60.0 (N = 20)	41.4 (N = 29)	–

Note
*Means a significant diff. to the first DC; (BiBB/BAuA comparative value for persons below the age of 45, 19 %, and above the age of 45 years, 22 %)

The great variation between the CS group and L group in plant A1 is striking: while from a longitudinal perspective no significant change can be detected, the cross-sectional analysis reveals a distinct negative development (different group structures!).

Upper Back

The upper back shows a slight to strong negative development in all assembly units when comparing to the first data collection. It should be noted here that in plant A1 the frequency of complaints declined during the second data collection. This is probably due to ergonomic improvements that were implemented in plant A1 between the first and second data collection (see section Ergonomic Evaluation of the Work Systems). However, plant B shows a distinct trend towards increased discomfort in the upper back (Table 9).

Lower Back

The lower back analysis gives the same picture as the upper back analysis, but with an even higher incidence of complaints. The findings in plant A1 are characterised by strong fluctuations. The improvements seen here between 2.DC and 3.DC can be ascribed to the ergonomic action taken (introduction of lifting aids). The extent of complaints registered in plant B resembles that of plant A2 (Table 10).

Table 9 Longitudinal (L) and cross-sectional (CS) view of physical complaints (%)—upper back

		1.DC	2.DC	3.DC
Plant A1	L	29.41 (N = 17)	52.94* (N = 17)	35.29 (N = 17)
	CS	26.53 (N = 167)	41.04 (N = 173)	32.79 (N = 161)
Plant B	L	38.98 (N = 59)	42.37 (N = 59)	47.46 (N = 59)
	CS	36.82 (N = 201)	35.77 (N = 137)	48.98*/** (N = 98)
Plant A2	L	44.44 (N = 9)	55.56 (N = 9)	–
	CS	43.75 (N = 15)	50.00 (N = 27)	–
Plant A3	L	44.44 (N = 9)	44.44 (N = 9)	–
	CS	55.00 (N = 20)	27.59 (N = 29)	–

Note
*Means a significant diff. to the first DC
**Means a significant diff. to the second DC; (BiBB/BAuA comparative value for persons below the age of 45, 40 %, and above the age of 45, 42 %)

Table 10 Longitudinal (L) and cross-sectional (CS) view of physical complaints (%)—lower back

		1.DC	2.DC	3.DC
Plant A1	L	58.8 (N = 17)	82.4 (N = 17)	47.1** (N = 17)
	CS	52.3 (N = 174)	61.5 (N = 179)	61.0 (N = 59)
Plant B	L	67.1 (N = 70)	65.7 (N = 70)	77.1 (N = 70)
	CS	71.1 (N = 225)	68.3 (N = 145)	79.8** (N = 104)
Plant A2	L	71.4 (N = 7)	71.4 (N = 7)	–
	CS	76.5 (N = 17)	75.0 (N = 24)	–
Plant A3	L	80.0 (N = 10)	80.0 (N = 10)	–
	CS	71.4 (N = 21)	50.0 (N = 28)	–

Note
*Means a significant diff. to the first DC
** Means a significant diff. to the second DC; (BiBB/BAuA comparative value for persons below the age of 45, 40 %, and above the age of 45, 42 %)

Knee

Strain on the knees remains constant and at a relatively high level in plants A3 and B (compared with the BiBB/BAuA sample). By constrast, plant A1 shows a slight increase of complaints. A distinct improvement in plant A2 can be observed in the longitudinal analysis, yet this is not reflected in the cross-sectional data (group structure). The improvement in plant A2 is largely due to the use of shoes with high shock absorbing and cushioning properties and compensatory movements as part of the job rotation (material supply) (Table 11).

Summary of Physical Complaints

Despite attempts to move towards more ergonomic work design (implementation of lifting aids, steps for height adjustment, changes in the line balances, shoes with high shock absorbing properties, etc.), the changes made to the working conditions did not lead to a significant reduction of physical complaints. The reasons for this are other changes aimed at optimised performance and cycle time reduction; the omission of hidden micro-breaks and compensatory movements (going, fetching parts); the standardisation of sequence of movements at very high repetitive levels; and the low proportion of time allocated to secondary tasks (material supply, maintenance and repairs, quality assurance, optimisation of workflow, etc.). This is particularly true in plant B, where (the model of) a value-oriented production system has been adopted and the concept of participative teamwork has been dismissed.

Table 11 Longitudinal (L) and cross-sectional (CS) view of physical complaints (%)—Knee

		1.DC	2.DC	3.DC
Plant A1	L	26.7 (N = 15)	33.3 (N = 15)	46.7 (N = 15)
	CS	28.4 (N = 169)	36.9 (N = 176)	40.7 (N = 59)
Plant B	L	44.1 (N = 59)	49.2 (N = 59)	50.9 (N = 59)
	CS	48.3 (N = 211)	42.0 (N = 138)	50.0 (N = 92)
Plant A2	L	62.5 (N = 8)	25.0* (N = 8)	–
	CS	50.0 (N = 16)	48.2 (N = 27)	–
Plant A3	L	44.4 (N = 9)	44.4 (N = 9)	–
	CS	27.3 (N = 22)	34.5 (N = 28)	–

Note
*Means a significant diff. to the first DC; BiBB/BAuA comparative value for persons below the age of 45, 18 %, and above the age of 45, 21 %

As the listed physical complaints generally increase with age (cf. BiBB/BAuB data), the new work concepts (featuring integrated production systems) are likely to worsen physical complaints and decrease the likelihood of employees performing this work until reaching the statutory retirement age.

Work Ability Index

The Work Ability Index (WAI) (Ilmarinen and Tempel 2002) is subject to an age-related decline that can be moderated through better working conditions. Table 12 reveals how differently WAI scores developed in the various plants (over the course of the study). A significant downward trend can be observed in the longitudinal analysis of plant B, and this trend is also visible in the cross-sectional data. Plants A1 and A2 display a similar trend during the first two data collections. However, in the third data collection, plant A1's WAI scores show a slight improvement. In constrast, plant A3 is the only assembly system that features a significant improvement between both data collections. These findings support the hypothesis of Ilmarinen and Tempel (2002), who argue that work design can keep a WAI score constant and even improve it in some cases. Plant A3 is such an exception. This plant launched a number of alterations in the work process, such as replacing physically strenuous welding tasks with welding systems, replacing very loud engraving machines with low-noise embossing machines, stepping up teamwork using continuous improvement activities and adopting an employee-oriented leadership style. Overall it can be said that the WAI scores can be graded as "good". This corresponds with the moderate rate of absenteeism. These rates fluctuate over the course of the study, in plant A1 between 3 and 5 %, in plant B

Table 12 Longitudinal changes in the WAI

Plant	1.DC	2.DC	3.DC	N
A1	39.1	38.3	38.9	18
B	38.8	37.3*	36.7*	79
A2	36.1	35.7	–	10
A3	29.5	37.8*	–	11

*Means a significant diff. to the first DC; (WAI scores: 49–44 very good, 43–37 good, 36–28 moderate, and below 27 poor)

between 4.2 and 5 %, in plant A2 between 4.1 and 9.3 %, and in plant A3 between 4.8 and 5.9 % (Table 13).

When looking at the cross-sectional WAI scores (plant B) in the three age groups, a significant difference can be observed (see Table 14).

A different picture emerges from a longitudinal perspective (Table 15), with a steady deterioration in the youngest and the middle age group and constancy or an improvement in the oldest age group. Here, in line with the healthy-worker effect, it can be assumed that only persons in relatively good health are employed at the assembly lines.

Job Satisfaction

In accordance with Ilmarinen and Tempel (2002) we assume a positive correlation between work ability and job satisfaction. Consequently, it is no surprise that scores for job satisfaction are subject to corresponding, comparable changes. The job satisfaction results are positive (the scale goes from 1 meaning low to 7 meaning high)—except for plant B, where a significant deterioration in job satisfaction can be observed between 1.DC and 3.DC. This deterioration is both longitudinal and cross-sectional: 4.7(L)/4.7 (CS) (1.DC) fell to 4.5 (L)/4.6 (CS) (2.DC) and then to 4.2 (L)/4.1(CS) (3.DC). Within the group of employees aged 45 years and above, job satisfaction remained stable (4.7), whilst the greatest decline occurred in the group of 36–45 year-olds (from 4.8 down to 4.1, N = 29).

Subjective Work Analysis (*SALSA*)

Only selected findings from the subjective work analysis that were considered particularly relevant are presented here. Employee-oriented behaviour of supervisors is viewed more critically in plant B than in plant A1. Employee-oriented

Table 13 Cross-sectional changes in the WAI

Plant	1.DC	2.DC	3.DC	N1/N2/N3
A1	38.8	37.7	39.3	185/186/65
B	37.6	36.9	36.2	242/153/109
A2	36.5	35.9	–	18/29
A3	30.6	38.5*	–	22/35

*Means a significant diff. to the first DC; (WAI scores: 49–44 very good, 43–37 good, 36–28 moderate, and below 27 poor)

Table 14 Cross-sectional WAI scores (plant B); significant diff. between: 1 and 2; 1 and 3; 2 and 3

	Below 35 (N = 93)	36–44 (N = 77)	45+ (N = 66)
WAI	39.92	37.42*	34.83*/**

*Means a significant diff. to the first DC
**Means a significant diff. to the second DC

Table 15 Longitudinal WAI scores according to age groups (plant B)

	Below 35 (N = 37)	36–44 (N = 28)	45+ (N = 14)
1.DC	41.00	37.61	35.36
2.DC	39.76	35.11	35.07
3.DC	38.27	34.79	36.57

leadership behaviour in plant A1 showed a positive development over the course of the study. In the longitudinal analysis, ratings deteriorate when there is a direct supervisor change during the survey period. No such phenomenon can be observed in the cross-sectional data. This means that those persons who experience a change over time are more critical than those who merely rate their current situation. However, this assumption requires further investigation.

An interesting aspect of the findings from the subjective work analysis is the rating of task identity (see Table 16). As the assembled products greatly differ in terms of complexity, it is understandable that the appraisal of task identity varies. Completing a component (handbrake lever, plant A2) and assembling the gearbox (plant A1) is certainly experienced as more holistic than assembling a wiring harness (plant B) or assembling a component for the exhaust system (plant A3).

A significant decrease is noted in plant B. Presumably this is due to the limitation of different cycles (approximately 7–8) per employee, depending on their wage group.

Task diversity (see Table 17) rises in plants A1 and A3, while it decreases in plants A2 and B. In other words, standardisation, reduced cycle times and avoidance of waste (especially in plant B) happen at the expense of task variety. In plant A2, the decrease can be explained due to a reduction in technical faults. Technical measures and retrofitting caused malfunctions in individual systems to decline and output to rise, while the input required for maintenance, repairs and troubleshooting fell. However, now that everything is running smoothly and employees can focus on their core tasks, the effect of the short cycle time of 25 s is perceived to be far more intense.

Within the framework of her strain-biography study, Weichel (2012) used the 1.DC–plant B sample to show that task diversity at the workplace is the best predictor of adaptive performance. This means that high task diversity enables employees to handle new challenges more efficiently.

Limited task diversity also minimises chances of participation in decision-making and decision-shaping processes (opportunities for participation). The generally weak

Table 16 Task identity

		1.DC	2.DC	3.DC
Plant A1	L	3.4 (N = 18)	3.2 (N = 18)	3.2 (N = 18)
	CS	3.2 (N = 182)	3.1 (N = 199)	3.6*/** 9N = 65)
Plant B	L	2.7 (N = 77)	2.6 (N = 77)	2.4*/** (N = 77)
	CS	2.6 (N = 235)	2.6 (N = 153)	2.4 (N = 108)
Plant A2	L	3.6 (N = 10)	3.8 (N = 10)	–
	CS	3.8 (N = 18)	3.4 (N = 29)	–
Plant A3	L	2.9 (N = 11)	2.8 (N = 11)	–
	CS	3.0 (N = 23)	3.0 (N = 34)	–

Note
Scale 1 (not true) to 5 (very true)
*Means a significant diff. to the first DC
**Means a significant diff.to the second DC

Table 17 Task diversity

		1.DC	2.DC	3.DC
Plant A1	L	2.9 (N = 19)	3.4* (N = 19)	3.1 (N = 19)
	CS	2.8 (N = 184)	2.7 (N = 203)	3.1 (N = 65)
Plant B	L	2.7 (N = 81)	2.6 (N = 81)	2.5 (N = 81)
	CS	2.7 (N = 241)	2.6 (N = 155)	2.5 (N = 109)
Plant A2	L	2.6 (N = 10)	2.3 (N = 10)	–
	CS	2.3 (N = 18)	2.1 (N = 29)	–
Plant A3	L	2.1 (N = 11)	2.4 (N = 11)	–
	CS	2.4 (N = 23)	2.5 (N = 35)	–

Note
Scale 1 (not true) to 5 (very true)
*Means a significant diff. to the first DC

Table 18 Opportunities for participation

		1.DC	2.DC	3.DC
Plant A1	L	2.8 (N = 18)	2.8 (N = 18)	2.6 (N = 18)
	CS	2.7 (N = 183)	2.7 (N = 200)	2.7 (N = 65)
Plant B	L	2.5 (N = 80)	2.2 (N = 80)	2.3 (N = 80)
	CS	2.6 (N = 239)	2.3* (N = 154)	2.3* (N = 109)
Plant A2	L	2.5 (N = 10)	2.4 (N = 10)	–
	CS	2.4 (N = 18)	2.4 (N = 19)	–
Plant A3	L	2.6 (N = 11)	2.8 (N = 11)	–
	CS	2.7 (N = 23)	2.9 (N = 34)	–

Note
Scale 1 (not true) to 5 (very true)
*Means a significant diff. to the first DC

values (see Table 18) are weakest in plant B. Here it becomes obvious that the applied teamwork concept allows very little room for participation. It is striking that while the objective differences between plants A and B regarding opportunities for participation (see Fig. 2, teamwork) are large, the subjective employee ratings do not reflect these differences. This means that although chances of participation in plant A are objectively much higher due to certain organisational structures, they are not subjectively rated as such. For example, employees in plant A2 worked out design proposals regarding the colour and interior organisation that were later implemented. No such phenomenon can be found in plant B.

The least scope (between 1.2 and 1.7) is conceded to personal design of the workplace. In this regard, the plants barely differ from each other. The lowest scores (1.2–1.5) are recorded in plant B.

These findings are particularly problematic in terms of the aging workforce, as the prevailing planning philosophy does not officially permit individual leeway (e.g. individual work piece buffer, special storage space for parts, seating, etc.). Personal latitude clashes with endeavours to maintain standardised processes.

Personality Inventory

Data from the Big Five Inventory (BFI) and the Irritation Scale reveal nothing unusual. As expected, the BFI values change little over the survey period (the correlations vary between 0.49 and 0.86). The same can be said for cognitive and emotional irritation, where scores for overall irritation range from 2.8 to 3.8. Substantial differences between the individual plants were not detected.

Concerning the survey data, it should be mentioned that the participants made an effort to complete the questionnaires very conscientiously. One indicator for this is the accuracy of values given for body height. Correlations for body height are between 0.98 and 0.99 for the first, second and third data collection.

Work Design in Interaction with the Research Team

Cooperation between the researchers, the management, the workers' council and the employees differed greatly in the two companies. In plant B, neither the necessary data feedback nor discussions concerning direct interventions at the assembly line could be realised. All that was adopted there was the idea of implementing a strain-oriented, company-wide IT-based system (see Stanic 2010) for facilitating of job-rotation. Technical implementation of the IT has largely been fine-tuned, but the practical launch is not yet complete (as of 2011). Suggestions regarding the short cycle times, potential pre-assembling of modules and alterations to the product (e.g. separate tailgate assembly) were registered but not discussed in a serious manner.

In contrast, data feedback in plant A (A1, A2 and A3) was utilised to discuss feasible suggestions for improvement. Specific actions were derived from the discussions and partially implemented:

In A1: implementation of lifting aids, implementation of lifting devices to facilitate the retrieval of parts from a pallet box, avoidance of direct glare at the assembling stations, improved material supply.

In A2: colouring in the entire assembling and recreational area, improvement of material supply by means of suitable stepladders, job rotation between material supply and assembly, installation of an assembling station for physically disabled persons, optimisation of maintenance and repairs, application of therapeutic exercises to relieve complaints in the neck, shoulder, hand and arm region, etc.

In A3: noise-level reduction through implementation of new low-noise stamping press [below 40 db (A)] instead of the very loud needle press [up to 95 db (A)], implementation of welding systems to avoid awkward body postures, ergonomic optimisation of material supply, testing area and welding appliances, plus floor mats and lighting at certain workstations, etc.

The different approach towards the realisation of actions in the two companies is obvious. A key reason for the unwillingness to change at plant B is that modifications on the final assembly line bear a high economic risk due to their linkage with other assembling areas. In contrast, risks entailed by changes in the aggregate assembly can be controlled more easily.

Conclusion

The aim of the project "Age-differentiated work design in the automotive assembly" was to determine the age-related impact that typical assembling tasks have on employees. Various tasks in final vehicle assembly and equipment assembly were examined at two automotive companies. These tasks are characteristic of the automotive and supplying industry.

Within the project period, the automotive industry faced substantial economic fluctuations, the consequences of which are visible in the collected data.

In the first phase of the project, the two companies' interest in "demographic questions" was comparable. Company A's response to the subject matter was a comprehensive collective agreement and company B initialised a project to deal with the aging workforce. While company A's interest in the project grew over the years, backed by the management and the worker's council, company B's interest in the matter steadily dwindled. This adverse development affected data collection and the handling and treatment of the results.

During the project it became clear that differentiated work systems (plant A2 and A3) are better suited for implementing and evaluating measures of work design customised for an ageing workforce than the rather extensive work systems found in the final vehicle assembly area (plant B). Strong support from the worker's council in plant A was a prerequisite for the success of the project. The worker's council in plant B supported data collection but they did not follow up on the findings.

The comparable methods and instruments of value-oriented production systems found in both companies lead to an increase in productivity in conjunction with similar effects on the employees.

Cycle times are reduced, hidden breaks are omitted, and non-value adding activities (e.g. material supply) are eliminated, increasing efficiency. These rationalisations culminate in the examined Chaku–Chaku work systems, which are characterised by very short cycle times. The ergonomically motivated measures of work design that have been implemented to "improve added value" did not lead to any relief for employees, because shortening the cycle time resulted in a concomitant intensification of work performance. This was the case in plant B and plant A2 in particular.

In the four analysed work systems, physical strain rose during the study period as a consequence of one-sided strain. These strains worsen with age. A similar relationship exists between age and work ability, except that work ability decreases with age. Task diversity and the extent to which work tasks are holistic have significantly decreased in the perception of the employees. The same applies for opportunities for participation and personal latitude.

Overall job satisfaction did not change. Satisfaction with the company is high in both plants, which means that the perceived deterioration of working conditions (especially in plant B) did not lead to dissatisfaction with the company as a whole.

Job rotation measures alone are not sufficient when customising work systems to meet the demands of an aging workforce. The recommendations listed under Lessons Learned (see section Lessons Learned) are essential.

Acknowledgments The authors thank the German Research Foundation (DFG) for funding this research (FR 560/8-3). We would also like to thank all cooperation partners from industry and academia for their support and sharing data and experiences. Especially, we would like to thank the employees of the industrial engineering department, the work council, the factory doctor, the foremen as well as the employess of the human resources department of the introduced automobile manufacturers.

References

Bongwald O, Luttmann A, Laurig W (1995) Leitfaden für die Beurteilung von Hebe- und Tragetätigkeiten: Gesundheitsgefährdung, gesetzliche Regelungen, Messmethoden, Beurteilungskriterien und Beurteilungsverfahren. HVBG, Sankt Augustin

Bundesanstalt für Berufsbildung und Bundesanstalt für Arbeitsschutz und Arbeitsmedizin (BiBB/BAuA) (eds) (2005/2006) Arbeit und Berufe im Wandel, Erwerb und Verwertung beruflicher Qualifikationen, Erwerbstätigenbefragung 2005/2006 Dortmund

BMW Annual Report (2011) http://geschaeftsbericht2011.bmwgroup.com. Accessed Aug 2012

Conti R, Angelis J, Cooper C, Farager B, Gill C (2006) The effects of lean production on worker job stress. Int J Oper Prod Manage 26(9):1013–1038

Ellegast RP (2005) Fachgespräch Ergonomie 2004 Zusammenfassung der Vorträge, gehalten während des Fachgespräches Ergonomie am 15/16. November 2004 in Dresden. BGIA, Sankt Augustin (BGIA-Report)

Enríquez Díaz JA (2012) Arbeitswissenschaftliche Evaluation ganzheitlicher Produktionssysteme für die Aggregatemontage in der Automobilindustrie. Dissertation, RWTH Aachen, Fakultät für Maschinenwesen

Freiboth M (1998) Empirischer Vergleich von Arbeitsorganisationsformen in der Automobilindustrie. Dissertation, Universität Kassel

Frieling E, Sonntag KH (1999) Lehrbuch Arbeitspsychologie. Hans Huber, Bern, pp 438–443

Ilmarinen J, Tempel J (2002) Arbeitsfähigkeit 2010 Was können wir tun damit Sie gesund bleiben? VSA-Verlag, Hamburg

Jürgens U (2007) Warum Toyota so lange so stark ist. IG Metall Baden-Württemberg, Stuttgart

Kono H (2004) Fundamental principles and side-effects for effective manufacturing revolution. In: Proceedings of the 5th Asia Pacific industrial engineering and management systems-conference, 23 May 2010

Kotzab D, Enríquez Díaz JA, Sytch A, Frieling E (2011) Changes in employee attitudes towards work activity in a chaku–chaku assembly line—a follow up study in a German automotive manufacturing company. In: ODAM–X (ed) Human factors in organizational design and management: II-81–II-86. IEA Press, Santa Monica

Kuorinka I, Jonsson B, Kilbom A, Vinterberg H, Biering-Sørensen F, Andersson G, Jørgensen K (1987) Standardised Nordic questionnaires for the analysis of musculoskeletal symptoms. Appl Ergonomics 18(3):233–237

Landau K (2011) Arbeitsgestaltung für alternde Belegschaften. ASU-protect Das Magazin für das Arbeitsschutzmanagement, Gentner, Stuttgart, pp 30–33, 1 Nov 2011

Neuberger O, Allerbeck M (1978) Messung und Analyse von Arbeitszufriedenheit. Erfahrungen mit dem Arbeitsbeschreibungsbogen (ABB). Huber, Bern

Neuhaus R (2010) Evaluation und Benchmarking der Umsetzung von Produktionssystemen in Deutschland. Books on den demand, Norderstedt

Nübling M, Hasselhorn HM, Seitsamo J, Ilmarinen J (2004) Comparing the use of the short and the long disease list in the work ability index questionnaire. In: Proceedings of the 2nd international symposium of work ability. ICOH, 18–20 Oct 2004. Verona, p 74

Ohno T (2009) Das Toyota-Produktionssystem: Das Standardwerk zur Lean Production. Campus, UK Frankfurt

Rammstedt B, John OP (2005) Kurzversion des Big Five Inventory (BFI-K): Entwicklung und Validierung eines ökonomischen Inventars zur Erfassung der fünf Faktoren der Persönlichkeit. Diagnostica 51:195–206

Rimann M, Udris I (1997) Fragebogen zur persönlichen Situation im Beruf und Betrieb. Salsa. In: Strohm O, Escher OP (eds) Mensch, Technik, Organisation: Unternehmen arbeitspsychologisch bewerten. Ein Mehr-Ebenen-Ansatz unter besonderer Berücksichtigung von Mensch, Technik und Organisation. vdf Hochschulverlag, Zürich, pp 281–298

Schumann F (1911/2011) Die Arbeiter der Daimler-Motoren-Gesellschaft Stuttgart-Untertürkheim. Herausgegeben von IG Metall Baden-Württemberg. Original: Duncker–und Humblot. Wiederabdruck: INFO IDEE Medienverlag, Ludwigsburg

Stanic S (2010) Fahrzeugendmontage—Herausforderungen für den demografischen WandelSchriftenreihe. Personal–und Organisationsentwicklung. Kassel University Press, Kassel

Sytch A, Kotzab D, Enríquez Díaz JA, Nöring R, Frieling E (2011) Einstufung von industrieller Teamarbeit—Eine Klassifikation ausgewählter Montagesysteme. In: GfA Mensch, Technik (eds) Organisation—Vernetzung: Tagungsband der GfA Frühjahrskonferenz GfA-Press, Dortmund, pp 631–634

Verband der Deutschen Automobilindustrie—VDA (2011) Jahresbericht. Frankfurt a.m

Waters TR, Putz-Anderson V, Garg A, Fine LJ (1993) Revised NIOSH equation for the design and evaluation of manual lifting tasks. Ergonomics 36(7):749–776

Weichel J (2012) Ressourcen und Belastungen in der Berufsbiografie. Kassel University Press GmbH, Kassel

Wimmer E, Schneider MC, Blum P (2010) Antrieb für die Zukunft—Wie VW und Toyota um die Pole-Position ringen. Schäffer/Poeschel, Stuttgart

Woll A, Tittlbach S, Schott N, Bös K (2004) Diagnose körperlich-sportlicher Aktivität. Fitness und Gesundheit. Dissertation.de–Verlag im Internet, Berlin

Yagyu S (2007) Das synchrone Managementsystem: Wegweiser zur Neugestaltung der Produktion auf Grundlage des synchronen Produktionssystems. mi-Fachverl., Landsberg am Lech

Capability Related Stress Analysis to Support Design of Work Systems

Holger Rademacher, Ralph Bruder, Max Bierwirth, Dorothee Müglich, Andrea Sinn-Behrendt and Kurt Landau

Lessons Learned

Two complementary studies have been conducted to analyze the effects of the demographic change in the automotive industry. First, physical exposures due to work tasks in production sites of automotive industry have been analyzed. In this study 159 workplaces were evaluated using the European-Assembly-Worksheet (EAWS). Second, physical capabilities of workers in automotive industry were examined. Several tests that simulate the complex load situations in ordinary work situations of automobile workers were developed and conducted with 106 male workers (aged 20–63 years) in the automotive industry. The test performance of each individual was rated and potential age dependencies of industry relevant capabilities were statistically analyzed.

The results prove the existence of age dependencies of several specific physical capabilities with significant practical relevance. Capabilities concerning manual

H. Rademacher (✉) · R. Bruder · M. Bierwirth ·
D. Müglich · A. Sinn-Behrendt · K. Landau
Institute of Ergonomics, Technische Universität Darmstadt, Darmstadt, Germany
e-mail: rademacher@gesamtmetall.de

R. Bruder
e-mail: bruder@iad.tu-darmstadt.de

M. Bierwirth
e-mail: bierwirth@iad.tu-darmstadt.de

D. Müglich
e-mail: d.mueglich@iad.tu-darmstadt.de

A. Sinn-Behrendt
e-mail: sinn@iad.tu-darmstadt.de

K. Landau
e-mail: landau@iad.tu-darmstadt.de

C. M. Schlick et al. (eds.), *Age-Differentiated Work Systems*,
DOI: 10.1007/978-3-642-35057-3_10, © Springer-Verlag Berlin Heidelberg 2013

materials handling and maintaining awkward body postures show statistically significant differences between age groups. These types of exposures were also found to be the most frequent ones in workplace assessments in the automotive industry. Based on our consolidated findings, we give the following five recommendations for an age-related analysis and design of work systems:

- Assess and evaluate the capabilities of workers in a production plant with a differentiated view
- Assess work load situations and identify "critical" workplaces in industrial practice
- Be aware that capabilities related to manual materials handling and working with the trunk strongly bent forward diminish significantly with age
- Use ergonomic work design to lower physical demands and thereby reduce adverse age-related effects on workers' health and performance
- Develop a database to match standardized capability profiles of the actual workforce and work task demand profiles based on assessed load situations.

Introduction

Most industrial nations are experiencing a continuous change in the structure of their workforce in manufacturing industry. For decades the percentage of older workers has been rising. In order to maintain their future competitiveness in the global market manufacturing companies may encounter difficulties because of the demographic change (see Age-differentiated Work Systems: Introduction and Overview to a Six-Year Research Program in Gemany, Schlick et al.). One possibility to overcome these difficulties is to take this development into account in the design of production systems. In order to avoid adverse effects like musculoskeletal disorders, evidence-based knowledge about the characteristics of physical capabilities and their relation to age is required.

The aim of this long-term research project was therefore to gather information about influences of age on workers' physical capabilities. This should allow an age-differentiated view on an individual's capabilities as well as on exposures due to work tasks that are frequently found in industrial practice. Thus, the field study is focused on assessing mechanical exposures at "typical" workplaces in automotive industry. The second study is a laboratory study designed to evaluate individual physical capabilities of workers, who work at these workplaces. Therefore, the second study is conducted directly in plants of an automotive company. The automotive industry is among those with the highest percentage of manual work, especially in assembly areas, and faces big challenges because of the demographic change. However, production system design offers potential to effectively enable ergonomic work design.

With profound knowledge about existing work demands and workers' capabilities, age-differentiated work design and employee assignment can be effectively supported. This might help to prevent work-related (musculoskeletal) disorders, reduce early-retirement as well as to improve motivation and job-performance of older workers. One assumption is that age has influences on some, but not all capabilities to cope with mechanical work-related exposures (see Age-differentiated Work Systems: Introduction and Overview to a Six-Year Research Program in Gemany, Schlick et al.). Another assumption is that some of these capabilities are needed at workplaces in the automotive industry and are, therefore, "age-critical" (Rademacher et al. 2011).

Literature Review

The scientific discussion about aging workers and age-specific capabilities in industrial settings is relatively diversified. Regarding recent research, there is a lack of profound scientific studies of workload exposures as well as of worker capabilities.

There are no studies published which give detailed quantitative information about type and extent of mechanical exposures found at workplaces in automotive manufacturing. However, some studies give qualitative statements about the appearance of specific physical risk factors for musculoskeletal disorders (Kahya 2007; Landau et al. 2008). Other studies include only a sample of workplaces with specific tasks or types of exposures (Enríquez Díaz et al. 2010; Punnett et al. 2004).

Concerning effects of workers' age on physical capacities (e.g. strength and endurance), several studies can be found. In an extensive literature review Kenny et al. (2008) describe studies that examine age-related physiological functions (e.g. maximum muscle strength or maximum oxygen uptake) and age-related changes of the musculoskeletal system (e.g. flexibility or stability of the spine). Furthermore, they evaluated theoretical influences of the age process on job performance. Hence, their conclusion is that habitual physical activities delay the decrease of physical capacity of workers as well as further risks of work-related disorders. Moreover, physical training lengthens the career of older workers.

In a recent study Hamberg-van Reenen et al. (2009) examined the muscle capacity of approximately 1,800 workers. They found that the older workers on average had the highest static muscle endurance in the neck and shoulder region. When they examined the back muscles they had converse results, because on average the maximum muscle power was detected at an age of 36 years. Furthermore, best results of muscle capacity are determined among young workers, who performed sports activities more than 3 h per week. In contrast, they found the highest muscle capacity among old workers, who moderately participated in sports (less than 3 h per week).

The relationship between productivity and employment of older workers was studied by Shephard (2000). He described age-related trends in personal characteristics, capabilities and skills. As a result, he developed proposals for worksite modification, employee wellness and training programs to help maintaining productivity.

Savinainen (2004) took random samples in order to evaluate the influence of aging upon productivity in manufacturing. He observed the development during a follow-up period of 10–16 years. A deficit model about age resulted from this study. However, in order to explain the decreasing physical capacities due to age Savinainen refers to the concept of "synergistic capacity" by Kumashiro (2000). "A synergistic capacity can be described as a general capacity to perform tasks supported by a wealth of experience and knowledge" (Kumashiro 2000, p. 1009).

A long-term study by Bildt Thorbjörnsson et al. (2000) examined influences of work stress, subjectively perceived strain and gender on lumbar spine symptoms. Although they demonstrated significant influence related to gender, sedentary work, whole body vibrations, working overtime and other psycho-social factors, no statement about influences of age was made.

Aittomäki et al. (2005) examined the difference of work stress, age and gender. Results suggest a linear trend of less physically demanding work in older than in younger age groups. However, there was an interaction between advancing age and poor functioning in physically demanding work. The participants of the study were employees of the city of Helsinki (Finland). A transfer of these results to assembly workers in automobile manufacturing, therefore, may be questionable.

Most studies only test basic physiological functions of workers to demonstrate their age-related decrease or increase, but do not assess the potential age-dependency of "work-related" capabilities. Regarding capabilities of workers only one empirical survey is known by the authors, where capacities were tested in complex load situations that simulate ordinary work situations of the workers.

Gall and Parkhouse (2004) examined the changes in physical capacities as a function of age in power line technicians. The participants had to perform work-related tests that simulated physical workload power line technicians normally experience during their work. They compared younger ($\leq$39 years) and older (>50 years) workers, a total of 40 test persons. Results (for six out of nine test variables) show no statistical differences between the mean values of the different age groups. The group of older participants achieved significantly lower results in the "aerobic capacity test", "one-handed pull down", and both "right and left standard handgrip tests". However, Gall and Parkhouse (2004) demonstrated older workers were able to cope with their job demands. They concluded that heavy manual work seems to maintain task-specific physical capacities as age progresses.

The literature presents no satisfying findings to the question which work-related capabilities of workers are significantly influenced by age. The field of possible influences of age on industry relevant worker capabilities is barely evaluated. Petrenz (1999) and a literature review by Silverstein (2008) show that existing studies on effects of aging only focus on changes in maximum capabilities. Whether these changes are critical in practice, for coping with certain long term work-related musculoskeletal risk factors, has not been thoroughly discussed in these reviews. The results of this literature review underline the importance of our approach to examine individuals' capabilities using specifically developed tests. The results of the workplace assessments, the questionnaire and the capability tests are presented in the following.

Methods

Assessment of Typical Mechanical Exposures in Automotive Industry (Study A)

To get an overview about the typical mechanical exposures of workers in the automotive industry, workplaces in two different companies were analyzed: Firstly, 117 workplaces at an automotive supplier that produces car and motorcycle tires; secondly, 42 workplaces at the final assembly line and the paint shop of a car manufacturer. The sample covered cycle times from 5 to 180 s and a large variety of manual work tasks, typical for the automotive industry such as assembly of car interiors, exterior parts, carrying and lifting of parts, short cycled assembly and machine control tasks.

To assess mechanical exposures at all 159 workplaces the "European Assembly Worksheet" (EAWS, Schaub et al. 2008) was used. The EAWS is a screening method based on observation and collection of workplace data such as distances, weights and forces. The EAWS is a valid screening tool for the assessment of physical work load in the automotive industry (Rademacher et al. 2006). The EAWS analyzes risk factors for musculoskeletal disorders in five sections concerning different types of physical workload:

- Static working postures
- Action forces of the whole body and finger-hand system
- Manual materials handling
- (High) repetitive loads on upper limbs
- "Extra points" for specific exposures that are not covered in previous sections.

For all risk factors load duration and intensity were assessed. A score of "load points" is allocated to each risk factor present at the individual workplace and a total score is calculated. However, for "(high) repetitive loads on upper limbs" a separate score is calculated. Afterwards the calculated scores are rated with a three zone traffic light scheme (DIN EN 614-1 2006); low risk (green), possible risk (yellow) and high risk (red).

Based on the scores allocated to each section, the EAWS analysis delivers a detailed overview of the main risk factors present at a workplace and allows a comprehensive risk assessment concerning the physical workload.

Assessment of Workers' Capabilities (Study B)

Subjects

The capability testing has been realized as a laboratory study under field study conditions directly in plants of a German vehicle manufacturer. The examination included 106 male workers of a production plant from this company. The age of the workers varied between 20 and 63 years (mean = 40.34 years; standard deviation SD = 11.72 years). In this study the workers were divided into two major age groups, "young workers" aged 20–35 years (mean = 28.78 years; SD = 4.35 years) and "older workers", who are 45–63 years old (mean = 50.66 years; SD = 3.94 years). Each group was further divided into three subgroups. The study was conducted to evaluate differences between younger and older workers' capabilities. Hence, the age group of workers between 36 and 44 years is at this state of the study not included. It is also presumed from the literature that this age group does not show any significant decrease or increase in physiological functions.

Persons, who were required for this study, needed work experience (including apprenticeship) of more than 5 years and must not be more than 6 months out of work practice (e.g. unemployed or retired). Exclusion criteria were reception of disability rent, present inability to work or extended periods of inability to work (more than 3 months in the last 12 months due to musculoskeletal symptoms including neurological dysfunctions).

Capability Testing

Besides the subjective view on limitations and work ability, objective tests of the capabilities of workers were used. The aim of the capability testing was to evaluate if there are significant differences ($p < 0.05$) in work-related capabilities between the groups of "younger workers" (20–35 years) and "older workers" (45–63 years).

Based on the results of the workplace assessments in this study and the experience of the assessment of more than 200 workplaces done by the authors in another study (Landau et al. 2008; Rademacher et al. 2008), the tests were designed that they represent relevant work tasks with common mechanical exposures in modern production facilities. The aim was to cover as many stressful work-related tasks as possible performed in assembly areas of the automotive industry. For research results concerning electronics industry and manual dexterity see chapter Influence of Age and Expertise on Manual Dexterity in the Work Context: The Bremen-Hand-Study@Jacobs, Voelcker-Rehage et al.

For measuring the functional physical ability of a person to perform work-related tasks, the "Isernhagen Work Systems Functional Capacity Evaluation" (IWS-FCE, Isernhagen 1988) is a commonly used method (e.g. in rehabilitation) which has shown satisfying validity and reliability (Chen 2007; Gouttebarge et al. 2004). Due to the fact that capabilities of active workers should be analyzed on-site, only a very limited amount of time for the entire assessment of an individual's capabilities was available. The IWS-FCE includes 29 tests and takes about two days to be completed. That is why the IWS-FCE could not be used in the study. Instead 8 different tests developed by Rademacher et al. (2008) were used, which could be completed in one hour per subject and are based on the methodical approach of the IWS-FCE.

For all capability tests the age of the worker and the test set-up are determined as the independent variable. The independent and dependent variables of the capability testing are listed in Table 1. The controlled variables are the gender of the participants and the surrounding conditions.

The set up of each test used in this study is explained in the following. All tests are standardized, but can be adapted to the individual anthropometry of a participant.

Cardiopulmonary Load Test

Cardiopulmonary capacity is evaluated for each participant to get a rough estimate of the general fitness of the worker. Based on this test the "Pulse Performance Index" (PPI) can be calculated. The PPI is a parameter for the physical condition

Table 1 Overview of variables

Tests	Independent variables	Dependent variables
Cardiopulmonary load test	• 20 squats/10 steps 10 times	• Time needed for completing the test • Heart rate (before and after the test) • Blood pressure (before and after the test) • Time until resting heart rate is reached again
Manual materials handling	• Body posture • Load weight • Number of repetitions	• Time needed for completing the test • Heart rate (before and after the test) • Individual work intensity • (Preliminary expert rating)
Performing tasks in static postures	• Body posture • Arm position • Time	• Heart rate (before and after the test) • Total repetitions of completed screwing/ number of finished repositioning operations • Grades of physical limitation • (Preliminary expert rating)

and the capacity of the cardiovascular system (Meißner-Pöthig 1997). A PPI score of less than 1.00 is an indicator of a below average capacity of the cardiovascular system. If the PPI score is higher than 2.00 a good (above average) cardiovascular condition is provided (Seibt et al. 2007).

The test demands 20 squats performed in a steady pace. In the case of individual contraindications, e.g. knee or hip disorders, ascending and descending 10 steps with 10 repetitions is used as an alternative test.

The time needed to complete the test, heart rate and blood pressure before and after the test were measured. The time to reach the resting heart rate after the test is also noted. To calculate the PPI the difference between the heart rate after test performance and resting heart rate is divided by the measured test duration.

Manual Materials Handling

To assess the capabilities for manual materials handling, in particular lifting and repositioning of loads, two tests were developed: One for lifting between floor and waist level, a second for lifting from floor to head level.

Executing the test was explained and demonstrated by the test instructor. The participants were advised to bend their knees, to keep their back upright und to put down the box between their feet.

Lifting between floor and waist level

The participant stands in front of a shelf, which is adjusted to his body height. The task is to lift a wooden box slightly from the shelf to the floor while doing a body rotation of 90°. Afterwards the box is lifted up again on the shelf (initial position). The worker has to repeat the action with different load weights from 5 to 20 kg 10 repetitions each. The load weight increases in steps of 5 kg.

Lifting between floor and head level

In contrast to the previous test the participant lifts the box from floor to waist level (without rotation), places the box briefly on the shelf, then lifts it on a second shelf at shoulder level (adjusted to workers' body height, handles are at head level when box is on the shelf) and again places it there briefly. Finally, the individual has to lift the box from the upper shelf and place it on the floor again (initial position). Depending on the results of the previous lifting test the test instructor decides whether the worker has to perform 5 or 10 repetitions (15 or 30 repositions) at each load weight. The load weight is increased incrementally in three steps of 5 kg from 5 to 15 kg. The test instructor may assign an additional, higher load weight depending on his assessment of the workers' capability (but max. 35 kg).

The measured dependent test variables are the heart rate before and after carrying each load weight and the time for completing the test. Additionally, the individual work intensity is subjectively evaluated by the test instructor on a scale from "light" (1) to "maximal" (8) based on the observation of the individual's test performance (expert rating).

Performing Tasks in Static Postures

The capabilities to perform tasks in static postures were assessed with five different tests, which are described below. The focus while performing the tests was on trunk postures particularly trunk bending as well as on different elevated arm levels.

Screwing hex nuts manually with both hands above head level

This test was used in order to rate the capabilities of a person coping with the demands "upright standing in a posture without body support", "static arm postures in the overhead area" and "high-frequent manual screwing movements". The participant has to loosen and fasten hex nuts, which are coupled with a screw through a link chain. The link chain is fixed across a shelf, approximately 5 cm above the head of the participant. The participant starts with his dominant hand at one end of the chain and moves the hex nuts with their screws inversely and sequentially to the opposite end of the chain (from left to right or from right to left). After 3 min the worker continues loosening and fastening the hex nuts with the non-dominant hand for another 3 min. The non-screwing hand only grasps the head of the screw. Both arms are kept at chain height during the whole test.

High-frequent repositioning at shoulder level

The test delivers information about the capability to perform high-frequent pinch grips with two fingers (thumb and index finger) combined with strong shoulder abduction (<80°) and dorsal extension (>45°) in the wrist. The participant has to loosen and fasten pegs between two cords. In the experimental setup there are 2 parallel cords with 10 pegs on each side of the lower string. The strings are on a horizontal line at participants' chest level, with a distance of 5 cm between each other. The pegs have to be transferred sequentially to the opposite side of the upper string and vice versa. The instruction is to fully squeeze each peg and move it from one end of the first string to the other end of the second string (from left to right or from right to left). The pegs shall be kept fully squeezed while moving them. The worker starts with his dominant hand at one end of the lower string with pegs no. 1 to 10. After 3 min the worker changes to the non-dominant hand and continues loosening and fastening the pegs no. 11–20 for another 3 min.

Screwing a semi-filled water bottle with one hand above head level

This test delivers supplementary data to the hand-arm testings before concerning the capabilities to maintain specific body postures (standing upright at one position and maintaining one arm at shoulder level). The capabilities to perform (high) frequent pinch grips with five fingers as well as strong pronation and supination (>60°) in the forearms are considered with this test, too. The participant screws in and out a cylindric object (e.g. a plastic bottle) filled with 500 ml of water by gripping its bottom (Ø ~ 80 mm). The basic position of the worker for this test is 90° flexion in his shoulder and elbow. The use of only one hand (not the fingers) is allowed to screw the bottle and the participant has to perform a maximal pro- and supination in order to fix the bottle with the first turn. The participant starts with his dominant hand. After 3 min the worker continues with the non-dominant hand for another 3 min.

Use of screwdriver (top-down) while trunk strongly bent forward

The capability assessed in this test is performing and maintaining strong trunk bending in addition to stretching one arm away from the body. The participant uses a cordless screwdriver (Bosch PSR 18–2, weight 2 kg) to screw completely in (towards the floor) and out (towards the ceiling) 20 hexagon screws, which are arranged in two lines with 10 screws respectively. The metal plate which contains threads for the screws is located vertically on the middle of the participants' thighs. An adjustable support for the legs of the participant is provided to enable a trunk bending of about 60° in order to screw with the right or left arm (dominant hand). The active arm is stretched out moderately from the body (elbow flexion between 45° and 60°). The participant holds the screwdriver with the dominant hand in a pistol grip and the other hand grasps the battery unit of the screwdriver.

Use of screwdriver (bottom-up) while trunk strongly bent forward and rotated

In addition to the test described before, we examined the capabilities to maintain a moderate to strong static trunk rotation (~20°), a moderate static lateral flexion of the head (>10°) and to exert a static force with the arms. The participant has to screw in and out 10 hexagon screws, which are arranged in one line. In addition the worker is instructed to keep eye contact to the screw. There is no possibility to support the own body with the screwdriver on the head of a screw as the participant could in the other mechanical screwing task. The metal plate with threads is located vertically between the middle and lower part of the participants' sternum. The participant has an adjusted support around his legs to push against. The test set up is the same as in the previous test.

The measured test variables are the heart rate before and after the test and the total repetitions of completed screwing respectively repositioning operations within the test duration (6 min). In addition, the level of physical limitation is subjectively evaluated by the test instructor on a scale from "no limitations" (0) to "severe limitations" (4) based on observed difficulties or limitations of the individual participant such as number and duration of interruptions while screwing (expert rating). As the individual's limitations, not the performance, were assessed with this expert rating, a 5-step ordinal scale was considered as satisfactory instead of a more detailed 8-step scale.

Subjective Assessments (Questionnaires)

Additionally to the tests, participants fill out a standardized questionnaire named "Detection of workplace-related disorders" developed by the authors. The questionnaire is based on the "Body Part Discomfort Scale" of Corlett and

Bishop (1976) and the "Nordic Questionnaire" (Kuorinka et al. 1987). The interviewee gives information about joint, back or muscle disorders, their location and an individual rating of intensity [on an ordinal scale "no disorders" (0) to "severe disorders" (7)]. Furthermore, we ask questions about the duration of disorders, therapies, duration of work incapacity and personal countermeasures against the disorders. The questionnaire also contains questions about sports activities and nicotine consumption.

Finally, the interviewee answers the questions of the 'Work Ability Index' (WAI, short version) (Tuomi et al. 1998) in order to do a self-assessment of his work ability. This questionnaire contains 7 items, each rated on a 7-point scale from 1 to 7 points (max. 49 points). The items are dealing with:

- Current work ability compared with the lifetime best
- Work ability in relation to the demands of the job
- Number and type of current diagnosed diseases
- Self-assessed work impairment due to diseases
- Sick leave during the past year (12 months)
- Self-prognosis of work ability two years from now
- Mental resources.

Rating of Capabilities

To determine an individual's capabilities related to manual materials handling and working in static postures the results of the cardiopulmonary load test (the PPI), the measured test variables and the expert rating of the individual's performance are assessed and combined in specific rating algorithms developed by Rademacher (in preparation). The rating of capabilities related to manual materials handling, for example, is based on the values of the following variables:

- Results of cardiopulmonary load test (PPI)
- Results of questionnaire "detection of workplace-related disorders"
- Measured heart rate before and after test
- Measured time for completing the test
- Expert rating of individual work intensity of the test
- Test dropout/abort.

Through the algorithms the capabilities of each individual are rated on a 5–step ordinal scale, from "no limitation of capability" (0) to "strong limitation of capability" (4). The results of the workplace assessments, the questionnaire and the capability tests are presented in the following.

Results

Assessment of Typical Mechanical Exposures in Automotive Industry (Field Study A)

The assessments conducted in company 1 show that work tasks are mainly characterized by manual materials handling (70.49 %) and static postures (27.44 %). Concerning manual materials handling, especially the task frequency represents the most critical risk factor. Awkward postures and a higher load weight (on average 11.5 kg) are the main contributors at high risk workplaces.

Regarding all 42 workplaces assessed in company 2, awkward postures have the strongest impact in terms of mechanical exposures (53.80 %)—specifically, working with arms at or above shoulder level as well as bended trunk postures. A significant amount of extra points (20.23 %) for automobile assembly specific exposures e.g. working on moving objects, constricted accessibility and visibility was assigned in the assessments. Manual materials handling is also an issue in car manufacturing (18.43 %). Action forces contribute very little to mechanical exposures (on average 7.55 %).

The results of the workplace assessments in the two companies are shown in Fig. 1.

Assessment of Workers' Capabilities (Study B)

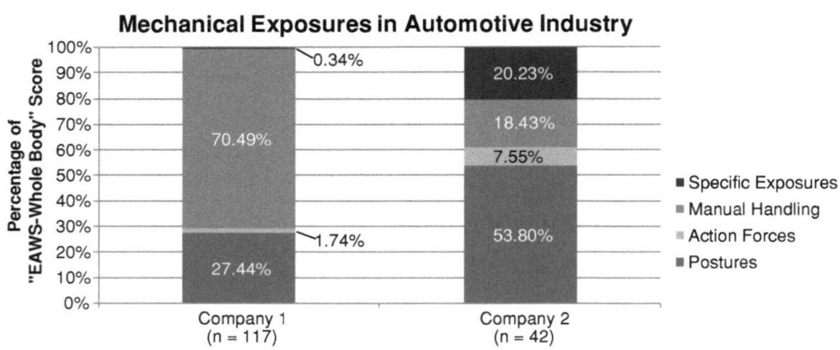

Fig. 1 Results of the workplace assessments

Capabilities of Manual Materials Handling

The capabilities for manual materials handling were assessed for 101 participants. Five workers had contraindications for lifting and therefore could not perform these tests. First, the capability to lift a load between floor and waist level and

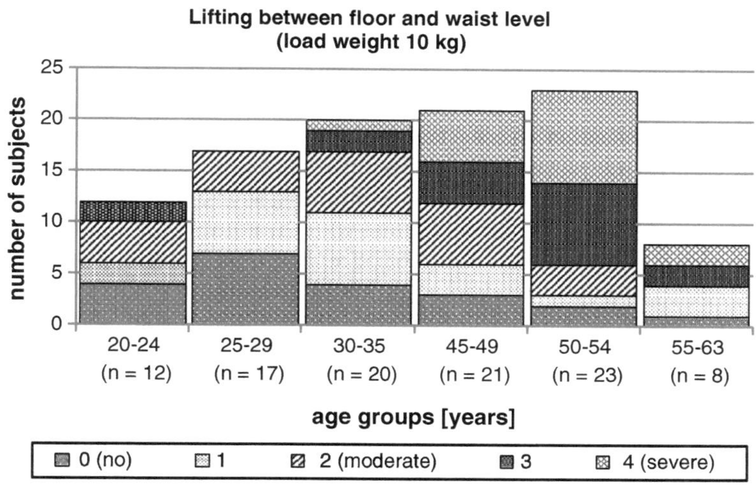

Fig. 2 Assessed grade of limitation of the capability to lift a load weight (10 kg) between floor and waist level

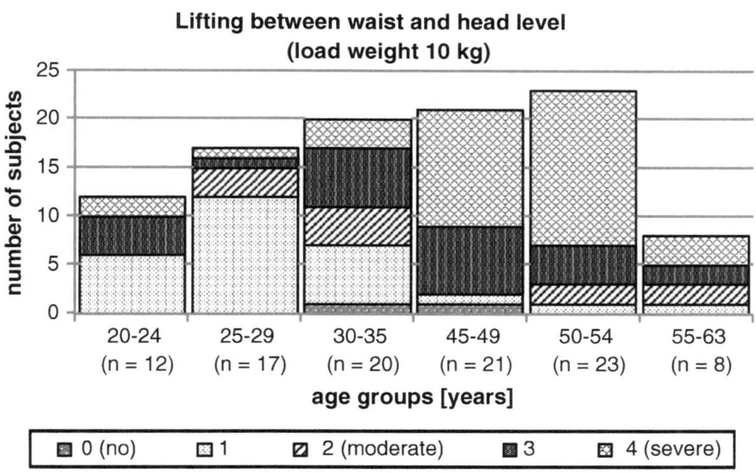

Fig. 3 Assessed grade of limitation of the capability to lift a load weight (10 kg) between waist and head level

second the capability to lift a load from waist to head level was evaluated. The results are depicted in numbers of workers according to the grades of limitation per age group. The results for a load weight of 10 kg are shown in Figs. 2 and 3 below.

Lifting Between Floor and Waist Level

The lowest limitations can be found in the age group of workers from 25 to 29 years (median $\tilde{x} = 1.0$; mean absolute deviation from the median $MD\tilde{x} = 0.65$). Beginning with the age group of 30–35 years, a growing number of workers with severe grades of limitation can be found. In the group of workers aged 55–63 years there were only 8 workers, who were available at the company to participate in this study. The age group with the highest grades of limitation is the group of 50–54 year-olds. The median is at 3.0 with a deviation of 0.87. In comparison, the group of workers aged 55–63 years has a median of 2.0 with a deviation of 1.38, which are about the same as the group of 45–49 years old workers ($\tilde{x} = 2.0$; $MD\tilde{x} = 1.10$) and the overall values ($\tilde{x} = 2.0$; $MD\tilde{x} = 0.93$).

Statistical analyses with the H-Test by Kruskal and Wallis (1952) (Bortz et al. 2000; Kruskal and Wallis 1952) show that there is a highly significant difference ($p < 0.001$) in the capabilities to lift a load of 10 kg from floor to waist level between "younger workers" (20–35 years) and "older workers" (45–63 years). A weak but highly significant correlation with age has also been found ($\tau = 0.355$; $p < 0.001$).

Concerning the other load weights used in this test (5, 15 and 20 kg) similar results have been found. As expected, lower limitations are assessed in all age groups for a load weight of 5 kg (overall median = 1.0; $MD\tilde{x} = 0.68$). Nevertheless, a highly significant ($p < 0.001$) difference between "younger" and "older workers" exists also for a load of 5 kg. The correlation with age is also highly significant ($p = 0.001$) but weak ($\tau = 0.252$).

The overall values for a load of 15 kg (median = 3.0; $MD\tilde{x} = 0.76$) show that higher grades of limitation are to be found with that higher load weight. The group of "older workers" has a high median of 3.5 ($MD\tilde{x} = 0.81$) whereas "young workers" have a median of only 2.0 ($MD\tilde{x} = 0.71$). This proves to be a highly significant difference ($p < 0.001$) as well. Like in the case of a load of 10 kg, the group of 50–54 year-olds show more severe grades of limitation ($\tilde{x} = 4.0$; $MD\tilde{x} = 0.65$) than the group of workers aged 55–63 years ($\tilde{x} = 3.0$; $MD\tilde{x} = 1.25$). A weak but highly significant correlation with age has also been found for lifting a load of 15 kg from floor to waist level ($\tau = 0.328$; $p < 0.001$).

For a load of 20 kg hardly any difference in capabilities to a load of 15 kg can be found for the group of "young workers" with a median of 3.0 ($MD\tilde{x} = 0.78$). For "older workers", however, the median is very high (4.0) with a very small deviation ($MD\tilde{x} = 0.60$). Nevertheless, the correlation with age is weak ($\tau = 0.289$; $p < 0.001$).

Lifting Between Waist and Head Level

Lifting loads from waist to head level produces slightly different results than lifting from floor to waist level (see Fig. 3).

In contrast to lifting 10 kg from floor to waist level, higher grades of limitation are found among all age groups for lifting 10 kg from waist to head level. Already among the youngest (age group of 20–24 years old workers) two workers have severe limitations. The lowest limitations are again to be found among the workers from 25 to 29 years of age ($\tilde{x} = 1.0$; MD$\tilde{x} = 0.47$). The strongest limitations were found among the age group from 50 to 54 years ($\tilde{x} = 4.0$; MD$\tilde{x} = 0.48$) and also among 45–49 year-olds ($\tilde{x} = 4.0$; MD$\tilde{x} = 0.67$). Accordingly, the difference between main age groups of "younger workers" ($\tilde{x} = 1.0$; MD$\tilde{x} = 0.86$) and "older workers" ($\tilde{x} = 4.0$; MD$\tilde{x} = 0.62$) is highly significant ($p < 0.001$). However, the correlation between age and lifting a load weight of 10 kg to head level is weak ($\tau = 0.383$; $p < 0.001$).

A stronger correlation with age ($\tau = 0.405$; $p = 0.001$) was found for a load weight of 5 kg. The difference between "younger" and "older workers" is also highly significant ($p < 0.001$). However, lower limitations were found among the oldest workers ($\tilde{x} = 1.5$; MD$\tilde{x} = 1.38$) than among the 45–49 year-olds ($\tilde{x} = 3.0$; MD$\tilde{x} = 0.95$) and the 50–54 year-olds ($\tilde{x} = 3.0$; MD$\tilde{x} = 0.83$).

For lifting a load of 15 kg from waist to head level, the correlation is again weaker ($\tau = 0.361$; $p < 0.001$). However, higher values are found among all age groups. Although the group of the youngest workers (20–24 years) has almost the same level of limitations ($\tilde{x} = 4.0$; MD$\tilde{x} = 0.67$) as each sub-group (45–49 years: $\tilde{x} = 4.0$; MD$\tilde{x} = 0.24$; 50–54 years: $\tilde{x} = 4.0$; MD$\tilde{x} = 0.13$; 55–63 years: $\tilde{x} = 4.0$; MD$\tilde{x} = 0.25$) of the main group "older workers", a highly significant difference ($p < 0.001$) between "young workers" ($\tilde{x} = 3.0$; MD$\tilde{x} = 0.71$) and "older workers" ($\tilde{x} = 4.0$; MD$\tilde{x} = 0.19$) has been found.

Capabilities to Maintain Static Postures

The capabilities to work in a static posture are divided into working with static arm postures and with static trunk bending; each with two different intensities. In total, 100 workers participated in these tests. Due to contraindications for back bending six workers did not perform these tests. As above, the results are depicted in numbers of participants according to the grade of limitation per age group.

Working at Shoulder Level

Figure 4 shows the results for the participants' limitations working with one hand at shoulder level. On the left, the results are depicted for the individual dominant hand and on the right, for the individual non-dominant hand.

Results for the dominant hand show only minimal differences between age groups. The lowest limitations are found among the oldest workers ($\tilde{x} = 0.0$; MD$\tilde{x} = 0.43$). However, no significant differences between the age groups and no

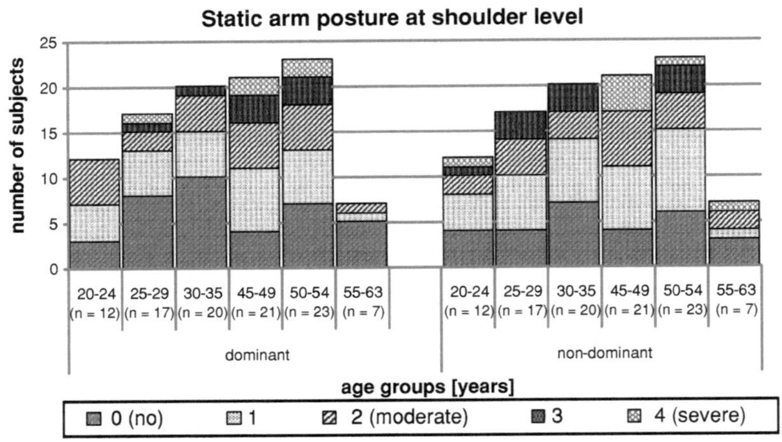

Fig. 4 Assessed grade of limitation of the capability to work in a static posture with one arm at shoulder level

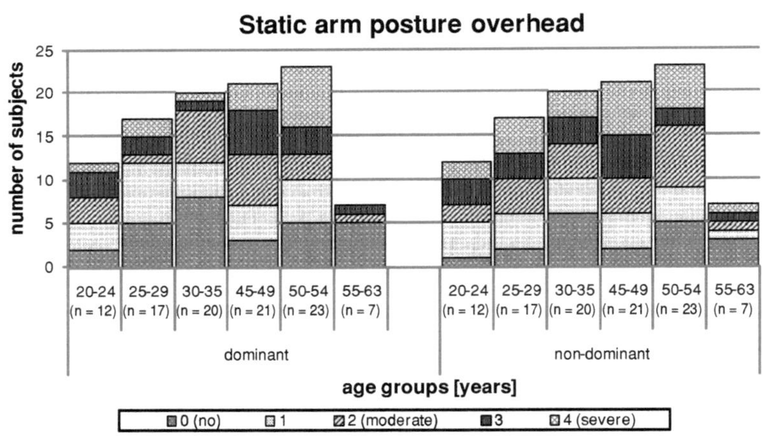

Fig. 5 Assessed grade of limitation of the capability to work in a static posture with one arm above head level

significant correlation between age and working with the dominant hand on shoulder level can be found.

The same is true for the non-dominant hand. However, in this case the least limitations are found among the age group of 30–35 year-olds. A Wilcoxon signed-rank test with a level of significance of $\alpha = 0.05$ revealed that there is no significant difference between the dominant and non-dominant hand.

Working Above Head Level

For performing tasks above head level (see Fig. 5), again, the oldest workers show the lowest limitations using the dominant hand ($\tilde{x} = 0.0$; $MD\tilde{x} = 0.71$). The highest limitations are found among the 50–54 year-olds ($\tilde{x} = 2.0$; $MD\tilde{x} = 1.39$) and the 45–49 years-old workers ($\tilde{x} = 2.0$; $MD\tilde{x} = 1.00$). However, no significant differences between the age groups and no significant correlation with age could be identified.

Using the non-dominant hand, the overall values are higher ($\tilde{x} = 2.0$; $MD\tilde{x} = 1.16$) than using the dominant hand ($\tilde{x} = 1.0$; $MD\tilde{x} = 1.05$). The highest limitations in average are found in the group of 45–49 year-old workers ($\tilde{x} = 3.0$; $MD\tilde{x} = 1.14$) and the lowest among the 55–63 year-olds ($\tilde{x} = 1.0$; $MD\tilde{x} = 1.29$). As in the case of working at shoulder level, no significant differences between the age groups and no correlation with age could be found. In contrast to working at shoulder level, a significant difference was identified between working above head level with the individual dominant or the non-dominant hand ($p = 0.004$).

Trunk Moderately Bent Forward

Assessed limitations for working with moderately forward bent trunk are shown in Fig. 6. The lowest limitations are found among the group of workers from 25 to 29 years of age ($\tilde{x} = 0.0$; $MD\tilde{x} = 0.65$) and 55–63 years-old workers ($\tilde{x} = 0.0$; $MD\tilde{x} = 0.57$). Among the group of the youngest workers (20–24 years) 3 participants with severe limitations were identified which leads to worse results ($\tilde{x} = 1.5$; $MD\tilde{x} = 1.58$) than the group of 50–54 year–old workers ($\tilde{x} = 1.0$;

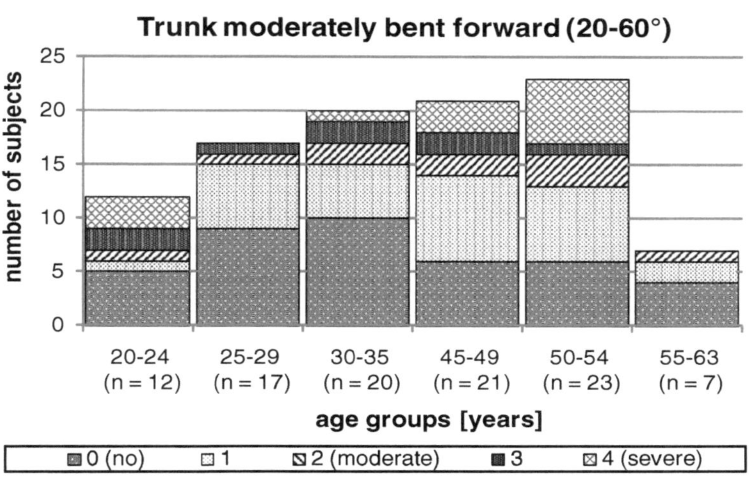

Fig. 6 Assessed grade of limitation of the capability to work in a static posture with trunk moderately bent forward

MD$\tilde{x}$ = 1.26). No significant differences between the age groups and no correlations with age were found for working with moderately bent trunk.

Trunk Strongly Bent Forward

Figure 7 shows the results for working with the trunk strongly bent forward. Once again, more limitations are found among the group of the youngest workers ($\tilde{x} = 2.5$; MD$\tilde{x} = 1.50$) than among the 50–54 year-olds ($\tilde{x} = 2.0$; MD$\tilde{x} = 1.22$) or among the 55–63 year-olds ($\tilde{x} = 2.0$; MD$\tilde{x} = 0.71$). The most limitations are found in the group of workers from 45 to 49 years of age ($\tilde{x} = 3.0$; MD$\tilde{x} = 1.19$) and the least among the 25–29 year-old workers ($\tilde{x} = 1.0$; MD$\tilde{x} = 0.53$). A significant difference between "younger workers" (20–35 years) and "older workers" (45–63 years) was identified ($p = 0.027$). However, no significant correlation between age and the capability of working with the trunk strongly bent forward was found.

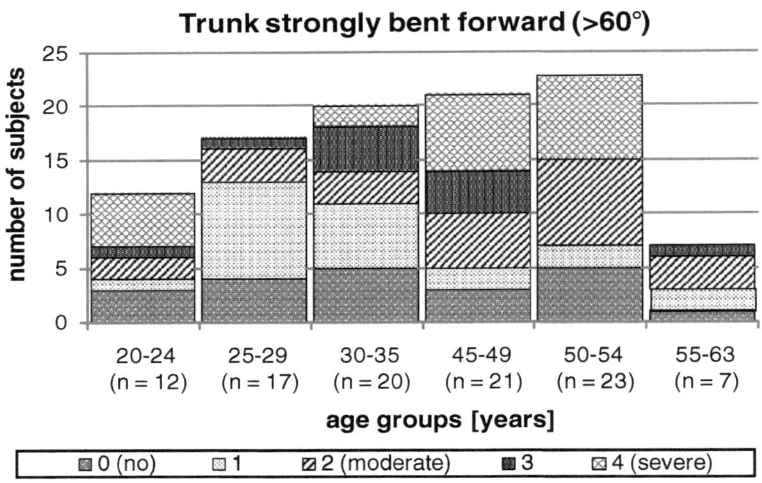

Fig. 7 Assessed limitations working with trunk strongly bent forward

Questionnaire Data

Work Ability Index (WAI)

The short version of the Work Ability Index was used to assess the subjective work ability of the workers. The WAI score was calculated by summation of all single item scores and then classified into four categories:

- Poor: 7–27 points
- Moderate: 28–36 points
- Good: 37–43 points
- Excellent: 44–49 points.

Higher scores of the WAI indicate better work ability.

Data of 104 participants could be analyzed. Two participants did not complete the WAI-questionnaire. The results are presented in Fig. 8.

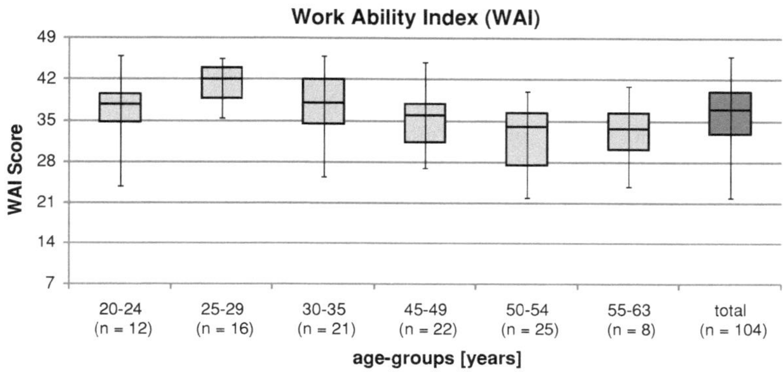

Fig. 8 Results from WAI-questionnaire

The Box-Whisker-Plots in Fig. 8 show the median, the lower and upper quartile as well as the sample minimum and maximum for each age group (numbers of participants are noted on the x-axis).

The overall median of the WAI score was 37 ($MD\tilde{x} = 4.08$). This result represents a good self-reported work ability of workers in a production plant in the automotive industry.

The age group of 25–29 years shows the highest median WAI score ($\tilde{x} = 42$; $MD\tilde{x} = 2.69$). The age group of 30–35 years ranks second with a median WAI score of 38 ($MD\tilde{x} = 4.9$). Both groups have good self-reported work ability.

The lowest median WAI score with 34 points is found among the workers of 50–54 years and 55–63 years. Their work ability is only on a moderate level. Overall, the "older workers" have a lower mean WAI score than the "younger workers". This was also statistically calculated with the U-Test and highly significant ($p < 0.001$).

An interesting aspect is that the age group of 25–29 years got the most ratings of excellent (31.25 %) and good (56.25 %) work ability. None of the participants in this age group show poor self-assessed work ability. In contrast the workers between 50 and 54 years have the highest percentage (28 %) of poor work ability. Concerning age-related differences in experienced strain see chapter Age-Related Differences in Critical Driving Situations: The Influence of Dual-Task Situations, S-R Compatibility and Driving Expertise, Aschersleben et al.

Discussion

Cross-Case Analysis

It is evident from the results that physical capabilities concerning specific body-postures show significant differences between "younger" (20–35 years) and "older workers" (45–63 years). However, relatively few significant differences ($p < 0.05$) can be identified in our study.

Considerable differences between "younger" and "older workers" are found for capabilities related to manual materials handling. Highly significant are the differences between the age groups of 20–35 years and 45–63 years. However, the comparison of age subgroups also shows several significant differences.

Though, higher grades of limitation can be also found among "younger workers", and the oldest workers (55–63 years) in our sample are rarely the group with the highest grades of limitation. Instead, this is often the group of 50–54 year-old participants. The good results of the 55–63 year-old workers may suggest a Healthy-Worker-Effect (Wen et al. 1983), but only 8 workers in this age group were available to participate. Also, psychological aspects to show their fitness at this age could have influenced the results of this group (see chapter Age Differences in Motivation and Stress at Work, Hertel et al.) but were not measured in this study. Therefore, no substantial evidence concerning these two effects could be gathered in this study.

Overall, only in the case of lifting a load of 5 kg between floor and waist level no limitations were found only in the group of 25–29 year-olds. In all other cases limitations were detected in all age groups; with a median growing roughly by 1.0 with each load level (5, 10, 15, and 20 up to 45 kg). At a load weight of 20 kg lifted between floor and waist level as well as lifting a load weight of 15 kg between waist and head level the median reaches its maximum of 4.0.

In almost all cases the group with the least limitations is the group of 25–29 year-old workers. The youngest workers (20–24 years) have stronger limitations than the participants of the consecutive two age groups from 25 to 35 years of age.

The assessments of capabilities to work in static postures show less influences of age: Any correlation or significant differences between the age groups were found for working on shoulder or above head level. The oldest workers were in both cases even the group with the least limitations. In contrast to age, the use of the dominant or non-dominant hand showed influence on limitations (at least in the case of working above head level).

Working with bent trunk, significant differences between the age groups exist only in the case of a strongly bent trunk (>60°). Solely the group of workers, who are in the age of 45–49 years show higher but no significant differences ($p < 0.01$) compared to 25–29 year-old workers in their capability to strongly bend their trunk forward. Here, "older workers" (45–63 years) have significantly stronger limitations than "younger workers" (20–35 years). However, in the case of moderate trunk bending the youngest workers (20–24 years) show the most limitations, but no significant differences can be identified.

In most significant cases the "older workers" reveal more severe grades of limitation of their capabilities than the group of "young workers". All in all, only for manual materials handling and working with strongly bent trunk significant (negative) effects of age were found.

The results of the cardiopulmonary load test show that the general fitness of workers is in "normal range" (PPI mean values between 1.0 and 2.0). However, a highly significant ($p < 0.01$) difference between "younger" and "older workers" could be identified. "Older workers" achieve in average slightly lower PPI values. As the calculation of the PPI is a combination of individual strain (measured in the interval between maximum and normal heart rate) and performance level (measured time to complete the 20 squats), no clear conclusion concerning age-related effects on performance or strain level can be drawn.

Results from the WAI questionnaire back the results from the objective assessments conducted in this study. Only the group of the oldest workers (55–63 years) estimate their work ability with lower scores as the results of the objective assessments suggest. As the participants were available only for a very limited amount of time, the capability tests had only a relative short duration. To reach measurable effects relatively demanding tasks were chosen for the test setting. Tests that use less demanding tasks, but in return a longer time duration (e.g. an 8 h shift) may produce different results. As no longitudinal study was conducted, selection effects may be present. Furthermore, the question arises, if physical exposures from the workers' actual and past jobs affect the results.

Transfer to Industrial Practice

Looking at the results of this study, it can be stated that workers' capabilities seem to be influenced by age. But it is evident, that not all capabilities significantly diminish with age. Only specific capabilities decrease. Namely, manual materials handling and working with strongly bent trunk show the strongest age dependence in our sample.

The assessments of more than 150 workplaces in the automotive industry conducted in this study together with experiences from workplace assessments in other studies (see also chapter Assembly Tasks in the Automotive Industry: A Challenge for Older Employees, Frieling et al.) clearly show that manual materials handling and working in static body postures are capabilities of high relevance for workers in the automotive industry.

A serious problem for workers is the combination of load weight and/or awkward body postures while performing work tasks, which are located in a vehicle's interior. Thus, appropriate actions should be initiated.

Ergonomic work design offers opportunities to reduce the need for these "age-critical" tasks. Lifting aids and the reduction of the individual load weight as well as avoiding tasks that incorporate lifting above waist level can eliminate these adverse age-related effects on the worker's health and performance. Anthropometric work design should be utilized to avoid any strong trunk bending while performing work tasks. In cases where technical improvements cannot be realized, organizational measures like job rotation may help to reduce duration of exposures and adverse effects. These measures should not only address "older workers", who already show limitations of their capabilities, but "young workers" must be prevented from critical exposures. Furthermore, individual training and compensation strategies can be used to sustain a constant capability "level" (see also chapter Influence of Age and Expertise on Manual Dexterity in the Work Context: TheBremen-Hand-Study@Jacobs, Voelcker-Rehage et al.).

Outlook

The number of "older" workers in industry nations is rising. From the results of this study there is a need for action, as we know, that specific work-related capabilities of workers diminish significantly with age and these capabilities are needed to cope with physical demands on workplaces in automotive industry.

To keep the automotive industry competitive on the global market it is necessary to successfully manage an aging workforce (see chapter Effects of an Ageing Workforce on the Performance of Assembly Systems, Zülch et al.). Different possibilities exist to prevent a decrease of worker capabilities and to sustain work-related capabilities. This includes the redesign of workplaces, changing work organization (e.g. job rotation) and integration of behavioral ergonomics, as mentioned previously.

Furthermore, it has to be postulated that ignoring age-related work design will have negative effects on workers as well as on the competitiveness of manufacturing companies. Negative consequences for workers could be excessive physical (as well as mental) strain, work-related (musculoskeletal) disorders and early retirement because of disabilities. It will also dissipate and squander valuable resources e.g. experience of "older workers" (see Age-differentiated Work Systems: Introduction and Overview to a Six-Year Research Program in Gemany, Schlick et al.).

To achieve a healthy and productive workforce, knowledge and systematic data collection about exposures on specific workplaces as well as work-related capabilities of workers, especially of those who are "age-critical" (e.g. manual materials handling and working with trunk strongly bent forward), are required. This would allow developing and using a large pool of data concerning capability profiles of workers. Some companies already have available data about capabilities of their workers; especially about those, who already have disorders. However, in most cases this data is not systematically collected and thoroughly analyzed.

As a next step, the presented results of this study will be used to develop a database with standardized work-related capability-profiles of workers. Data about capabilities gathered in this study will be compared with data available in a company to develop a common data structure. To be able to do so, there is a need for standardized work-related capability testing for all workers in production plants and a subsequent systematic transformation into a database. Therefore, a standardized capability testing procedure will be developed, which can be used efficiently in practice in order to be used in companies.

Eventually, the collected data can be used for a matching of worker capabilities with the work-related exposures, especially with age-critical tasks in a company. This allows to detect possible risks for workers' health and to design workplaces suitable to all age groups (age-robust workplace design).

A further aim could be that multiple companies (e.g. automotive manufacturers) use one common database as a source of information for an "age-robust" design of workplaces. In this database information about age-critical tasks and design aspects, like the comprehensive literature database concerning aspects of age and vision developed by Keil (2011) could be added.

In the long term, exposure biographies of workers could be generated. The comparison of individual exposure biographies and capability ratings may deliver additional information in order to create work task demand profiles.

Another approach could be to develop a screening method for an age-differentiated assessment of mechanical exposures. The screening of existing or planned workplaces could make age-related ergonomic deficits in work system design transparent. Consequently, guidelines for age-differentiated ergonomic work design could be developed.

Acknowledgments The authors thank the German Research Foundation (DFG) for funding this research (LA 286/16-1, BR 3544/2-3). We would also like to thank all cooperation partners from industry and science for their support and sharing data and experiences.

References

Aittomäki A, Lahelma E, Ross E, Leino-Arjas P, Martikainen P (2005) Gender differences in the association of age with physical workload and functioning. J Occup Environ Med 62:95–100

Bildt Thorbjörnsson C, Alfredsson L, Frederiksson K, Michelsen H, Punnett L, Vingard E, Torgen M, Kilbom A (2000) Physical and psychosocial factors related to low back pain during a 24 year period. Spine 25:369–375

Bortz J, Lienert GA, Boehnke K (2000) Verteilungsfreie Methoden in der Biostatistik. Springer, Berlin

Chen JJ (2007) Functional capacity evaluation and disability. The Iowa Orthopaedic Journal 27:121–127

Corlett EN, Bishop RB (1976) A technique for assessing postural discomfort. J Ergon 19:175–182

Enríquez Díaz JA, Shiekirka S, Markus A, Frieling E, Straeter O (2010) Chaku–Chaku assembly system—A new trend in the automotive industry: First results of the ergonomic evaluation on a case study. In: Karwowski W, Salvendy G (eds.) Proceedings of the 3rd international conference on applied human factors and ergonomics. USA Publishing, Miami

Gall B, Parkhouse W (2004) Changes in physical capacity as a function of age in heavy manual work. J Ergon 47:671–687

Gouttebarge V, Wind H, Kuijer PP F, Frings-Dresen MH (2004) Reliability and validity of functional capacity evaluation methods: a systematic review with reference to Blankenship system, Ergos work simulator, Ergo-Kit and Isernhagen work system. International archives of occupational and environmental health 77:527–537

Hamberg-van Reenen HH, van der Beek AJ, Blatter BM, van Mechelen W, Bongers PM (2009) Age-related differences in muscular capacity among workers. Int Arch Occup Environ Health 82:1115–1121

Isernhagen SJ (1988) Functional capacity evaluation. In: Isernhagen SJ (ed) Work injury: management and prevention. Aspen, Gaithersburg, pp 139–174

Kahya E (2007) The effects of job characteristics and working conditions on job performance. Int J Ind Ergon 37:515–523

Keil M (2011) Konsequenzen des demographischen Wandels für zukünftige Produktions- und Technologieabläufe am Beispiel der altersbedingten Veränderung der Fähigkeit des Sehens. Dissertation. Wissenschaftliche Schriftenreihe des IBF, Heft 91, Chemnitz

Kenny GP, Yardley JE, Martineau L, Jay O (2008) Physical work capacity in older adults: implications for the aging worker. Am J Ind Med 51:610–625

Kruskal WH, Wallis WA (1952) Use of ranks in one-criterion variance analysis. J Am Stat Assoc 47:583–621

Kumashiro M (2000) Ergonomics strategies and actions for achieving productive use of an ageing workforce. J Ergon 43:1007–1018

Kuorinka I, Jonsson B, Kilbom A, Vinterberg H, Biering-Sørensen F, Andersson G, Jørgensen K (1987) Standardized nordic questionnaires for the analysis of musculoskeletal Symptoms. Appl Ergon 18(233):237

Landau K, Rademacher H, Meschke H, Winter G, Schaub K, Grasmück M, Mölbert I, Sommer M, Schulze J (2008) Musculoskeletal disorders in assembly jobs in the automotive industry with special reference to age management aspects. Int J Ind Ergon 38:561–576

Meißner-Pöthig D (1997) Vitalitätsdiagnostik nach Pöthig®. In: Meißner-Pöthig D, Michalak U (eds) Vitalität und ärztliche Intervention: Vitalitätsdiagnostik. Grundlagen—Angebote—Konsequenzen, Stuttgart, pp 64–72

Petrenz J (1999) Alter und berufliches Leistungsvermögen. In: Gussone M (ed) Ältere Arbeitnehmer—Altern und Erwerbsarbeit in rechtlicher, arbeits- und sozialwissenschaftlicher Sicht. Bund, Frankfurt am Main, pp 63–99

Punnett L, Gold J, Katz JN, Gore R, Wegman DH (2004) Ergonomic stressors and upper extremity musculoskeletal disorders in automobile manufacturing: a one year follow up study. Occup Environ Med 61:668

Rademacher H (in preparation) Engpassorientierte Analyse physischer Arbeitsbelastungen und Mitarbeiterfähigkeiten als Grundlage zur altersdifferenzierten Arbeitsgestaltung am Beispiel der Automobil- und Zulieferindustrie. Dissertation an der TU Darmstadt

Rademacher H, Bruder R, Sinn-Behrendt A, Landau K (2011) Identifying demographic bottlenecks for musculoskeletal risks in production areas—implications for the design of industrial workplaces and assignment of workers. In: 10th International symposium on human factors in organisational design and management, Grahamstown

Rademacher H, Sinn-Behrendt A, Landau K, Bruder R (2008) Ergonomic analysis and design of industrial workplaces with special reference to age. In: Karwowski W, Salvendy G (eds.) Conference proceedings of the 2008 applied human factors and ergonomics conference, Las Vegas

Rademacher H, Sinn-Behrendt A, Schaub K, Landau K (2006) Ein Verfahren zur Ermittlung altersbedingter Engpässe hinsichtlich muskuloskelettaler Belastungen. Z Arbeitswissenschaft 60:230–244

Schaub K, Landau K, Bruder R (2008) Development and application of methods for stress analysis in manufacturing. In: Karwowski W, Salvendy G (eds.) Conference proceedings of the 2008 applied human factors and ergonomics conference, Las Vegas

Seibt R, Dutschke D, Hübler A, Scheuch K (2007) Handlungsanleitung zur Durchführung einer individuellen Vorsorgediagnostik für Lehrkräfte. In: Seibt R, Scheuch K (eds) Verbundprojekt Lange Lehren. Selbstverlag der Technischen Universität Dresden, Dresden

Shephard RJ (2000) Aging and productivity: some physiological issues. Int J Ind Ergon 25:535–545

Savinainen M (2004) Physical capacity and workload among ageing workers. Dissertation, Faculty of Medicine, Tampere, University of Tampere

Silverstein M (2008) Meeting the challenges of an aging workforce. Am J Ind Med 51:269–280

Tuomi K, Ilmarinen J, Jahkola A, Katajarinne L, Tulkki A (1998) Work ability index (2nd edn). Finnish Institute of Occupational Health, Helsinki

Wen CP, Tsai SP, Gibson RL (1983) Anatomy of the healthy worker effect: a critical review. J Occup Environ Med 25:283–289

Standards

DIN EN 614-1:2006 Safety of machinery. Ergonomic design principles. Terminology and general principles

Field Study of Age-Critical Assembly Processes in the Automotive Industry

Kerstin Börner, Christian Scherf, Bianca Leitner-Mai and Birgit Spanner-Ulmer

Lessons Learned

Based on an increasing proportion of aging employees, and thereby a rise in the personnel's mean age, companies are currently being forced to face the new challenges posed by the aging of their workforce. It remains to be proven whether older employees exhibit the same performance results as their younger counterparts in a short-cycled assembly system in which a high demand is placed upon sensory motoric skills and physical abilities. Despite research exhibiting an average performance loss of 1.5 % per decade in physical prowess from the age of 30 (Fries 1989), tangible observations are still lacking. A variety of factors constitute this general loss in performance which can, however, be compensated for through the use of individual adaptation processes. We developed a test design complimented by field studies in which employees were directly observed at their workplace on an assembly line. This observation stage was documented with the use of up to eight simultaneously running cameras that recorded each detail of the assembly tasks asked of the participants. The outcome of the observation stage of the study was approximately 300 min of video material showing the basic motions of each individual subject of the study. Concurrent to the video data that was

K. Börner (✉) · C. Scherf · B. Leitner-Mai · B. Spanner-Ulmer
Chair for Ergonomics at the Institute of Industrial Management and Factory Systems, Chemnitz University of Technology, Chemnitz, Germany
e-mail: kerstin.boerner@mb.tu-chemnitz.de

C. Scherf
e-mail: christian.scherf@mb.tu-chemnitz.de

B. Leitner-Mai
e-mail: bianca.leitner-mai@mb.tu-chemnitz.de

B. Spanner-Ulmer
e-mail: awi@tu-chemnitz.de

C. M. Schlick et al. (eds.), *Age-Differentiated Work Systems*,
DOI: 10.1007/978-3-642-35057-3_11, © Springer-Verlag Berlin Heidelberg 2013

collected during the observation stage, data pertaining to biological effort on the body (heart rate and breathing rate) were also collated allowing for well-founded conclusions to be made concerning how the load of both younger and older assembly workers change during an 8 h shift. On the basis of our investigations, we have been able to draw the following observations:

- Assembly workers, regardless of age, exhibited a similar strain on the basis of subjective appraisal and the differences in pulse measured.
- Partly hard, physical work in 8 h shifts is accomplishable for older male assembly line workers (45–60 years) at a steady state.
- Sensory motoric demanding work in 8 h shifts is also achievable for older female assembly line workers (44–57 years) at a steady state.
- Older (46–60 years) and middle-aged (31–42 years) male assembly line workers show a constant work speed compared to younger workers (21–29 years).
- Older workers showed significantly slower assembly times during a given 8 h shift in two assembly processes observed.

Introduction

Due to oft cited and well noted contemporary demographic changes, companies are currently being forced to maintain senior employees within their workforce. As a result, and to make such an action feasible, manufacturing companies, in particular, have found it necessary to redesign production and work processes in order to meet the pressing implications of an ageing workforce.

Despite a large field of studies related to an age-induced decline of basic physical and mental abilities (e.g., Allen 2002; Schaie 1996; Spirduso et al. 2005) it has been found that the current design of workplaces and production processes fail to completely incorporate human factors and ergonomics (Keil et al. 2010). Production lines with directed motions and repetitive movement sequences, frequently required within companies' strict short cycle times, often result in the strained posture of the employee. Such factors place high physical demands on the employees as well as demands from a production perspective. On account of the necessity of short cycle times on an assembly line the chance to individually regulate the speed of work and/or work process are reduced to a minimum. Moreover, the dictate of productivity seeks further cycle time reductions and an increase in productivity of work content (Frieling et al. 2006). In light of this increased physical and mental load on employees, calls of further investigation exist surrounding the principle of humane work (Spanner-Ulmer et al. 2009).

Current approaches aimed at dealing with an aging workforce are somewhat lacking (Frieling et al. 2006; Spanner-Ulmer et al. 2009). The automotive industry, for example, has implemented new workplaces and workstations for those physically restricted employees in an attempt to alleviate the pressures of their aging workforce (Bös 2007). Whilst this has been an effective solution to date, when considering future projections of work commitments it becomes clear that it will not be feasible for companies to reassign all employees of age-related deteriorating performance to soft-workplaces or exempt them from the rotation system. Furthermore, such a division of labor implies possible controversial implications. On the one hand, it gives rise to potential irritations regarding the apparent different performance requirements of the employees resulting in a clear separation between younger and older age groups (Keil et al. 2010). On the other hand, it remains questionable whether those often content-free tasks comply with the ergonomic standards for health- and personality-enhancing working conditions.

A further critical point is the strong causality between performance restrictions and senior employees. The often one-sided discussion ascribes older employees a general performance decline which is not found when complex tasks are considered (Hacker 2003; Wild-Wall et al. 2009). Considering the current extent of research, the impact of performance decline, load duration and age-critical basic motions on performance in assembly systems remain unknown (Frieling et al. 2006; Keil et al. 2010) whilst the positive influence of work experience, automated motion sequences, and performance regulation on the work performance has only partially been analyzed (Henry et al. 2004; Park 1994). The extent to which these factors can be implemented into a strictly clock-dependent assembly system should be the subject of future investigation.

The research project, completed within the framework of this priority program, aims to answer the fundamental question of whether senior employees in assembly line positions are physically strained more than younger employees. By extension, such a project also implies the analysis of the possibilities of operational individualization on the assembly line. The chapter's key point of interest therefore aims to explore the possible variance between younger and older employees within this assembly process field. In order to achieve this, we conducted two field studies, the first observed 23 female assembly workers, and the second, 31 male assembly workers both from the automotive sector.

Background

Rohmert (1984) developed and established the ergonomic stress-strain-concept constituting our point of departure for this project. By integrating the four steps of analysis, measurement, assessment and redesign, it is possible to plan and design work according to humane and business rational points of view. The concept is based on a simple cause-effect-relationship of stress and strain.

Stress, measured objectively, results from a given work system and can be assessed according to its extent and duration. Depending on an individual's constitution, disposition, qualification and adaptation features, this stress appears as subjective strain on the working staff. Accordingly, being a subjective measurement, strain expresses the individually perceived amount of the exterior occurring stress on the body. This severe cause-effect-relationship between "objective" stress and "subjective" strain only rarely occurs in reality. On the one hand, exogenous effects of a work system, considered under the term stress, are not always objectively measurable (e.g., posture, work climate). On the other hand, activity regulation allows for a variation of implementation conditions (e.g., physical effort and working speed) and subject-related behavior (e.g., fatigue-related evasive movement) (Rohmert 1984).

Although, there is a clear dominance of physically demanding work over intellectual tasks when working in short-cycled assembly systems, a certain amount of activity regulation can be assumed. Even regarding simple motor tasks, such as lifting and carrying loads, posture and load application points are individually variable parameters. Figure 1 shows a modification of the age-differentiated stress-strain-concept introduced by Keil et al. (2010) and is based on the advanced stress-strain-concept (Schlick et al. 2010). The classical age-differentiated stress-strain-concept has been expanded by activity regulation and system reaction (Scherf et al. 2010).

It remains questionable whether decreasing physical performance ability, supported by various studies (e.g., Allen 2002; Spirduso et al. 2005), has any effect on the work process itself, or whether the work performance can be sustained by

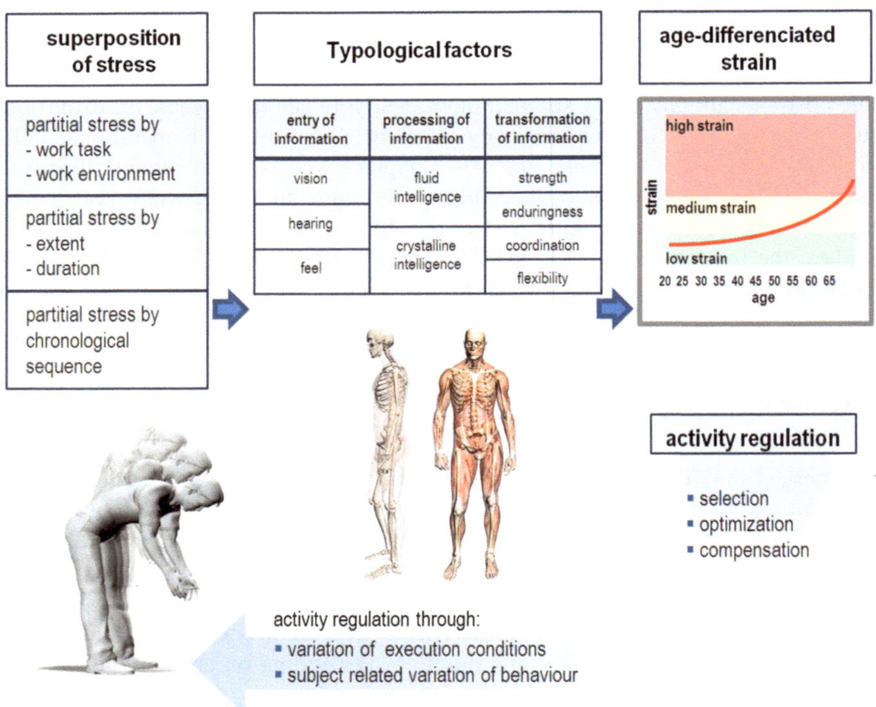

Fig. 1 Advanced age-differentiated stress-strain-concept [according to Keil et al. (2010) and Rohmert (1984)]

compensatory effects (activity regulation). Scientific evidence, however, is available for the assessment of decreasing physical performance with increasing age for the variables of strength (e.g., Runge et al. 2004; Viitasalo et al. 1985) and flexibility (e.g., Barnes et al. 2001; Chateauroux and Wang 2008).

Furthermore, other studies substantiate an age-related performance decline in neuromuscular performance ability (Fozard et al. 1994; Pierson and Montoye 1958). This approach in assessing neuromuscular performance ability is gaining greater importance in contemporary work processes due to rising demands through process automation, miniaturization and shorter cycle times (Heuer and Hegele 2006). Pierson and Montoye (1958) were among the first to differentiate between reaction and movement time within the stimulus-response time. Movement time is recorded at its lowest at the age of 18, and thereafter constantly rises in conjunction with age. More recent research by Fozard et al. (1994), as well as by Stelmach et al. (1986), confirm this relation and also point to the impact of the complexity of the task (e.g., stimulus discrimination and control of directed movements).

Within the framework of the priority program "Age-differentiated work systems" Heuer and Hegele (2006) were able to prove that precise target movements become slower with older age, this is especially the case in relation to higher ranges and increased accuracy. Moreover, a performance decline in movements with spatially divided execution and observation in older workers can be assumed. It has been shown in more recent research by Hegele and Heuer (2010) that conscious-independent adaptation processes are age-stable, whereas conscious-dependent components of this adaptation are reduced in older subjects. Further studies by Bützler et al. (2011) also show, alongside the angle influence of movements of the hand-arm-system, a significant age influence. Here, older subjects show increased movement times independent of angle and target distance.

As far as recent research is concerned, no substantial answers with regard to an age-differentiated design of tasks and work processes can be found (Frieling et al. 2006). Furthermore, the literature reveals no sufficient results showing which basic assembly motions underlie an age-critical influence. The studies mentioned above, often conducted under laboratory conditions, attest to the fact that older people have a general performance loss in the areas of strength, flexibility and neuromuscular performance ability. It should however be clearly emphasized, that speed-centered and peak force-grounded tests do not allow conclusions to be made in relation to job performance (Hacker 2003). Moreover, productivity, in relation to total output and work input, is less determined by the physical assignment of the employee than rather the degree of automation and the efficiency of processes (Shepard 1999). Such an example is that of older subjects have been observed to exhibit better work performance results than younger people regarding realistic task planning (Hacker et al. 1999). These statements also support the assumptions of the expanded stress-strain-concept by Rohmert (1984). Although, the possibilities for activity regulation through variation of the mode of operation, assembly time and assembly quality are low in strictly cycled assembly systems, it can be assumed that the employees do use the minimal regulation scope (Scherf et al. 2011). To what extent this scope of action is used, is the subject of this research chapter.

Furthermore, the extent to which the duration of stress influences short-cycled assembly systems remains to be fully explored (Keil et al. 2010). It has been shown that if workload exceeds a certain limit, daily reoccurring performance is not possible (Schmidtke 1993). This critical value of workload is known as the "continuous performance limit" and marks the threshold of heart rate frequency and the extent to which a constant performance, without muscular fatigue, can be delivered (Müller 1950). Stays a worker's heart rate below this continuous performance limit he works within a "steady state". An age-critical basic motion is not, however, defined only by the single load level, but also by the load duration (Scherf et al. 2011). The load duration factor is considered in different ergonomic evaluation procedures; how far these data are validated by longitudinal studies in field tests remains unproven. In addition, age has only been entered as an influencing factor in a limited number of evaluation procedures (Keil et al. 2010).

In summary, field tests analyzing strains of younger and older assembly line workers in short-cycled assembly systems are needed in order to make validated statements as to the way in which stress duration, age-critical basic motions, as well as stress change, have an influence on the strain of younger and older employees and to make suggestions for future enhancements in the manufacturing industry.

Methodology

We conducted two similar field studies (denoted henceforth as 'A' and 'B') in two factories from within the automotive industry. Due to the unequal distribution of men and women in the factories chosen, we focused on the respectively larger group of employees. In field study A, we observed female workers during the working shift time of 6:00a.m. to 2:45p.m. at the single workstation at which belt buckles were assembled. In field study B, we studied the actions of male workers between the times of 6:00a.m. and 2:00p.m. as they worked on the assembly line for 4 h assembling flywheels (with its upper-middle load requirements) followed by 4 h at a station with low load. In both studies we used the same instruments for observation.

Subjects

We recruited the participants for field study A at a first-tier supplier in the automotive industry, and for field study B at an OEM plant for car engines, both in Saxony (eastern Germany). We invited all available assembly line workers from the chosen workplaces to informative introductory meetings during their working time and asked them to participate voluntarily in the study. A prerequisite for the study was that the subjects concerned had to have been trained at the workplace on the assembly line for which they would be examined. After workers had given their written consent to be a part of the study, they were sent for a medical examination in which, amongst others, the following values were recorded: resting pulse rate, Pulse Performance Index-test (PPI), Body Mass Index (BMI) as well as notification of any discomfort of shoulder, arm and hand. Any potential participant exhibiting a medical condition or taking medication that could possibly affect heart frequency were excluded from the sample. Ultimately, 23 well-trained female workers aged between 27 and 57 years (M = 43.9 years, SD = 6.82 years) took part in the field study A. For field study B, we obtained 31 male participants between 21 and 60 years (M = 38.45 years, SD = 11.78 years).

Workplace

As noted above, the field study observations were conducted at the workstation "belt buckle assembly" (field study A) for observation within a plant of the automotive supply industry (factory A) and at the workplace "flywheel assembly" in an automotive engine factory (factory B) for field study B.

Many of the performance tasks involved at the two workstations studied included basic motions of the hand-arm-system. We analyzed the assembly process using the Methods-Time Measurement (MTM); a preset motion-time-system, used primarily in industrial settings, to analyze the methods used to perform any manual task. A basic MTM-cycle consists of the reach, grasp, move, position and release procedure (Bokranz and Landau 2006). Thereafter, we analyzed these motions in relation to age-related changes using the "Chemnitz Age Database" (Keil et al. 2009).

The single-work-stations in factory A were arranged in two groups. Each group consisted of four workplaces for belt buckle assembly and was surrounded by a constantly moving common assembly line to collect completed pieces. Together, subjects of one group were obliged to assemble a designated number of belt buckles within an 8 h shift, they were, however, free to determine their individual working speed. We simultaneously observed four workplaces by use of video recording equipment and two scientific observers. We chose two employees from each group to ensure that the observed assembly workers were not required to restock the basic components required to complete task. The employees responsible for the restocking of assembly parts were not included in the observation that day. The research technique chosen did not interfere with the production process. Nine different models of belt buckle were assembled at identical workplaces, however only one type was assembled at any given time. For each belt buckle model between nine and eleven individual pieces had to be assembled manually. Workers either sat or stood in front of an assembly desk on which several small load carriers with the individual pieces to be assembled were laid out. Stock was refilled following the Kanban-System, as well as the pre-mounted basic modules provided to the right of each worker. Kanban is a manufacturing process developed in Japan that works in accordance with the Pull-principle whereby cards are used to reorder materials in a just-in-time system (Wiendahl 2010).

Each belt buckle was assembled in approximately 35 s and was then placed on the continuously moving conveyor-belt to the left of the worker. Two cameras were attached to each workplace to record the motions of the hand-arm-system. Four workplaces were equipped with cameras giving a total of eight cameras recording simultaneously.

To assess the physical load of the investigated workplace we used the European Assembly Work Sheet (EAWS) (see chapter Capability Related Stress Analysis to Support Design of Work Systems, Rademacher et al.). EAWS allocates points for postures, forces, manual handling and repetitive motions, wherein the incorporation of short-cycle loads is characteristic (Schaub and Ahmadi 2007). The results are then categorized according to load and risk: 0–25 denoting low load and low health risk (green), 26–50 for middle load and possible health risk (yellow) and >50 for high load and high health risk (red) (Otto and Scholl 2011). For the belt buckle assembly we found a load of 15.5 for sitting and 19.7 for a mixed calculation of sitting and standing. These results indicated a low load placed on the participants.

The focus in factory B was on the workstation concerning "flywheel assembly" in which subjects mounted flywheels, weighing between five and twelve kilograms (kg), on a motor block on an assembly line for 4 h. The average cycle time was less than 1 min. The flywheels were supplied to the worker on palettes of eight flywheels in a stack of eight palettes (i.e. a total of 64 flywheels per stack), which were brought to the assembly line by a forklift operator. These stacks could then be lifted electrically to maintain a constant working height for the workers. Furthermore, the workers were able to manually rotate the palettes for easier access. The mounting process we analyzed included the following actions: reach for the flywheel, lift, place on the motor block, locate and fix with the first screw. The internal company rating of the workstation based on EAWS was calculated at 48 indicating an upper-middle load rating. A second workstation, calculated to have 13 EAWS points (i.e. low load), was chosen as compensation for the fact that subjects had operated at a relatively high load for 4 h when under normal conditions they are required to do this for only 2 h. Both workstations required the subject to adopt a standing position in order to complete their tasks.

Data Collection

Objective Measurements

We used the Aquila-Complete-System for motion recording a system consisting of two linked computers accessing eight synchronized cameras. The software was developed specifically for this study by FusionSystems. A time schedule, distributed evenly over the shift, was defined in order to record the sequences of operation; employee breaks were also considered. Recordings of 3 min were taken 30 times per shift in field study A whilst a total of 48 four-minute recordings were captured in field study B. Additional manual recording was also possible. The videos documented the mode of operation and showed whether the production process was running smoothly.

The subject's heart rate was logged during the entire morning shift at intervals of once per second using a wireless Bioharness^{TM} chest strap from Zephyr^{TM}.

Two researchers were present during the entire 8 h shift and completed a number of protocols which included information about changes of the belt buckle type and flywheel type. Furthermore, we recorded all the activities of the subjects over the course of the shift: conversations, strenuous movements, operating position (sitting or standing) and any time away from the workstation. Temperature was documented every hour in field study A and every 10 min in field study B. Additionally, we used a calibrated Brüel and Kjaer hand-held sound level meter (version 2250-L) to collate the average noise levels of a 5-min-period on an hourly basis.

Questionnaire Data

Different questionnaires were used to collect subjective data. The Work Ability Index was determined by the short-form of WAI (Hasselhorn and Freude 2007) and either completed at home and brought to the medical investigation, or filled in during the medical investigation. The maximum possible score was 49 what equates to a very good ability to work. By means of strain-ratings (Richter et al. 2002) and NASA Task Load Index (TLX) (Hart and Staveland 1988), the individual's subjective strain was measured on the day of observation; these strain-ratings included 12 questions which were answered shortly before and again after the shift, as well as between any rotations of workstations. Subjects assessed their mood (e.g., insecure, alert, energetic, bored) on a range of 1–6 (1 meaning to disagree strongly and 6 to agree strongly). The TLX contained 7 questions which the participants were also asked after the shift and between rotations. The answers formed a measurement for the strain experienced during the observed shift (scale 0–20 in which 0 equals lowly strained and 20 highly strained). The individual chronotype was detected with the Munich Chronotype Questionnaire (MCTQ) (Roenneberg et al. 2003) which was completed by the workers outside of the working environment.

Procedure

Procedure Field Study A

Data for field study A were collated between January and March 2011 during ongoing production at factory A. For all subjects, observation and measurements were carried out during complete morning shifts (6:00a.m. to 2:45p.m.) without disturbance of the production process. The subjects completed the strain-ratings

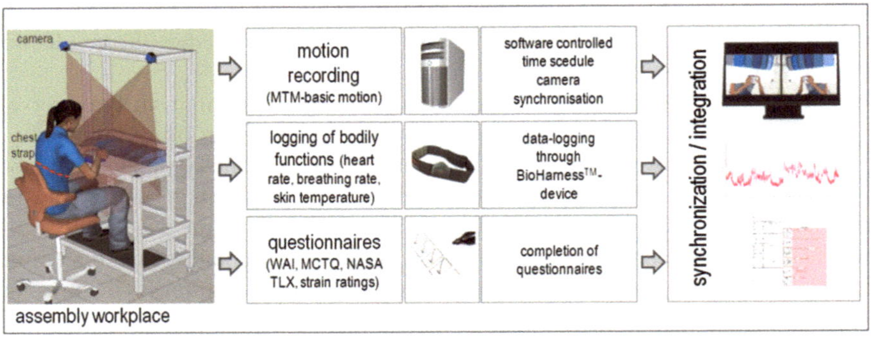

Fig. 2 Schematic experimental design

shortly before the shift in a separate room. Thereafter, the aforementioned chest strap was activated and applied around the subject's chest underneath their clothes. The subject then proceeded to their work stations and completed their normal assembly tasks. On account of the chest strap's receiver's battery life being insufficient to continue recording each second for 8 h, the receiver was replaced with a fully charged unit and recalibrated after 4 to 4.5 h of recording. Movements of the hand-arm-system were recorded with the cameras at the 30 predefined intervals distributed over the whole shift. After the shift the chest strap was detached, the data was transferred to a computer and straps were washed for reasons of hygiene. The subjects were asked to repeat the strain-ratings, as well as the TLX ratings. The WAI and the MCTQ were completed away from the working environment and at a later time. The experimental design is shown schematically in Fig. 2.

Procedure Field Study B

Between May and November 2011 we conducted field study B, also during ongoing production at factory B. We observed two almost identical workstations for flywheel assembly at two assembly lines during the morning shift (6:00a.m. to 2:00p.m.). Nine men formed the daily team responsible for a certain section of the assembly line. Usually, the working places within the team were rotated every 2 h however, for our study two men were exempt from this regular rotation on the given research day. Instead they were assigned for the two selected stations with upper-middle and low load requirements for 4 h each. After 4 h these subjects changed their work stations among themselves. Ideally, we sought to observe the workers over an entire 8 h shift at one workstation in order to record the total load duration. On account of the upper-middle EAWS load rating of flywheel production, this, however, was not permitted by the factory concerned and as a result the 4 h rotation procedure was agreed upon as a feasible alternative.

Again, we provided the workers with the chest strap as in study A and the male subjects were also requested to complete identical questionnaires as their female counterparts in the study A. Additionally however, participants in field study B also completed the strain-ratings and the TLX midway during of the shift when rotating their workstations.

We monitored the chosen workplaces with four cameras 48 times per shift for durations of 4 min each. Each subject was observed on two different observation days: once as they began their shift on the "flywheel assembly" station (upper-middle load) and once again as they started their shift at the station with the low load task.

Analysis

Values of the participant's heart frequency recorded with the chest strap were transferred in an Excel chart and synchronized with the recorded checkpoints. We were then able to determine the corresponding working pulse difference for the observed assembly processes. Furthermore, we were also able to calculate an "average working pulse difference" per minute for each subject when regular assembly (assembly without distraction, e.g., talking with a colleague) was completed. We then calculated the mean pulse difference for 10 min intervals for the age groups we had formed, making sure that at least 75 % of the subjects of each age group were represented. Video data were played in VirtualDub-1.6.19 and assembly times achieved were determined manually. We calculated the BMI of the subjects with the standard formula: [kg/(height in m)^2]. Observation protocols were used to verify the balanced working conditions referring to temperature and noise for all of the periods of observation.

Statistical analysis of the data was undertaken using the SPSS®. Predictive Analytics Software (PASW® Statistics 18.0.0). The *t* test was used to find significant differences, and correlation coefficients (r) were calculated. In addition, partial correlation, Pearson Chi square-test and Fisher's exact test were also employed. All *p* values are reported as significant at less than 0.05.

Analysis of Field Study A

During each 3-min video recorded during field study A, an average of between five and six belt buckles were assembled. From each motion recording, showing regular work without distraction, the assembly of one belt buckle was randomly selected for analysis. As the assembly of the different belt buckle types showed some variation, we therefore analyzed the two largest denominators (denoted hereafter as Period No. 1 and Period No. 2) of the belt buckle assembly and compared them between the two age groups (see Table 1 for details). Period No. 1 covered seven basic MTM cycles, Period No. 2 two basic MTM cycles with the

Table 1 Mean values (±SD) of subjects, Y = young, M = middle, O = old group of participants (field study A and B)

Field study group	A		B		
Age group	Y-Women	O-Women	Y-Men	M-Men	O-Men
Total subjects (N)	13	10	10	9	12
Age (years)	39.7 ± 4.4	49.4 ± 5.3	25.4 ± 2.1	36.0 ± 4.5	51.2 ± 5.0
BMI (kg/m^2)	24.3 ± 4.2	27.5 ± 5.2	24.4 ± 2.1	24.9 ± 3.0	28.4 ± 4.3
WAI (points)	37.5 ± 5.4	38.6 ± 5.0	43.1 ± 2.9	41.4 ± 5.7	40.6 ± 4.4
Assembly time (s)	14.94 ± 2.30 5.91 ± 1.20	17.04 ± 3.05 6.42 ± 1.15	12.38 ± 1.75	11.89 ± 2.41	13.02 ± 2.30
Pulse difference (min^{-1})	17.84 ± 1.51	16.55 ± 1.29	17.40 ± 8.80	21.03 ± 7.07	13.66 ± 7.45

basic motion pressing. To determine the achieved assembly time for Period No. 1 and Period No. 2 on the belt buckle assembly line, video records were used. Identification of individual variance in the assembly's mode of operation sequence, arrangement of load carriers and other noticeable behavior, such as storing two different parts in one container, was also obtained from the video data.

Within industrial sciences, it is an accepted practice to determine strain using the difference in pulse, as to eliminate the individually wide range of the resting and working pulse (Hettinger 1991). To determine the difference in pulse whilst at work (= pulse difference), we subtracted the resting pulse (measured in a lying position) from the working pulse, measured in a sitting position. Heart frequencies in a standing working position were corrected with −3.242 beats/min which we ascertained as the process conditioned difference (based on over 6,000 values gained in our field study).

PPI is an indicator for the cardiovascular performance and reflects the individual's fitness level. It is calculated with the following formula:

$$PPI = \frac{Pulse\ after\ 20\ squats\ (BPM) - Resting\ pulse\ (BPM)}{Time\ in\ seconds\ required\ for\ 20\ squats}$$

A PPI-value of <1 indicates a low cardiovascular load capacity (i.e. a low fitness level) whereas PPI-values of >2 indicate reduced stress on the cardiovascular system and therefore higher fitness levels (Seibt et al. 2007). We used the resting pulse measured in a lying position.

Analysis of Field Study B

For field study B we analyzed the subjects at the workstation at which flywheels were assembled. As noted above, we observed the workstation for 4 h at the beginning of a shift. We excluded non-regular work (e.g., no assembly, assembly with distraction) from our analysis. During the 4-min motion-recordings of regular work we documented assembly of between four and six flywheels. We randomly selected one complete section of the mounting process (reach for the flywheel, lift, place on the motor block, located and fix with the first screw) for each predetermined video interval and analyzed the required time for assembly and corresponding heart frequency. We also obtained information about the variances of mode of operation from the video data, i.e. how subjects fixed the first screw, if they used the vertical adjustment and rotating of palettes as well as using a device bolt. For comparative purposes, we divided the 31 male subjects in the three age groups (see Table 1 for details).

A complete attainment of participants' resting pulse in a lying position was not possible due to organizational reasons, we therefore determined the pulse difference at work by subtracting the lowest heart frequency at the workstation (standing at the workplace) during a process of conditioned intermission (e.g., waiting for a motor block) from the measured working pulse (Schlick et al. 2010). For analysis we divided the weight of the mounted flywheels into two groups: "below 10 kg" and "over 10 kg".

Results

Field Study A

First results of field study A are already published (Börner et al. 2012a, b). In this field study A, 23 women between 27 and 57 years were accepted to take part in the study. We divided them in two age groups (young: M = 39.7 years, SD = 4.4 years; old: M = 49.4 years, SD = 5.3 years) (see Table 1). The subjects had no significant difference referring to their BMI, the mean of which was calculated at 24.3 kg/m^2 (SD = 4.2 kg/m^2) for the younger age group and 27.5 kg/m^2 (SD = 5.2 kg/m^2) for the elder group (see Table 1).

The WAI for the younger women (M = 37.5, SD = 5.4) and for the older women (M = 38.6, SD = 5.0) was defined as "good" (good = 37–43; Hasselhorn and Freude 2007) with no significant difference within the groups (see Table 1). The younger group achieved a significantly higher PPI-value (M = 1.58, SD = 0.30) in comparison to the older group (M = 1.27, SD = 0.31; $p = 0.023$).

When reviewing the data of the questionnaires it became apparent that both the total score of the TLX, and the score for the single items, showed no significant differences between the younger and the older female groups (see Fig. 3). For all subjects, temporal demand (M = 10.04, SD = 3.28) was perceived as higher than mental demand (M = 6.61, SD = 3.487; $p = 0.001$). Physical demand (M = 9.09, SD = 3.99) was also experienced at levels higher than mental demand (M = 6.61, SD = 3.49; $p = 0.006$).

In reference to the strain-ratings, significant differences between all subjects pre- and post-shift were found. All women felt less energetic ($p = 0.031$), less distracted ($p = 0.025$), less alert ($p < 0.001$) and more exhausted ($p = 0.014$)

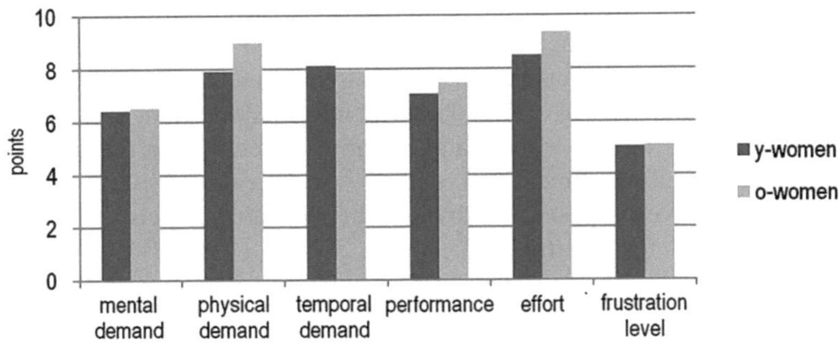

Fig. 3 Age-related Task Load Index (TLX), a comparison of post-shift findings (field study A, y-women: younger women, o-women: older women) (Börner et al. 2012a)

after the shift. Two of the 12 items in the strain-ratings displayed a significant difference in relation to age. First, the older women stated to being more insecure (M = 1.80, SD = 0.92) at the beginning of the shift than the younger women (M = 1.15, SD = 0.56; $p = 0.048$). Second, although the energy level before and after the shift remained stable for older women (4.1), it declined for the younger from 3.9 to 3.1. Therefore, after the shift the younger women (M = 3.08, SD = 1.12) felt less energetic than the older women (M = 4.10, SD = 0.57; $p = 0.015$) did.

Most of the subjects (75 %) sat during the measurement and observation process, however, based on our results, we found no statistical evidence to suggest a difference in sitting or standing between either younger or older women. Some differences in the mode of operation were documented including variations in the sequence of assembly, differing arrangement of load carriers and individual storage organization (e.g., using one's own lap to temporarily store basic modules). However, there was no significant difference in the mode of operation that could be attributed to the age of the participant.

On average the younger women (M = 14.94 s, SD = 2.30 s) needed significantly less assembly time for Period No. 1 than the older women required (M = 17.04 s, SD = 3.05 s; $p < 0.001$) (see Table 1). Figure 4a shows the age-dependent average assembly times for Period No. 1 during the moments of video recording in the morning shift . Results also confirm similar difference in assembly time for Period No. 2 as older women (M = 6.42 s, SD = 1.15 s) were recorded as needing significantly more time for the assembly of this process than the younger women (M = 5.91 s, SD = 1.20 s; $p < 0.001$).

Therefore, a correlation between age and average assembly time can be found ($r = 0.592, p = 0.003$) as well as between age and MCTQ ($r = -0.52, p = 0.011$). A partial correlation of age and average assembly time with the covariate MCTQ showed a significant correlation ($r = 0.567, p < 0.01$). Fisher's exact test was used to show the distribution of chronotypes between younger and older women. There is no correlation between average assembly time and WAI, BMI, arm discomfort or shoulder discomfort, however a correlation between hand discomfort and average

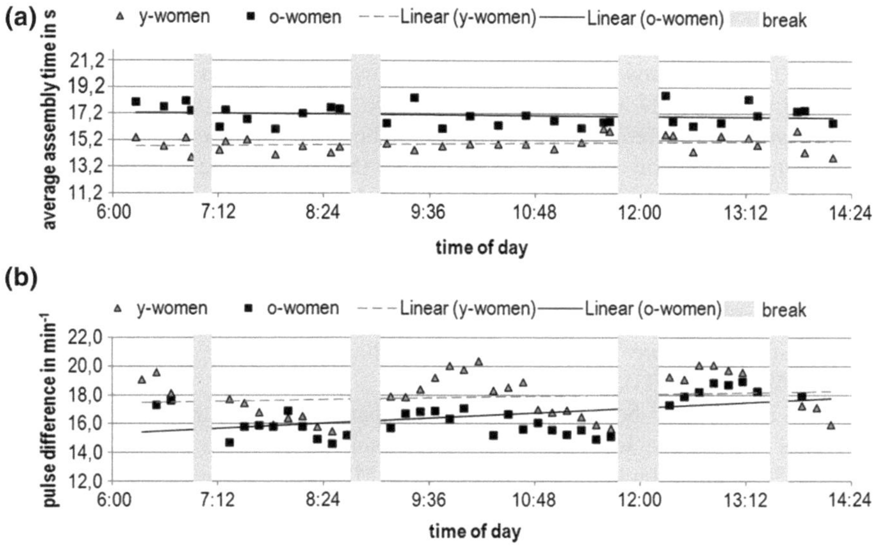

Fig. 4 **a** Average assembly time for Period No. 1 (field study A). **b** Average pulse difference for intervals of 10 min (field study A) (Börner et al. 2012a)

assembly time was found to be present ($r = 0.532$, $p = 0.009$). Fisher's exact test stated the equal distribution of hand discomfort of younger and older women. Hand pain, therefore, is not accountable for an age-difference in assembly time. Interestingly, there is a negative correlation between average assembly time and nonworking time within the group of younger women ($r = -0.0616$, $p = 0.025$), but no such correlation within the group of older women.

The average pulse difference (between working and resting pulse) is plotted in Fig. 4b as a function of time in which the younger and older participants are represented by grey triangles and black quadrats respectively. As Fig. 4b shows a slightly upward linear trend applies for both age groups, whilst the pulse difference of the older women rises more steeply. With regard to all of the observed measurements, there is a significant pulse difference between the younger ($M = 17.84\ min^{-1}$, $SD = 1.51\ min^{-1}$) and older women ($M = 16.55\ min^{-1}$, $SD = 1.29\ min^{-1}$; $p < 0.001$). The older women's group shows the lower pulse difference. There is no significant correlation between pulse difference and assembly time when averaged over all measured time points.

Field Study B

In field study B, 31 men between 21 and 60 years of age took part in the investigation. On account of more even distribution compared to the sample group in field study A, we were able to form three age groups: young, middle and old, with a mean age of 25, 36 and 51, respectively (see Table 1). The younger group (M = 24.4 kg/m^2, SD = 2.1 kg/m^2) and middle group (M = 24.9 kg/m^2, SD = 3.0 kg/m^2) showed a significantly lower BMI than the older group (M = 28.4 kg/m^2, SD = 4.3 kg/m^2) (see Table 1).

No significant differences in reference to the WAI were found (see Table 1). The younger men showed a mean WAI of 43.1 (SD = 2.9), the middle group 41.4 (SD = 5.7) and the older group 40.6 (SD = 4.4). Regarding the strain-ratings, there was no significant difference within the three age groups at the beginning of the shift. However after 4 h at the flywheel workstation the older group (M = 1.75, SD = 0.97) was significantly more insecure than the younger group (M = 1.1, SD = 0.32; $p = 0.046$) and the younger group (M = 3.9, SD = 0.88) felt significantly more alert after the shift than the middle group (M = 2.56, SD = 1.59; $p = 0.033$).

When all subjects' boredom levels are compared it is clear that, independent of age, after 4 h on the workstation all participants were more bored (M = 2.45, SD = 1.52) than at the beginning of the shift (M = 1.68, SD = 1.17; $p = 0.002$). After the entire 8-h shift they recorded feeling more unchallenged (M = 3.1, SD = 1.42) than at the beginning of the shift (M = 2.32, SD = 1.05; $p = 0.007$). Furthermore, they mentioned to be more bored at the end of the shift (M = 2.29, SD = 1.44) than to the beginning (M = 1.68, SD = 1.17; $p = 0.005$). The analysis of the total score of the TLX showed no significant difference after 4 h with upper-middle load compared to following 4 h of low load. Yet, the middle group (M = 13.78, SD = 4.66) referred significantly higher physical demand than the older group after the 4 h on the workstation (M = 7.83, SD = 5.29; $p = 0.015$). In comparison, the younger group were calculated to have a mean TLX point index of 10.3 (SD = 4.79) (see Fig. 5a).

Figure 5b shows the six items of the TLX, independent of age, after 4 h of upper-middle load and after 4 h of low load, respectively. Significant differences among the items after upper-middle load and among the items after low load are noticeable, for example after the 4 h on the flywheel workstation, the workers exhibited independent of age a higher physical demand (M = 10.35, SD = 5.38) than temporal demand (M = 8.26, SD = 5.60; $p = 0.02$). The temporal demand in turn, was appraised as being higher than the mental demand (M = 5.29, SD = 4.26; $p = 0.002$). There are no significant differences between the upper-middle load and low load workstations, however, the items 'physical demand' and 'effort' show a falling trend of stress level at the different workplaces.

In analyzing the difference in working and resting pulse, we found highly significant differences between the three age groups ($p < 0.001$) at the flywheel workstation assembly point. The older men showed the lowest pulse difference with M = 13.66 min^{-1} (SD = 7.45 min^{-1}). The younger men exhibited a mean of 17.40 min^{-1} (SD = 8.80 min^{-1}), followed by the middle group with a mean of 21.03 min^{-1} (SD = 7.07 min^{-1}). The 10-min means of the pulse differences can be seen in Fig. 6. While the pulse differences for the younger and older group rose during the 4 h shift, the pulse differences for the middle group can be seen as declining.

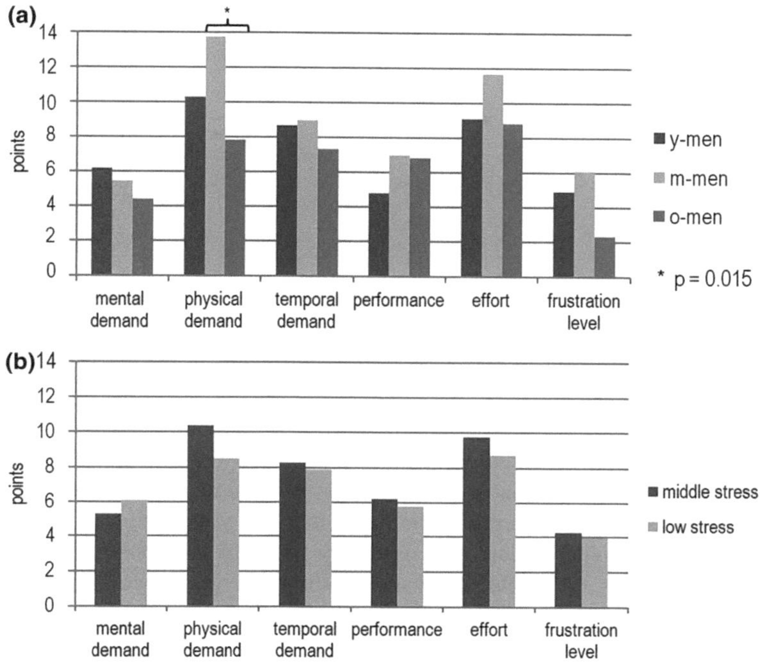

Fig. 5 **a** Age-related Task Load Index (TLX) after middle stress, mid-shift (field study B, y-men: younger men, m-men: middle-aged men, o-men: older men). **b** TLX-items between middle and low stress independent on age, mid-shift (field study B)

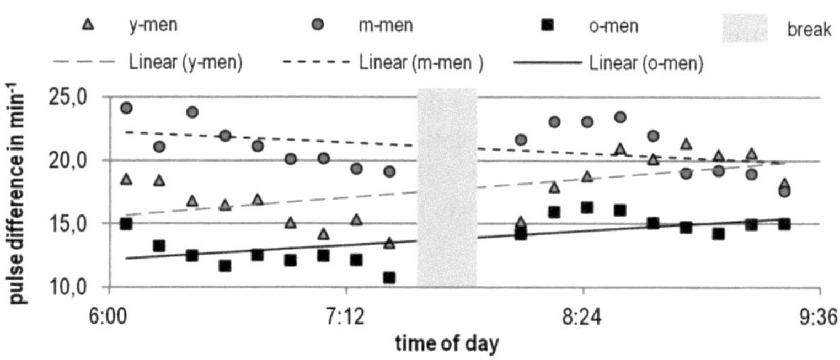

Fig. 6 Average pulse difference between working and resting of younger, middle and older men for intervals of 10 min (field study B)

Concerning to the average assembly times for the selected process, the younger group needed significantly less time (M = 12.38 s, SD = 1.75 s) than the older group (M = 13.02 s, SD = 2.41 s; $p = 0$ .001), but they needed significantly more time than the middle group (M = 11.89 s, SD = 2.31 s; $p = 0.02$). At five given measuring intervals, this difference between the three age groups was significant

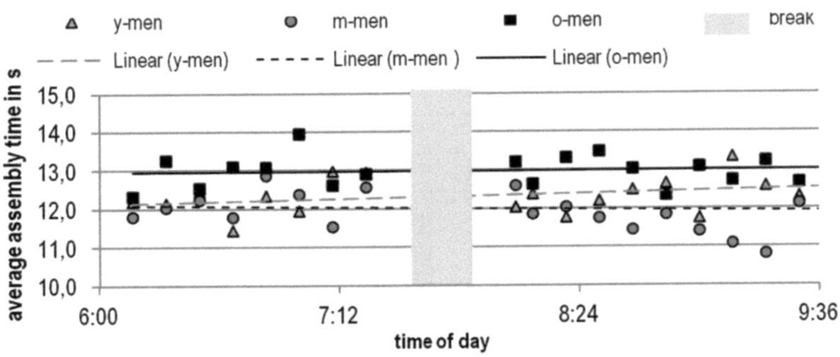

Fig. 7 Average assembly time for younger (y-men), middle (m-men) and older men (o-men) (field study B)

and for a further four measured points it was marginally significant. As a result we can state with confidence that the middle age group assembles the observed flywheel component on the assembly line faster than the younger group, who in turn are faster than the older group. The assembly times in the course of the 4 h are shown in Fig. 7. As the linear trend shows, the older group works at a constant (slow) speed, the middle group starts fast and maintains this tempo, whilst the younger group starts fast but becomes slower over the course of the shift.

As a mode of operation, the workers chose between using a bolt as help in mounting the flywheel or not. For loose fixing of the first screw they used a powered screwdriver or, alternatively, their fingers. There was no correlation between the assembly time and the handling with the screws. The participants also had the option of turning the palettes on which the flywheels were stored as well as lifting the palettes electronically; however, only 26 % of all subjects turned the palettes whilst 87 % lifted them. Although more of the older subjects turned the pallets, we found no strong correlation between age and the use of these available aids.

There is, however, a positive correlation between the weight of the flywheel and use of the device bolt ($r = 0.456$, $p < 0.001$), but no correlation between weight and assembly time. Furthermore, we found no correlation between BMI and assembly time.

Discussion

Field Study A

Literature classifies a "good ability to work" when WAI levels are judged to fall between the values of 37 and 43 (Hasselhorn and Freude 2007). The women of the two age-groups observed in this chapter exhibited such an ability scoring 37.5 and 38.6, admittedly vor this being at the lower end of the WAI classification. The results of BMI and PPI indicated similar physical conditions between the two subject groups. Despite the differences a PPI > 1 is still good for both groups (Seibt et al. 2007). All the subjects studied were well trained at the "belt buckle assembly" workstation and were physically fit with no sign of cardiovascular problems. Any hand, arm or shoulder discomforts had been noted prior to the test. The distribution, however, was independent of age and only factor of hand discomfort was shown to have had an influence on the assembly time.

Results of the TLX in field study A showed a similar subjective appraisal of work as no significant difference between the younger and the older groups were found. For all subjects, temporal and physical demand was perceived to be higher than mental demand, yet all values are located in the lower half of the TLX scale. The results of strain ratings, both before and after the observed shifts, indicate the influence of work on all subjects. Independent of age, all women felt less energetic, less alert, less distracted and more exhausted at the end of the work day. Although, the older women (1.8) stated to be significantly more insecure at beginning of the shift than the younger women (1.15) it is clear that based on the 1–6 scale, and the low ratings returned, that both groups were not really insecure as 1 means "disagree strongly".

The older subjects' stable energy level, both stated in the strain ratings before and after the shift, may be attributed to the slower speed of work. Seen statistically, this difference equates to a 14 % slower assembly time when compared to the younger participants. This relative faster speed exhibited by the younger subjects of the study may explain why they felt less energetic after the shift ended compared to their older counterparts. The differences in assembly time could also be ascribed to the fact that there was no externally implemented time pressure in the rate of assembly throughout the work cycle. Thus, the women were free to determine their own individual working speed. The correlation between average assembly time and non-working times among the younger subjects may indicate that they use their increased working speed in order to acquire additional work breaks. We cannot make this same conclusion for the older women.

Heart rate monitoring was used as an indicator for the objective strain as it was possible to implement and register data without influencing the assembly process: an essential requirement in conducting the field study. The average pulse difference between work and rest is significantly higher for the younger women (17.8 min^{-1}) as opposed to the older women (16.5 min^{-1}). A possible explanation for the lower pulse difference in the older subjects may be due to the slower assembly performance; however, despite this significantly difference in pulse differences between the age groups, it remains too small to allow for clear interpretations to be drawn. As shown in Fig. 4b, it is obvious that the pulse difference during the shift rises only slightly. Thus, we may assume that subjects of each age group operate within a steady state. The expected continuous performance limit for the pulse difference under the given conditions was 20 min^{-1} (Hettinger 1991). Obviously, it is not possible to deduce a higher strain for a certain age group from the means of the pulse differences alone. The measurable slight increase of the pulse difference through the 8 h-shift for both age groups proves the subjective strain experienced by our subjects. According to the strain ratings all subjects declared being more exhausted, less energetic and less alert after the shift.

Based on the model of selective optimization with compensation (Baltes and Baltes 1990) we expected an influence on the assembly process. We therefore examined the arrangement of parts and load carriers on the subjects' workplace. Although some workers changed the arrangement of the load carriers and/or disposed of parts in advance, there were no significant differences of younger and older workers in the examined aspects. Possible explanations could be the low number of subjects and the organization of the highly standardized workplace which offers only a handful of options of individual activity regulation.

The research design involved the constant 8 h monitoring of well trained, healthy assembly workers. The maximum number of female workers that became subjects of our study, and were made available to us by our cooperation partner, was 23. It is critically noted that the sample with 23 women is statistically small and the resulting age distribution was not ideal: many of the women were recorded to be in their mid-forties and no younger women under the age of 27 were found for the investigation thereby skewing the age balance and only allowing for the representation of two age groups.

Heart frequency taken of the subjects and the ensuing findings were not as meaningful as had been predicted before the investigation. It is possible that the physical demand at the workplace in field study A was too low for this purpose. We investigated physical strain of the hand/arm system using small parts of a low weight in field study A. Clearly, the physical stress of the work on a "belt buckle assembly" workplace is relatively low, whilst the stress for the sensory motoric system is comparatively high. Based on the Chemnitz Age Database, we assumed that both visual and sensory motoric demands would have a higher influence on strain within the older age group. However, the results do not confirm our assumption. The subjects indicated that their physical strain, when compared to the NASA TLX, was within a medium range.

Field Study B

Based on the experiences gathered in field study A, we searched systematically for a workplace displaying tasks with a higher physical load for field study B in order to more fully test our hypotheses. The flywheel assembly workstation was rated at 48 EAWS points, denoting it as upper-middle load. No suitable workstation in the automotive plant was found to have a higher load which we could have tested without interfering with the production process, therefore the flywheel assembly workstation was chosen.

In pursuit of better comparative data, the nature of the flywheel assembly line also ensured that all the chosen subjects would be working from the same standing position during the course of our observation. A further advantage of this workstation was the distribution of the age of the workers. With 31 participating men aged between 21 and 60, we were able to create and study three different age groups. These three age-groups of men exhibited a good working ability with no significant differences in the results of the WAI. Results however, confirmed that performance decreased with age. Decreasing values for the WAI in the older demographic have been shown elsewhere (Prümper and Richenhagen 2011) and the results of our survey confirm these findings.

A performance rating (WAI) of 43.1 was recorded for the younger male participant. The middle group scored 41.4 and the older men 40.6. Confirming the EAWS rating of 48, the flywheel assembly station was also adjudged to be particularly physically demanding (TLX) by the workers of the factory.

The BMI for the older group of men was significantly higher than the BMI of the middle and younger age groups. While it has been noted in literature that BMI rises with age (Müller and Bosy-Westphal 2008), no correlation between a high BMI and slow performance could be found. Although the older group (1.75) was significantly more insecure than the younger group (1.1), after 4 h at the flywheel workplace, both groups were quite secure as the value of 1 means "disagree strongly" on a scale from 1 to 6. The high self-confidence exhibited in the strain ratings confirmed that all workers were highly trained and familiar with the process.

Rather surprisingly, it was the middle group that achieved the best (i.e. quickest) assembly times in field study B being 0.5 s faster than the younger group, and approximately 1.2 s faster than the older group.

Within the items of the TLX, the middle age group also estimated physical stress to be the highest whilst the older group judged the physical stress significantly lower with the younger subjects occupying the middle value. This leads to the question of whether a shorter assembly time is achieved through a distinctly intensified work input.

The comparison of the differences in pulse over a 4 h period clearly supports the participants' subjective appraisals: shorter (i.e. faster) assembly times are achieved through a higher work input. This fact is exemplified in a significantly higher pulse difference for the middle group as compared with the older group. Further interest is also in the noticeable relative higher working pulse rate of the younger group when compared to that of the older group. Whilst this higher pulse rate may be attributed to a higher workload carried out by the younger and middle group, the data collected comparing the pulse differences of the three age groups was unable to confirm this assumption. The heart frequency for all three groups leveled to a steady state. The higher pulse differences and the higher subjective appraisals of the physical strain were not substantiated in the objective data.

Outlook

This field study clearly documents the effects of stress and strain of short-cycled, repetitive performances in the automotive and in the automotive supplier industry. The results of the investigation were based on the data of our real-time observations, the videos records and the heart frequency readings taken throughout the duration of an 8 h shift the participants were subjected to. These objective measurements have been complemented by the acquisition of subjective data taken from several questionnaires. For the first time, the measurements of work performance in the value-added process, in combination with the subjective strain and objective stress gathered here, allow us to make verifiable statements about the extent to which older assembly workers endure more strain than younger assembly workers.

Our test design deliberately avoided the use of experimental, non-work related and not knowledge based experimental designs. We therefore directly measured the performance in the value-added process and were able to observe an entire 8 h morning shift. With the data gathered, we have been able to deliver not only a snapshot of the participants' performance, but also, valuably, we have been able to measure the endurance stress in the assembly process and explain this graphically. Our results will allow concrete conclusions to be made for the considered work process regarding the performance of the younger and older assembly workers. In doing this, we would like to take this opportunity to clarify that senior assembly workers are equally able to fulfill their sensory motoric tasks as well as their physically demanding work, as their younger assembly co-workers. On closer inspection of the assembly times, a noticeable difference between the older and the younger assembly worker can be observed. Nevertheless, the difference is relatively low, equaling a mere 14 % extra assembly time in field study A (with self-determined working speed), and 5 % in field study B (with assembly line-determined working speed).

Interestingly the middle group in field study B exhibited the lowest average assembly time. This, as assumed, resulted in a significantly more pronounced pulse difference. Furthermore, we found references that the model of selective optimization with compensation, suggested by Baltes and Baltes (1990), would also apply when a strict and standardized short cycle regime is used. On average, the group of the older participants had slower assembly times than the sample of younger participants, however, the older sample in turn worked at a more constant pace over the 4 h period. Furthermore, approximately 35 % of the older workers used the possibility to rotate the pallet holding the flywheels simplifying selection and lifting. In comparison, only 20 % of the younger participants did the same, a further indicator of selective optimization procedures.

Our results have been verified with long-time measurements and therefore, we believe, are well-founded. However, it would be problematic to attempt to generalize these results in the current form since our study focused on the two, very specific assembly processes. Thus, the extent to which selection effects have an influence on, for example the "Healthy Worker Effect", cannot be fully clarified. Despite this however, we are of the opinion that our research has been a success as, for the first time, it offers a fundamental understanding, and the conviction to state that, at least for the selected assembly processes, that older assembly workers are not significantly more strained compared to their younger counterparts.

Nevertheless this chapter provides scope and a point of departure for necessary further research. In field study A, it was impossible to document the production output of a single individual, a necessary requirement in attempting to make statements about the productivity. Production in field study B was observed at an assembly line, therefore the individual output of pieces was not dependent on age of the participant, but rather the availability of parts to be mounted.

It should also be noted that there are limits attributed to the cardiovascular load measurement utilized in this study that was only able to measure the strain on the cardiovascular system . Whilst this allows us to make conclusions on the physical strain of heavy, manual work or fatigue loading on large muscle groups it is not sufficient to measure the strain placed on small muscle groups and strain on the sensory channel. To solve this obvious limitation, future research projects could find the use of an Electromyography (EMG) measurement beneficial. That said however, EMG measurements are difficult to conduct if one does not wish to interrupt the production process. Our desire not to hinder the production process on the assembly line during our investigation led to our decision to use this cardiovascular load measurement (explained above) to assess the heart rate owing to its ease of implementation and evaluation.

To conclude, whilst our study has been a success in the two instances in which it was implemented, and we have been able to make, with confidence, a number of statements referring the stress and strain on a given range of worker's age groups, our studies need to be replicated at other institutions and organizations involved in manual production in order to obtain a larger sample to generate more distinct and noteworthy conclusions.

Acknowledgments Financial support by the German Research Foundation (DFG) under its priority program SPP 1184, Age Differentiated Work Systems "(grant SP 970/3-1)" is gratefully acknowledged.

References

Allen P (2002) Reaction time. In: Ekerdt DJ (ed) Encyclopedia of aging, vol 4. Macmillan Reference USA, New York, pp 1165–1169

Baltes PB, Baltes MM (1990) Psychological perspectives on successful aging: the model of selective optimization with compensation. In: Baltes PB, Baltes MM (eds) Successful aging: perspectives from the behavioral sciences. Cambridge University Press, New York, pp 1–34

Barnes CJ, Van Steyn SJ, Fischer RA (2001) The effects of age, sex, and shoulder dominance on range of motion of the shoulder. J Shoulder Elbow Surg 10:242–246

Bokranz R, Landau K (2006) Produktivitätsmanagement von Arbeitssystemen. MTM-Handbuch, Schäffer-Poeschel, Stuttgart

Börner K, Leitner-Mai B, Scherf C, Spanner-Ulmer B (2012a) Feldstudie zur altersabhängigen Beanspruchung von Montagemitarbeitern am Beispiel der Automobilzulieferindustrie. In: GfA (ed) Gestaltung nachhaltiger Arbeitssysteme: Wege zur gesunden, effizienten und sicheren Arbeit, Frühjahrstagung der Gesellschaft für Arbeitswissenschaft, 22–24 Feb 2012. GfA-Press, Kassel, pp 299–303

Börner K, Scherf C, Leitner-Mai B, Spanner-Ulmer B (2012b) Field study of age-differentiated strain for assembly line workers in the automotive industry. Work: A J Prevention Assess Rehabilitation 41:5160–5166

Bös G (2007) Strategisches Management alternder Belegschaften bei der Audi AG. In: Holz M, Da-Cruz P (eds) Demografischer Wandel in Unternehmen, Herausforderungen für die strategische Personalplanung. Gabler, Wiesbaden, pp 199–212

Bützler J, Vetter S, Jochems J, Schlick CM (2011) Refinement of Fitt's law for large touch screens: age differentiated analysis of the motion angle in bivariate pointing task. In: Göbel M, Christie C, Zschernack S, Todd A, Mattison M (eds) Research for the missing link. 10th international symposium on human factors in organizational design and management, vol 2. IEA Press, Santa Monica, CA, pp 93–99

Chateauroux E, Wang X (2008) Effects of age, gender, and target location on seated reach capacity and posture. Hum Factors 50:211–226

Fozard JL, Vercruyssen M, Reynolds SL, Hancock PA, Quilter RE (1994) Age differences and change in reaction time: the Baltimore longitudinal study of aging. J Gerontology: Psychological Sci 49:179–189

Frieling E, Buch M, Wieselhuber J (2006) Alter(n)sgerechte Arbeitssystemgestaltung in der Automobilindustrie: die demografische Herausforderung bewältigen. Zeitschrift für Arbeitswissenschaft 60:213–219

Fries JF (1989) Erfolgreiches Altern: Medizinische und demografische Perspektiven. In: Baltes MM, Kohli M, Sames K (eds) Erfolgreiches Altern. Hans Huber, Bern

Hacker W (2003) Leistungsfähigkeit und Alter. In: IABColloquium, Praxis trifft Wissenschaft, Titel: Eine Frage des Alters: Herausforderungen für eine zukunftsorientierte Beschäftigungspolitik, 20–21 Oct 2003 in der Führungsakademie der BA in Lauf

Hacker W, Auerbach I, Hagendorf E, Rudolf M (1999) Erfüllen zurückgestellter zeitbezogener Aufträge: selbstinduziertes Nutzen von Abrufhilfen. Zeitschrift für Experimentelle Psychol 46:81–106

Hart SG, Staveland LE (1988) Development of a NASA-TLX (task load index): results of empirical and theoretical research. In: Hancock PA, Meshkati N (eds) Human mental workload. North-Holland, Amsterdam, pp 139–183

Hasselhorn HM, Freude G (2007) Der Work ability Index: ein Leitfaden, Bundesanstalt für Arbeitsschutz und Arbeitsmedizin (ed) Wirtschaftsverlag NW, Bremerhaven

Hegele M, Heuer H (2010) Anpassung an visumotorische Transformationen unter indirekter Sicht in frühem und spätem Erwerbsalter. Zeitschrift für Arbeitswissenschaft 64:265–275

Henry JD, MacLeod MS, Phillips LH, Crawford LR (2004) A meta-analytic review of prospective memory and aging. Psychol Aging 19:27–39

Hettinger T (1991) Handhabung von Lasten: Ergonomische Gesichtspunkte. Hanser, München

Heuer H, Hegele M (2006) Bewegungen mit indirekter Sicht in Altersverlauf. Zeitschrift für Arbeitswissenschaft 60:221–229

Keil M, Spanner-Ulmer B, Scherf C (2009) Änderungen der menschlichen Leistungsfaktoren in Alter: Entwicklung eines interdisziplinären Altersmodells. In: GfA (ed) Arbeit, Beschäftigungsfähigkeit und Produktivität in 21. Jahrhundert, Frühjahrstagung der Gesellschaft für Arbeitswissenschaft, 4–6 Mar 2009. GfA-Press, Dortmund, pp 145–148

Keil M, Hensel R, Spanner-Ulmer B (2010) Fähigkeitsgerechte Prozessmodellbausteine zur Generierung altersdifferenzierter Beanspruchungsprofile. Zeitschrift für Arbeitswissenschaft 3:205–215

Müller EA (1950) Ein Leistungs-Pulsindex als Maß der Leistungsfähigkeit. Arbeitsphysiologie, Bd 14:271–284

Müller MJ, Bosy-Westphal A (2008) Energiehaushalt und Ernährung. In: Speckmann EJ, Hescheler J, Köhling R (eds) Physiolgie. Urban and Fischer, München, pp 583–614

Otto A, Scholl A (2011) Incorporating ergonomic risks into assembly line balancing. Eur J Oper Res 212:277–286

Park DC (1994) Aging, cognition, and work. Human Perform 7:181–205

Pierson WR, Montoye HJ (1958) Movement time, reaction time, and age. J Gerontol 13:418–421

Prümper J, Richenhagen G (2011) Von der Arbeitsunfähigkeit zum Haus der Arbeitsfähigkeit. Der Work Ability Index und seine Anwendung. In: Seyfried B (ed) Ältere Beschäftigte: Zu jung, um alt zu sein. Konzepte: Forschungsergebnisse: Instrumente. Bielefeld, Bertelsmann, pp 135–146

Richter P, Debitz U, Schulze F (2002) Diagnostik von Arbeitsanforderungen und kumulativen Beanspruchungsfolgen am Beispiel eines Call Centers. Zeitschrift für Arbeitswissenschaft 56:67–76

Roenneberg T, Wirz-Justice A, Merrow A (2003) Life between clocks: daily temporal patterns of human chronotypes. J BiolRhythms 18:80–90

Rohmert W (1984) Das Belastungs-Beanspruchungs-Konzept. Zeitschrift für Arbeitswissenschaft 4:193–200

Runge M, Rittweger J, Russo CR, Schiessl H, Felsenberg D (2004) Is muscle power output a key factor in the age-related decline in physical performance? A comparison of muscle cross section, chair rising test and jumping power. Clin Physiol Funct Imaging 24:335–340

Schaie KW (1996) Intellectual development in adulthood. Cambridge University Press, New York

Schaub K, Ahmadi K (2007) Vom AAWS zum EAWS: ein erweitertes Screening-Verfahren für körperliche Belastungen. In: GfA (ed) Kompetenzentwicklung in realen und virtuellen Arbeitssystemen, Bericht zum 53. Arbeitswissenschaftlichen Kongress vom 28 Feb–2 Mar 2007, GfA Press, Dortmund, pp 601–604

Scherf C, Leitner-Mai B, Börner K, Spanner-Ulmer B (2010) Versuchsdesign zur Generierung altersdifferenzierter Beanspruchungsprofile. In: Müller E, Spanner-Ulmer B (eds) Nachhaltigkeit in Planung und Produktion. Tagungsband 4. Symposium Wissenschaft und Praxis and 8. Fachtagung Vernetzt planen und produzieren: VPP2010, Wissenschaftliche Schriftenreihe des Institutes für Betriebswissenschaften und Fabriksysteme, Sonderheft 16. IBF, Chemnitz, pp 1–12

Scherf C, Börner K, Leitner-Mai B, Spanner-Ulmer B (2011) Identifying age differentiated strain profiles for assembly line workers on the basis of age critical reference processes. In: Göbel M, Christie C, Zschernack S, Todd A, Mattison M (eds) Research for the missing link. 10th international symposium on human factors in organizational design and management, vol 2. IEA Press, Santa Monica, CA, pp 87–92

Schlick C, Bruder R, Luczak H (eds) (2010) Arbeitswissenschaft. Springer, Heidelberg

Schmidtke H (1993) Ergonomie, 3rd edn. Hanser, München

Seibt R, Dutschke D, Hübler A, Scheuch K (2007) Handlungsanleitung zur Durchführung einer individuellen Vorsorgediagnostik für Lehrkräfte. In: Seibt R (ed) Das Dresdner Modell. Selbstverlag der TU Dresden, Dresden

Shephard RJ (1999) Age and physical work capacity. Exp Aging Res 25:331–343

Spanner-Ulmer B, Frieling E, Landau K, Bruder R (2009) Produktivität und Alter. In: GfA (ed) Produktivität im Betrieb, Herbsttagung der Gesellschaft für Arbeitswissenschaft, 23–25 Sept 2009. GfA-Press, Dortmund, pp 81–117

Spirduso WW, Francis KL, MacRae PG (2005) Physical dimensions of aging, 2nd edn. Human Kinetics, Champaign

Stelmach GE, Goggin NL, Garcia-Colera A (1986) Movement specification time with age. Exp Aging Res 13:39–46

Viitasalo JT, Era P, Leskinen AL, Heikkinen E (1985) Muscular strength profiles and anthropometry in random samples of men aged 31–35, 51–55 and 71–75 years. Ergonomics 28(11):1563–1574

Wiendahl HP (2010) Betriebsorganisation für Ingenieure, 7th edn. Hanser, München

Wild-Wall N, Gajewski P, Falkenstein M (2009) Kognitive Leistungsfähigkeit ältere Arbeitnehmer. Zeitschrift für Gerontologie Geriatrie 4:299–304

Age-Related Differences in Critical Driving Situations: The Influence of Dual-Task Situations, S-R Compatibility and Driving Expertise

Gisa Aschersleben, Katrin Arning and Jochen Müsseler

Lessons Learned

In the near future, there will be a significant increase in the number of older people who hold driving licenses. Maintaining mobility by driving a car is a key factor for elderly people as mobility is a prerequisite for social contacts, social integration and an autonomous lifestyle. Contrary to the stereotype that older drivers are risky and constitute a threat to general traffic safety accident statistics show that fatality rates are the highest in young drivers and that rate declines to a minimum for drivers of the age of 40–60 years. This can partly be explained by compensatory driving strategies: older drivers prefer to drive shorter distances, avoid critical situations like rush hours and drive preferably during daytime. Nevertheless, there is an increased accident-risk of older drivers in complex driving situations resulting from decreasing driving practice and, more importantly, age-related changes in information processing capacities (see chapter Age-Differentiated Systems: Introduction and Overview to a Six Year Research Program in Germany, Schlick et al.). The aim of the project was to analyze age-related differences in the applied context of critical driving situations. Findings reported in the literature suggest a decreased performance of elderly people in laboratory situations as well

G. Aschersleben (✉)
Developmental Psychology, Saarland University, Saarbrücken, Germany
e-mail: aschersleben@mx.uni-saarland.de

K. Arning
Human-Computer Interaction Center, RWTH Aachen University, Aachen, Germany
e-mail: arning@comm.rwth-aachen.de

J. Müsseler
Psychology Department, RWTH Aachen University, Aachen, Germany
e-mail: muesseler@psych.rwth-aachen.de

C. M. Schlick et al. (eds.), *Age-Differentiated Work Systems*,
DOI: 10.1007/978-3-642-35057-3_12,

as in a natural driving context. However, it is still an open question, whether this is also true for dual-task requirements in a driving context and whether expertise might compensate for age-related effects. In two experiments, we first studied the influence of age on spatial compatibility effects in natural scenes and, then, the effects of a secondary task as well as the effects of driving expertise were examined.

The results of the experiments indicate that in the driving context, reaction time in an avoiding situation (e.g. avoiding to drive into a person appearing unexpectedly on the scene by turning a steering wheel away from the person) was shorter than a turn of the steering wheel towards the stimulus. These results speak against an automatic response tendency elicited by the position of the stimulus, as assumed often by compatibility theories (Kornblum et al. 1990). Although older drivers (mean age 62; young drivers mean age 23) showed an inferior driving performance with slower reactions, the compatibility effect was not increased in older drivers, which contradicts the general pattern of results obtained in the literature. As one possible explanation for this observation, the influence of expertise was examined. In addition, the influence of a secondary task was tested by presenting an acoustical message by the navigation system. Participants reacted faster when the direction announced in the navigation message was contralateral to the pedestrians' side of appearance. Again, older drivers showed slower reactions, but there was no increased interference effect. Moreover, two groups of professional drivers (young, mean age 28, and old drivers, mean age 56) were tested: Older professional drivers reached a comparable performance in reaction times as younger professional drivers and driving experience resulted in a reduced vulnerability to interfering stimuli—probably because professional drivers are more experienced in the primary task (driving a car, or, more specifically, reacting in dangerous situations while driving a car) and because they are more used to dual-task situations while driving a car (using a mobile phone or listening to a message of the navigation system while driving a car).

Based on these findings the following recommendations can be derived:

- When designing electronic driver assistant systems (e.g., collision avoidance system, lane assistant) that support the driver to increase safety, it should be taken into account that every additional information given to the driver not only demands the driver's attention in general but that there are also specific cross task effects (e.g., a lateral signal from the collision avoidance system can have an interfering effect on a left/right reaction with the steering wheel).
- Age-related changes in information processing capacities can be compensated for by training. As it is not always possible for elderly people to avoid complex driving situations, older drivers should undergo driving courses regularly, especially if their driving practice decreases, and specifically train driving in complex situations (lane-changing manoeuvres, turning of at intersections, right of way decisions and driving situations with multiple task demands).

The Older Driver

Over the next thirty years there will be a significant increase in the number of older people who hold driving licenses. According to the US Institute of Highway Safety, by 2029 one out of four licensed drivers is anticipated to be an older driver (age 50+). This trend will not only be caused by demographic change but also by a more mobility-prone lifestyle in western societies. Many older drivers will have access to a car for all their lives, moreover the proportion of older women holding driving licenses will increase. Maintaining mobility by driving a car is a key factor for elderly people as mobility is a prerequisite for social contacts, social integration and an autonomous lifestyle (OECD 2001).

Common stereotypes perceive older drivers as risky and as a threat to general traffic safety. However, accident statistics contradict this stereotype: fatality rates are the highest in young drivers and that rate declines to a minimum for drivers of the age of 40–60 years to slightly increase again to a maximum for those aged 75 and older (Statistisches Bundesamt 2005). Research in traffic and accident analyses shows that older drivers are particularly prone to have accidents resulting from inadequate handling of complex traffic situations such as lane-changing manoeuvres, turning of at intersections, right of way decisions and in driving situations with multiple task demands (e.g. driving a car and using a mobile phone; cf. Mcgwin and Brown 1999; Cox and Cox 1998; Hancock et al. 2003). Contrary to that, older drivers are "under-represented" in crashes involving loss of control or collisions due to speeding, risky overtaking or driving under the influence of alcohol. The higher accident-risk of older drivers in complex driving situations can be explained by two factors: (1) decreasing driving practice and (2) age-related changes in information processing capacities. Regarding decreasing driving practice, older drivers tend to use compensatory driving strategies. They prefer to drive shorter distances, avoid critical situations like rush hours and drive preferably during daytime which leads in sum to a lower annual mileage and to a reduction of driving practice (Hartenstein 1995; Stutts 1998). Referring to age-related changes in information processing capacities, a number of physiological and psychological functions decline with increasing age, which negatively affects driving abilities and performance. A more detailed description of those changes will be given in the following section.

Dual-Task Demands and Cross-Task Compatibility in Driving

Age-related changes in the information processing system comprise changes in perceptual, motor and cognitive abilities, which are gradual and vary widely from individual to individual (Park and Schwarz 1999). Age-related changes in the *perceptual system*, which are relevant in driving, refer to a decline in auditory and visual abilities of older adults (Kline and Scialfa 1997). Changes in the auditory system comprise an increase in the auditory threshold (especially for high-frequency tones), difficulties in the perception of speech and in the discrimination and the spatial location of tones. Apart from an ongoing decrease in visual acuity, the eyes' ability to accommodate and adapt to light changes declines and the sensitivity to glare and reflections increases. Older adults therefore have higher difficulties seeing and determining the speed and distance of the traffic they need to merge with. *Motor abilities*, that is the ability to execute fast and accurate movements, are also declining over the life span (Vercryssen 1997). Moreover, there is a decline in muscle power, nerve sensitivity, skeletal strength and flexibility, which may limit driving operations and makes it difficult to move the head to see traffic at junctions. The most relevant changes in the context of the present study refer to age-related declines in *cognitive functions*—for instance, the decline in working-memory capacities, the slowing of processing speed and the reduced ability to distinguish relevant from irrelevant information (Park and Schwarz 1999). The most prominent change is the slowing of information processing, i.e. a slowing in stimulus detection, decision making, response selection, and response execution (Salthouse 1996). Processing capacity is also reduced over the lifespan, with increased problems while simultaneously processing information such as talking and carrying out some driving manoeuvre. The ability to discriminate relevant from irrelevant information is most impaired in elderly people if they have to perform the task under time pressure (Plude and Hoyer 1986). Many studies provided evidence for age-related declines in information processing abilities, especially with increasing task complexity (Kliegl et al. 2003; Li et al. 2004).

Dual task demands are typically studied in a dual-task paradigm, where participants are required to react as fast as possible to two stimuli being presented in close succession. For example, participants are instructed to press a left/right key (R1) in response to a high/low tone (S1) and to say "blue" or "yellow" (R2) in response to a blue or yellow stimulus (S2). Under conditions with a short time interval (stimulus-onset asynchrony, SOA) between the presentation of S1 and S2 (i.e. S2 is presented when response preparation and production of R1 has not yet been completed), dramatic increases in reaction time and errors in the secondary task (S2–R2) are observed. These performance impairments were explained by capacity limitations at the response selection stage (for an overview see Pashler 1998). It was assumed that two reactions could not be selected simultaneously because the central stage of response selection was limited to process one event at a time. As a consequence, response selection of the secondary task had to wait until response selection of the primary task was completed. More recent dual-task studies found evidence that processing S1 not only impairs the selection of R2, but also impairs perceptual encoding of S2. Using a go/no-go manipulation in the primary task, Müsseler and Wühr (2002) demonstrated that performance in the secondary task is impaired even under conditions under which only identification of the stimulus but no response selection is required in the primary task (see also Jolicoeur 1999). Moreover, by introducing compatibility relationships between tasks (Lien and Proctor 2002), it was shown that compatibility information could successfully bypass the response-selection bottleneck (Hommel 1998; Müsseler et al. 2005, 2006). Therefore, cross-task compatibility is a further important factor, which has to be considered in dual task situations.

Cross-task compatibility describes whether the information presented in the second task provides information, which is compatible or incompatible to the primary task. While driving a car a driver often has to immediately react to visual information that is spatially localized (e.g. avoiding obstacles that suddenly appear at the right side of the road). From compatibility research it is well known that speed as well as accuracy of a spatial reaction is influenced by the spatial location of the imperative stimulus. A stimulus that is presented ipsilaterally (at the same side at which the reaction is required) results in faster responses and less errors than a stimulus that is spatially non-corresponding with the required reaction. In classical compatibility theories, this pattern of results is explained by an automatic activation of the corresponding reaction that in the case of an incompatible situation has to be inhibited first before the correct response can be performed (Kornblum et al. 1990).

Up to now, the question about the effects of compatibility and dual task demands in driving situations has been mostly ignored in research. Moreover, it is important to note that the increase of age-related competencies such as driving expertise and more defensive and anticipatory driving style cannot fully compensate for the age-related decrease in the information processing system. Older drivers do not necessarily drive worse; however, as it is not always possible to avoid problematic and demanding traffic situations elderly people drive with higher cognitive load. Thus, it is important to study the influence of this fact on reactions in critical situations in which a coordination of different tasks is required. Therefore the present study aimed for an investigation of the effects of age, compatibility and dual task demands in an applied context of critical driving situations. We were interested to address the following questions:

- Do older drivers show an inferior driving performance in critical driving situations?
- Does age increase interference effects of dual task demands and cross-task compatibility in critical driving situations?
- Does driving expertise compensate for the effects of age?

Experiment 1: Spatial Compatibility in Natural Scenes: The Effect of Age

The purpose of Experiment 1 was twofold: First, we examined whether driving responses in a taxi-driver scenario would follow the assumptions of spatial compatibility or if we are able to replicate the findings presented in Müsseler et al. (2009). There we found that the typically observed advantage of spatially compatible responses was reversed for dangerous situations in natural scenes, that is, for critical driving situations the spatially incompatible response was faster than the compatible response. Classical compatibility theories would imply that first an automatic activation of the corresponding reaction (i.e. in the direction of the dangerous stimulus) takes place, which then has to be inhibited before the correct response can be selected. The question was whether this automatic activation of the ipsilateral response was the same in driving situations. Second, we analyzed compatibility effects in critical driving situations in an older sample. As interference effects in compatibility tasks increase in elderly people (Pick and Proctor 1999), we assumed that compatibility effects in an applied driving situation would be more pronounced in an older sample.

A taxi-driver scenario was realized in the experiment, where younger and older participants watched short movies from the driver's perspective. The movies showed a street scene, in which participants approached an intersection and a pedestrian suddenly entered the street from the left or the right side, either calling a taxi by waving with the arm or causing a critical situation by carelessly entering the street. Participants were instructed to react as fast as possible by turning a steering wheel clockwise or counterclockwise either towards the location of the person (in order to pick this person up) or away from the person (in order to avoid hitting the person).

Based on the findings presented in Müsseler et al. (2009), we expected faster reactions when participants had to turn the steering wheel away from the stimulus (avoiding reaction; spatially incompatible condition) in comparison to a turn of the steering wheel towards the stimulus (spatially compatible condition). Moreover, as interference effects in compatibility tasks increase in elderly people, we expected the group of older adults to show not only slower reactions but also an increased compatibility effect.

Method

Participants

A total of 30 participants, 15 older adults with a mean age of 61.7 years (range 56–65 years) and 15 younger adults with a mean age of 24.2 years (range 21–30 years) took part in the experiment. A valid driver license was required to participate in the experiment. Younger adults held their driver license, on average, for 6 years (range 3–10 years) whereas older adults held their driver license for more than 41 years (range 37–45 years). Average amount of driving amounted to 9,600 km per year in the younger adults and 12,100 km per year in the older adults. All participants of the present and the subsequent experiments were right-handed and reported normal or corrected-to-normal vision. Older participants were recruited by advertisements in a local newspaper, younger participants were students, which fulfilled a course requirement or were recruited through the social network of the experimenter.

Fig. 1 Experimental setup: participant in the seat box in front of the screen

Apparatus and Stimuli

The experiment was carried out in a dimly lit and soundproof room and was controlled by a Macintosh computer with Matlab using the Psychophysics Toolbox (Brainard 1997; Pelli 1997). Stimuli were projected with a beamer (Epson EMP-82 with a resolution of 1024 × 768 pixel) on a screen (146 × 208 cm). Participants sat on a car seat in a seat box 240 cm in front of the screen (see Fig. 1). They used a steering wheel (Logitech Formula GP Racing Wheel, 25 cm in diameter) to react to the stimuli. A 45°-deviation of the steering wheel from the home position (straight ahead) determined reaction times. Response directions were also collected to analyze accuracy. Participants started a trial by stepping on an accelerator pedal, which was placed in front of the car seat positioned according to participants' length of legs.

As visual stimuli movies showing traffic scenes were presented. They were projected with a size of 110 × 150 cm on the screen and showed a straight-ahead driving scene in a one-way street from the participants' perspective. The movies were based on digitized photos of a street scene and the zooming-in function of a video software (iMovie, Apple Computer Inc.) was used in order to evoke a driving impression. After 2 s the movie ended and a picture of a pedestrian (39 × 14 cm) was superimposed into the street scene 6° to the left or the right of the screen center. The pedestrians' pictures were mirrored in order to equally enter the street scene from the left or right hand side. The pedestrian either waved and turned towards the participant holding eye contact; or was carelessly entering the street without keeping eye contact.

Procedure

First, the car seat was adjusted to participants' seating preferences and participants were instructed to grab the steering wheel with their dominant hand on the top. In this position, the hand movement or direction of the steering-wheel rotation corresponded to the driving direction (for different results with regard to the hand position (see Guiard 1983; Stins and Michael 1997). Participants were told to act like taxi-drivers, which were exclusively driving on one-way streets (in order to allow turning to the left side of the street without considering the oncoming traffic). Each trial started with a static street scene and the taxi drive was started by stepping on the accelerator pedal. After 2 s a pedestrian (either waving or carelessly entering the street) appeared to the left or right hand side of the street and participants had to react with a steering-wheel movement. As taxi-drivers they were instructed to turn towards waving pedestrians to pick them up (compatible S–R relationship) or to turn away from pedestrians stepping on the street to avoid hitting them by moving the steering wheel in the opposite direction (incompatible S–R relationship). Participants were instructed to react as fast and correct as possible.

After participants' steering-wheel response the screen cleared and participants had to realign the steering wheel into a straight-ahead position. If participants gave a wrong steering wheel response or if reaction times exceeded 1,500 ms, an auditory error feedback was given. The next trial started with a step on the accelerator pedal. Each participant accomplished 16 practice trials in order to ensure an understanding of the experimental procedure. The experiment lasted about 60 min.

Design

The experiment had a two-factorial design with age (younger vs. older adults) as between-subject factor and spatial compatibility (compatible vs. incompatible) as within-subject factor. Participants performed 256 trials, with a short break after half of trials. As dependent variables reaction times of steering-wheel responses and percentage errors (proportion of incorrect response directions) were recorded.

Results and Discussion

Median reaction times and percentage of errors of each participant were entered into 2 (age) $\times$ 2 (spatial compatibility) analyses of variance (ANOVA). The analysis of reaction times revealed highly significant main effects of age ($F(1, 28) = 19.8$, $MSE = 16607.14$, $p < 0.001$) and spatial compatibility ($F(1, 28) = 19.6$, $MSE = 1652.03$, $p < 0.001$). The interaction between age and spatial compatibility missed statistical significance ($p > 0.10$). The main effect of age indicated longer reaction times in older participants ($M = 743$ ms, $SD = 116$) in comparison to younger participants ($M = 595$ ms, $SD = 57$). The spatial compatibility effect referred to

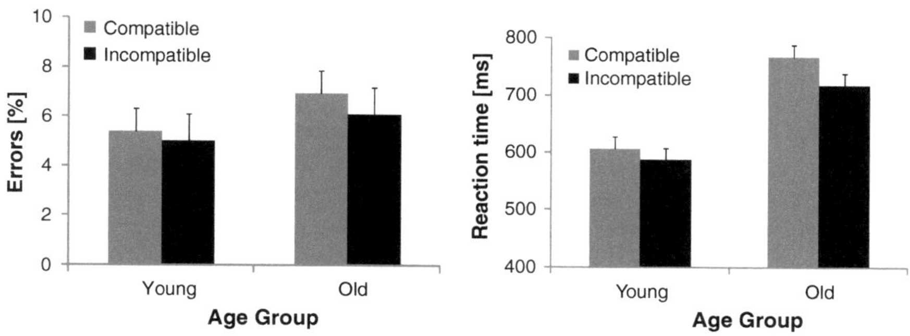

Fig. 2 Mean reaction times (*left*) and percentage errors (*right*) of younger and older drivers in compatible and incompatible situations in Experiment 1

faster reaction times in spatially incompatible conditions (M = 646 ms, SD = 120) than in spatially compatible conditions (M = 692 ms, SD = 121), i.e. participants reacted 47 ms faster when they had to avoid a collision with a pedestrian by turning the steering wheel in the opposite direction than picking up the pedestrian by turning the steering wheel towards them (Fig. 2).

The missing interaction between age and spatial compatibility indicates that the spatial compatibility effect was not increased in the group of older drivers. Reaction times of younger participants were 30 ms faster in the incompatible condition; older participants reacted 63 ms faster in the incompatible condition. The analysis of percentage errors did not reveal any significant effects or interactions. For older and younger participants as well as for spatially compatible and incompatible conditions no appreciable differences were observed in error rates.

Contrary to assumptions of classical compatibility theories, reaction time was faster when drivers had to turn the steering wheel away from the stimulus (avoiding reaction; spatially incompatible condition) in comparison to a turn of the steering wheel towards the stimulus (spatially compatible condition). Apparently, driving performance (as measured in reaction time and errors) in a natural setting of younger and older drivers was affected by the affective meaning of stimuli leading to reversed compatibility effects for dangerous situations— which replicates our findings in a younger sample (Müsseler et al. 2009, Exp. 1). Older drivers showed an inferior driving performance with slower reactions. No interaction between age and compatibility was found—contrary to previous findings in laboratory situations with neutral stimuli—which indicates, that the compatibility effect was not increased in older drivers. One possible explanation is, that older drivers' higher driving expertise moderated compatibility effects. This explanation will be tested in Experiment 2b.

Experiment 2a: Spatial Compatibility in Natural Scenes: The Effect of a Secondary Task

Experiment 2 further examined the effects of spatial compatibility and age in driving in a dual task situation. The same driving scenario was used as in the previous experiment. Younger and older participants watched short movies showing a street scene where pedestrians suddenly entered the street. They were instructed to avoid a collision with the careless pedestrian by turning the steering wheel as fast and correct as possible in the opposite direction (primary task). Additionally, an auditory driving direction message of a navigational system was presented to which participants had to react in an unspeeded manner (secondary task). Research using dual-task paradigms has not only shown unspecific impairments in reaction times and errors but also so-called cross-task compatibility effects. Reaction times in the primary task were impaired when an incompatible stimulus was presented in the secondary task (Müsseler et al. 2005, 2006). Thus, we expected an effect of cross-task compatibility between the primary task of driving and the secondary task of responding to the message of the navigation system. Moreover, as the influence of a secondary task has been shown to increase with age (see chapter Age-related Changes of Neural Control Processes and their Significance for Driving Performance, Hahn et al.), we examined the effects of cross-task compatibility in older drivers, expecting enlarged interference effects across tasks in older drivers.

Method

Participants

A total of 34 new participants, 17 older adults with a mean age of 60.7 years (range 56–65 years) and 17 younger adults with a mean age of 23.2 years (range 21–30 years) took part in the experiment.

Apparatus and Stimuli

The same experimental setup was used as in Experiment 1 with two exceptions. The experimental procedure was controlled by a Syntron AMD PC running on Windows XP. The auditory driving direction message ("turn left"/"turn right") was recorded and added as sound track to the movies and presented via speaker boxes (Logitech R-10), which were placed 50 cm to the left and to the right of the participant. Moreover, only the careless pedestrian entering the street without keeping eye contact was presented.

Procedure

As primary task, participants were required to react with an avoiding response to a pedestrian, who suddenly entered the street from the left or right side. Just before the presentation of the visual stimulus, participants heard a spoken driving direction message of a navigation system ("turn left/turn right"). The secondary task required participants to remember this message and to give an unspeeded response by turning the steering wheel according to the driving direction message to the left or the right after the primary task was accomplished.

As in Experiment 1 each trial started with the presentation of a static street scene and the movie was started by stepping on the accelerator pedal. Before the pedestrian appeared the spoken message of the navigation system was presented. In order to assess the baseline driving performance in the primary task, in 1/3 of the trials no navigation message was presented. Moreover, in order to reduce the frequency of dangerous driving situations, in 1/3 of the trials no pedestrian appeared and participants only had to respond to the secondary task. The experiment lasted about 90 min.

Design

The experiment had a two-factorial design with age (younger vs. older adults) as a between-subject factor and spatial correspondence (corresponding vs. non-corresponding) as within-subject factor. Spatial correspondence refers to the spatial location of the visual stimulus in the primary task (pedestrian appearing at the left or at the right side of the scene) and the spatial information given in the acoustical message of the navigation system ("turn left", "turn right"). Note, that under spatial correspondence between the stimuli in the primary and secondary task, the required steering wheel reaction is incompatible to the spatial location of the person and the spatial information given in the acoustical message of the navigation system and vice versa.

Participants performed 252 trials (84 trials per condition) in randomized order. As dependent variables reaction times of steering-wheel responses and percentage errors (proportion of incorrect response directions) were recorded.

Results and Discussion

Median reaction times and percentage of errors of each participant were entered into 2 (age) $\times$ 2 (spatial correspondence) ANOVA. The ANOVA on reaction times revealed a highly significant main effect of age, with older drivers showing slower reactions than younger drivers [$F(1,32) = 42.5$, $MSE = 17391.49$, $p < 0.001$)] and a highly significant main effect of correspondence ($F(1,32) = 12.6$, $MSE = 192.38$, $p = 0.001$), with faster reactions in non-corresponding situations (Fig. 3).

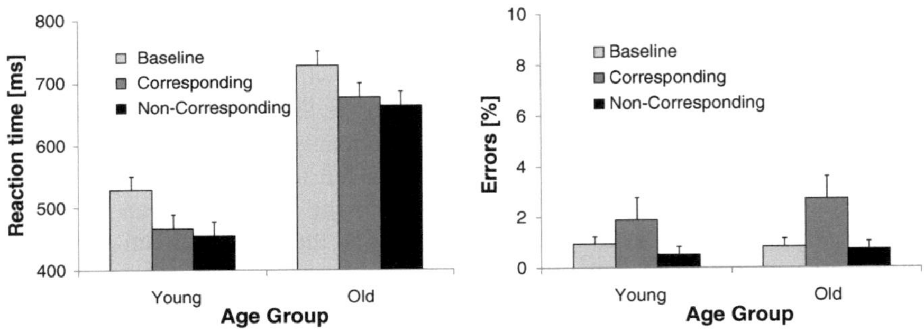

Fig. 3 Mean reaction times (*left*) and percentage errors (*right*) of younger and older drivers in the baseline condition and in corresponding and non-corresponding situations in Experiment 2a

The interaction between age and correspondence again missed statistical significance ($p > 0.20$). Contrary to expectations, older drivers did not show higher correspondence-related interference effects. The main effect of age implies, that older participants showed slower reactions (M = 669 ms, SD = 99) in comparison to younger adults (M = 461 ms, SD = 88). Regarding the main effect of correspondence, reactions were faster in non-corresponding conditions (M = 559 ms, SD = 140) in comparison to corresponding conditions (M = 571 ms, SD = 141). The analyses of percentage errors closely mirrored the pattern of results obtained in reaction times.

Unexpectedly, in the single task condition (baseline) reaction times were significantly slower than in the dual task condition, both in the group of young participants (single task: M = 513 ms, SD = 82, dual task: M = 448 ms, SD = 71; $p < 0.001$) as well as in the group of old participants (M = 652 ms, SD = 126, dual task: M = 583 ms, SD = 137; $p < 0.001$). This might be explained by the fact that the auditory signal from the navigation system (stimulus in the secondary task) always directly preceded the visual stimulus in the primary task (pedestrian suddenly appearing on the scene). Therefore, the stimulus in the secondary task probably served as a cue for the stimulus in the primary task thus resulting in reduced reaction times under dual task conditions.

Findings show that participants' driving performance was superior in non-corresponding situations, i.e. when the driving direction message of the navigation system was contralateral to the pedestrians' side of appearance. In other words, participants reacted faster and with lower error rates when the direction of the navigation message corresponded to the escape direction when avoiding the collision with the pedestrian. As expected, we found slower reactions in older drivers. Contrary to expectations, there was no increased interference effect in older drivers. One possible explanation for that finding is, that older drivers' expertise again compensated for age-related interference effects. Therefore the factor driving expertise was included in Experiment 2b.

Experiment 2b: Spatial Compatibility in Natural Scenes: The Effect of Driving Expertise

In Experiment 2b, the effect of driving expertise on age and compatibility in a dual-task driving situation was examined. For this purpose, a group of professional drivers was tested using the identical setup that was applied to non-professional drivers in Experiment 2a and the performance of both groups was compared directly. We assumed that the age-related decrease in driving performance in the dual-task situation is compensated by the greater amount of driving expertise in the elderly people.

Method

Participants

A total of 34 new participants, 17 older adults with a mean age of 55.7 years (range 50–65 years) and 17 younger adults with a mean age of 27.9 years (range 21–35 years) took part in the experiment. Driving expertise was operationalized by professional driving of at least 1,000 km per month regular driving experience in the last two years, at minimum. On average, professional drivers spent 32 h per week driving as compared to 5 h in the non-professional drivers.

Stimuli, Procedure and Design

Stimuli, procedure and design were identical to those in Experiment 2a.

Results and Discussion

For the sake of concision, we directly compared the results of professional with non-professional drivers. Median reaction times and percentage of errors of each participant were entered into 2 (age) $\times$ 2 (expertise) $\times$ 2 (spatial correspondence) ANOVA.

The analysis of reaction times revealed highly significant main effects of age ($F(1,64) = 38.0$, $MSE = 16389.72$, $p < 0.001$), expertise ($F(1,64) = 20.3$, $MSE = 16389.72$; $p < 0.001$) and interactions between age and expertise ($F(1,64) = 11.1$, $MSE = 16389.72$; $p < 0.005$) as well as between correspondence and expertise ($F(1,64) = 15.7$, $MSE = 2764.13$; $p < 0.001$). The main effect of correspondence, the interaction between correspondence and age and the triple interaction of age, expertise and correspondence did not reach statistical significance.

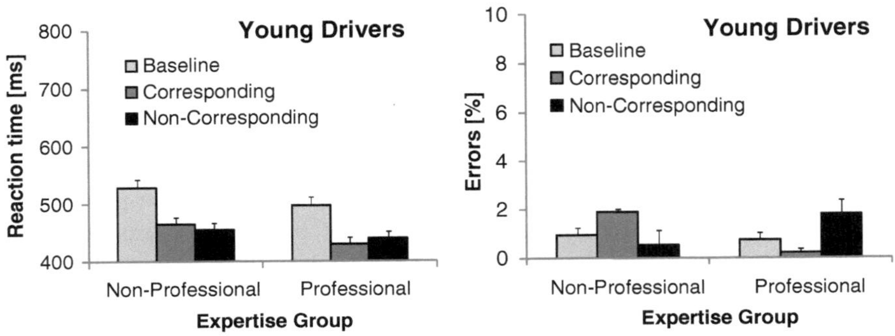

Fig. 4 Mean reaction times (*left*) and percentage errors (*right*) of young professional and nonprofessional drivers in the baseline condition and in corresponding and non-corresponding situations in Experiment 2b

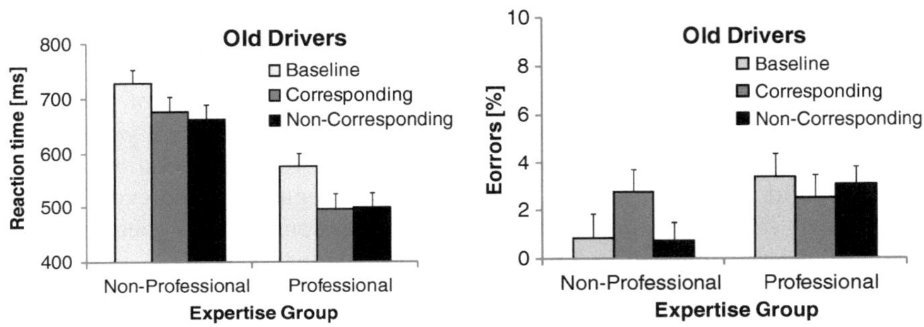

Fig. 5 Mean reaction times (*left*) and percentage errors (*right*) of old professional and nonprofessional drivers in the baseline condition and in corresponding and non-corresponding situations in Experiment 2b

The main effect of age indicates, that older participants showed 128 ms slower reactions (M = 583 ms, SD = 136) in comparison to younger adults (M = 455 ms, SD = 75). Regarding the main effect of expertise, non-professional drivers (M = 565 ms, SD = 140) were 93 ms slower than professional drivers (M = 472 ms, SD = 93). The interaction between age and expertise indicates that age-related differences in driving performance were larger in non-professional drivers than in professional drivers (Figs. 4 and 5). Older non-professional drivers showed 209 ms slower reactions (M = 669 ms, SD = 98) in comparison to younger non-professional drivers (M = 461 ms, SD = 88; $p < 0.001$). In professional drivers this difference between younger (M = 435 ms, SD = 48) and older drivers (M = 497 ms, SD = 115) was only 62 ms but still significant ($p < 0.05$).

Regarding the interaction between expertise and correspondence, correspondence effects were only found in non-professional drivers, whereas professional drivers reacted comparably fast in corresponding and non-corresponding situations. As reported in Experiment 2a, non-professional drivers' reactions were

12 ms faster in non-corresponding conditions (M = 559 ms, SD = 140) in comparison to corresponding conditions (M = 571 ms, SD = 141; $p < 0.05$). However, for professional drivers no significant differences in reaction times between corresponding (M = 463 ms, SD = 94) and non-corresponding conditions (M = 469 ms, SD = 91; $p > 0.05$) were found.

Again, when comparing reaction times in the dual-task condition with the baseline condition (where no navigation message was presented), faster reaction times were obtained in the dual-task condition ($F(1,67) = 199.5$, MSE = 761.59; $p < 0.001$). Participants reacted 67 ms faster with the navigation message (M = 516 ms, SD = 128) than in the baseline condition without navigation message (M = 583 ms, SD = 126).

The analysis of percentage errors revealed a significant main effect of age ($F(1, 64) = 5.0$, MSE = 9.03; $p < 0.05$) and an interaction between expertise and correspondence ($F(1,67) = 200.9$, MSE = 761.59; $p < 0.001$). No other reliable effects were found. The main effect of age indicates that older drivers had higher error rates (M = 2.26 %, SD = 2.79) than younger drivers (M = 1.09 %, SD = 1.09). Regarding the interaction between expertise and correspondence, non-professional drivers had a 1.68 % higher error rate in corresponding (M = 2.31 %, SD = 2.76) than in non-corresponding situations (M = 0.63 %, SD = 1.07; $p < 0.001$). Contrary to that, professional drivers' error rates were lower in corresponding (M = 1.37 %, SD = 3.05) than in non-corresponding conditions (M = 2.42 %, SD = 2.91; $p < 0.05$; Figs. 4 and 5). The presence or absence of the navigation message did not affect participants' error rates.

Older drivers showed a considerably reduced driving performance, but contrary to typical findings in the literature no increased age-related interference effects were found, which hint at compensating resources in older drivers. As a potential compensating factor investigated in this study driving expertise moderated the effects of age in driving performance, with older professional drivers reaching a comparable performance in reaction times as younger professional drivers. Moreover, driving expertise co-acted with correspondence-mappings in the driving task: for non-professional drivers a reversed correspondence-effect was found with superior performance in the spatially non-corresponding condition, which is probably a result of valence correspondence overruling spatial correspondence (Müsseler et al. 2009). However, in professional drivers no effect of cross-task-compatibility was obtained, suggesting that expertise not only reduces age-related performance differences but also affects compatibility-related interference effects. Finally, the inferior performance in the baseline condition, where no navigation message was given, suggests that the navigation message had a cue-function, which facilitated driving reactions in the primary task.

General Discussion and Conclusion

The aim of the present study was to analyze age-related differences in the applied context of critical driving situations. Findings reported in the literature suggest a decreased performance of elderly people in laboratory situations as well as in a natural driving context. However, it is still an open question, whether this is also true for dual-task requirements in a driving context and whether expertise might compensate for age-related effects. In two experiments, we first studied the influence of age on spatial compatibility effects in natural scenes and, then the effects of a secondary task as well as the effects of driving expertise were examined.

In Experiment 1, we replicated and extended the findings obtained by Müsseler et al. (2009). They demonstrated that the typically observed advantage of spatially compatible responses is reversed for dangerous situations in natural scenes. In the driving context, reaction time in an avoiding situation (e.g. stimulus: an unexpected person appearing on the right side of the scene, reaction: turning a steering wheel to the left) was shorter than a turn of the steering wheel towards the stimulus. These results speak against an automatic response tendency elicited by the position of the stimulus, as assumed often by compatibility theories (Kornblum et al. 1990). The advantage of an incompatible stimulus–response coupling was not only observed in young subjects but also in a group of elderly participants. Although older drivers showed an inferior driving performance with slower reactions, no interaction between age und compatibility was found. Thus, different to the general pattern of results obtained in the literature, the compatibility effect was not increased in older drivers. As one possible explanation for this observation, the influence of expertise was examined in Experiment 2.

In Experiment 2, the effects of a secondary task as well as the effects of driving expertise were examined. Recent studies using the dual-task paradigm demonstrated so-called cross-task-compatibility effects, that is, performance in the primary task was not only impaired by the presence of a secondary task in general, but there were also specific impairments based on the compatibility relation between the stimulus in the secondary task and the reaction in the primary task (Müsseler et al. 2005, 2006). Results obtained in Experiment 2 showed that participants' driving performance was superior in non-corresponding situations, however, in this case a cross-task compatibility effect was obtained. Participants reacted faster when the direction announced in the navigation message was contralateral to the pedestrians' side of appearance. Again, older drivers showed slower reactions but there was no increased interference effect.

One possible explanation for the absence of increased interference effects in the group of elderly participants might be the fact, that age of participants is confounded with driving experience. The group of young participants had on average 5 years of driving experience, whereas the group of older drivers had 40 years of driving experience on average. Therefore, two factors might be at work producing these results. On the one hand, with increasing age, susceptibility to interference effects increases with increasing age, as research on spatial compatibility as well as dual-task situations has demonstrated (e.g. Pick and Proctor 1999). On the other hand, with increasing driving experience the influence of interfering stimulation (e.g. the acoustical message of a navigation system) should be reduced.

To test this explanation, two groups of professional drivers (young and old drivers) were tested with the identical experimental setup in Experiment 2b. In accordance with our hypotheses, expertise moderated the effects of age in driving performance. First, older professional drivers reached a comparable performance in reaction times as younger professional drivers. Second and more importantly, driving expertise co-acted with compatibility mappings in that in professional drivers no effect of cross-task-compatibility was obtained. Thus, driving experiences result in a reduced vulnerability to interfering stimuli probably because professional drivers are more experienced in the primary task (driving a car, or, more specifically, reacting in dangerous situations while driving a car) and because they are more used to dual-task situations while driving a car (using a mobile phone or listening to a message of the navigation system while driving a car).

One unexpected result was obtained in the comparison of single-task vs. dual-task situations. In all four groups of participants (young and old, professional and non-professional drivers) increased reaction times were observed in the single-task situation (baseline) as compared to the dual-task situation. A typical result, however, would have been that the secondary task requires processing capacity and due to limited processing capacity reaction time increases in dual-task conditions. In the present study, the stimulus in the secondary task (acoustical signal of the navigation system) was always presented immediately before the stimulus of the primary task (person suddenly appearing on the scene). As the reaction in the primary task required participants to react as fast as possible with a turn of the steering wheel, the acoustical signal probably served as a cue for the appearance of the visual stimulus and thus, facilitated the reactions in the primary task, even though it was only predictable in 66 % of the trials (in 1/3 of the trials no visual stimulus followed).

The results obtained in the present project do not only have to be considered in research but also have practical consequences for the design of electronic driver assistant systems. Although further research is needed to validate the findings in different applied situations (e.g., with information given from a collision warning system), the present results already indicate that it should be taken into account that every additional information given to the driver not only demands the driver's attention in general (see chapter Age-related Changes of Neural Control Processes and their Significance for Driving Performance, Hahn et al.) but that there are also specific cross task effects. Moreover, the results indicate that age-related changes in information processing capacities can be compensated for by training. As it is not always possible for elderly people to avoid *complex* driving situations, older drivers should undergo driving course regularly, especially if their driving practice decreases.

Acknowledgments We wish to thank Denise Veit, Dagmar Cora und Antonia Windrisch for carrying out the experiments and Oliver Gärtner for programming Experiment 2. This research was supported by a grant from the Deutsche Forschungsgemeinschaft to Jochen Müsseler and Gisa Aschersleben (DFG MU 1298/5).

References

Brainard DH (1997) The psychophysics toolbox. Spat Vis 10:433–436

Cox AB, Cox DJ (1998) Compensatory driving strategy of older people may increase driving risk. J Am Geriatr Soc 46:1058–1059

Guiard Y (1983) The lateral coding of rotations: a study of the simon effect with wheel-rotation responses. J Mot Behav 15:331–342

Hancock PA, Lesch M, Simmons L (2003) The distraction effects of phone use during a crucial driving maneuver. Accid Anal Prev 25:501–514

Hartenstein W (1995) Das "Älterwerden" der Autofahrer-Population: Größenordnungen, Fahrgewohnheiten, Einstellungen, Auswirkungen. In: Lindlacher J (Ed) Ältere Menschen im Straßenverkehr. Schriftenreihe Straßenverkehr, Bd. 34. ADAC, München, pp 16–24

Hommel B (1998) Automatic stimulus-response translation in dual-task performance. J Exp Psychol Hum Percept Perform 24:1368–1384

Jolicœur P (1999) Dual-task interference and visual encoding. J Exp Psychol Hum Percept Perform 25:596–616

Kliegl R, Krampe RT, Mayr U (2003) Formal models of age differences in task complexity effects. In: Staudinger U, Lindenberger U (eds) Understanding human development: life-span psychology in exchange with other disciplines. Kluwer, Dordrecht, pp 289–313

Kline DW, Scialfa CT (1997) Sensory and perceptual functioning: basic research and human factors implications. In: Fisk AD, Rogers WA (eds) Handbook of human factors and the older adult. Academic Press, San Diego, pp 27–54

Kornblum S, Hasbroucq T, Osman A (1990) Dimensional overlap: cognitive basis for stimulus-response compatibility: a model and taxonomy. Psychol Rev 97:253–270

Li S-C, Lindenberger U, Hommel B, Aschersleben G, Prinz W, Baltes P (2004) Transformations in the couplings among intellectual abilities and constituent cognitive processes across the life span. Psychol Sci 15:155–163

Lien MC, Proctor RW (2002) Stimulus-response compatibility and psychological refractory period effects: implications for response selection. Psychon Bull Rev 9:212–238

McGwin G Jr, Brown DB (1999) Characteristics of traffic crashes among young, middle-aged, and older drivers. Accid Anal Prev 31:181–198

Müsseler J, Wühr P (2002) Response-evoked interference in visual encoding. In: Prinz W, Hommel B (eds) Attention and performance XIX: common mechanisms in perception and action. Oxford University Press, Oxford, pp 520–537

Müsseler J, Aschersleben G, Arning K, Proctor R (2009) Reversed effects of spatial compatibility in natural scenes. Am J Psychol 122:325–336

Müsseler J, Koch I, Wühr P (2005) Compatibility effects with irrelevant locations in dual task: a cross-task simon effect. Eur J Cogn Psychol 17:708–772

Müsseler J, Wühr P, Umiltà C (2006) Processing of irrelevant location information under dual-task conditions. Psychol Res 70:459–46

OECD (2001) Ageing and transport; mobility needs and safety issues. Organisation for Economic Co-operation and Development, Paris

Park DC, Schwarz N (1999) Cognitive ageing. Buchanan, Philadelphia

Pashler H (1998) Attention. Psychology Press, Hove

Pelli DG (1997) The VideoToolbox software for visual psychophysics: transforming numbers into movies. Spat Vis 10:437–442

Pick DF, Proctor RW (1999) Age differences in the effects of irrelevant location information. In: Scerbo MW, Mouloua M (eds) Automation technology and human performance: current research and trends. Lawrence Erlbaum, Mahwah, pp 258–261

Plude DJ, Hoyer WJ (1986) Age and the selectivity of visual information processing. Psychol Aging 1:4–10

Salthouse T (1996) The processing-speed theory of adult age differences in cognition. Psychol Rev 103:403–428

Statistisches Bundesamt [German Federal Statistical Office] (2005) Unfallgeschehen im Straßenverkehr [Accidents in road traffic 2005]. Pressestelle, Wiesbaden. http://www.destatis.de/jetspeed/portal/cms/Sites/destatis/Internet/DE/Presse/pk/2006/Unfallgeschehen/PressebroschuereUnfallgeschehen05,property=file.pdf

Stins JF, Michaels CF (1997) Stimulus-response compatibility is information-action compatibility. Ecol Psychol 9:25–45

Stutts JC (1998) Do older drivers with visual and cognitive impairments drive less? J Am Geriatr Soc 46:854–861

Vercruyssen M (1997) Movement control and speed of behavior. In: Fisk AD, Rogers WA (Eds) Handbook human factors and the older adult. Academic Press, San Diego, pp 55–86

Age-Related Changes of Neural Control Processes and Their Significance for Driving Performance

Melanie Hahn, Nele Wild-Wall and Michael Falkenstein

Lessons Learned

For an increasing number of older people driving is the main source of mobility and therefore a significant factor for quality of life and independence concerning social and working life. Driving is a complex task involving visual, motor and cognitive skills which underlie age-related changes (Anstey et al. 2005). While some of the visual and motor deficits can be compensated e.g., by visual aids and driver assistance systems, changes in cognitive functions are more challenging. Besides the increasing technical improvements in cars there are more and more training programs for older drivers. To develop successful programs or technical aids that improve driving for older people it is necessary to know the specific problems in driving that come along with increasing age and related cognitive changes.

We conducted a series of three experiments in order to test several hypotheses concerning the relevance of age-related changes in cognitive functions for car driving. All experiments consisted of driving like tracking tasks and secondary cognitive tasks which occasionally required manual or verbal responses. In the first experiment with a more artificial setup we also assessed neurophysiological data

M. Hahn (✉)
Leibniz Research Centre for Human Factors and Working Environment (IfADo), Dortmund, Germany
e-mail: hahn@ifado.de

N. Wild-Wall
Rhein-Waal University of Applied Science, Kamp-Lintfort, Germany
e-mail: Nele.Wild-Wall@hochschule-rhein-waal.de

M. Falkenstein
Leibniz Research Centre for Human Factors and Working Environment (IfADo), Dortmund, Germany
e-mail: falkenstein@ifado.de

C. M. Schlick et al. (eds.), *Age-Differentiated Work Systems*,
DOI: 10.1007/978-3-642-35057-3_13,

via the electroencephalogram (EEG). The EEG allows us to examine cognitive processing with very high time resolution and in situations where no overt performance can be measured. The second and third experiments were conducted in a driving simulator because older people benefit from more realistic settings in experimental situations and in order to increase the external validity of our results.

We could show that the tracking performance of older people was especially hampered after responding manually to relevant stimuli in a secondary task. This motoric interference could not be reduced by changing the response modality of the secondary task from manual to verbal responses. With the aid of the EEG data we could show that older people seem to have a deficit in differentiating between relevant and irrelevant secondary stimuli. The undifferentiated investment of attentional resources to relevant and irrelevant stimuli may have significant effects on the response to further, possibly critical stimuli while driving. Indeed, as we could show in a third experiment, older compared to younger people respond more slowly and with more errors to critical events that occur immediately after irrelevant stimuli. Based on our findings the following recommendations for the development of experimental settings, in-vehicle information systems and traffic environment can be given:

- Older people benefit from naturalistic settings in experimental setups (see also chapter Age-Related Differences in Critical Driving Situations: The Influence of Dual-Task Situations, S-R Compatibility and Driving Expertise, Aschersleben et al.).
- As older people showed problems with the coordination of driving-like tracking and a secondary manual response, situations should be reduced which involve parallel secondary manual responses.
- Verbal responses seem to be no appropriate alternative to manual responses in a secondary task while driving.
- Because the differentiation between relevant and irrelevant stimuli is reduced with increasing age, less relevant and definitely irrelevant information should be avoided in the proximal and distal traffic environment. This claim is especially important, if the less relevant stimuli occasionally offer response options.
- In order to compensate for the general age-related slowing, any relevant information should be given as early as possible.

Introduction

Mobility is highly important for participating in social and working life. Especially driving is a key index of mobility and an increasing number of older people are willing to maintain driving as a main source of mobility. According to statistical predictions in Germany, the distance travelled by drivers 65 years and older will almost double from 76 to about 150 billion kilometers from 2002 to 2050 (Statistisches Bundesamt

2008). For older people, driving is a significant factor for quality of life and for being independent (Oxley and Whelan 2008). Also before retirement age, there is a strong relationship between participating in road traffic and occupational engagement. On the other hand, increasing age is a main predictor for driving cessation. This may not only reduce mobility but also has negative impact on many areas of life for older people who stop driving. It has been shown that driving cessation is related to a decrease in out-of-home activity levels and social isolation (Marottoli et al. 2000) or to reduced health status and increased depressive symptoms (Ragland et al. 2005).

Driving itself is a highly complex task, which asks for simultaneous processing of multiple stimuli and to generate rapid and accurate responses. Several skills are related to safe driving like visual, motor, and also cognitive skills (Anstey et al. 2005). These skills underlie age-related changes even during normal aging (e.g., presbyopia in the visual domain, reduced muscle force in the motor domain, or reduced attention, speed of processing or working memory in the cognitive domain). Many of these age-related changes may be compensated for, e.g., by eyeglasses in order to compensate for presbyopia, by easy entry aids in order to compensate for orthopedically problems or by power assisted steering devices in order to compensate for reduced muscle force. Others of them, especially changes in cognitive functions, need to be more acknowledged as potential problems and challenge. This was the case in the last years: Age-related cognitive changes have been more recognized as being relevant for driving performance which has led to increased research effort in this field (see Anstey et al. 2012 for an overview) and also efficient training programs for older drivers like e.g., the speed of processing training by Ball and colleagues (e.g., Edwards et al. 2009). At the same time, technical equipment like e.g., in-vehicle information systems became established, comprising more and more functionality and complexity. Studies suggest that driving performance can decrease when secondary tasks, e.g., the handling of an in-vehicle information device, have to be performed while driving (e.g., Horberry et al. 2006). This may be a result of increased distraction and inattention due to more or less relevant stimuli and response options, which are secondary to driving but offered by the technical devices. Distraction and inattention are main causes of traffic accidents. Considering the just mentioned normal age-related changes in cognitive functioning, this may be especially relevant for older drivers.

In general, the absolute number of accident involvements of older drivers is not higher than in any other age group. Reasons for this are the lower proportion of older drivers and that they tend to drive less or use compensation strategies like e.g., taking only familiar routes or stopping night driving. However, if referenced to the kilometers driven, increasing age bears a higher risk of being involved in a car accident (Preusser et al. 1998) and of being injured (Eberhard 2008). In addition, situations may occur in which some compensation strategies (e.g., driving only at daylight) are not appropriate or cannot be utilized in order to fulfill occupational or social requirements.

As to the given facts, effort is necessary in order to support safety and mobility of an aging society. This includes also a solid knowledge about age-related cognitive changes and their potential relevance for driving performance. Such knowledge may lead to recommendations for designing technical devices in the proximal and distal driver environment in order to support safe driving, and thus, mobility and independence.

Age-Related Cognitive Changes and Driving

Today, in social and working life, the resources and potentials of aging people are recognized and underlined. By now, this view displaced the deficit model of aging. However, aging, even in health, is undisputable related to changes in perceptual, motor and cognitive functioning. This is underlined by many laboratory studies, which examined younger and older participants in cross-sectional or longitudinal experimental designs (for an overview see Bialystok and Craik 2006). Although age-related decline in several cognitive functions occurs with high inter-individual variability, changes need to be recognized as potentially relevant for fitness and safety to drive in order to maintain mobility.

Many experimental studies include paradigms where fast and accurate responding is required. A reduction of response speed with increasing age is the most prevalent observation which is especially pronounced in complex experimental tasks or when fast decision between response options is required (Salthouse 2000; chapter Age-Related Differences in Critical Driving Situations: The Influence of Dual-Task Situations, S-R Compatibility and Driving Expertise, Aschersleben et al.). In an own study by Yordanova and colleagues (2004) we showed that such reduction of response speed in some conditions may be explained by the slowing of motor processes but not the decision stage or earlier perceptual stages. Also the preparation of fast motor responses to cues which only partially predict specific response requirements, was shown to be compromised in older compared to young participants (Sterr and Dean 2008). Further, the production of accurately timed manual responses shows a decline with increasing age (Wild-Wall et al. 2009). The just mentioned age-related changes concerning the preparation, speed and timing of sensorimotor processes may eventually not affect normal driving situations. However, delays of milliseconds may make a difference in non-standard, unexpected and dangerous situations.

Next to the just mentioned changes in rather basic sensorimotor processes, age-related changes especially concern so called executive functions (see e.g., Bialystok and Craik 2006). Executive functions can be understood as higher order cognitive functions which modulate the operation of other cognitive processes and coordinate goal directed mental activity. Although until now, there is no widely accepted unitary concept of executive functions, prominent examples are linked to working memory and comprise for instance executive attention, switching and dividing attention, the selection of relevant information and the inhibition of irrelevant information or inappropriate and premature responses. Imagine, for instance, the situation waiting at a red light for the straight-ahead lane. Suddenly, the light for the left-turning vehicles switches to green. In such situation, executive functions such as inhibition prevent us from accelerating, even if the gas pedal is slightly touched. Taken together, executive functions are highly important in complex situations where fast and accurate decisions are necessary. For such situations, car driving is a key example.

Executive functions are closely linked to the integrity of frontal brain areas. Interestingly, these brain areas are one of the earliest to comprise age-related changes, starting already in in the fourth decade of life (Salat et al. 2009). Therefore it is not surprising that changes in executive functions are one of the most reported age-related changes of cognitive functioning in research (see Bialystok and Craik 2006 for an overview). Such changes often go hand-in-hand with altered electrophysiological brain responses, so called event-related potentials (ERPs), which can be extracted from the electroencephalogram (EEG). ERPs are correlates of specific cognitive functions like e.g., perception, attentive stimulus processing or response selection and preparation. They can be characterized by scalp location, amplitude and latency and they are modulated by specific task demands and also by structural and/or functional age-related brain changes.

We and others have shown that the ability to inhibit the processing of irrelevant information is compromised with increasing age which is also reflected in increased brain activation following irrelevant stimuli (e.g., Wild-Wall and Falkenstein 2010, Gazzaley and D'Esposito 2007). On the other hand, the ability to select relevant information was shown to be intact in higher age (Gazzeley and D'Esposito 2007). Also the abilities to maintain multiple task rules in working memory and to switch between tasks without external cue decline with increasing age (Gajewski et al. 2010). These changes may be explained by reduced working memory resources and are also reflected in altered electrophysiological brain responses.

Specific task parameters (e.g., complexity of the stimuli) and the task context (e.g., the general complexity of the task) are important aspects which determine if the age-related changes in executive functions become evident in behavior or not. Critical are especially situations in which the coordination of two or more tasks is required and processing resources need to be flexibly allocated to the different tasks. In fact, age-related performance decline in laboratory dual task situations is well documented in many studies (Verhaeghen et al. 2003). Age-related performance decline was particularly observed in dual tasks which combine sensorimotor and cognitive task demands, for instance when walking and solving a cognitive task (Li and Lindenberger 2002).

Age-related changes in executive functions, that is, in working memory, dual tasking, task switching and inhibition, are not only well documented in laboratory studies, but are also highly relevant for car driving. At all times during driving, it is necessary to steer the car and frequently to react to the current traffic situation and to traffic signs. Constantly, the driver has to select relevant information and inhibit irrelevant and distracting stimuli in the proximal and distal environment. Considering the potential problems of older people due to the age-related changes in executive functions, driving performance may suffer especially in highly complex situations with multiple task demands where the flexible assignment of processing resources is necessary. In addition, in-vehicle information systems and technical devices eventually require the processing of secondary stimuli and motor responses which may further increase the demands of a normal driving situation.

Although until now there is manifold evidence from laboratory studies which point to an age-related performance decline in dual-task situations, such studies often fail to show ecological validity. In addition, studies are rare which investigate dual-task situations including continuous (driving-like) sensorimotor tasks and secondary task demands of varying relevance or modality (see chapter Age-Related Differences in Critical Driving Situations: The Influence of Dual-Task Situations, S-R Compatibility and Driving Expertise, Aschersleben et al.).

In order to fill these gaps, the present line of research picked up the above-mentioned age-related changes in cognition and traced several questions. First of all, cognitive and particularly executive processes should be identified which are compromised with increasing age and which are especially relevant for driving performance. Here, we focused on situations with secondary task demands which include the processing of and the response to secondary visual stimuli as for instance offered by in-vehicle technical devices. Related to this we examined (a) if the presentation of secondary stimuli with varying degree of relevance differently affects the performance of younger and older drivers. (b) By means of event-related potentials we intended to examine the specific neurophysiological mechanisms which underlie the processing of secondary stimuli in dual-task situations especially if no responses are required. (c) As we observed evidence that the coordination of driving-like tracking with a secondary manual response is especially defacing for older people, it was questioned whether the possibility of a verbal response affects driving-like tracking performance of older people to a lower degree. In addition, by using also a driving simulator, our results allow us (d) to estimate, if older participants benefit especially from more realistic task setups.

Methods and Results

To answer these questions we conducted three different experiments. The first experiment can be seen as a pre-experiment with a more controlled but less ecological valid set-up. It should identify the relevant conditions in a driving-like dual task which are most affected by aging. The second and the third experiment were conducted in a driving simulator and are based on the results of the first experiment.

Experiment 1

In our first experiment participants should perform a driving-like dual task that consists of a compensatory tracking task and a visual attention task on a computer. While performing both tasks, the electroencephalogram (EEG) was recorded.

The dual task was intensively practiced the day before the experimental session. In the experimental session, younger and older participants performed six experimental blocks (11 min each), which lasted more than one hour.

Tracking Task

As a primary visuo-motor task we used a compensatory tracking task, where participants had to keep a horizontal green bar in the center of a limited space on the computer monitor by using a PC steering wheel with both hands. In accordance with the instruction the participants should compensate the horizontal movement of the side wind-simulating bar by rotating the steering wheel. The bar was moving according to a complex signal of eight different superimposed and phase-delayed sine waves. If participants kept the bar in the middle of the defined space, the bar remained green. The bar changed its color into red, if the deviance to the middle of the space exceeded a certain limit. Participants were instructed to keep the bar in the middle of the space as exactly as possible and to avoid the red color.

Visual Attention Task

In the secondary task participants had to respond to two target letters with the left or right hand according to predetermined rules, e.g., letter A—right button/hand and letter X—left button/hand. Target letters were randomly chosen from the alphabet and changed after each block, which always contained 120 trials.

The presentation order of the target letters was once randomly determined and then presented to all participants in the same order. Previous targets and all remaining letters from the alphabet were used as non-target letters.

Each trial started with a fixation circle (1,000 ms) and was followed by two stars in a virtual diagonal as a cue (500 ms) that indicate relevant locations for the subsequent stimuli. 2,500 ms after cue onset two letters were presented (500 ms)—either at the relevant (cued) or irrelevant (non-cued) positions (see Fig. 1). If one of the two instructed target letters appeared at a relevant position, which was the case in 40 % of the trials, participants had to press one of two buttons corresponding to the target letter. The buttons were affixed to the steering wheel. Participants should not respond if target letters were presented at irrelevant positions (irrelevant target-likes: 20 %) or if non-target letters (at relevant or irrelevant positions, 20 % each) were presented. After each response and non-response participants received a feedback about the correctness and the speed of their response. They were instructed to respond as fast and correctly as possible. Generally participants were instructed to pay attention to both tasks, however, they should emphasize the tracking task (primary task) and disregard the visual attention task (secondary task) if necessary.

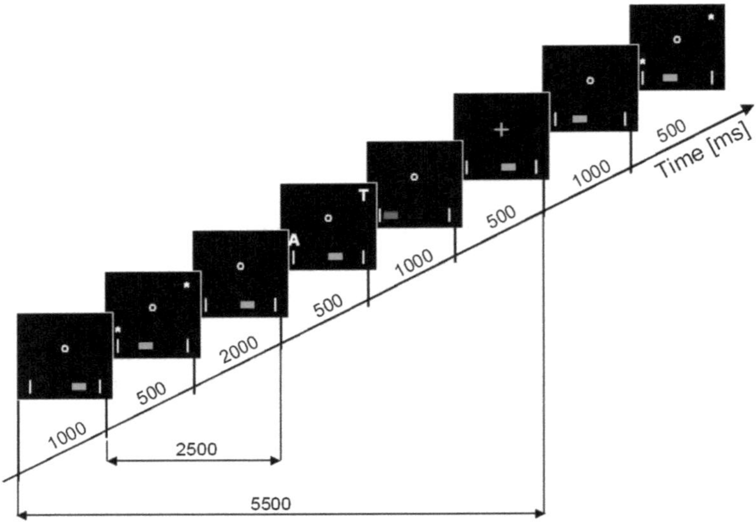

Fig. 1 Timing of the events that occur in one trial of the dual task consisting of the visual attention task and the tracking task. In this example the participant responded correctly with a right hand button press to target letter *A* and received a positive feedback (+)

Data Acquisition and Analysis

We assessed the performance in the visual attention task by measuring the response time and different types of error rates (wrong button presses, misses and false alarms). We calculated the individual overall tracking performance (individual baseline tracking performance over all six experimental blocks) as average difference between the center of the moving bar and the center of the limited space, assessed by the root-mean-squared error (RMSE), and calculated as percentage relative to the maximum possible deviation. In order to examine the influence of the processing of secondary stimuli on the tracking task we assessed the tracking performance (a) in two intervals of 1,000 ms directly following the onset of the cue and from 1,200 to 2,200 ms after cue onset and (b) in the critical time interval of 1,500 ms after the onset of relevant or irrelevant (non-) target stimuli in the visual attention task. The deviation in these intervals was measured as RMSE and was adjusted by the individual baseline tracking performance (100 %), thus given as proportion. Because the critical interval contained a motoric response (a button press) in case of the relevant targets, we additionally analyzed the interval between stimuli onset and response and the interval between response onset and 500 ms after the response separately.

To obtain correlates of invested attentional resources in time segments without overt responses and hence performance data, we used event-related potentials (ERP). Here, we measured the Contingent Negative Variation (CNV) to the secondary cue stimuli. The CNV is a vertex negative potential which is especially prominent under cognitive effort and in anticipation of upcoming and potentially relevant events as signaled e.g., by cues. We also measured the so-called P3 potential in relation to relevant or irrelevant (non-)target stimuli. The P3 is a positive deflection in the ERP in the time range of about 300–500 ms after the onset of auditory or visual stimuli. It consists of two subcomponents: While the P3a, which has a fronto-central scalp maximum, is assumed to reflect the orientation to a stimulus, the following P3b, having a parietal maximum, is assumed to reflect the invested amount of processing resources in a certain condition.

In the present experiments the amplitude of the P3b was measured as maximum positive value within the time window of 300–900 ms after (non-)target onset. The P3 latency was determined as the time interval between stimulus onset and P3 peak. The performance in the visual attention task and the EEG-data were averaged across all correct trials. All statistical analyses (RT, error rates, tracking performance, EEG-data) were conducted by repeated measures analyses of variance (ANOVAs). We used the between-factor group (young versus old) and the additional within-factor type of stimulus (relevant target, relevant non-target, irrelevant target-like, irrelevant nontarget) in the analyses of error rate and tracking performance. In the analyses of EEG-data we also included the within-factor electrode (Fz, Cz, Pz). For all p-values the level of significance was set at .05. Data acquisition and analysis were the same in all three experiments.

Subjects

The sample consisted of 25 young participants (20–33 years; M = 25.2, SD = 3.0; 13 women) and 24 older participants (57–70 years; M = 64.5, SD = 3.5; 12 women). All participants were healthy, had no history of neurological or psychiatric disorders and were not taking any drugs affecting the central nervous system. They had normal or corrected to normal vision and hearing and were experienced and active drivers. These demands also apply to the subjects in the following experiments.

Results

Older people showed significantly ($p < 0.05$) slower response times and committed more false alarms and missings in the secondary visual attention task than young people. Both groups did not differ in their error rates to targets (wrong button presses after relevant targets).

Moreover, the older group showed generally a larger absolute tracking error over the whole course of the experiment and particularly in the interval (1,500 ms) after the presentation of secondary stimuli. The tracking performance of the older compared to the young group was decreased mainly after the presentation of a relevant target. This may be due to the processing of the relevant target itself or to the required button press. Analyses of the time interval before and after the button press revealed no age differences in tracking performance before, but clear age differences after the button press to relevant stimuli. This suggests that the decreased tracking performance of older people in this time interval was likely not caused by the presence of a relevant target per se but seems to be the result of motoric interference between the sensorimotor tracking task and the manual button press.

Analysis of the relative tracking performance in the two time intervals of the preparation interval revealed no differences between the groups in the interval directly following the cues. Interestingly, the relative tracking performance of the older compared to the younger group was even better in the retention interval from 1,200 to 2,200 ms after cue onset.

Analyses of the P3b potential (Fig. 2), derived from the EEG as neurophysiological response to the stimuli in the visual attention task, revealed significantly longer latencies in NoResponse-trials (relevant non-targets or irrelevant non-targets and target-likes). Older people had generally longer P3b latencies after relevant non-targets and irrelevant target-likes and non-targets.

In addition, we obtained significantly smaller amplitudes of the P3b in responsetrials (i.e., after relevant targets) for older than for younger people (see Fig. 2). Young people showed the expected differences in P3-amplitude after the different stimulus types, that is, a significantly larger P3-amplitude after relevant targets than after relevant non-targets and irrelevant target-likes or non-targets. In contrast, there were only small and not significant differences of the P3b amplitude between the four types of stimuli for the older group.

These results of the first experiment seem to reflect an age-related decrease in the ability to differentiate between relevant and irrelevant stimuli. Apparently older people seem to invest a comparable amount of processing resources for relevant and irrelevant stimuli (cf. Hahn et al. 2011). In addition, the performance data suggest higher difficulty of older participants to coordinate a button press with the sensorimotor tracking task. However, in neutral time intervals as for instance in the preparation interval following the cues, older people showed even a relatively better tracking performance and task coordination. This was paralleled by

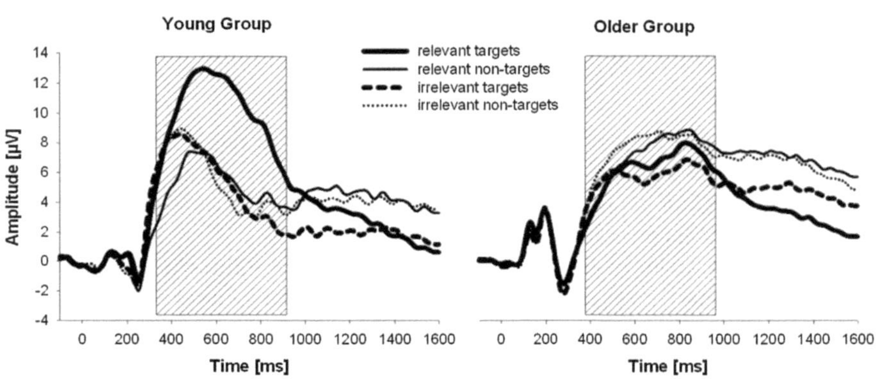

Fig. 2 Grand average ERPs for each group and stimulus type at electrode position Pz

increased cognitive and preparatory effort as indexed by the CNV (cf. Wild-Wall et al. 2011).

A shortcoming of the first experiment is that we did not test the primary and secondary task separately in single task conditions. Therefore we conducted the second experiment with separate single task and dual task conditions.

Experiment 2

Based on the results of the first experiment, the second experiment should answer the question, if the apparent motoric interference between the sensorimotor tracking task and the manual button press can be decreased by changing the manual response of the secondary task into a verbal response. Additionally we differentiated between task performance in dual task and single task conditions.

Single and Dual Task

In the second experiment participants had to perform a slightly modified version of the dual task used in the first experiment. Instead of moving a bar with a PC steering wheel the dual task was performed in a driving simulator in which participants had to navigate their simulated car on a road. In the single task conditions they had either to drive or to perform the visual attention task with manual (left or right button press) or verbal responses (saying "left" or "right). In the dual task condition the visual attention task had to be performed while driving.

Subjects

The sample of the second experiment consisted of 20 young participants (25–35 years, M = 28.9, SD = 2.8, 10 women) and 20 older participants (60–69 years, M = 64.6, SD = 2.6, 10 women).

Results

Concerning response speed in the visual attention task (Fig. 3, left panel) we replicated the results of the first experiment: Generally, older compared to younger people showed significantly slower responses to relevant target stimuli. Both groups needed more time to respond in the dual task than in the single task and they responded more slowly in the verbal than in the manual condition. The older group committed significantly more wrong responses and more missings than the young group. Independent of age, both groups omitted more targets in the verbal than in the manual condition (see Fig. 3, right panel).

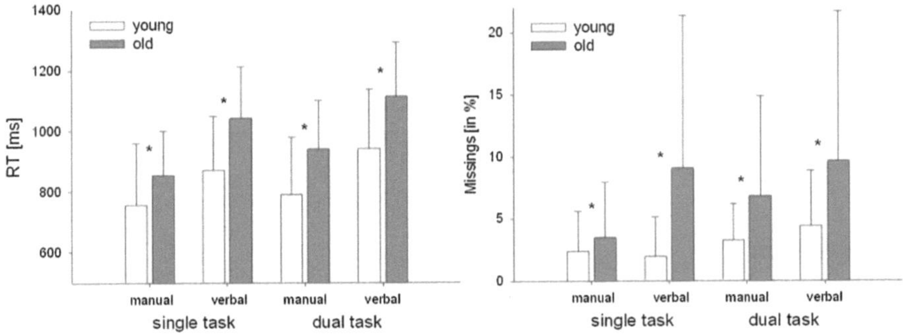

Fig. 3 Response time and missing rate in the visual attention task for each age group and task condition; * $p < 0.05$

In contrast to the first experiment there was no age difference in false alarm rates: In both groups, most false alarms were committed in the dual task with manual response option. Also contradicting the first experiment, the overall tracking performance of the older group was comparable to the young group. In both groups the overall tracking performance in the dual task with verbal responses was significantly impaired. Considering the critical time interval of 1,500 ms after the presentation of secondary visual stimuli in the dual task situation, older people showed the same or even a better tracking performance after relevant or irrelevant targets in the dual task situation than young people.

Taken together, the results of the second experiment underline that older participants benefit from more realistic settings in experimental setups. Importantly, independent of age, verbal responses seem not to be a better alternative compared to secondary manual responses.

Experiment 3

The third experiment was conducted to examine if the lack of differentiating between relevant and irrelevant stimuli in older people (see section Experiment 1) may have an impact on the response to critical events if they occur closely after irrelevant stimuli.

Single and Dual Task

In the third experiment participants had to follow a car in the driving simulator while performing the visual attention task including manual responses (see section Experiment 2). Participants had to perform three task types: (a) a combination of the tracking task and the visual attention task, (b) a combination of the tracking task and infrequent responses to the brake light and (c) a combination of the tracking task, the visual attention task and infrequent responses to the brake light. In the last two conditions the back brake light of the car in front of the participants flashed up after 50 % of the three stimulus conditions which required no response, that are, the relevant non-targets and irrelevant non-targets or target-likes. Participants had to respond to the brake light with a press on the brake pedal with their right foot as fast as possible. The time interval between onset of the irrelevant stimuli and onset of the brake light was varied (stimulus onset asynchrony, SOA = 10 or 450 or 900 ms). In addition, it has to be noted that in the task type (b), including the tracking and the brake light task, the stimuli of the visual attention task were also presented, however, without any task instruction and response requirement.

Subjects

The sample of the third experiment consisted of 20 young participants (25–31 years, M = 27.3, SD = 1.9, 10 women) and 20 older participants (60–69 years, M = 63.9, SD = 3.1, 10 women).

Results

There was no age-related difference in overall tracking performance in the three task types. Both groups showed the best tracking performance in blocks in which they just had to track and respond to the brake light. The tracking performance decreased significantly in the blocks that included all three subtasks (tracking, visual attention task and responding to the break light). Surprisingly, tracking performance got worst in the blocks consisting of tracking and the visual attention task.

In both conditions which included the visual attention task the response time of the older group was significantly longer than that of the young group. There were no differences in missings in the visual attention task between the two age groups nor between the two task types. Moreover, older participants showed significantly more wrong button presses after relevant targets and more false alarms than young participants in both task types.

Both groups showed significant greater response times to the brake lights in the most difficult condition which combined all three subtasks (tracking, visual attention task and brake light task) than in the easier condition consisting only of the tracking and the brake light task. In the most difficult condition and independent of age, the longest response times to brake lights were observed in the shortest SOA condition, that is, if the brake light occurred 10 ms after irrelevant stimuli. Only in this condition the older participants responded significantly slower to the brake lights than the young.

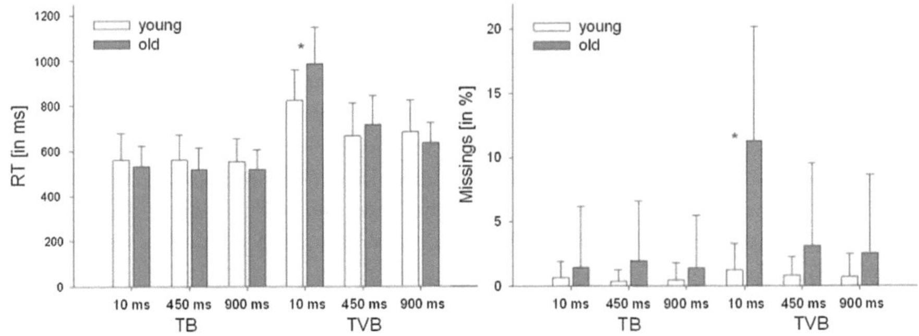

Fig. 4 Response time and missing rate of the subtask "responding to the brake light" after irrelevant stimuli in the visual attention task for each age group and SOA, separated for the two task combinations tracking and brake light (*TB*) and tracking, visual attention and brake light task (*TVB*); * $p < 0.05$

In addition the older participants omitted significantly more brake lights than the young group. However, as can be seen in Fig. 4, the two age groups differed in their missing rate only in the difficult task with the shortest SOA of 10 ms (old > young). Young people omitted more brake lights when they occurred shortly after irrelevant stimuli compared to longer intervals in the difficult condition. Comparable results were obtained for the older group, however, the rate of omitted brake lights decreased further from the second longest to the longest interval between irrelevant stimuli and brake light.

To summarize, independent of age, the tracking performance was most hampered when participants had to perform a secondary visual attention task. In addition, the fast and accurate responding to a critical event, i.e., the brake light, was most affected shortly after irrelevant stimuli. Most importantly, this performance impairment was more pronounced in the older compared to the young group and especially, when the visual stimuli were potentially response relevant.

Discussion

In all of the conducted experiments older people showed longer response times and committed more errors (missings and false alarms) in the visual attention task than younger people. The age-related general slowing is in accordance with several other studies analyzing motor responses in single and dual tasks (e.g., Falkenstein et al. 2006; Verhaeghen et al. 2003; Yordanova et al. 2004). Presumably the slowing is at least partly related to an alteration of motor responding in complex tasks (Voelcker-Rehage et al. 2006) and may also have contributed to the overall worse tracking performance of the older compared to the young participants in the less realistic setting of the first experiment.

The increased error rate of missings and false alarms in older versus young people leads to the assumption that the identification of relevant targets seemed to be more difficult for older people than for young people. The performance pattern in the older group is in line with other studies which also found increased rates of misses and false alarms in a continuous performance task (e.g., Braver et al. 2001; Mani et al. 2005). The P3 analysis supports the presumption of an insufficient differentiation between relevant and irrelevant stimuli in the secondary task as the older people in contrast to the young people did not show a greater parietal P3-amplitude after relevant targets compared to irrelevant target-likes and non-targets (for details see Hahn et al. 2011). In other words, older people seemed to invest the same amount of processing resources for all stimulus types, irrespective of the relevance of the stimulus (see e.g., Czigler and Balázs 2005 for comparable results). The lack of differentiation between relevant and irrelevant stimuli could be explained by an age-related deficient context representation (Braver and Barch 2002) that can cause the further processing of irrelevant information. It is also possible that older people have generally selective problems to suppress the processing of irrelevant information (Gazzaley and D'Esposito 2007; Hasher and Zacks 1988). Possibly, the relatively improved tracking performance in the retention interval following the cue may have also contributed to the increased rate of missings and false alarms to the subsequent visual stimuli: In this interval, the older people may have focused on the tracking task which on the other hand may have hampered the retention of the cue information (for details see Wild-Wall et al. 2011).

The comparable rate of target errors, i.e., the confusion of response side (left versus right), in the young and older group suggests that older people were able to process target stimuli as effectively as young people, after they identified them correctly as targets (cf. Gazzeley and D'esposito 2007). This was further supported by the comparable P3-latencies after relevant targets in both groups.

The generally decreased tracking performance of the older group only in the first experiment is probably due to the complexity of the dual task paradigm, to a general age-related decline in motor performance or maybe also to the less realistic task set-up. Because older participants showed specific problems in their tracking performance after the button presses, but not after stimulus onset and before button presses, it can be suggested that the tracking performance of older people after relevant targets is decreased due to motoric interference and not to attentional limitations in the processing of relevant stimuli (cf. Hartley and Maquestiaux 2007).

The results of the first experiment led to two questions: (1) Can we enhance the tracking performance of older people by using a different response modality for the secondary task (e.g., verbal responses) in order to reduce motoric interference? (2) Does the age-related lack of differentiation between relevant and irrelevant stimuli, especially the unnecessary investment of resources to irrelevant stimuli in older people, have an impact on the response to critical events while driving?

In the second experiment we examined in a more realistic setting of a driving simulator if the change to another response modality (verbal responses) could enhance the tracking performance of older people by reducing motoric interference. Other studies found no differences between driving performance in a driving simulator and on the road (e.g., Shechtman et al. 2009), thus, support the external validity of a driving simulator. Moreover, we underlined that especially older people benefit from a more realistic context as we could show here that their tracking performance was comparable to that of the younger group. A similar phenomenon is known from memory research: Older people show significantly better prospective memory performance in naturalistic tasks than younger adults, but displayed deficits in laboratory-based prospective memory tasks in contrast to young people (Aberle et al. 2010).

In accordance with Brouwer et al. (1991) we expected to enhance older people's tracking performance by changing the response mode of the secondary visual attention task from manual into verbal in order to decrease motoric interference. Surprisingly verbal responses did not enhance the performance of older people. Therefore, for both groups, verbal responses seem not to be an appropriate alternative to manual responses in dual task situations like driving. Apparently, verbal responses generally need more attentional resources and also longer production times than manual responses which is reflected by the higher rate of missings and longer response times in the verbal condition.

The third experiment, also conducted in a driving simulator, should answer the question if the lack of differentiation between relevant and irrelevant stimuli, especially the comparable amount of attentional resources used for irrelevant stimuli in older people, may result in delayed or wrong responses to critical events (here: brake lights of a car).

First of all, there was no age difference in tracking performance: Both groups showed the best tracking performance if the tracking was combined with the brake light task, a decreased performance if combined with the brake light and the visual attention task and surprisingly the worst performance if combined only with the visual attention task. Trying to explain this surprising result, it is possible that performing the brake light task improves tracking performance by focusing attention to the brake lights and therefore also to the road. However, when the visual attention task is involved, this may detract attention from the road and thus from lane-keeping.

In both groups the response times decreased with increasing interval between the irrelevant stimuli and the brake light, especially in the difficult task where they showed longer response times and omitted more often the response to the brake light (cf. Levy et al. 2006). Importantly, this effect was more pronounced in older people. This is in line with other studies that show larger age-dependent delays of processing a second stimulus in situations where two response-relevant stimuli are presented in short succession (psychological refractory period or PRP-effect; e.g., Allen et al. 1998).

Interestingly, older people showed this effect only in the difficult task with a very short interval after the irrelevant stimuli. In contrast, there was no age difference in the easier task or after longer intervals between irrelevant stimulus and critical event (break light). Therefore it is suggested that the decreasing performance of older people in responding to the brake lights in the difficult task is due to the amount of attentional resources they invest in the processing of the just preceding irrelevant stimuli, especially, when such stimuli are occasionally response relevant.

To sum up, with the present experiments we found evidence for an age-related deficit in the differentiation between relevant and irrelevant stimuli in a dual task situation which is especially pronounced, when the irrelevant stimuli are potentially response relevant. Moreover, this lack of differentiation seems to result in an age-related significant slowing and increasing error rates to critical events that occur shortly after irrelevant stimuli. The simultaneous execution of two motor demands (tracking and button press) had particular effects on the tracking performance of the older group, but these problems could not be avoided by changing the response modality from manual in verbal response in one of the subtasks. Rather we could show that verbal responses—at least in complex dual task situations as we used here—are no appropriate alternative to manual responses as they result in higher response times, more missings of relevant targets and a poorer tracking performance in both age groups.

Outlook

The results of the presented experiments provide information about problems of older drivers in more realistic multi-task situations and may have helpful implications for the age- and user-friendly design of in-vehicle information systems and traffic environments in general. For example—especially for older people, who seem to have problems to differentiate between relevant and irrelevant information and respond more slowly and with more errors to events occurring shortly after irrelevant information—it may be important to avoid irrelevant information as far as possible, and present only relevant information. Since older people generally respond more slowly, relevant information should be presented at the earliest possible point in time. Although there may be some motoric interference if (especially older) drivers have to respond manually to secondary stimuli e.g., of an in-vehicle information system, the implementation of verbal responses does not seem to be an appropriate alternative solution in a complex dual task situation as we used here.

The present results stimulate further questions for research and for application. Further research in more realistic task settings is necessary in order to identify other cognitive processes, which underlie age-related changes and are relevant for driving performance. For instance, the age-related decline in working memory performance, as is evident in many laboratory studies, may have an impact especially in very complex traffic situations. However, applied research concerning this question is missing.

The knowledge about relevant cognitive processes and their age-trajectories needs to be included in a model which allows drawing predictions about human performance in new and hypothetical driving environments. Such model may serve more applied research. Most important, it may also guide the development, design and evaluation of technical innovations in the proximal and distal driver environments.

Acknowledgments The research was funded by the German Research Foundation (DFG) according to the Priority Program No. 1184, Age-differentiated Work Systems (WI 3105/2). We thank Ludger Blanke, Christiane Westedt, Trung-Tin Pham, Ines Fries, Janina Roskamp, Ines Mombrei, Christine Mertes and Sabrina Wueller for their help in setting up and conducting the experiments, and all subjects for their participation.

References

Aberle I, Rendell PG, Rose NS, McDaniel MA, Kliegel M (2010) The age prospective memory paradox: young adults may not give their best outside of the lab. Dev Psychol 46:1444–1453

Allen PA, Smith AF, Vires-Collins H, Sperry S (1998) The psychological refractory period: evidence for age differences in attentional time-sharing. Psychol Aging 13:218–229

Anstey KJ, Wood J, Lord S, Walker JG (2005) Cognitive, sensory and physical factors enabling driving safety in older adults. Clin Psychol Rev 25:45–65

Anstey KJ, Horswill MS, Wood JM, Hatherly C (2012) The role of cognitive and visual abilities as predictors in the multifactorial model of driving safety. Accid Anal Prev 45:766–774

Bialystok E, Craik FIM (2006) Livespan cognition. Oxford University Press, New York

Braver TS, Barch DM (2002) A theory of cognitive control, aging cognition, and neuromodulation. Neurosci Biobehav Rev 26:809–817

Braver TS, Barch DM, Keys BA, Carter CS, Cohen JD, Kaye JA et al (2001) Context processing in older adults: evidence for a theory relating cognitive control to neurobiology in healthy aging. J Exp Psychol Gen 130:746–763

Brouwer WH, Waterink W, van Wolffelaar PC, Rothengatter T (1991) Divided attention in experienced young and older drivers: lane tracking and visual analysis in a dynamic driving simulator. Hum Factors 33:573–582

Czigler I, Balázs L (2005) Age-related effects of novel visual stimuli in a letter-matching task: an event-related potential study. Biol Psychol 69:229–242

Eberhard J (2008) Older drivers' "high per-mile crash involvement": the implications for licensing authorities. Traffic Inj Prev 9:284–290

Edwards JD, Myers C, Ross LA, Roenker DL, Cissell GM, McLaughlin AM, Ball KK (2009) The longitudinal impact of cognitive speed of processing training on driving mobility. Gerontologist 49:485–494

Falkenstein M, Yordanova J, Kolev V (2006) Effects of aging on slowing of motor-response generation. Int J Psychophysiol 59:22–29

Gajewski PD, Wild-Wall N, Schapkin SA, Erdmann U, Freude G, Falkenstein M (2010) Effects of aging and job demands on cognitive flexibility assessed by task switching. Biol Psychol 85:187–199

Gazzaley A, D'Esposito M (2007) Top-down modulation and normal aging. Ann N Y Acad Sci 1097:67–83

Hahn M, Wild-Wall N, Falkenstein M (2011) Age-related differences in performance and stimulus processing in dual task situation. Brain Res 1414:66–76

Hartley AA, Maquestiaux F (2007) Success and failure at dual-task coordination by younger and older adults. Psychol Aging 22:215–222

Hasher L, Zacks RT (1988) Working memory, comprehension, and aging: a review and a new view. In: Bower GH (ed) The psychology of learning and motivation, vol 22. Academic Press, San Diego, pp 193–225

Horberry T, Anderson J, Regan MA, Triggs TJ, Brown J (2006) Driver distraction: the effects of concurrent in-vehicle tasks, road environment complexity and age on driving performance. Accid Anal Prev 38:185–191

Levy J, Pashler H, Boer E (2006) Central interference in driving: is there any stopping the psychological refractory period? Psychol Sci 17:228–235

Li KZH, Lindenberger U (2002) Relations between aging sensory/sensorimotor and cognitive functions. Neurosci Biobehav Rev 26:777–783

Mani TM, Bedwell JS, Miller S (2005) Age-related decrements in performance on a brief continuous performance test. Arch Clin Neuropsychol 20:575–586

Marottoli R, Mendes de Leon C, Glass T, Williams C, Cooney L, Berkman L (2000) Consequences of driving cessation: decreased out-of-home activity levels. J Gerontol B Psychol Sci Soc Sci 55:S334–S340

Oxley J, Whelan M (2008) It cannot be all about safety: the benefits of prolonged mobility. Traffic Inj Prev 9:367–378

Preusser DF, Williams AF, Ferguson SA, Ulmer RG, Weinstein HB (1998) Fatal crash risk for older drivers at intersections. Accid Anal Prev 30:151–159

Ragland D, Satariano W, MacLeod K (2005) Driving cessation and increased depressive symptoms. Gerontological Soc Am 60A:399–403

Salat DH, Lee SY, van der Kouwe AJ, Greve DN, Fischl B, Rosas HD (2009) Age-associated alterations in cortical gray and white matter signal intensity and gray to white matter contrast. NeuroImage 48:21–28

Salthouse TA (2000) Aging and measures of processing speed. Biol Psychol 54:35–54

Shechtman O, Classen S, Awadzi K, Mann W (2009) Comparison of driving errors between on-the-road and simulated driving assessment: a validation study. Traffic Inj Prev 10:379–385

Statistisches Bundesamt (2008) Verkehr—Unfälle von Senioren im Straßenverkehr 2007, Wiesbaden

Sterr A, Dean P (2008) Neural correlates of movement preparation in healthy ageing. Eur J Neurosci 27:254–260

Verhaeghen P, Steitz DW, Sliwinski MJ, Cerella J (2003) Aging and dual-task performance: a meta-analysis. Psychol Aging 18:443–460

Voelcker-Rehage C, Stronge AJ, Alberts JL (2006) Age-related differences in working memory and force control under dual-task conditions. Aging Neuropsychol Cogn 13:366–384

Wild-Wall N, Willemssen R, Falkenstein M (2009) Feedback-related processes during a time-production task in young and older adults. Clin Neurophysiol 120:407–413

Wild-Wall N, Falkenstein M (2010) Age-dependent impairment of auditory processing under spatially focused and divided attention: an electrophysiological study. Biol Psychol 83(1):27–36

Wild-Wall N, Hahn M, Falkenstein M (2011) Preparatory processes and compensatory effort in older and younger participants in a driving-like dual task. Hum Factors 53(2):91–102

Yordanova J, Kolev V, Hohnsbein J, Falkenstein M (2004) Sensorimotor slowing with ageing is mediated by a functional dysregulation of motor-generation processes: evidence from high-resolution event-related potentials. Brain 127:351–362

Integrating Training, Instruction and Design into Universal User Interfaces

Michael Sengpiel, Malte Sönksen and Hartmut Wandke

Lessons Learned

Elderly people are often considered unable or unwilling to learn novel things or to acquire new skills. The saying "You can't teach an old dog new tricks" has many equivalents in other languages. However, we know from many findings in basic research and from common experience (seniors can even learn to juggle, to play the piano or to speak Chinese) that this widespread assumption is simply wrong. Yet it exists, and it may contribute to the so-called digital divide in society and especially in the working environment, when older employees are considered *too old* for learning to use new computer-based technologies.

In our studies, we were able to show that older people can learn very well to interact with new digital devices. Our main finding is that through appropriate user interface design and close integration of training and instruction into the interaction process, differences between young and old users can be reduced tremendously and in some cases even completely eliminated.

The theoretical foundation for our experiments is the SOC-Theory developed by Baltes and Baltes (1990), in which the acronym SOC stands for Selection, Optimization and Compensation. In an extension of this theory we shifted the interpersonal SOC-processes to extra-personal entities i.e. to computer programs.

We performed seven experiments in which we varied several variables, which are assumed to support users' learning. We reduced training time stepwise (from

M. Sengpiel (✉) · M. Sönksen · H. Wandke
Humboldt-Universität zu Berlin, Berlin, Germany
e-mail: michael.sengpiel@hu-berlin.de

M. Sönksen
e-mail: malte.soenksen@hu-berlin.de

H. Wandke
e-mail: hartmut.wandke@psychologie.hu-berlin.de

C. M. Schlick et al. (eds.), *Age-Differentiated Work Systems*,
DOI: 10.1007/978-3-642-35057-3_14, © Springer-Verlag Berlin Heidelberg 2013

90 min to less than 3 min) and changed the user interface from a traditional function-oriented/menu-based style to a task-oriented/wizard-based style.

The following findings can be derived from the series of seven experiments:

- The influence of age on performance disappears when computer literacy is taken into account as a control variable.
- In the literature, there are many recommendations for the layout of training procedures for elderly users. It is possible to develop a well-functioning and effective training program by following these recommendations. We called this "basic training". We were able to show, that elderly people can profit from basic training to a great extent. However, the already very positive results of basic training can still be improved when Guided Error Training components are included. Older users are sometimes worried about breaking a program or causing unknown system states and Guided Error Training offers a good prophylactic measure.
- Training effects can be further improved by integrating "how" and "why" issues into the procedure. Usually, elderly people tend to learn only action sequences (step by step procedures). They reach a better understanding and interaction performance when it is explained to them how action steps are related to each other.
- Contrary to our expectations, the introduction of role models (video-based elderly instructors) did not lead to further improvements. However, a very short instructional video (2:37 min) showing only a hand and providing some basic spoken explanations can reduce performance differences between old and young subjects.
- With a redesign of the user interface (wizard) the differences between young and old disappeared completely; it is possible to improve the performance of older users significantly without impairing performance of young users. In all experiments, both young and old participants benefited from training and design changes, yet the gain for older users was always disproportionately greater than the gain for young users. Old and young people could perform equally well regarding effectiveness and efficiency in terms of interaction steps. Regarding time on task, older users showed significant improvements but did not reach the speed of young users.

Introduction

Since the early seventies of the last century, the scientific literature (e.g. Baltes and Nesselroade 1970; Schaie 1970) has shown that the traditional, popular and widely-applied conception that ageing is always a steady process of decline is a misconception. In a huge number of experiments many researchers (see Schmiedek et al. 2010 for a comprehensive overview) have shown that elderly people can learn new skills and can improve their basic cognitive abilities (like memory span, spatial thinking, reasoning etc.). Of course there are limitations, especially at high age (80 years and more) and there are great inter-individual differences. Even if a normal state of health is assumed, some individuals show worse performance data in their sixties than some who might be 90 years and older. Besides some genetic aspects, the concept of successful ageing provides a good theoretical foundation to explain the differences between individuals.

Especially the SOC-Theory, which stands for Selection, Optimization, and Compensation (Baltes and Baltes 1990; Li et al. 2001), gives a lot of explanations how the natural decline of basic cognitive abilities in the area of fluid intelligence (sometimes also called the "mechanics of the mind") can sometimes be offset by conscious, systematic and controlled strategies or by unconscious, spontaneous and quasi-automated behaviour. An example for the first may be the decision to do "brain jogging" exercises; an example for the latter may be to avoid driving during night-time or rush hour.

While the scientific picture of ageing changed in the last decades of the previous century, public opinion is still determined by the traditional prejudice that the elderly suffer from continuous and unchangeable decline of skills and performances. We call this the "age stereotype". Staufer (1992) carried out several interviews with older white-collar workers, their superiors, and Human Resources managers in various German enterprises. He found that this age stereotype is not only common, but is also accepted as true by the elderly individuals themselves. That older people themselves take over the traditional stereotype was shown by Marqui et al. (2002). In their laboratory experiment, young and old subjects had to make a judgment whether they would be able to answer a set of questions in various areas of knowledge. For each question a pre-answer rating scale of feeling-of-knowing and a post-answer rating scale of (un) certainty had to be filled in. The authors compared the performance (percentage of correct answers) and rating data (how do subjects evaluate their own knowledge) in two areas of knowledge: (1) information technology and (2) culture and history. Naturally, the older subjects had less knowledge of information technology. However, even in cases where the knowledge of younger and older subjects was objectively equal, the older participants felt "less knowledgeable" regarding the IT sector. Comparable findings are also available in the real world, as the Mori study (1999) pointed out: When it comes to the causal attribution of disturbances in the use of computers (programs crash, freeze, or cannot be used), older workers (>55 years) believed significantly more often than their younger colleagues that they themselves "were to blame" for the occurrence of the computer trouble.

Conceptual Framework

From the existing literature it is evident that human–machine interfaces need special attention when the machines are intended to be accessible for users of all age groups. Our research program on how to remove the barriers for older users in particular started with a number of basic considerations:

(1) It should be recognized that there is a gap between state-of-the-art basic research in cognitive ageing and the application of the findings to the everyday treatment of elderly employees in industry and administration. Based on the wrong but still very popular deficit model of ageing, elderly employees are considered less able to work under pressure, less able to learn and less knowledgeable with regard to computer and IT-based processes. Based on this, at least partly wrong, age stereotype, elderly employees do not get enough training in new technologies, they are not adequately considered when new computer-based work places are introduced, and they are more or less excluded from technological innovations. In professional jobs, younger colleagues often replace them. Popular slogans like the distinction between digital natives and digital immigrants (Prensky 2001), although originally intended as an appeal to overcome the distinction, are often used to explain existing differences as natural and unalterable. This is quite typical for a self-fulfilling prophecy. Therefore, one essential aspect of our research was to demonstrate that elderly people are able to improve their performance not only in elementary memory tasks, but also in everyday activities using computer technology.

(2) The SOC-Theory has mainly been used to explain why elderly people are able to master various tasks in laboratory settings and in real life, although several perceptual, motoric, and cognitive functions do indeed deteriorate. The focus of the traditional application of SOC is to assume an adaptive process within the person, e.g. people may decide to avoid buying a ticket from a computerized ticket vending machine (TVM) and to buy this ticket at a counter from a human employee. They select a place and time for buying at which counters are available and open. Or they optimize their use of a ticket vending machine by watching other people operate the TVM, by asking service employees or other passengers, and by trying out keys or touch screen actions. Finally, they apply a compensating strategy e.g. by allowing for extra time or even by buying the ticket a day before departure. Although in the original version of SOC some external resources are mentioned, especially with regard to sensory (eyeglasses) and motor (walking stick) function, the power of computer technology to support elderly people has not been considered sufficiently. One of the goals of our research project was to relocate selection, optimization, and compensation into the technology. No longer must only humans adapt to technology, but technology should be adapted to humans as well. Applied to the usage of a TVM, that means the system should offer not all but a limited subset of options (e.g. tickets). Optimization means that the system should guide the user step by step through the purchasing process. Compensation means that the TVM provides additional information when the user has some lack of knowledge (e.g. how to proceed or what ticket types and tariffs are available).

(3) While basic research is mainly interested in age-related changes, we focused our studies on cohort effects. In contrast to age-related decline, which is either irreversible or can be delayed or stopped only temporarily, cohort effects can

be completely eliminated. Many supposed deficiencies of elderly employees are not caused by their age but by their working environment and by organizational measures. Especially with regard to information and communication technologies (ICT), elderly employees may lack the necessary knowledge ("computer literacy") to effectively use computers and computerized machinery because they have performed their jobs with traditional tools and machines for many years. A later cohort of employees will have started their job experience with ICT. Moreover, some basic interaction knowledge may have been acquired at different life stages. In the decades before 1970, there was no ICT at the workplace (apart from a few very specialized jobs in computer centres). In the eighties of the last century, special training courses were offered for the use of desktop computers in offices. Beginning with the Internet Revolution, computer services became relevant for many spheres of life and learning was more and more based on self-organized training on the job. Since the year 2000, ICT has been integrated into nearly all traditional technologies (pervasive computing, ubiquitous computing). Pre-school children acquire computer literacy nearly effortlessly by playing with toys and gadgets. On the other hand, elderly employees try hard to learn, e.g. how to operate a smart phone. The great impact of lacking computer literacy both on job performance and job satisfaction as well as the optimistic view of overcoming this lack (in contrast to pure age-based deficiencies) were decisive for the choice of our subject of research.

(4) While basic research in changes with ageing is focused on perceptual, motor and cognitive changes, our approach was also directed to attitudes, emotions, and motivational factors. Generally, elderly people avoid risks and uncertainty. They have a greater need for security. When beginning something new, e.g. trying out a new electronic device or a new computer program, they may fail. Therefore elderly people avoid such situations. They show greater fear to make an attempt to use an unknown or less familiar system and if they do try to use it, they often fail indeed. In contrast to younger people, who often argue that the system is ill-designed or a program has flaws, elderly people often believe that a failure has been caused by their own behaviour. Therefore, during all training sessions and in all design efforts, special efforts were made to assure that the elderly users always felt fearless, comfortable, and relaxed. Additionally, the emotional attitude towards computer systems had to be controlled. We also considered self-efficacy as a dependent variable in training sessions and experiments. One of our basic hypotheses was, that elderly people not only are able to use interactive systems successfully but that they also reach a significant gain in self-efficacy.

(5) A main characteristic of our approach was to combine training and design in order to show that the "digital gap" between generations can be attenuated or even eliminated. With the help of training, we addressed the adaptation on the side of the user. With the help of design, we addressed the adaptation on the side of the system, which, effectively, is an adaptation on the side of the developer of the system (the programmer, designer, ergonomist etc.). If there

were a silver bullet, it would obviously have to be on the side of design activities. If a system selects and offers only functions which the user is able to understand and use, if it optimizes the path from an initial state to the goal state by reducing the mental workload, and if it compensates the lack of computer literacy on the user side, then the need for training should be greatly reduced.

However, the need for training can almost never be reduced to zero. Therefore, the training itself became the subject of design. The most essential principle for training design was *integration*. All training procedures were embedded into the (simulated) interactive device and the trainees could easily switch into the users' role without leaving the system and switch back again just as easily.

Finally, our training procedures should work according to the SOC-Theory: Selecting training tasks which were suited to the actual knowledge of the users, optimizing the training by providing video-based role models (virtual trainers also of higher age), who demonstrated and explained the interaction principles, and compensating the lack of users' knowledge by explaining the various functions, interaction elements and data objects.

Experimental Studies

Within the context of the project ALISA, a ticket-vending machine (TVM) of the local public transportation provider in Berlin (BVG) was selected as the technical system. The choice of this system as an experimental subject is based mainly on three aspects:

- As BVG ticket offices are replaced by TVMs, their use becomes increasingly inevitable, requiring the acquisition of the necessary user skills. However, older people currently rarely use TVMs.
- There are numerous studies that examine the use and acceptance of TVMs by older people (e.g. Tacken et al. 2005).
- The TVM of the BVG is a well-structured system with sufficient complexity to provide tasks of varying difficulty. It can be simulated reasonably well and learning content can be modelled easily.

Just like the original TVM, the simulated TVM was equipped with a touch screen. This kind of input device does not require spatial transformations for eye-hand coordination between hand and cursor movement and has proven to be better suited for inexperienced and older users than a computer mouse (see Schneider et al. 2008). The focus of the research project was not to develop a special user interface and training for this specific TVM, but rather to derive generalized training and design principles for older users, which can be applied to other interactive systems.

For an overview of the dependent variables and the samples, all conducted studies are listed in Table 1. These studies are presented along a continuum from three variations of exhaustive training to minimal video instruction and a wizard redesign of the user interface without additional instruction.

Four Variations of Exhaustive Training

Based on a comprehensive task analysis (see Struve and Wandke 2009), a video-model-based training for the simulated TVM was developed. This training was structured into an introduction and six lessons with increasing difficulty and complexity, and with different learning objectives. The content of each lesson was divided into smaller sequences. Thus, the learners' quick success due to the resulting step-by-step procedure should have a positive impact on motivation and especially on the perceived self-efficacy. Each lesson was divided into the following sections:

In the introduction, the topic of each lesson and the ticket types contained were named first, providing a first motivational support especially for older learners wanting to know why they should learn. The learning process began with the presentation of a single ticket and its most important tariff information and conditions. Subsequently, the necessary interaction steps to acquire a ticket were shown by the model using the TVM, while simultaneously providing verbal

Table 1 Overview of the experimental studies conducted in the ALISA-project

Study	Dependent variables	Experimental group		Control group	
		N	Age	N	Age
Guided error training	Errors, completion time, additional steps/grey clicks (in reproduction and transfer tasks), declarative & procedural knowledge, cognitive load, perceived self-efficacy, motivation	40	M = 66.77 SD = 3.63	40	M = 23.85 SD = 3.11
Worked examples		40	M = 66.95 SD = 3.65	40	M = 25.63 SD = 2.93
Training without model		20	M = 69.55 SD = 3.53	20	M = 66.85 SD = 4.60
Sustainability of training effects		19	M = 67.0 SD = 3.06	–	–
Minimal instruction and wizard redesign of the user interface	Effectiveness, efficiency (steps), efficiency (time), Satisfaction	62	M = 68.2 SD = 4.8	62	M = 24.5 SD = 4.14

N sample size
M mean
SD standard deviation

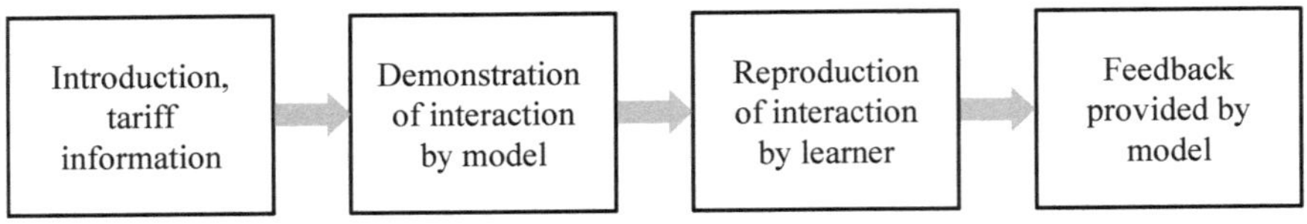

Fig. 1 Schematically illustrated sequence of the learning process within the training

instructions. This section was followed by a reproductive phase in which the learner was asked to practice the solution steps shown before. If she deviated from the presented solution, an error feedback was given, which asked her to watch the instructional video again. After a successful completion of the reproduction task, the learner received positive feedback from the model. In Fig. 1 the process of learning within the operator training is illustrated schematically.

After this procedure was repeated for all the tickets of the lesson, three transfer tasks followed in which the mediated procedural knowledge and knowledge about the tariffs had to be applied to new but similar contexts. These tasks were formulated as real-world scenarios in which learners were asked to acquire the tickets not only for themselves, but also for friends and family members. Here, reviewing the instructional videos was no longer possible, so the tasks could be regarded as an indicator for learning success. In both sequences, the learning process and the testing of learning success, the time to complete the tasks and the number of errors were recorded as indicators of effectiveness of training. Figure 2 shows screenshots of the different sections of operator training.

In the didactic design of the interactive trainings, numerous aspects were considered.

Cognitive Load Theory

Within the Cognitive Load Theory (Chandler and Sweller 1991), learning materials shall be designed to maximize the resources available for information processing of the content (germane load) and to optimize the use of working memory. Therefore, the lessons were designed to be neither too easy (mental underload) nor too difficult (mental overload) and the computer-based realization of the training aimed to provide this optimal workload by adapting the complexity of the training content to the knowledge of the learner.

Worked Examples Effect

All learning content within the lessons was expressed in the form of action- and problem-orientated examples. When using the Worked Examples Effect (Sweller et al. 1998) of the Cognitive Load Theory, the newly obtained knowledge should be organized and presented so that it can be integrated into already existing

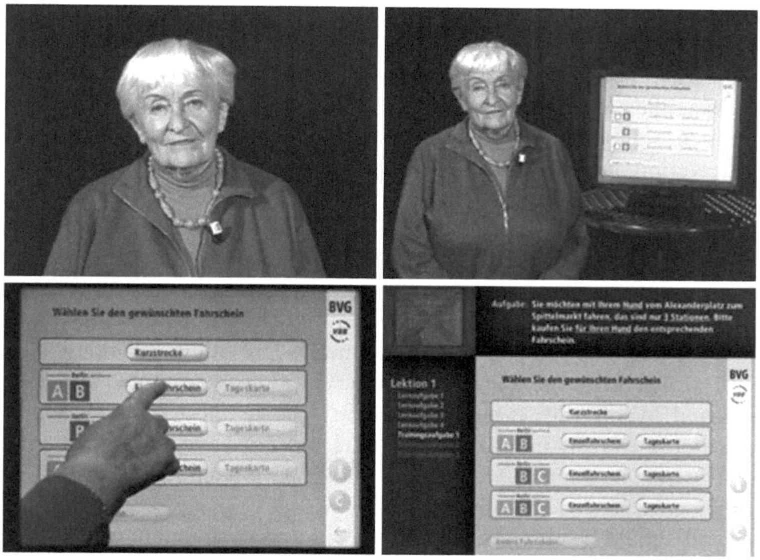

Fig. 2 Screenshots of the operator training; *top left* introduction; *top right* explanation of the interaction steps; *bottom left* demonstration by model; *bottom right* reproduction task

knowledge structures from previous lessons. To this effect, no more than five interaction steps were taught per sequence to ensure the target-aimed processing into the working memory. Longer sequences were divided into two segments.

Training Wheels

Furthermore, the interface of the simulation was based on the approach of the Training Wheels of Carroll and Carrithers (1984), where all the buttons of an interactive system that had not been learned yet were greyed out and disabled. This approach addressed quick learning success, which had been shown to be effective for older people in various studies (Bruder et al. 2007a, b).

Parts-Whole-Sequences

Globally required concepts (e.g. interaction elements of the TVM and information about the tariff zones) were presented to the subjects before the first lesson. This principle of the Parts-Whole-Sequences (Mayer and Moreno 2003) was also used within the lessons, while the scheme of the learning process described above was implemented.

Consideration of SOC

In its construction of the operator training and the consideration of important didactic principles, the ALISA project succeeded in combining aspects of interface design and design of training and in implementing the model of optimization through selection and compensation (SOC-model, Baltes and Baltes 1990). Here, the processes of the model were transferred from the person to the technology. For example, *selection* was achieved through the described functional reduction and subsequent gradual construction, considering the knowledge of the user and her previous opportunities for interaction with the system. *Optimization* was addressed by structuring the learning material, allowing integration of new knowledge into existing knowledge structures. Instead of merely providing information, the formation of an adequate mental model of the TVM was promoted. *Compensation* was achieved by offering external guidance. The necessary interaction steps to acquire a ticket were demonstrated so users did not have to work them out for themselves, freeing cognitive capacity. The SOC-model can be found again in many other measures of the training design.

Finally, it should be emphasized that a focal area of an operator training can be seen in the socio-motivational support of the learner. The real ticket purchasing situation could induce stress, for example due to other waiting buyers. However, all training sessions could be handled at a self-determined speed. Also, the demonstration by a video model, which was of the same age as the participants, and the feedback given by the model encouraged the identification and familiarity between the learner and the model and strengthened motivation and the learning process. A detailed description of the development of operator trainings, as well as other educational and motivational aspects can be found in Struve and Wandke (2009). All descriptions thus far have related to the standard version of the developed operating training. Studies in different versions, which varied particularly in their didactical implications, were implemented during the entire project. These studies, together with their most important differences, will be described in detail now.

Guided Error Training

This study used a training based on the concept of Guided Error Training by Ivancic and Hesketh (1995). During the demonstration of every interactional step the video model made mistakes. When discrepancies compared to expected system states or in the final entry check were noticed, they were immediately corrected. Apart from the original procedure, the model verbalized the reason for the mistake as well as the applied solution strategy. The interactional steps demonstrating mistakes were chosen according to the task analysis and the results of a user test (Dittberner 2007).

Worked Examples

The standard training version is a product-oriented Worked Examples Training (Sweller et al. 1998), where the model presents task goals and appropriate solution steps, demonstrating *how* to solve a task correctly. In a different study, a training identical to the product-orientated version was designed to use a process-orientated Worked Examples Effect, where the model additionally explained *why* it had chosen and used specific interactional steps, making implicit knowledge accessible to the learner and enabling a better understanding of the interactional steps and machine functions. Both the studies about Guided Error Training and the Worked Examples Effect used a younger sample as a control group.

Training Without Model

The third study was intended to explore the impact of the visual presence of the model on socio-motivational effects. Thus the training was modified by substituting the model with a static picture of the ticket-vending machine (see Fig. 3). Additionally, the spoken text was adapted so that the model did not use the first person singular. Instead the learner was trained by being addressed directly. However, socio-motivational cues in the form of personalization and language were still given according to Social Agency Theory (Mayer 2005) and the close-up remained unmodified so that at least the model's hand was visible.

Sustainability of Training Effects

In a fourth study, the sustainability of training without visual presence of the model was analysed. The selected version of the training was based on the results of previous studies showing that the visual demonstration of a model created no additional benefits in most of the explored variables. The selection is useful for deriving generally accepted training principles and accommodating the desire for an economical solution. To evaluate a sustainable effect, the participants' abilities

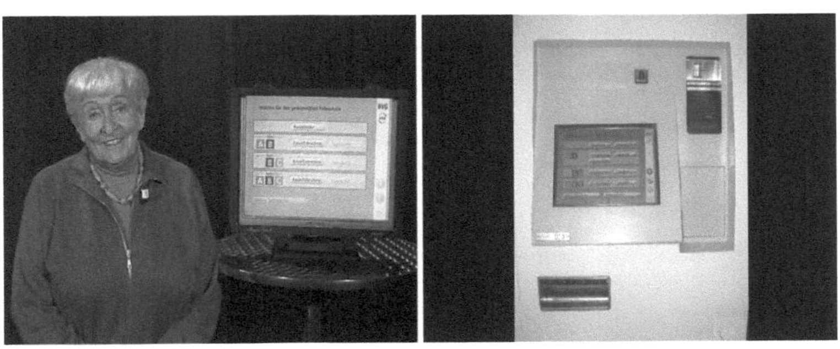

Fig. 3 Screenshots of the training with (*left*) and without (*right*) visual presence of the model

to use the ticket-vending machine were documented two weeks before they had completed the training. Another five weeks later the participants' knowledge and their handling of the TVM were documented once again to demonstrate the sustainability of the effect.

The same variables were used to evaluate the different versions of the training. These include the performance in the reproduction tasks and the analogue transfer tasks, which was measured as number of errors, completion time and deviation from the optimal approach (additional steps). Also, so-called grey clicks were recorded, which count the attempts to use inactive ("greyed out") or already selected buttons. The declarative as well as the procedural knowledge of the participants was assessed. Learning itself was measured as resulting cognitive load, perceived self-efficacy and motivation. Table 1 provides an overview of all studies.

Minimal Instruction and Redesign of the User Interface

Participants for these studies were recruited in two age groups: the older comprised of a total of n = 62 adults (M = 68.2 years) and the younger of n = 62 younger adults (M = 24.5 years). The age groups were split evenly into three experimental setups. In the control setup, participants were asked to solve eleven tasks using a simulated TVM. In these eleven tasks, participants had to select tickets for purchase using a simulated ticket vending machine (TVM), which was presented on a 19 touch screen monitor (see Fig. 4).

The eleven tasks were designed to be realistic (i.e. Please purchase a single ticket for Berlin ABC, reduced fare) and to vary in difficulty. Figure 5 shows the frequency of correctly solved tasks in the order they were presented. On average, the younger group solved 84 % and the older group 52 % of them correctly.

In the video setup, participants watched a short instructional video immediately before they solved the same eleven tasks. In the wizard setup, participants received no video instructions and solved the same eleven tasks using a modified GUI (Graphical User Interface) that had been designed to require less computer literacy and followed a wizard paradigm.

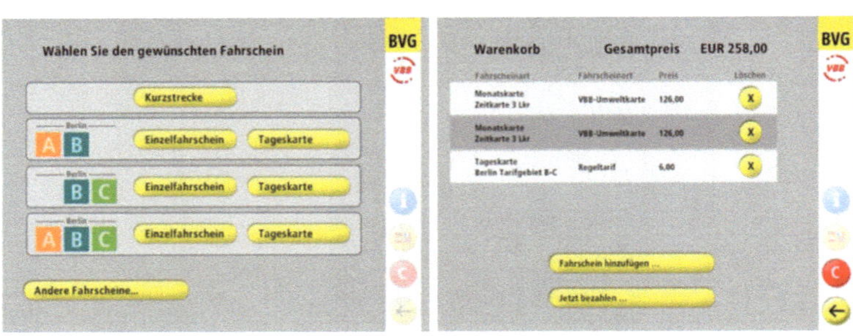

Fig. 4 Screenshots of the simulated ticket vending machine (TVM)

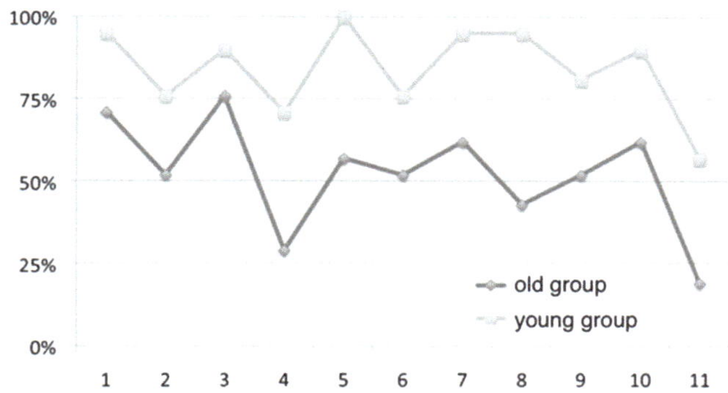

Fig. 5 Frequencies of correctly solved tasks for the young and the old group in the control setup using the original TVM

Minimal Video Instruction

In this study, a potential lack of interaction knowledge was compensated by a brief instructional video (2:37 min), in which a narrator's voice provided only basic interaction knowledge for the use of the ticket vending machine (e.g. "This is a button. When it is pressed, the color changes from yellow to green…") and a hand pointed to the objects of reference on the GUI, which was designed to resemble the TVM simulation. To avoid concomitant teaching of domain knowledge, all ticket button text had been removed. Figure 6 shows screenshots from the video. The video was played back in a loop and participants were instructed to touch the screen to stop the video and start with the tasks when they saw fit.

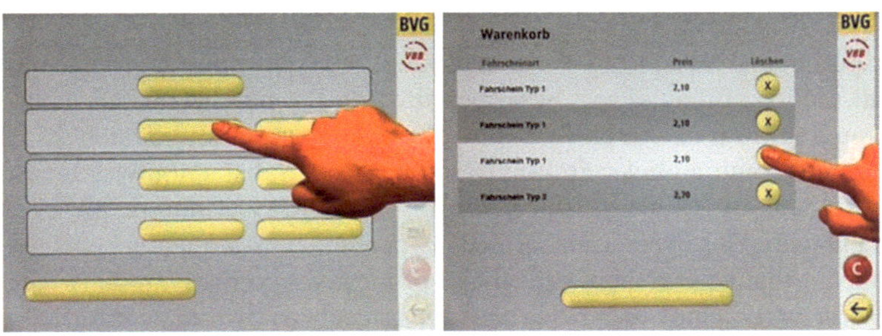

Fig. 6 Screenshots from the minimal video instruction

Wizard-Redesign of the User Interface without Additional Instruction

In the wizard study, the graphical user interface (GUI) of the TVM was redesigned to require less computer literacy (compensation) and to guide the user step by step through the ticket selection process (optimization). The wizard was not meant to anthropomorphize the machine, but refers to a design pattern that guides the user through complex tasks by decomposing them into a set of manageable steps (Folmer and Bosch 2003). This design pattern should benefit particularly older users by reducing cognitive requirements on visual search and working memory, which tend to decrease in old age (Schaie 1990). Since older and younger users have prior domain knowledge about the task of purchasing a ticket, e.g. at the ticket counter, the process of ticket selection was decomposed into four main questions from the user's perspective: (1) Who wants to go? (2) Where? (3) How long should the ticket be valid? (4) How many tickets are needed? The wizard was built to be functionally equivalent to the original TVM and differed only in the GUI. Figure 7 shows screenshots from the wizard GUI of the TVM simulation.

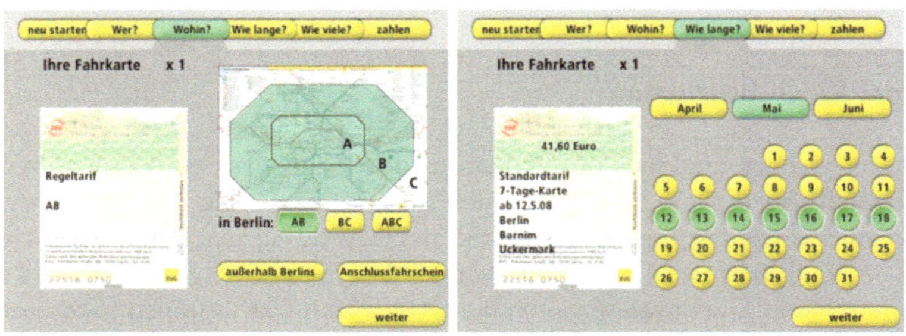

Fig. 7 Screenshots from the wizard-GUI of the TVM

Dependent Variables

The experimental conditions were compared regarding effectiveness, efficiency and satisfaction. Effectiveness was measured as the number of correctly solved tasks, ranging from one to eleven. Efficiency was measured separately as the time and the steps (button clicks) it took to solve a task. Satisfaction was measured as the sum score of 13 items selected from the QUIS (Chin et al. 1988). All dependent variables were transformed to percent. Consequently, each score of 50 % means that a participant solved half of the tasks correctly, took twice as long as the best participants and twice the steps necessary, while scoring half of the possible points on the satisfaction questionnaire, respectively. Figure 8 shows the number of necessary steps and the fastest times participants took to solve the eleven tasks, which provide the basis for the calculation of efficiency measures.

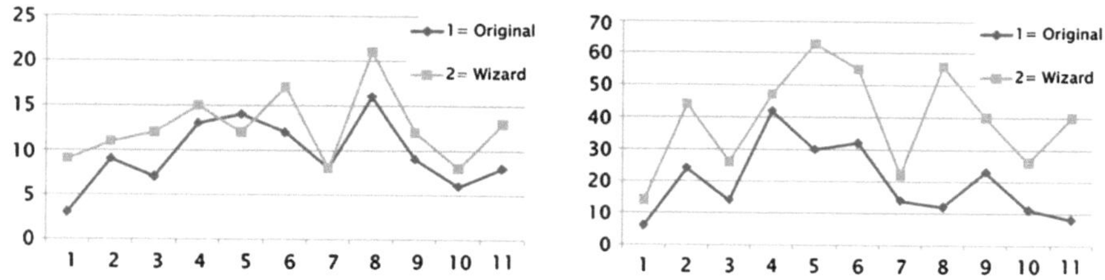

Fig. 8 Number of necessary steps (*left*) and fastest times (right, in seconds) participants took to solve the eleven tasks, providing the basis for the calculation of efficiency measures

Results

Training Studies

Global Results

At this point, results will be summarized which could be observed across all studies. These will be not presented in detail for the separate studies. In all studies with younger participants as a control group (Guided Error Training and Worked Examples), they generally committed fewer errors and solved tasks faster than older participants. This refers to the learning process as well as learning success (analog transfer tasks). In almost all tasks, younger participants needed fewer steps and committed fewer grey clicks. The questionnaires for measuring cognitive load indicate that neither the younger nor the older participants perceived the learning process with the operator training as particularly exhausting.

All participants experienced an increase in procedural and declarative knowledge after the operator training, yet it was significantly higher for older participants. Knowledge increase differences between the different versions of the operator training need not be attributed to the respective didactic design aspects, for they could be explained with the law of initial values, implying that a high initial value correlates negatively with its growth. In this respect, the findings of a preliminary study for the development of the operator training by Struve and Wandke (2009) are very interesting. Here, tasks for declarative and procedural knowledge were presented to ten older novices as well as to ten experts in using the TVM. The results of the novices corresponded roughly to the results of the presented studies. However, it was shown that even experts could only solve about 75 % of the tasks.

A similar pattern was found for the perceived self-efficacy, which also increased among participants in all studies, independent of age. As expected, the increase for the older participants was higher than for the younger ones. In addition, all subjects showed above-average values of self-efficacy, resulting in a ceiling effect. As expected, the initial baseline was higher among younger than among older participants.

Finally, among older participants, no difference between men and women were found. Both groups performed equally well when using the operator training.

For practical purposes, an operator training for interactive systems, such as a TVM, could enhance the efficiency and effectiveness in using the system. Furthermore the knowledge on how to use it would be enhanced.

Effects of Guided Error Training

A 2-factor ANOVA revealed one of the main effects of the first study: a significant interaction between training version and age regarding the number of errors in the transfer tasks. Younger participants committed a similar number of errors in the Guided Error Training compared to the basic training version, while older participants produced a significantly lower number of errors in the Guided Error Training. This effect could be found for the processing time and the error rate in the learning process (reproduction task) as well as for the processing time of the transfer tasks. The results therefore indicate that the errors which were described in the instructions lead to better performance in solving similar tasks. Figure 9 visualizes these results.

Regarding declarative knowledge, an ANOVA revealed another significant effect for the training version, regardless of the age of the participants. Subjects who were trained with the Guided Error Training reached higher test scores than those who learned with the basic training version. This effect was found for the procedural knowledge as well. In addition, there was an interaction between the moment of knowledge measurement, training version, and age: Older participants who used the Guided Error Training had a significantly higher gain in procedural knowledge than older subjects who learned with the basic training version. For younger subjects, this effect could not be found.

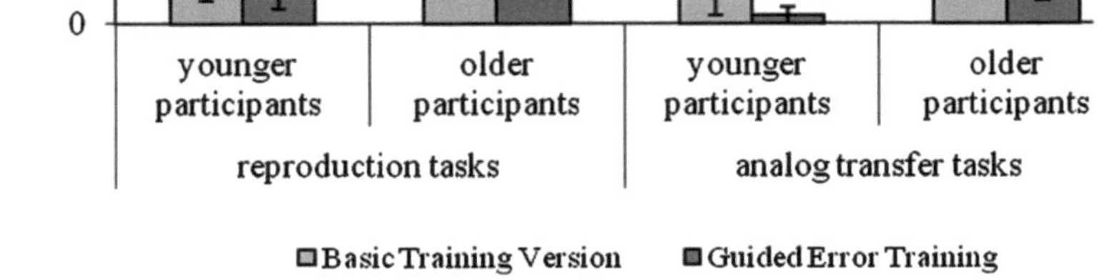

Fig. 9 Number of errors in reproduction tasks and analog transfer tasks for younger and older participants

In summary, learning with Guided Error Training had no effect on the learning process itself but seemed to have an effect on the knowledge and success of learning, especially for older subjects. Thus, presenting errors and solutions seems to result in deeper processing of learning material. A possible explanation could be that the fear of older people to break something or do something wrong was reduced by observing the model who committed errors without serious consequences. Thus, showing older users of TVMs how to correct mistakes they could theoretically commit in a preceding training could lead to a better performance in their real usage of the system and possibly to a faster buying process.

Effects of Worked Examples

For the process-oriented Worked Examples, only the global effects reported above, regarding the direct comparison of younger and older participants, were found in this study as well. An ANOVA revealed the first significant effect of the training version analyzing the gray clicks. First, the previously reported effect of age on gray clicks was found. Second, the older subjects committed significantly more gray clicks than younger subjects. Third, an effect of interaction between the training version and age could be detected: Older users of the process-oriented Worked Examples Training showed significantly less gray clicks than older users of the product-oriented basic training. This effect could not be found in the younger group. Figure 10 shows results for additional steps and gray clicks in comparison.

The acquired knowledge showed significant interaction effects between age, training and moment of knowledge measurement. Surprisingly, for declarative knowledge, older participants showed a significantly higher increase of knowledge

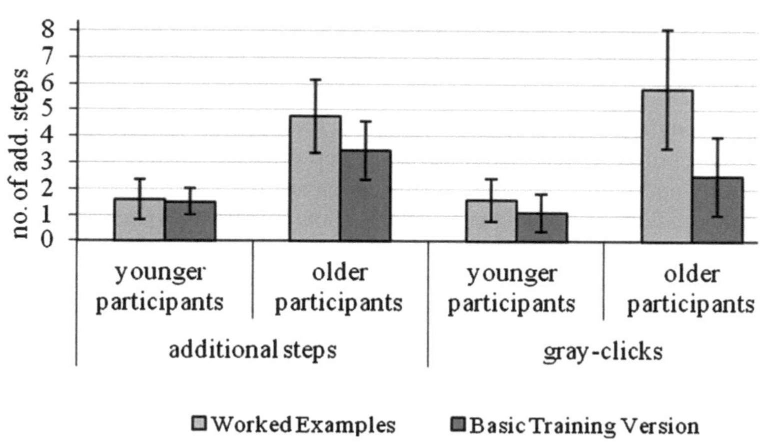

Fig. 10 Additional steps and gray clicks for older and younger participants in the basic training version and the worked examples training

in the product-oriented basic training version, than in the process-oriented Worked Examples Training, while this effect was reversed for the younger participants. For procedural knowledge, results regarding age point in the opposite direction. While younger subjects achieved higher scores in product-oriented basic training, older participants had a significantly higher increase of procedural knowledge in the process-oriented Worked Examples Training.

In summary, learning with the process-oriented Worked Examples Training did not lead to higher improvement in the learning process or learning success compared to the basic training version. Therefore, we assume that due to the already nearly optimal adjustment of the basic training version, additional instruction methods could not generate improvement effects (Struve and Wandke 2009). The expertise reversal effect (Kalyuga et al. 2003) could provide another explanation, stating that learning materials become dispensable for novices with an increase in their expertise. These materials could represent redundant information with respect to the Cognitive Load Theory (Chandler and Sweller 1991), indicating the unnecessary use of cognitive capacity, with its detrimental effect on the learning process. Since the effect did not occur in the group of younger subjects, they might have had sufficient cognitive capacities to compensate an adverse effect of redundant information. However, a practical implication is that learning materials should be adapted to the learner's level of knowledge and additional information should be reduced enough to allow for autonomous problem-solving without presenting solutions.

Training without Model

For the Training without Model, no significant effects could be reported for the learning process and the learning success. Regarding the acquired declarative and procedural knowledge and self-efficacy, an age effect was found in accordance with the global results described above.

One possible explanation for the lack of effects could be that one cannot assume that the narrator in the Training without Model was not perceived by the participants. Although the visual representation of the model was largely omitted, the participants could recognize both the voice and the hand of the model (see Fig. 2 bottom left) and could therefore easily anticipate the entire model. This would support the findings of Moreno et al. (2001) showing that the visual presentation of a speaker or a model has no additional positive impact on learning and motivation. The subjects could therefore have experienced a similar level of socio-motivational support in both versions of the training.

Due to the lack of significant results comparing the basic training version to the Training without Model, the data was used for an exploratory analysis. Figure 11 presents the results graphically. As a first result, the sample was divided into people with high and low self-efficacy. Accordingly, an ANOVA revealed a significant effect of the training version regarding error rate in the analog transfer tasks. Contrary to expectations, participants with low self-efficacy committed

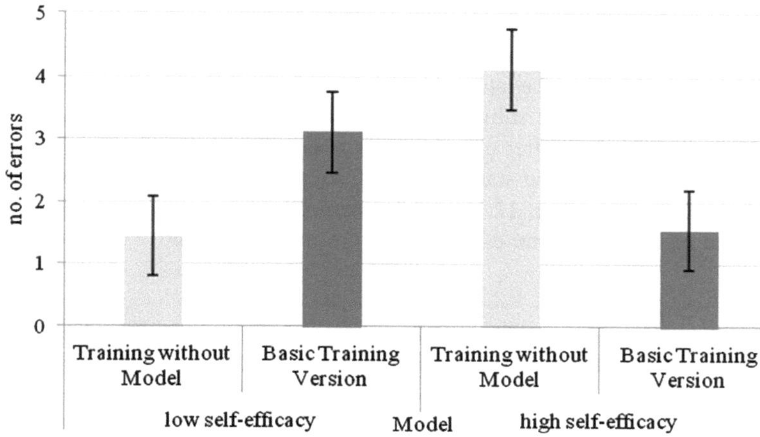

Fig. 11 Number of errors in the training without Model and the basic training version for persons with low and high self-efficacy

fewer errors in the Training without Model, while those with high self-efficacy produced fewer errors in the basic Training with Model.

Bandura (1997) could provide a possible explanation, stating that the self-efficacy expectations are strengthened especially by one's own successful actions. The high value of self-efficacy in the group without model could also have led to an overestimation of their own knowledge and thus, to an overestimation of their abilities. This might result in a significantly increased number of errors in this version of the training.

In summary, the visual presence of a model did not improve the learning process or learning success more than the simpler version. Therefore, since the creation of video footage is a time and cost intensive undertaking, it appears reasonable to abandon this form of representation in practical contexts.

Sustainability Effects

In this study, the results can be reported from two points of view. On the one hand, it was possible to detect the status quo of older participants using the TVM due to the design of the study. That is, the performance (errors and time) and the knowledge of the sample were measured without conducting the operator training. In addition, efficiency and knowledge were recorded immediately after completing the Training without Model, which allowed a direct quantification of improvements in both domains. An ANOVA revealed a significant effect, indicating that the older subjects committed more errors in the status quo measurement than after the training. A significant improvement of processing time from the status quo measurement to the point after training could also be reported.

Therefore, it appears that training can improve the use of the TVM significantly. To identify the sustainability of these effects, the sample received the tasks again five weeks after the status quo measurement. In this measurement, an ANOVA revealed that the subjects committed significantly more errors after five weeks than immediately after training, but the number of errors was still far below the number of errors in the status quo measurement. The data for the committed errors are presented in Fig. 12. The time required to complete the tasks showed a corresponding trend, but the differences were not significant.

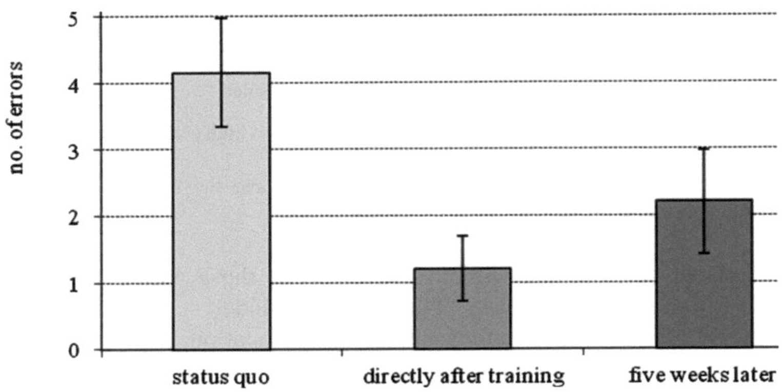

Fig. 12 Number of errors at the three points of measurement

Analysing declarative knowledge, similar results could be found. Again, an ANOVA revealed significant improvements from the status-quo measurement to the time after the training and to the moment five weeks later. Moreover, the decrease of declarative knowledge from the time directly after the training to the measurement five weeks later was not significant. It can therefore be described as relatively stable knowledge caused by the operator training. The same significant pattern was also found for the perceived self-efficacy. Finally, the same effect was found for the procedural knowledge.

In summary, it could be shown that the operator training supported knowledge acquisition and increased the performance in dealing directly with interactive systems, such as TVMs. The operator training also increased the perceived self-efficacy. These positive effects are relatively stable. However, they decrease over time. So for practical purposes, it is essential that the interactive system is used regularly after completing an operator training.

Design Studies

Computer Literacy and Experience

Since participants computer literacy and experience influence their performance using the TVM, the groups were tested for respective differences. To measure computer literacy (CL) as the "ability to understand and use computer related symbols, functional elements and interaction patterns", the Computer Literacy Scale (CLS) was administered at the beginning of the experiment as a paper and pencil questionnaire (Sengpiel and Dittberner 2008).

As expected, older participants had significantly lower scores in the knowledge part of the CLS than those in the younger group, yet participants using the original TVM GUI did not differ significantly in computer literacy from those who watched the video or from those who used the wizard. Hence, performance differences between the video, wizard and control group cannot be attributed to differences in computer literacy (CL).

Efficacy

Compared to using the original TVM, the older group benefited from watching the video and even more from using the wizard, while the young group had very high efficacy in all three setups. A MANOVA revealed that the older group did indeed benefit more from watching the video or using the wizard, so much so that they did

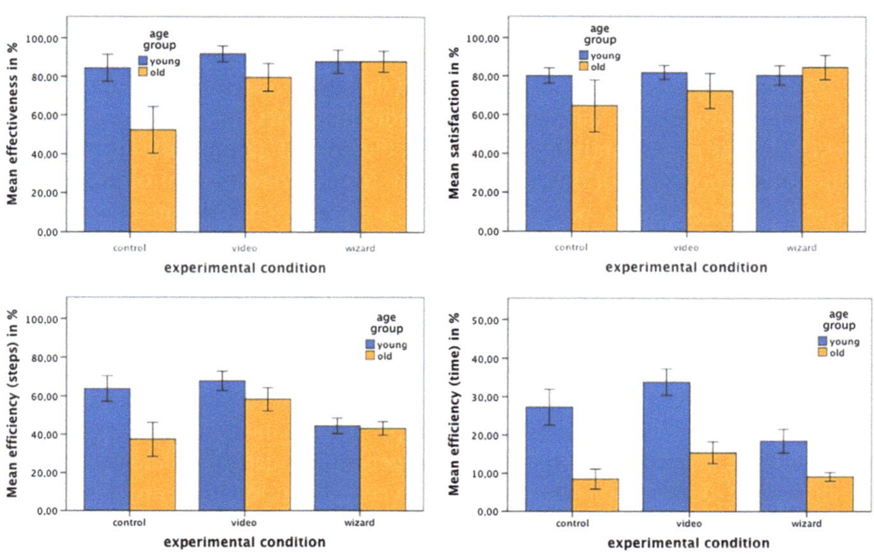

Fig. 13 Effectiveness, efficiency (measured in steps and time) and satisfaction in all six groups

not differ significantly from participants in the younger control group in the number of correctly solved tasks. Figure 13 gives an overview of the results for all six groups.

Efficiency Measured in Time

Both age groups benefited from watching the video in their efficiency measured in time; yet a MANOVA revealed that the older group did not benefit more than the younger group from watching the video or using the wizard. As expected, participants in the older group who watched the video or used the wizard still differed significantly in efficiency measured in time from participants in the younger control groups. This finding is consistent with research showing that the speed of information processing declines with age.

Efficiency Measured in Steps

In efficiency measured in steps, even the young group benefited from watching the brief video and again the older group benefited much more. In the wizard setup, both age groups were equally efficient in steps.

Satisfaction

Satisfaction ratings were rather high overall and the video had little effect on the users' satisfaction, but particularly for the old group, the wizard did. There was a significant main effect of the age group on satisfaction with the TVM, indicating that older participants were less satisfied, and there was a significant main effect of experimental conditions on satisfaction, indicating that for the wizard group, satisfaction was significantly higher than for the control group while for the video group, satisfaction did not differ significantly from wizard or control groups.

The interaction effects between the age group and the experimental setup on satisfaction was significant, indicating that the older group did indeed benefit more from watching the video or using the wizard.

Watching a brief instructional video or using a wizard to select the desired ticket proved to be beneficial for both age groups. Specifically, the experimental groups solved more tasks in less time and less steps than the control group and the older group benefited more from the video and the wizard than the younger group in effectiveness, efficiency measured in steps and satisfaction. As expected, the video and the wizard could not close the generation gap in efficiency in time, which largely depends on the decreased speed of information processing [see Czaja (1996) for an overview of relevant age-related changes in cognitive abilities].

For practical purposes, the younger generation might not want or need to switch to a wizard or to watch an instructional video for a simple machine like the TVM investigated. However, if the older generation had such a video available to them and actually watched it right before the tasks were performed or switched to the wizard, our findings suggest that they could use the TVM as effectively and efficiently (measured in steps) as the younger generation.

Discussion

If we consider the results of all experiments from a comprehensive perspective, it becomes evident that the principles underlying the basic training program and the various extensions proved to be successfully applicable in multimedia-based training and instructions for elderly users. This refers to the three columns of the SOC-Theory as well as the principles we have derived from applied training and design literature.

Selection (S) is a very powerful strategy if implemented in a multimedia-based training program. It reduces initial reservations and by adapting the number of options in a user interface to the actual knowledge of the users, it provides a good fit between task requirements and knowledge and skills.

Optimization (O) is in any case part of a training program; in short, we could even say that repeating training steps in order to make the trainee feel competent *is* optimization.

Compensation (C) is primarily a matter of instruction and design. Different strategies could be implemented in user interfaces and training programs. If they were clearly labelled, e.g. as a step-by-step safe way vs. an all-at-once short-cut, elderly users or all other users with temporary or permanent limitations in knowledge, abilities or skills could select a compensating strategy to reach their goals.

Beside these model-based principles several others have proved to be a good choice. One is the general use of touchscreen devices. The touchscreen worked very well and although we did not vary input and output devices, the results from the Aachen Group (see chapterErgonomic Design of Human–Computer Interfaces for Aging Users, Schlick et al.) in the joint research program support our decision in a very strong way (Schneider et al. 2008).

From the variations of exhaustive training, three important conclusions can be drawn: First, the general operator training proved to be useful, because it supports knowledge acquisition and increases the perceived self-efficacy and thus the performance in dealing directly with an interactive system. This positive effect is stable in most criteria, however it does decrease over time. Second, to maintain the learned skills it is essential that the corresponding device is used regularly and integrated into everyday life. This was confirmed by the findings of Bruder (2008), who examined the sustainability of training for mobile phones in one of her studies. She was able to show that only the older participants who used their mobile phones daily benefited from the training in the long term. However, a TVM is usually not used regularly and the design of optimal training and instructions is difficult to achieve. Therefore, the ALISA-project conducted parallel efforts to develop minimal instructions and a wizard redesign of the user interface, hoping that these universal design interventions result in a significant reduction of training time while still ensuring an effective and efficient use of the TVM. Third, the finding that the presence or absence of video models did not affect the training outcome should not be over-generalized. It may be valid for interactive machines and electronic devices. However, another group within the SPP 1184 research program found that video models could be very helpful when more generalized competencies were to be acquired. Time management competence is a very broad competence resulting in very different observable behaviour. In such cases video models are irreplaceable (Noefer et al. 2009).

Both design interventions, minimal video instruction and wizard redesign, proved to be suitable approaches to a "universal design", yet they are not equivalent: In particular, rarely used "walk-up-and-use-systems" in public spaces should be designed so that they can be used spontaneously by anybody and even a simple device like the TVM can pose a challenge for older users. Generally, they have less computer literacy that can guide them in the use and exploration of new technology and since the face of technology changes quickly, generational differences will persist.

However, our research shows that for the younger-old, many of the age differences in user success can be mitigated by clear instructions and that specifically a brief instructional video presented immediately before the use of a ticket vending machine can eliminate age differences in effectiveness and efficiency of use measured in steps. The advantages of such a brief video are twofold: they are easily produced and they can be integrated into many devices where they can provide help on demand precisely when needed, when the user is ready for the information and can practice right on the task she was motivated to do to begin with (Carroll 1990). For many devices, this concept could also be extended to complete training programs as shown for mobile phones by Bruder et al. (2007a, b).

Another way to compensate for a lack of computer literacy is to redesign the interface to require less computer literacy for successful use. One way to achieve this goal is to follow a wizard design pattern, and we were able to show that it eliminated age differences in effectiveness and efficiency of use measured in steps as well as or even better than the instructional video. Moreover, the wizard yielded significantly higher satisfaction than the original TVM, particularly for older users. The main advantage of such a wizard lies in the effectiveness of its use. While on average the older group was able to select the correct ticket with the original TVM in 53 % of the cases, the instructional video increased effectiveness to 80 % and the wizard even to 89 %, which is about the same effectiveness the younger group showed in all setups. It can be concluded, that using the wizard required little prior interaction knowledge and was easy enough to be used successfully by all age groups. This effectiveness comes at a price of decreased efficiency. While for the purchase of a single ticket the wizard is still faster than the instructional video plus the original TVM, the knowledge gained in the video can be transferred to future uses, while the knowledge gained through the use of the wizard cannot easily be transferred to the original TVM.

Outlook

For practical purposes, to support older users it seems appropriate to combine the approaches and their advantages. For tasks performed rather rarely that focus on effectiveness rather than efficiency, a wizard design pattern proved to be a good fit. A wizard and an instructional video could be integrated into existing machines to provide information and instructions on many levels to be available when and how they are needed. Many devices come with a high-resolution screen, a loudspeaker, and a line-out port for headphones. So it is relatively easy to integrate tutorial and training programs directly into the device which has to be learned to operate. Even the usage of purely mechanical devices could be explained and trained when a mobile computer (e.g. a smartphone) is available. This will be the case for almost everyone in a few years. Traditional barcodes or RFID-tags attached to devices and sensors or smartphone cameras could be used to download interactive training programs and to start demonstrations and exercises. These training units could alternate with phases of operations with the real device. We call this "ubiquitous training" in analogy to Ubiquitous Computing or "training just in time".

Finally, the less flexible or adaptable the user interface (perhaps for economic reasons) and the greater the personal importance and frequency of use of an interactive device, the more reasonable and likely it becomes to invest efforts in an exhaustive training program.

Two caveats shall be mentioned regarding the generalizability of the findings. First, the investigated TVM can be classified as a "walk-up-and-use-system" that anyone should be able to use without prior training and other, more complex interactive systems [such as project planning software (Schneider et al. 2008)] might pose problems for older users that are not as easily mitigated. Second, for experimental reasons the instructional video was designed to teach only interaction knowledge, while for practical purposes it will be of interest to teach domain knowledge as well, which could make the brief instructional video even more effective.

Acknowledgments This research was funded by the German Research Foundation (DFG) as part of the priority program "Age-differentiated Work Systems" (SPP 1184). We would like to thank all members of the ALISA research project: Doreen Struve, Franziska Becker, Lisa Dehne, Diana Dittberner, Laura Felten, Jochen Heyden, Marko Peucker, Anna Uhl, Sebastian Ziller.

References

Baltes PB, Baltes MM (1990) Psychological perspectives on successful aging: the model of selective optimization with compensation. In: Baltes PB, Baltes MM (eds) Successful aging: perspectives from the behavioral sciences. Cambridge University Press, New York, pp 1–34

Baltes PB, Nesselroade JR (1970) Multivariate longitudinal and cross-sectional sequences for analyzing ontogenetic and generational change: a methodological note. Dev Psychol 2:163–168

Bandura A (1997) Self-efficacy: the exercise of control. Freeman, New York

Bruder C (2008) Gestaltungsprinzipien für das Training älterer Benutzer elektronischer Geräte. Dissertation an der TU Berlin. online available http://opus.kobv.de/tuberlin/volltexte/2008/1889/ last access: Nov. 28, 2012

Bruder C, Beyerlein T, Blessing L (2007a) Transfer of training success into everyday use. In: Gomez Chova L, Marti Belenguer D, Candel Torres I (eds) International technology, education and development conference—proceedings, international association of technology, education and development IATED, Valencia, Spain, Mar 7–9

Bruder C, Blessing L, Wandke H (2007b) Training the elderly in the use of electronic devices. lecture notes in computer science LNCS (part 1) 4554:637–646

Carroll JM (1990) The nurnberg funnel: designing minimalist instruction for practical computer skill. MIT Press, Cambridge

Carroll JM, Carrithers C (1984) Training wheels in a user interface. Commun ACM 27:800–806

Chandler P, Sweller J (1991) Cognitive load theory and the format of instruction. Cognit Instr 8:293–332

Chin JP, Diehl VA, Norman, KL (1988) Development of an instrument measuring user satisfaction of the human-computer interface. In: O'Hare JJ (ed) CHI'88: Proceedings of the SIGCHI conference on human factors in computing systems, ACM, USA, May 15–19, pp 213–218

Czaja SJ (1996) Aging and the acquisition of computer skills. In: Rogers WA, Fisk AD, Walker N (eds) Aging and skilled performance: advances in theory and applications. Lawrence Erlbaum Associates, Mahwah, pp 201–221

Dittberner D (2007) Zum Einfluss von Computer Literacy auf die Bedienung eines Fahrscheinautomaten. Unveröffentlichte Diplomarbeit, Technische Universität Berlin

Folmer E, Bosch J (2003) Usability patterns in software architecture. In: Proceedings of the 10th international conference on human computer interaction HCI, crete, greece, June 22–27, pp 93–97

Ivancic K, Hesketh B (1995) Making the best of errors during training. Training Res J sci pract training 1:103–126

Kalyuga S, Ayres P, Chandler P, Sweller J (2003) The expertise reversal effect. Educ Psychol 38:23–31

Li KZH, Lindenberger U, Freund AM, Baltes PB (2001) Walking while memorizing: a SOC study of age-related differences in compensatory behavior under dual-task conditions. Psychol Sci 12:230–237

Marquié JC, Jourdan-Boddaert L, Huet N (2002) Do older adults underestimate their actual computer knowledge? Behav Inf Technol 21:273–280

Mayer RE (2005) Principles of multimedia learning based on social cues: personalization, voice and image principles. In: Mayer RE (ed) Cambridge handbook of multimedia learning. Cambridge University Press, Cambridge, pp 201–212

Mayer RE, Moreno R (2003) Nine ways to reduce cognitive load in multimedia learning. Educ Psychol 38:43–52

Mori (1999) Downloaded from: http://www.compaq.co.uk/press/releases/archive/temp/2e12.asp. Visited on: May 27, 1999

Moreno R, Mayer RE, Spires HA, Lester JC (2001) The case for social agency in computer-based teaching: do students learn more deeply when they interact with animated pedagogical agents? Cogn Instr 19:177–213
Noefer K, Stegmaier R, Molter B, Sonntag K (2009) A great many things to do and not a minute to spare: can feedback from supervisors moderate the relationship between skill variety, time pressure and employees' innovative behavior? Creativity Res J 4:384–393
Prensky M (2001) Digital natives digital immigrants. Horiz 9:1–6
Schaie KW (1970) A reinterpretation of age-related changes in cognitive structure and functioning. In: Goulet LR, Baltes PB (eds) Life-span developmental psychology: research and theory. Academic Press, New York, pp 485–507
Schaie KW (1990) Intellectual development in adulthood. In: Birren JE, Schaie KW (eds) Handbook of the psychology of aging. Academic Press, San Diego, pp 291–309
Schmiedek F, Bauer C, Lövdén M, Brose A, Lindenberger U (2010) Cognitive enrichment in old age. J Gerontopsychology Geriatr Psychiatry 23:59–67
Schneider N, Schreiber S, Wilkes J, Grandt M, Schlick C (2008) Investigation of innovative input devices for the age-differentiated design of human-computer interaction, 52nd annual meeting of the human factors and ergonomics society, pp 144–148
Sengpiel M, Dittberner D (2008) The computer literacy scale (CLS) for older adults—development and validation. In: Herczeg M, Kindsmüller MC (eds) Mensch and computer 2008: viel mehr interaktion. Oldenbourg Verlag, München, pp 7–16
Staufer M (1992) Ältere Beschäftigte und technischer Wandel. Hogrefe, Göttingen
Struve D, Wandke H (2009) Video modelling for training older adults to use new technologies. *ACM Trans Accessible Comput 2:1–24
Sweller J, van Merriënboer JJG, Paas FGWC (1998) Cognitive architecture and instructional design. Educ Psychol Rev 10:251–296
Tacken M, Marcellini F, Mollenkopf H, Ruoppila I, Széman Z (2005) Use and acceptance of new technology by older people. findings of the International MOBILATE survey: enhancing mobility in later life. Gerontechnology 3:126–137

Ergonomic Design of Human–Computer Interfaces for Aging Users

Christopher M. Schlick, Sebastian Vetter, Jennifer Bützler, Nicole Jochems and Susanne Mütze-Niewöhner

Lessons Learned

The demographic changes taking place in many industrialized countries and the increased use of information technology in the workplace are creating an urgent need to carefully consider how older users work with computing systems. The changes in perceptual, cognitive and motor skills that often accompany the aging process can have important implications for the design of the human–computer interface. We investigated human–computer interaction exemplarily on the basis of self-developed project management application software for two reasons. First, computer aided project management (CAPM) is important in many manufacturing and service industries; and second, older workers are often responsible for the coordination and controlling of complex projects due to their extensive experience and excellent communication skills. A stage model of human information processing served as a theoretical framework for age-differentiated analysis and design. The model distinguishes four major stages: perception, cognition, response selection and response execution. Although there is a certain body of knowledge

C. M. Schlick (✉) · S. Vetter · J. Bützler · N. Jochems · S. Mütze-Niewöhner
Institute of Industrial Engineering and Ergonomics, RWTH Aachen University, Bergdriesch 27, Aachen 52062, Germany
e-mail: c.schlick@iaw.rwth-aachen.de

S. Vetter
e-mail: s.vetter@iaw.rwth-aachen.de

J. Bützler
e-mail: j.buetzler@iaw.rwth-aachen.de

N. Jochems
e-mail: n.jochems@iaw.rwth-aachen.de

S. Mütze-Niewöhner
e-mail: s.muetze@iaw.rwth-aachen.de

C. M. Schlick et al. (eds.), *Age-Differentiated Work Systems*,
DOI: 10.1007/978-3-642-35057-3_15, © Springer-Verlag Berlin Heidelberg 2013

regarding age-related changes in sensory processing and motor abilities under exacting laboratory conditions without computer-supported tasks, little is known about the effects of aging on human–computer interaction and their implications for the ergonomic design of software and hardware. We therefore analyzed effectiveness, efficiency and user satisfaction in seven experimental studies involving a total of 445 participants aged between 20 and 77 years. Based on our consolidated findings, we derived the following five recommendations for the age-differentiated and age-robust design of the human–computer interface:

- Compensate decreasing visual acuity of aging computer users with a larger font size. In most cases, it is not necessary to carry out comprehensive measurements of visual acuity, but simply to enlarge angular font size based on age from 16 to 22 arc minutes.
- If possible, reduce complex graphical information to be displayed on-screen by using zoom functions. Eliminate or minimize scrolling.
- Draw network diagrams (or similar graphical models) horizontally from left to right. Regarding the spatial spread of elements in the diagram, it depends on the task requirements: if decision-making based on syntactic relations between the activities is more important, use a wide spread; if a semantic analysis of predecessor-successor relations dominates, choose a narrow spread.
- Use touch screens for aging computer users. For computer users with impairments in hand-eye coordination, eye-gaze input is an effective alternative.
- When designing software for large touch screens, use a refined Fitts' law to guide the design process and to predict a time-optimal angular position of buttons and interaction elements. Furthermore, the model can be used a priori to calculate the optimal proportions of the side lengths for buttons and interaction elements depending on the angle.

Most of these recommendations can be easily adapted to other types of application software such as computer-aided manufacturing and design systems.

Introduction

Due to the aging populations in many Western and some Eastern societies and growing reliance on information technology, it is necessary to acquire a deeper understanding of the interrelation between age-related changes in mental and physical abilities and the ergonomic design of work with computing systems.

We analyzed age-related changes exemplarily based on self-developed project management application software. Computer aided project management is particularly interesting because it requires complex cognitive and perceptual skills as well as coordinative-communicative abilities. According to the Project Management Institute, project management is "the application of knowledge, skills, tools and techniques to project activities to meet project requirements" (IEEE 2011). Project planning, scheduling and schedule management/control are important tasks for a project manager. To support project planning, different techniques such as Gantt charts and responsibility assignment matrices have been developed. These techniques, however, are unable to fully describe complex time-related interdependencies between activities (Kerzner 2009). To develop a master plan—for instance—that provides comprehensive guidance for all project phases, time-critical interdependencies must be identified and modeled explicitly. Industry and academia frequently use network diagrams to do this (Dawson and Dawson 1994). In general, network diagrams consist of activities (tasks) and their interdependencies. An activity refers to the element of work that must be accomplished. Each activity has an expected duration and is interconnected via predecessor-successor relations with other activities. Figure 1 shows an example of a network diagram that is based on the popular Metra Potential Method (MPM, Kerbosh and Schell 1975) and describes a machine development project including manufacturing and assembly.

For each activity, detailed time data are encoded in a 2 × 3 matrix. In the first row, the earliest possible start time, the expected duration and the earliest possible completion time are itemized. The second row contains information about the latest possible start time, the buffer time and the latest possible completion time. The time-critical dependencies between activities can be identified by analysis of the so-called critical path. The critical path is the longest path through the network between the initial and the last activity and therefore determines the shortest possible time to complete the project. In the example provided, the critical path covers the activities 1, 2, 5, 7 and 8 as these activities contain no buffer time. Therefore, a time delay in one of the activities will inevitably extend the total duration of the project.

At a purely syntactic level, a network diagram can be regarded as a directed acyclic graph. The network graph consists of nodes (activities), which are connected by directed edges (dependencies). During project planning and scheduling, the directed acyclic graph is enriched by semantic information (name of the activity, earliest possible start and ending time, etc.). Each activity usually represents a single work package. To describe the main work processes of a project, different sequences of activities are modeled by predecessor-successor relations and integrated into a comprehensive project plan.

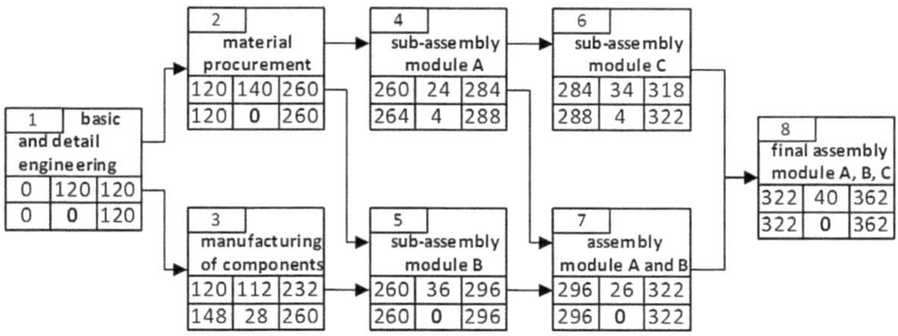

Fig. 1 Network diagram of a machine development project including manufacturing and assembly. The Meta Potential Method is used to model the activities of the project

The theoretical framework for the age-differentiated analysis and design of the human–computer interface for CAPM software is based on well-known stage models of human information processing (e.g. Wickens and Hollands 1999, see Fig. 2). The first stage covers the perception of different types of stimuli (e.g. visual, auditive or kinesthetic) through receptors in the human sensory system. In this stage, the information is processed by the central nervous system and the stimulus is recognized. The preprocessed information is then matched with information stored in working and long-term memory. Subsequently, the stored information content is categorized in either discrete states or continuous levels, which has direct implications for actions. After stimulus categorization, the person has to decide what to do. Decision making can be a rapid, nearly automatic process (e.g. the assignment of known duration to a specific activity) or a time-consuming critical reflection of pros and cons (e.g. decision making between different resource allocation policies). After an appropriate response has been selected, the final stage of information processing is the motor response execution (e.g. eye movement, speaking or grasping). Response execution is quite complex as it involves the appropriate timing and force of muscle commands to carry out the action.

In seven experimental studies with a total of 445 participants, we focused on analyzing the stages of visual perception, cognition and response execution regarding effective, efficient and satisfying human–computer interaction on an adequate level of mental workload.

In terms of human information processing, the initial stage in computer aided project management is the perception of text characters, graphical symbols and interconnected nodes in the different types of diagrams displayed on electronic information displays. It is well known that the ability to detect and differentiate nearby objects (visual acuity) naturally deteriorates as age increases and that the perception of complex information structures is potentially straining, especially for aging computer users. Therefore, we focused in our research on the effect of visual acuity on the perception of graphical symbols of different angular font sizes. Furthermore, we investigated theoretically derived visualization variants of complex network diagrams in a more applied setting.

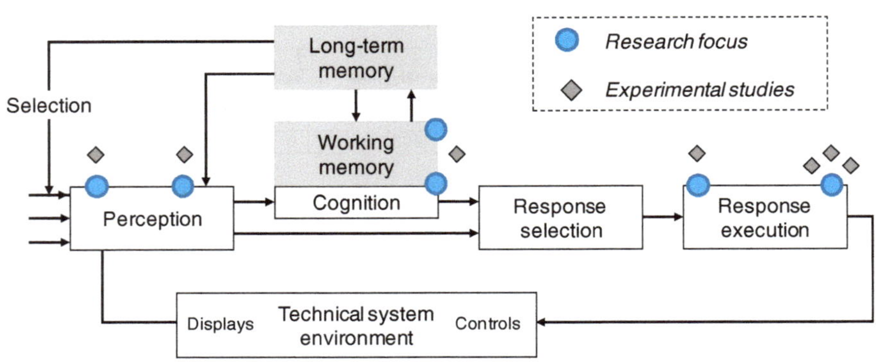

Fig. 2 Model of human information processing (adapted from Wickens and Hollands 1999)

Project planning, scheduling and schedule management/control are the main tasks of the project manager. These activities relate to the stage of cognition and decision making. Complex projects with many interrelated activities place high demands on cognitive processing, comprehension, memorization and interpretation abilities. In this context we focused on the structural layout of highly interconnected network diagrams in order to alleviate fast and reliable memorization and interpretation.

Once a decision based on syntactic and semantic relations in network diagrams has been made and the appropriate response has been mentally selected, the project manager may want to add or delete elements of the network diagram or change, update or adapt connections within the graph. It is well known that an age-related decrease in motor skills influences the speed and accuracy of providing data and control signals to a computer and potentially impedes the ease of use of conventional information input devices for graphical user interfaces. Therefore, we evaluated different direct and indirect information input devices regarding objective performance measures and user satisfaction. Furthermore, we analyzed ergonomic information presentation and input on large touch screens since we found that this interaction technique is especially beneficial for the performance and workload of aging computer users.

Current State of Research

Visual Perception

Eyesight is a highly important faculty for almost all work systems. Symptoms of an age-related eyesight degeneration process include a reduction of visual acuity, loss of power of accommodation and reduced color perception (Mouroulis 1999; Slataper 1950). A loss of visual acuity is often particularly critical with regard to performance and safety. Visual acuity describes the spatial resolution of the visual processing system. The unit of measurement is the reciprocal of individual distance in arc minutes at which two nearby points can be recognized as being separate. Visual acuity of 1.0 represents normal visual acuity, but it decreases with age. A 20-year-old normally has an acuteness of vision between 1.0 and 1.6 whereas an 80-year-old's visual acuity lies between 0.6 and 1.0. According to ISO 9241-303 (2008) angular font size should be at least 16 arc minutes. Beyond displaying text characters and symbols, the ergonomic visualization of complex graphical structures has become an important research area in human–computer interaction (Beard and Walker 1990; Card et al. 1999; Ware 2000; Federico et al. 2011). There are different approaches to support user navigation in complex network diagrams. Interfaces with an overview map show the details of a network simultaneously with a structural overview of the entire chart. By doing so, they can improve user satisfaction (North and Shneiderman 2000) and efficiency (Beard and Walker 1990; Cockburn et al. 2009). In interfaces providing a zoom function, either the entire chart or only a part of it is directly visible and can be manipulated by panning and zooming. Panning changes the area of the information space that is visible. Zooming changes the scale at which the information space is viewed. One can distinguish between linear and nonlinear panning and zooming. Linear panning and zooming is controlled with a mouse or keyboard and is based on a linear transfer function (Hornbaek et al. 2002). Nonlinear panning and zooming have been proposed in three different forms: (1) goal-directed zoom, in which direct zooming to a pre-defined size is supported (Woodruff et al. 1998), (2) combined zooming and panning, in which rapid panning automatically leads to a zoom out (Igarishi and Hinckley 2000) and (3) automatic zoom to objects, where a mouse click on the corresponding object automatically zooms to its center (Furnas and Zhang 1998; Ware 2000). But there are also negative aspects to "zoomable" graphical structures. Most notably, an increasing amount of information on-screen often leads to a decrease of the current field of view. Losing the spatial orientation of the entire information space can be one consequence (Pook et al. 2000).

Cognition

The effects of age on the speed and accuracy of memorizing syntactically connected graphical objects are ambiguous. Many authors have found evidence for an age-related decline in memorization performance as regards the spatial locations of objects (Sharps and Gollin 1987; Hoyer and Rybash 1992; Charness 1981; Light and Zelinski 1983; Moore et al. 1984; Pezdek 1983). Others, however, did not find significant age-related differences in recall performance of spatial object positions (Waddell and Rogoff 1981; Kirasic 1991). In most studies on memorization performance, distinct stimuli without logical dependencies between attributes are presented in a sequential order. To the best of our knowledge, no studies focusing on the memorization of complex graphical structures such as network diagrams were conducted before our work. Regarding the orientation of a graphical structure Winkelholz (2006) and Winkelholz and Schlick (2007) showed that memory performance is better for horizontal than for vertical structures. Furthermore, the results show that the memorization of narrow spaced objects can easily result in errors. Regarding the comparison of information, however, the proximity compatibility principle (Wickens and Carswell 1995), states that a close physical proximity on the display is particularly helpful.

Response Execution

Effective, efficient and satisfying human–computer interaction is strongly influenced by information input devices. Different input device characteristics place different requirements on human skills. Compatibility between device characteristics and a user's motor functions determines objective and subjective input performance to a large extent. Changes in motor abilities such as reduced muscle strength, reduced range of motion, and increased difficulty in executing fine movements are highly correlated with age (Walker et al. 1996). Therefore, aging computer users may need input devices with less complex sensorimotor transformations and lower requirements for precision motor control. Direct input devices such as touch screens and eye-tracking systems do not require a spatial transformation between the motor activity performed by the user and the calculated position of the cursor on the screen. While using indirect devices (e.g. the mouse), the plane of movement is usually rotated and shifted in relation to the plane of information output. Furthermore, large touch screens are a promising alternative in CAPM application areas where complex information needs to be presented and manipulated concurrently (e.g. interactive planning). Research has shown that this interaction technique can be beneficial for many user groups, particularly for aging computer users (Luczak et al. 2010; Murata and Iwase 2005; Rau and Hsu 2005). On large touch screens, effectiveness and efficiency of information input highly depends on the ergonomic design of the user interface. When redesigning project management software for large touch screens that was originally developed for traditional desktop computers, one needs to carefully choose the spatial characteristics of menus, buttons and icons as well as the size these elements should have in order to improve pointing performance and reduce human strain.

To determine 'optimal' target sizes and target positions in terms of pure motion time, Fitts' law (Fitts 1954) provides a highly satisfactory model. Fitts' law states that movement time (*MT*) is linearly dependent on the index of difficulty (*ID*) of a pointing task. According to the Shannon formulation of the law (Mackenzie 1989), the *ID* is defined as the dyadic logarithm of the quotient of movement amplitude (*A*) and horizontal target width (*W*):

$$MT = a + b \cdot ID \tag{1}$$

$$ID = \log_2\left(\frac{A + W}{W}\right) \tag{2}$$

The above formulation is also part of the ISO Standard (ISO 9241-14 2000). Fitts' original study only considered reciprocal movements along the axis of abscissa in Cartesian space. However, on large touch screens one has to deal with goal-directed movements over a half-plane and bivariate targets. In the literature, the two most prominent parameters to describe bivariate targets within the theoretical framework of Fitts' law are the W_{min} parameter, which represents the shortest length of the target sides, and the *W'* parameter which determines the target width in the direction of motion (ISO 9241-14 2000). Both parameters need to be evaluated experimentally in order to develop a comprehensive theoretical concept for ergonomic information input on large touch screens.

Methods and Results

Over six years of research, a total of 445 participants aged between 20 and 77 years participated in seven experimental studies. According to the experimental designs, the data was analyzed by full factorial analyses of variance with repeated measures (ANOVA), correlation analyses, and linear and nonlinear regression analyses. In the ANOVAs the age of the participants on their last birthday was categorized into one of three groups and served as a between-subjects variable. For each analysis, the level of significance was set to $p = 0.05$. All analyses reported are Greenhouse-Geisser corrected, if the assumption of sphericity was violated. If significant main effects occurred, the effect size ω^2 for repeated measures (Field 2009) was calculated as well. An effect size of $\omega^2 = 0.10$ corresponds to a small effect, $\omega^2 = 0.30$ to a medium effect and $\omega^2 = 0.50$ to a large effect. Only our main results are reported in the following.

Visual Perception

Optotypes Displayed on Electronic Information Displays

We investigated the age-induced change in visual acuity and its impact on performance in a target detection task in an experimental study (Vetter et al. 2010b) with 75 participants aged between 20 and 75 years. They were divided into three age groups (AG I: 20–39, AG II: 40–59, AG III: 60–75) with 25 persons in each group. Their visual acuity was measured in a pretest using the 'Rodatest 302' by Vistec. In this standardized eyesight test, Landolt ring optotypes (Fig. 3) were presented in a 5×5 matrix at two close distances of 40 and 55 cm and one far distance of 600 cm. The task was to detect, recognize and name the position of the ring opening.

After the pretest, a detection experiment based on Landolt rings of three different angular font sizes (12, 16 and 22 arc min) was carried out. The Landolt rings were displayed as visual targets on a 19-inch TFT LCD screen ($1{,}280 \times 1{,}024$ pixels). The viewing distance was kept constant at 70 cm. The participants were asked to press the space key as fast as possible when a Landolt ring with a previously defined position for the ring opening appeared on screen. Participants had to detect and

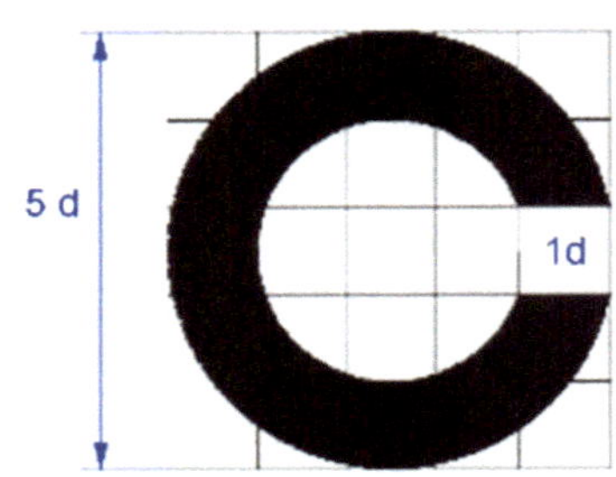

Fig. 3 The Landolt ring optotype used for testing visual acuity

recognize the previously specified Landolt ring out of a sequence of 120 Landolt rings (15 correct, 105 incorrect). In each sequence, the angular font size was constant. Each ring was shown for 1.5 s. After a five minute break, the procedure was repeated with the next font size permutation. The number of correctly detected Landolt rings (hits), the number of wrongly detected Landolt rings (false positive) and the response times for correct responses were analyzed using an ANOVA with repeated measures. The relationship between measured visual acuity, the participant's age in years and response time was investigated by correlation analysis.

The results of the pretest showed that—in conformity with our expectations—participant's aged between 20 and 39 years on average have greater visual acuity than participants aged between 40 and 59 years, who in turn have higher visual acuity values than the 60–75-year-olds. With regard to the distance of the Landolt ring optotypes, a longer distance reveals slightly lower visual acuity in the 20–39-year-olds, slightly higher visual acuity in the 40–59-year-old participants due to hypermetropia, and no clear tendency for the 60–75-year-olds.

Regarding the number of hits and false positives in the optotype detection task, overall performance was high, regardless of font size and age group. The results of inferential statistical tests show neither significant main effects nor significant interactions. Concerning the time of correct responses, the ANOVA reveals a strong main effect for font size ($\omega^2 = 0.73$) and for age group ($\omega^2 = 0.48$). There is a small but significant interaction effect for font size and age group ($\omega^2 = 0.10$). It is important to point out that the mean response time of users aged between 60 and 75 years correctly detecting ring optotypes with font sizes of 16 and 22 arc minutes is approximately the same as the response time of the 40–59-year-olds for optotypes with font sizes of 12 and 16 arc minutes, respectively. In other words, the participants from the older age group were on average as fast and accurate as participants from the middle age group working with the larger font sizes. Hence, age-related differences in response speed can be compensated by enlarging the angular font size from 16 to 22 arc minutes. All acquired values for visual acuity (corresponding to the distances of 40, 55 and 600 cm) show a significant negative correlation with response time in the target detection task: the lower the visual acuity, the slower the response. Overall, the strongest correlations with $r = 0.488$ to $r = 0.577$ exist between age of participants in years and response time in milliseconds.

Visualization of Network Diagrams

We analyzed visualization of complex network diagrams based on a sample of 90 participants aged between 20 and 75 years (Schneider et al. 2009). These participants were divided into three age groups (AG I: 20–39, AG II: 40–59, AG III: 60–75 years) with 30 persons each. The network diagrams included 14 activities that were represented on a 17-inch TFT LCD touch screen (1,280 $\times$ 1,024 pixels) that was also used for direct manipulation of the diagrams. Three different visualization variants, (1) overview map, (2) location-independent zoom function and (3) location-dependent zoom function, were tested against a de facto standard which can be found in a similar form in many commercial off-the-shelf software packages. Screenshots are shown in Fig. 4.

By touching an activity, detailed information was provided either at a fixed position in the upper left corner of the screen without reshaping the network (location-independent zoom function, Fig. 4, bottom left) or by zooming the activity at the requested position within the diagram and redrawing the surrounding elements (location-dependent zoom function). In the experiment, participants had to execute typical project management subtask like changing the start time or duration of an activity and connecting or deleting activities. Execution time, errors and subjective mental workload (according to the NASA Task Load Index, see Hart and Staveland 1988) were the dependent variables.

The results of the ANOVA of the execution time across all visualization variants show a significant main effect for age group with a large effect size of $\omega^2 = 0.66$. The visualization variant also has a significant effect on execution time. In this case, the effect size is particularly large ($\omega^2 = 0.91$). Furthermore, the

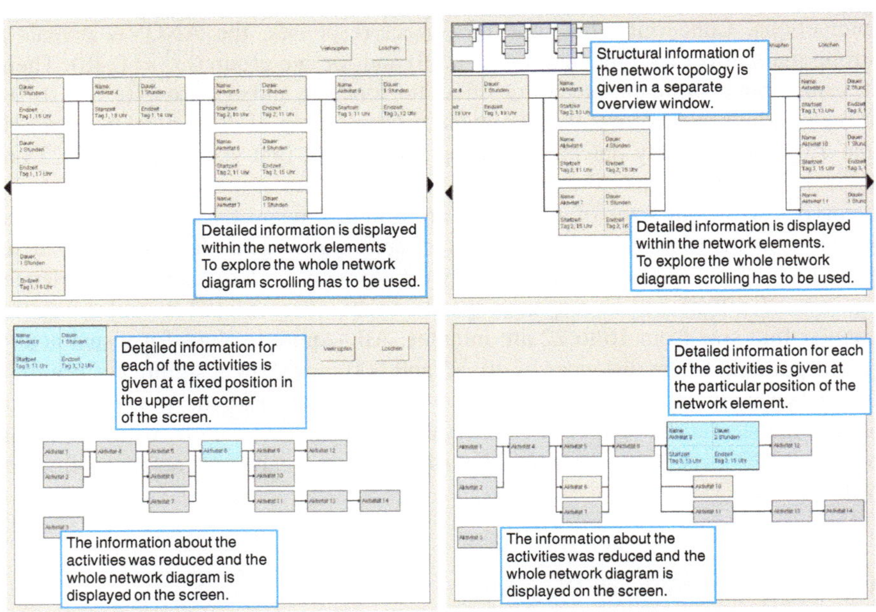

Fig. 4 Screenshots of investigated visualization variants: de facto standard form of visualization (*top left*), overview map (*top right*), location-independent zoom function (*bottom left*) and location dependent zoom function (*bottom right*). In the de facto standard form of visualization (*top left*) detailed information was given about the activities (name, earliest start time and duration) so the scroll bar had to be used to explore the whole network diagram. In the overview map, the standard map was enriched by the complete network topology, which was displayed in a separate overview window (*top right*). For the two other visualization variants, the information about the activities was reduced, the whole network diagram could be displayed and no scrolling was needed

data show that for 20–39 and 40–59-year-olds, the shortest average execution times occur when working with the location-independent zoom function. On the other hand, older participants (60–75-year-olds) worked fastest when using the location-dependent zoom function. Regarding the number of errors, the results of the ANOVA indicate significant effects for age group ($\omega^2 = 0.23$) and the visualization variant ($\omega^2 = 0.13$). The lowest error rate occurs for the visualization variants with zoom function. The subjective evaluation of mental workload showed that the location-dependent zoom function leads to the lowest workload and the de facto standard form of visualization leads to the highest workload.

Cognition

Memorization of Syntactic Relations in Network Diagrams

In the stage of cognition, the memorization of syntactic relations among elements in network diagrams was investigated with 90 participants aged between 20 and 75 years (Schneider et al. 2007). They were divided into three age groups (AG I: 20–39, AG II: 40–59, AG III: 60–75 years) with 30 persons each. Six layouts of abstracted network diagrams, depicted in Fig. 5, were displayed on a 19-inch TFT LCD screen (1,280 × 1,024 pixels) and had to be memorized by the participants. Each layout included a graphical structure with 25 elements. The elements were depicted as grey rectangles and did not include time data.

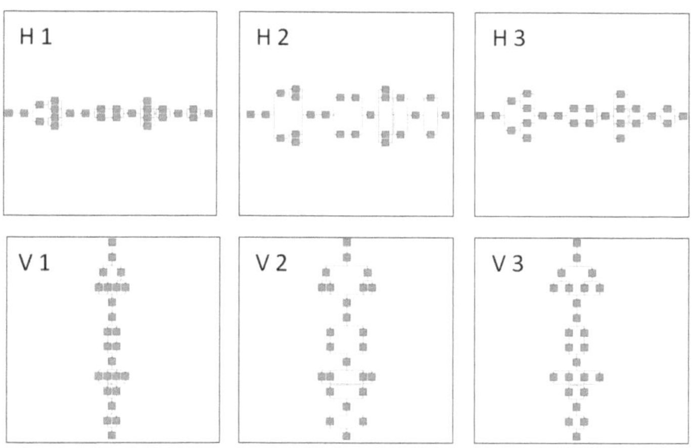

Fig. 5 Six investigated layouts of network diagrams. The diagrams differed regarding their orientation [horizontal (H) versus vertical (V)] and the spatial spread between activities [(1) narrow spread, (2) clustered, (3) uniformly spread]. The layouts *H2*, *V2* and *H3*, *V3* have a wide spread between the activities. In the layouts *H2* and *V2* parallel activities are clustered and in variant *H3* and *V3* all activities are uniformly distributed. Each network diagram consisted of 25 activities

After an acoustical signal, a random sequence consisting of five activities was highlighted in the given layout by sequentially changing the color of the five corresponding elements. Each activity was highlighted for two seconds. The end of a sequence was indicated by a second acoustical signal. Participants had to reproduce the sequence by clicking the activities in correct order with a mouse. A short break was obligatory before the next sequence was presented. Six sequences were randomly assigned for each layout. All the participants had to memorize the same six sequences corresponding to one layout. The order in which the layouts were presented was balanced. The dependent variables were the time to reproduce a single sequence correctly and the number of correctly reproduced sequences.

The ANOVAs confirmed significant age effects ($\omega^2 = 0.32$ and, $\omega^2 = 0.25$) for overall execution time and number of correctly memorized sequences. The highest memorization performance occurred in the group of 20–39-year-olds. The results revealed significant effects for spatial spread of activities in the network diagram on execution time ($\omega^2 = 0.15$) and the number of correctly repeated sequences ($\omega^2 = 0.68$). The layouts with a "clustered" and "uniformly spread" of activities (see Fig. 5) are beneficial for visual spatial memory (highest number of correctly repeated sequences, least required time). A layout with a small spread leads to significantly lower performance. Regarding the orientation of the network diagram, a horizontal layout is beneficial and leads to a significantly higher number of correctly repeated sequences ($\omega^2 = 0.70$) and the shortest required time ($\omega^2 = 0.20$).

Interpretation of Semantic Relations in Network Diagrams

The interpretation of semantic relations in complex network diagrams was investigated subsequently to the memorization study (Schneider et al. 2007). Therefore, the same sample was used. The six layouts of abstracted network diagrams (Fig. 5) were used. In the abstracted diagrams, only the earliest start and end times for each activity were displayed within the elements (see Fig. 6).

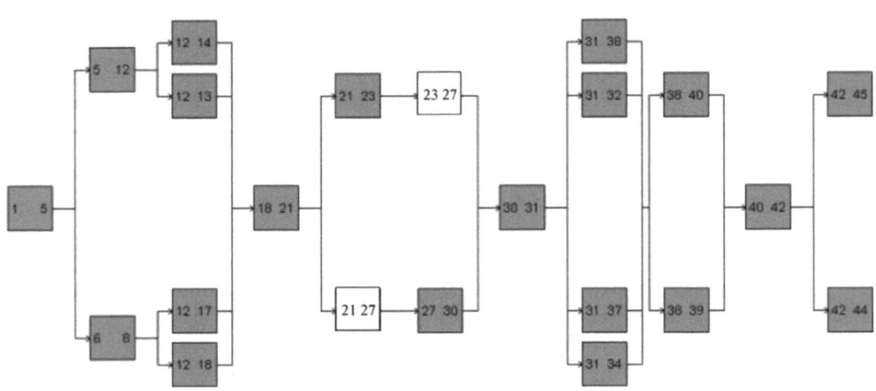

Fig. 6 Layout *H2* with spatial clustering of parallel activities

To solve the interpretation task, participants had to answer four questions for each layout. The questions were subdivided into two types. Questions of type 1 deal with temporal relations between activities [(1.1) "Which activity has a longer duration?", (1.2) "Which activity ends first?"], while questions of type 2 deal with predecessor-successor relations [(2.1) "How many direct predecessors does this activity have?", (2.2) "How many direct successors does this activity have?"]. After an acoustical signal, either one (questions 2.1 and 2.2) or two (questions 1.1 and 1.2) activities were randomly chosen and highlighted. After two seconds, one of these four questions appeared on screen and the participant had to answer the question by marking the correct activity or the correct number with a mouse click. Then the next activity was highlighted. Each participant had to answer identical questions in the same sequence per experimental condition. The order in which the various layouts were presented was balanced. The dependent variables were the execution time and the number of correctly answered questions.

Since there was very little variance in the number of correctly answered questions (almost no errors occurred), only the results for response time are presented. The ANOVA results reveal that the age group ($\omega^2 = 0.41$) and the spatial spread ($\omega^2 = 0.27$) of network activities both have significant effects on response time. Post-hoc paired comparisons of means show significant differences between all age groups. The mean response time significantly increased with age. For the spatial spread between activities, we found that the small spread of structures H1 and V1 leads to shorter response times than the alternative layouts "clustered" and "uniformly distributed" (see Fig. 5). Mean response time is shorter with a horizontal orientation of the network diagram, yet the effects are not significant.

Response Execution

Information Input Devices

Three information input devices, (1) a mouse, (2) a resistive touch sensitive screen surface and (3) an eye-gaze control system, were contrasted with one another on the basis of a two-dimensional pointing task (Schneider et al. 2008). A 17-inch TFT LCD screen (1,280 $\times$ 1,024 pixels) was used to display the pointing targets. A total of 90 subjects aged between 20 and 75 years participated in the study. They were divided into three age groups (AG I: 20–39, AG II: 40–59, AG III: 60–75 years) with 30 persons each. In the pointing task, the target width and the amplitude were varied systematically. The pointing movements had to be executed with each of the three input devices as fast and accurately as possible. While using the mouse, the cursor first had to be positioned at the home position indicated by a circle and then moved to the squared target. The home and target positions had to be confirmed with a mouse click. Using the touch screen, the pointing task involved touching the start position with the preferred index finger, then touching the target position with the same finger. With eye-gaze control, the task required participants to first visually fixate on a point inside the start circle and then fixate on the target square. The fixation dwell time that had to be exceeded to confirm a target object was set to 100 ms for all experimental conditions. Execution time was the primary dependent variable. Because of the different characteristics of the input devices, it was not possible to define a single objective error measure. In particular, when using the eye-gaze input, it is difficult to distinguish between consciously and not consciously controlled eye movements and the input errors that derived from them.

The ANOVA of the execution times across all input devices shows a significant main effect for age group ($\omega^2 = 0.55$). Post-hoc paired comparisons of means show significant differences between the 20–39-year-olds and the 40–59-year-olds as well as between the 20–39-year-olds and the 60–75-year-olds. The input device also has a significant effect on execution time ($\omega^2 = 0.51$). The touch screen leads to significant better performance than eye-gaze control and mouse input. Furthermore, there is a significant interaction effect between age group and input device. The data show that the average execution times of the 40–59-year-olds and the 60–75-year-olds differ significantly only for mouse input.

Performance Prediction Models for Large Touch Screens

Based on the paradigm of Fitts' law, human performance when executing goal-directed movements was analyzed on a large touch screen. The first study (Vetter et al. 2010a) was motivated by the fact that the definition of target width differs across previous human-computer studies. To answer the question of which definition of target width should be used, we analyzed the two most frequently used parameters of target width, W' and W_{min}, in a one-dimensional pointing task setting. W' determines the target width of an object in the direction of the approach angle. Conversely, the W_{min} parameter encodes the smaller value of either target height or width. The 30 right-handed participants were between 20 and 73 years old. A large touch screen with a projection area of 865 × 649 mm (4:3 ratio) was used. The participants were divided into two age groups (AG I: 20–32, AG III:

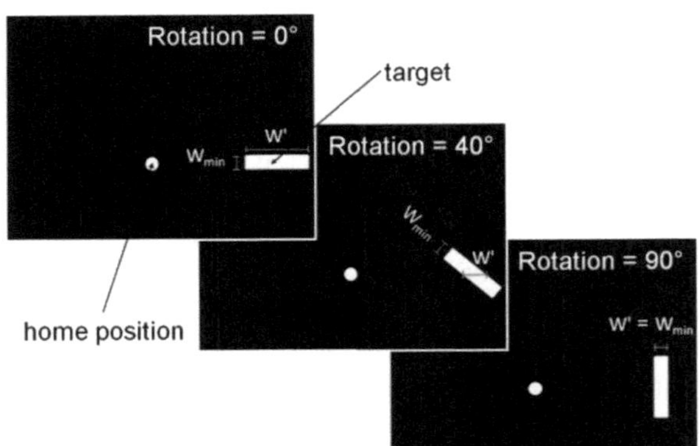

Fig. 7 Experimental setup exemplarily showing the target rotations of 0°, 40°, 90°

58–73 years) with 15 persons each. The experimental task was to point to rectangular target objects as fast and as accurately as possible. The targets were displayed under a constant angle of 0° (movement to the right, Fig. 7). By rotating the target objects around their center, the parameter W' was varied systematically, while W_{min} was kept constant.

The results of linear regression analysis clearly contradict the motion time prediction based on the W_{min} parameter, which proposes no change in MT when W_{min} is left unchanged. In contrast, the prediction based on the W' parameter is confirmed: a decreasing MT with increasing target width in the direction of motion was found. Furthermore, we found significant effects for the target height perpendicular to the direction of motion. This additional independent parameter is denoted by H'. The best fit between regression models and data was obtained by integrating an additional additive component into Fitts' formulation as:

$$MT = a + b\log_2\left(\frac{A}{W'} + 1\right) + c\log_2\left(\frac{A}{H'} + 1\right) \quad (3)$$

In the second study (Bützler et al. 2011), the influence of the motion angle was investigated with 30 participants aged between 21 and 77 years. The participants were divided into two age groups (AG I: 21–36, AG III: 58–77) with 15 persons each. The task was the same as in the first study. The angle between start and target object was varied systematically in 10° increments between 0° (movement to the right) and 180° (movement to the left). In order to eliminate directional influences from the target width, circular target objects were used (Fig. 8).

The results of the second study reveal that the average movement time on the large touch screen follows a sinusoidal function with the motion angle α as function argument (see Fig. 9).

Based on the results of the first and the second study, we were able to formulate the following integrative model for two-dimensional pointing and bivariate rectangular targets:

$$MT = a + b\log_2\left(\frac{A}{W'} + 1\right) + c\log_2\left(\frac{A}{H'} + 1\right) + d\sin(2\alpha). \quad (4)$$

A third study was conducted to validate the integrative model. For the validation study, 40 participants aged between 21 and 75 years had to execute the cited pointing task. The participants were divided into two age groups (AG I: 21–35, AG III: 51–75)

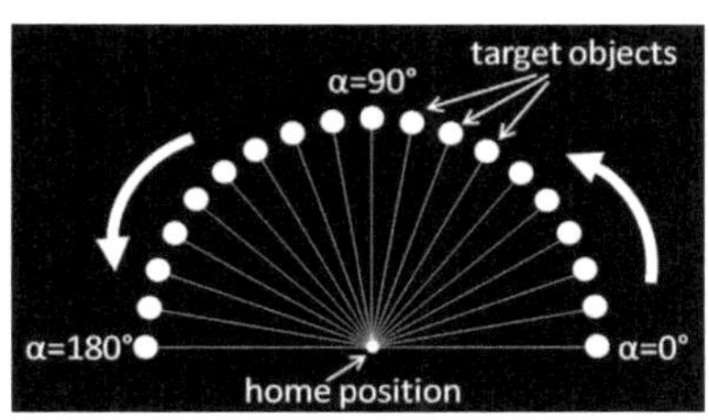

Fig. 8 Experimental setup showing the angular position of target objects

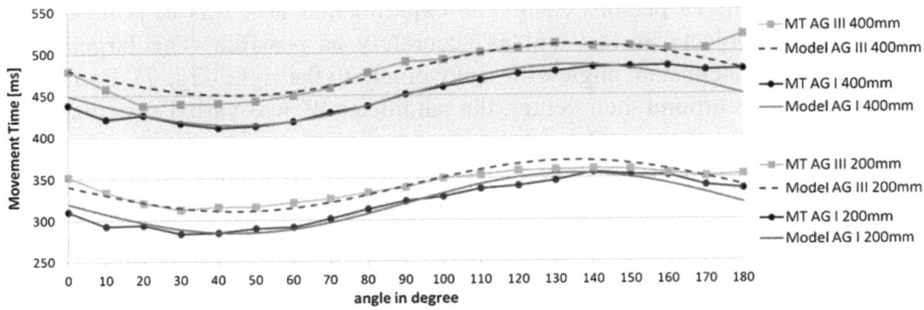

Fig. 9 Empirical average movement time and predicted movement time according to model from equation 4

with 20 persons each. 250 rectangular target objects had to be pointed. The target objects were displayed at a randomly chosen position on the screen (Fig. 10) covering a broad range of values in Fitts' famous index of difficulty ($ID \in [1.01; 4.88]$). The *ID* values were calculated according to Eq. 2.

We were able to successfully validate the model in the third study and found that compared to the original formulation of Fitts' law and to well-known refinements in the literature (e.g. Fitts 1954; Mackenzie and Buxton 1992; ISO 9241-9 2002; Accot and Zhai 2003; Murata and Iwase 2005; Appert et al. 2008; Yang and Xu 2010), our new model shows the highest predictive validity. The best goodness-of-fit was found when different regression coefficients depending on the age group of the participants were used.

For the 21–35-year-olds (AG I), the coefficient tuple is:

$$\theta_{\mathrm{AGI}} = \left(a = 25.003[\mathrm{ms}], b = 91.924\left[\frac{\mathrm{ms}}{\mathrm{bit}}\right], c = 42.415\left[\frac{\mathrm{ms}}{\mathrm{bit}}\right], d = -23.240[\mathrm{ms}]\right)$$

For the 51–75-year-olds (AG III), the coefficient tuple is:

$$\theta_{\mathrm{AGIII}} = \left(a = 23.980[\mathrm{ms}], b = 108.391\left[\frac{\mathrm{ms}}{\mathrm{bit}}\right], c = 50.966\left[\frac{\mathrm{ms}}{\mathrm{bit}}\right], d = -20.495[\mathrm{ms}]\right)$$

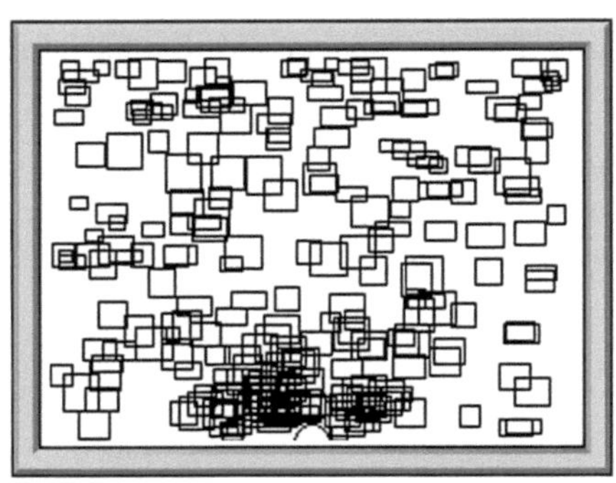

Fig. 10 Experimental setup in which all 250 randomly generated target objects are shown simultaneously

The resulting coefficient of determination is $R^2 = 0.937$ for the 21–35-year-olds and $R^2 = 0.924$ for the 51–75-year-olds.

Discussion

Visual Perception

For the visual perception of optotypes displayed on a computer screen, we found significant effects for visual acuity, age group and font size. The finding that differences in response time between the 40–59-year-olds and the 60–75-year-olds can be compensated by enlarging the font size from 12 to 16 or 16 to 22 arc minutes supports the concept of age-differentiated adaptation of the human-computer interface. The age of the participants was a better predictor of response time than visual acuity. This is probably due to lag times in neuromuscular responses. Therefore, we recommend a font size of at least 16 arc minutes and an adaptation of up to 22 arc minutes depending on the age of the participants.

To support visual perception of network diagrams, we developed different forms of visualization based on a literature review. The results reveal that all innovative visualization variants clearly outperform the de facto standard of many project management software systems. If execution time is put into focus, we recommend an age-differentiated design, since the 20–39-year-olds and the 40–59-year-olds worked fastest by using the location-independent zoom function (Fig. 4, bottom left), whereas the 60–75-year-olds benefited most from the location-dependent zoom function (Fig. 4, bottom right). Regarding error rate and mental workload, the location-dependent zoom function leads to the lowest average values in all age groups. We therefore recommend an ergonomic "Design for All" philosophy (Gregor et al. 2002) that uses a location-dependent zoom function.

Cognition

To analyze the central stage of human information processing, we varied the orientation and the spatial spread of activities in a network diagram and measured memory and interpretation performance based on syntactic and semantic relations. Participants in all age groups performed best when working with a horizontally oriented network diagram. Regarding the spatial spread of network activities, we detected an interesting trade-off: memorization based on syntactic relations was best for a wide spread, whereas interpretation based on temporal and predecessor-successor relations benefits from a small spread. These results are in accordance with the theory of Winkelholz and Schlick (2007), which predicts an improvement in human performance due to a smaller noise variance of visual spatial memory recall when representing the graph horizontally and with low ambiguity between interconnections of nodes. For the extraction of information on a semantic level, however, a small spread between activities has positive effects, corresponding to the proximity compatibility principle of Wickens and Carswell (1995). Since no interaction effects occurred, we recommend again a "Design for All" and the use of horizontally oriented network diagrams in conjunction with either a large or a small spread of activities, depending on the task.

Response Execution

To optimize response execution in human-computer interaction we compared a resistive touch screen and eye-gaze input with a conventional computer mouse in a simple pointing task setting. The results reveal highest average performance for touch screen input. In particular, aging computer users benefit from this kind of direct information input. When using a touch screen, older computer users can reach a performance level that is similar to young users working with the computer mouse. Although eye gaze input was especially unfamiliar for older users, they processed the task significantly faster than with conventional mouse input. We therefore strongly recommend touch screen input for aging computer users and eye-gaze input for motor impaired computer users.

Based on these results we analyzed touch screen input in depth by using Fitts' law. To accurately predict the execution time of goal-directed pointing movements on these screens, we extended Fitts' law to rectangular target objects and motions over a half plane. Based on two experimental studies, we found that rectangular target objects can be represented best by target width in direction of motion (W') and target height perpendicular to direction of motion (H'). Movement time follows a sinusoidal function depending on the angle α from the x-axis at which a point in the x–y-plane lies (polar angle). The parameter W' represents sensorimotor correction processes of movement amplitude. In a similar manner, H' reflects the time-dependent correction of movement direction. The sine term accounts for inertial anisotropy of the hand-arm system (Vetter 2012). The validity of the derived model was confirmed in a third study. The results of predictive performance modeling support an age-differentiated approach, since model parameters have to reflect age-related changes in movement planning and execution.

Outlook

In the near future, we plan to collaborate with a leading project management software company in Germany to transfer this scientific knowledge to industry and develop age-differentiated and age-robust software components for CAPM. By doing so, we will also be able to evaluate the impact of our findings on real project-planning and scheduling problems in industrial companies. For example, the trade-off between the spread of network elements concerning memorization of syntactic relations and interpretation of semantic relations can be investigated in a more practice-oriented experimental setting with senior project managers. Another interesting issue is an empirical evaluation of Gantt charts, which are highly relevant for project scheduling on a lower level of detail than network diagrams. Furthermore, we plan to assess the complexity of human–computer interaction when working with a combination of network diagrams and Gantt charts. Therefore, we will analyze sequences of discrete user-initiated events and discretized eye-movements as indicators of emergent complexity (Schlick et al. 2010). Through ongoing collaborations with industry, we will combine basic and applied research and prove our findings in ergonomic field studies. We hypothesize that the transfer of research findings to commercial software packages will not only improve the older project managers' efficiency and effectiveness; it will also reduce learning time and improve the user experience. In the future, predictive performance models such as Fitts' law can also help software designers to build anticipatory user interfaces that provide automatic adaptation of the layout and size of interface objects according to task demands as well as a user's behaviour and preferences (Schlick et al. 2006). Although all seven studies focused on CAPM, the significant effects can be interpreted as an indicator of a general need for age-differentiated and age-robust design in software systems. In light of the demographic changes taking place in many countries and growing reliance on computer technology, we also encourage the consideration of age-related differences in corresponding ISO standards, e.g. *human-centered design processes for interactive systems* (ISO 9241-210 2011) and *ergonomic requirements for office work with visual display terminals*—Part 14 menu dialogues (ISO 9241-14 2000) and guidelines.

Acknowledgments The authors would like to thank the German Research Foundation (DFG) for its kind support of research into age-differentiated design for the human–computer interface under Priority Program No. 1184, Age-Differentiated Work Systems (project funding reference number SCHL 1,805/1−3). We would also like to thank all the study participants for their contributions.

References

Accot J, Zhai S (2003) Refining Fitts' law models for bivariate pointing. In: Proceedings conference on human factors in computing systems, ACM CHI, Ft Lauderdale, pp 193–200

Appert C, Chapuis O, Beaudouin-Lafon M (2008) Evaluation of pointing performance on screen edges. In: Procedings ACM advanced visual interfaces, Napoli, Italy. ACM Press, New York, pp 119–126

Beard DV, Walker JQ (1990) Navigational techniques to improve the display of large two-dimensional spaces. Behav Inf Technol 9:451–466

Bützler J, Vetter S, Jochems N, Schlick C (2011) Refinement of Fitts' Law for large touch screens: age-differentiated analysis of the motion angle in bivariate pointing tasks. In:

Proceedings of human factors in organisational design and management X, 2, 2011, Grahamstown, South Africa: IEA Press, California, pp 93–99
Card SK, Mackinlay JD, Shneiderman B (1999) Readings in information visualization: using vision to think. The Morgan Kaufmann Series in Interactive Technologies, San Francisco
Charness N (1981) Visual short-term memory and aging in chess players. J Gerontol 36:615–619
Cockburn A, Karlson A, Bederson BB (2009) A review of overview + detail, zooming, and focus + context interfaces. ACM Comput Surv 41(1): 2:1–2:31
Dawson CW, Dawson RJ (1994) Clarification of node representation in generalized activity networks for practical project management. Int J Proj Manage 12(2):81–88
Federico P, Aigner W, Miksch S, Windhager F, Zenk L (2011) A visual analytics approach to dynamic social networks. In: Stefanie Lindstaedt, Michael Granitzer (eds.) Proceedings of the 11th international conference on knowledge management and knowledge technologies (i-KNOW '11), vol 47. ACM Press, New York, p 8
Field A (2009) Discovering statistics using SPSS. Sage Publications, London
Fitts PM (1954) The information capacity of the human motor system in controlling the amplitude of movement. J Exp Psychol 47:381–391
Furnas GW, Zhang X (1998) MuSE: A multiscale editor. In: Proceedings of the 11th annual ACM symposium on user interface software and technology UIST'98. ACM, New York pp 107–116
Gregor P, Newell AF, Zajicek M (2002) Designing for dynamic diversity: interfaces for older people. In: Proceedings of the 5th international ACM conference on assistive technologies, ACM, New York, pp 151–156
Hart SG, Staveland LE (1988) Development of NASA-TLX (Task load index): results of empirical and theoretical research. In: Hancock PA, Meshkati N (eds) Human mental workload. North Holland Press, Amsterdam
Hornbaek K, Bederson BB, Plaisant C (2002) Navigation patterns and usability of zoomable user interfaces with and without an overview. ACM Trans Comput Hum Interact 9(4):362–389
Hoyer WJ, Rybash JM (1992) Age and visual field differences in computing spatial rela-tions. Psychol Aging 7:339–342
IEEE (2011) IEEE guide—adoption of the project management institute (PMI(R)) standard a guide to the project management body of knowledge (PMBOK(R) guide) 4th (ed)
Igarishi T, Hinckley K (2000) Speed-dependent automatic zooming for browsing large documents. In: Proceedings of the ACM symposium on user interface software and technology(USIT), pp 139–148
Kerbosh J, Schell HJ (1975) Network planning by the extended METRA potential method report KS-l.l. University of Technology Eindhoven, Department of Industrial Engineering, Netherland
Kerzner H (2009) Project management: a systems approach to planning, scheduling, and controlling. Wiley, New Jersey
Kirasic KC (1991) Spatial cognition and behavior in young and elderly adults: implications for learning new environments. Psychol Aging 6(1):10–18
Light LL, Zelinski EM (1983) Memory for spatial information in young and old adults. Dev Psychol 19:901–906
Luczak H, Schlick C, Jochems N, Vetter S, Kausch B (2010) Touch screens for the elderly: some models and methods, prototypical development and experimental evaluation of human-computer interaction concepts for the elderly. In: Haftor D, Mirijamdotter A (eds) Information and communication technologies, society and human beings: theory and framework. IGI Global, Hershey, pp 116–135
Mackenzie IS (1989) A note on the information-theoretic basis for fitts' law. J Mot Behav 21:323–330
Mackenzie IS, Buxton W (1992) Extending fitts' law to two-dimensional tasks. In: Bauersfeld W, Bennett P, Lynch J (eds) Proceeding SIGCHI conference on human factors in computing systems, ACM, Monterey, pp 219–226

Moore TE, Richards B, Hood J (1984) Aging and the coding of spatial information. J Gerontol 39:210–212
Mouroulis P (1999) Visual instrumentation handbook: optical design and engineering principles. Mc Graw-Hill, New York, pp 101–159
Murata A, Iwase H (2005) Usability of touch-panel interfaces for older adults. Hum Factors 47:766–776
North C, Shneidermann B (2000) Snap-together visualization: evaluating coordination usage and construction. Int J Hum Comput Stud 53:715–739
Pezdek K (1983) Memory for items and their spatial locations by young and elderly adults. Dev Psychol 19:895–900
Pook S, Lecolinet E, Vaysseix G, Barillot E (2000) Context and interaction in zoomable user interfaces. In: AVI 2000 conference proceedings, pp 227–231
Rau P, Hsu J (2005) Interaction devices and web design for novice older users. Educ Gerontol 31(1):19–40
Schlick C, Winkelholz C, Motz F, Luczak H (2006) Self-generated complexity and human-machine interaction. In: Proceedings of the 2006 IEEE transactions on systems, man and cybernetics—part a: systems and humans, pp 220–232
Schlick C, Winkelholz C, Motz F, Duckwitz S, Grandt M (2010) Complexity assessment of human-computer interaction. Theor Issues Ergon Sci 11(3):151–173
Schneider N, Schreiber S, Wilkes J, Grandt M, Schlick C (2007) Age-differentiated adaptation of the human-computer-interface. In: Toomingas A, Lantz A, Berns T (eds) Proceedings of the 8th international conference on work with computing systems, Stockholm Sweden, pp 40–44
Schneider N, Vetter S, Kausch B, Schlick C (2009) Age-differentiated visualization of network plans for project management software, In: Proceedings of the IEA2009—17th world congress on ergonomics, Beijing, pp 1–8, 09–14 August 2009
Schneider N, Wilkes J, Grandt M, Schlick C (2008) Investigation of input devices for the age-differentiated design of human-computer interaction. In: Proceedings of the human factors and ergonomics society, 52nd annual meeting, New York. Hrsg: Human Factors And Ergonomics Society, Mira Digital Publishing, New York, pp 144–148, 22–26 Sept 2008
Sharps MJ, Gollin ES (1987) Memory for object locations in young and elderly adults. J Gerontol 42(3):336–341
Slataper F (1950) Age norms of refraction and vision. Arch Opthalmology 43:466–481
Vetter S (2012) Theoretische und experimentelle Analysen zum Fitts'schen Gesetz bei der Interaktion mit großflächigen berührungssensitiven Bildschirmen University, German
Vetter S, Bützler J, Jochems N, Schlick C (2010a) Untersuchungen zum Fitts̓schen Gesetz anhand eines großflächigen Touch-Displays, In: Innovative Interaktionstechnologien für Mensch-Maschine-Schnittstellen, 52. Fachausschusssitzung Anthropotechnik der DGLR Lilienthal-Oberth e.V, Berlin, DGLR e.V, M Grandt, A Bauch, (eds) Bonn, pp 147–158 (in German)
Vetter S, Jochems N, Kausch B, Mütze-Niewöhner S, Schlick C (2010b) Age-induced change in visual acuity and its impact on performance in a target detection task with electronic information displays. Occup Ergon 9(2):99–110
Waddell KJ, Rogoff B (1981) Effect of contextual organization on spatial memory of middle-aged and older women. Dev Psychol 17:878–885
Walker N, Philbin DA, Spruell C (1996) The use of signal detection theory in research on age-related differences in movement control. In: Aging and skilled performance. Advances in theory and applications. Rogers WA, Fisk AD, Walker N (eds) Lawrence Erlbaum Associates, pp 45–64
Ware C (2000) Information visualization: perception for design. Morgan Kaufmann (Academic Press), San Francisco, p 1
Wickens CD, Hollands JG (1999) Engineering psychology and human performance, 3rd edn. Prentice Hall
Wickens CD, Carswell M (1995) The proximity compatibility principle: its psychological foundation and relevance to display design. Hum Factor 37(3):473–494

Winkelholz C (2006) Analyse, Modellierung und Simulation der visuellräumlichen Kognition bei der Mensch-Maschine-Interaktion, Rheinisch-Westfälische Technische Hochschule Aachen, Fakultät für Maschinenwesen, Dissertation. (in German)
Winkelholz C, Schlick C (2007) Modeling human spatial memory within a symbolic archi-tecture of cognition. In: Barkowsky T (ed) Spatial cognition V, lecture notes in artificial intelligence. Springer, Berlin, pp 229–248
Woodruff A, Landay J, Stornebreaker M (1998) Goaldirected zoom. In: Summary of the ACM conference on human factors in computing systems CHI'98, Los Angeles, pp 305–306
Yang H, Xu X (2010) Bias towards regular configuration in 2D pointing. In: Proceedings of conference on human factors in computing systems, ACM, CHI, pp 1391–1400

Standards

ISO 9241-14 (2000) Ergonomic requirements for office work with visual display terminals (VDT)s—part 14 menu dialogues, ISO 9241-14: 2000 (E), 2000. International organization for standardization
ISO 9241-9 (2002) Ergonomic requirements for office work with visual display terminals (VDTs)—part 9-requirements for non-keyboard input devices (Vol February 15, 2002). International organization for standardization
ISO 9241-210 (2011) Ergonomics of human-system interaction—Part 210: Human-centred design for interactive systems, ISO 9241-210: 2011 (E), 2011. International organization for standardization
ISO 9241-303 (2008) Ergonomics of human-system interaction. Requirements for electronic visual displays. BS EN ISO 9241-303:2008. International organization for standardization

Age-Related Variations in the Control of Electronic Tools

Herbert Heuer, Mathias Hegele and Miya Kato Rand

Lessons Learned

Electronic tools, such as a computer mouse, are highly flexible with respect to the visuo-motor transformations they implement. Different from mechanical tools, they lack mechanical transparency. Several studies noted difficulties of older users with computer-mouse operations, but also with more complex visuo-motor transformations as found in laparoscopic surgery. These difficulties may partly result from generalized slowing, but partly also from reduced learning capabilities. We explored the nature of the latter kind of change in a series of experiments.

In all experiments reviewed in this chapter, participants learned novel visuo-motor transformations, in particular novel relations between hand movements and motions of a cursor on a monitor. Novel transformations were gain changes, rotations, and complex transformations as found in laparoscopic instruments. Different tests served to assess implicit and explicit knowledge that was acquired from learning. These were tests of after-effects (implicit knowledge), judgements of visually presented amplitudes and directions of "correct movements" (explicit knowledge), and tests of adaptive shifts (combination of explicit and implicit knowledge). The major changes at older working age can be summarized:

H. Heuer (✉) · M. K. Rand
Leibniz Research Centre for Working Environment and Human Factors, Dortmund, Germany
e-mail: heuer@ifado.de

M. K. Rand
e-mail: rand@ifado.de

M. Hegele
Justus Liebig University, Gießen, Germany
e-mail: mathias.hegele@sport.uni-giessen.de

C. M. Schlick et al. (eds.), *Age-Differentiated Work Systems*,
DOI: 10.1007/978-3-642-35057-3_16,

- Adaptive shifts decline.
- After-effects remain unaffected (sometimes they increase).
- Explicit knowledge declines.

These changes are found for sufficiently complex or large visuo-motor transformations, for which explicit knowledge is utilized for strategic corrections. For less conspicuous transformations, age-related changes can be absent at older working age or be restricted to a decline in explicit knowledge.

The age-related changes of learning novel visuo-motor transformations are quite robust across different conditions of learning. The absence of explicit learning has further consequences for dual adaptation and for the generalization across the workspace. Counter-measures against the age-related changes seem to require special efforts of improving explicit learning at older working age.

Problems in the Control of Electronic Tools at Older Working Age

When one uses a mechanical tool such as a hammer, one can feel the interaction forces and see the location of the impact of its head. The movements are governed by the laws of mechanics, and their consequences can be sensed directly. In contrast to mechanical tools, the laws of mechanics do not constrain electronic tools. The most widely used type is the computer mouse. With this tool, the transformation of hand movements into motions of the cursor on the computer monitor is mediated by an electronic device. This results in high flexibility of the definition of visuo-motor transformations at the cost of lacking natural interaction forces. For example, haptic information is related to the mouse movement only, but not to the cursor motion. When the cursor reaches an icon, there is no felt impact. Some observations suggest that older users can encounter difficulties with typical mouse operations such as cursor positioning or clicking an icon (Charness et al. 2004; Lee et al. 2012; Sandfeld and Jensen 2005; Smith et al. 1999; Walker et al. 1996, 1997).

Perhaps the most challenging electronic tools can be found in laparoscopic surgery. First, there is a complex mechanical transformation of the hand movements into the movements of the tip of the laparoscopic instrument (Gallagher et al. 1998; Heuer and Sülzenbrück 2009). Second, the image of the tip of the instrument and its surroundings is picked up by an endoscope and presented on a monitor at some place roughly in front of the surgeon. Therefore, the visual information is quite different from what the surgeon would see in direct view. Again there are indications that older surgeons encounter particular difficulties with these new surgical techniques (Neumayer et al. 2005; Risucci et al. 2001).

A part of the difficulties described could be due to generalized age-related slowing that can be found in humans (e.g., Li et al. 2001; Myerson et al. 1990; Proctor et al. 2005; Seidler and Stelmach 1995; Welford 1981) and in other animals, for example in mice (e.g., Fouquet et al. 2011). Age-related motor slowing is an incremental process that begins in early adulthood and continues throughout working life (Szafran 1951; Teeken et al. 1996; Yan et al. 1998). In addition the fairly general age-related decline of manual dexterity (see chapter Influence of Age and Expertise on Manual Dexterity in the Work Context: The Bremen-Hand-Study@Jacobs, Voelcker-Rehage et al.) could play a role. However, another part of the difficulties might be due to an age-related decline of visuo-motor plasticity of the brain (e.g., Sawaki et al. 2003). This might affect its capacity to learn to use electronic tools that introduce new relations between self-produced movements and their perceived consequences.

In this chapter, we start with an outline of a theoretical framework of the processes involved in learning to master visuo-motor transformations as they are inherent to electronic tools. Subsequently, we describe the basic experimental paradigm by which age-related changes of these processes can be studied. Based on these foundations, we present an overview of experimental findings. The existing data show that the acquisition of explicit knowledge of tool transformations as well as its usage for intentional (or strategic) corrections decline at older working age. These changes have proven to be astonishingly robust against some potential counter-measures. Furthermore, these age-related changes have negative consequences for the generalization and differentiation of adaptation to visuo-motor transformations across the workspace.

Theoretical Framework

Goal-directed movements serve to reach action goals. To start a program on a computer (action goal), for example, one needs to move the cursor on the respective control icon (movement goal). To achieve this, one has to specify the appropriate parameters of the hand movement (amplitude, direction) that will move the cursor to the desired location. Thus, one has to adjust the hand movement to the tool transformation T that translates (relative) mouse position **x** into (relative) cursor position **y**. A simple example is the transformation $y = T(x) = k \cdot x$, where **x** is the distance of the mouse from its initial position and **y** the distance of the cursor from its initial position.

The mastery of such a transformation can principally be achieved in multiple ways. Figure 1 illustrates a theoretical framework identifying the processes by which an inversion of the transformation T can be accomplished (cf. Heuer and Massen 2012). It is a combination of feed-forward (open-loop) control including an internal model of the transformation $M(T^{-1})$—more precisely: an internal model of the inverse transformation—and feed-back (closed-loop) control related to a feed-back controller. Both control routes approximate the inverse of the transformation T that is needed to produce a movement of the mouse with such an amplitude and direction that the cursor reaches its target. In addition, there are

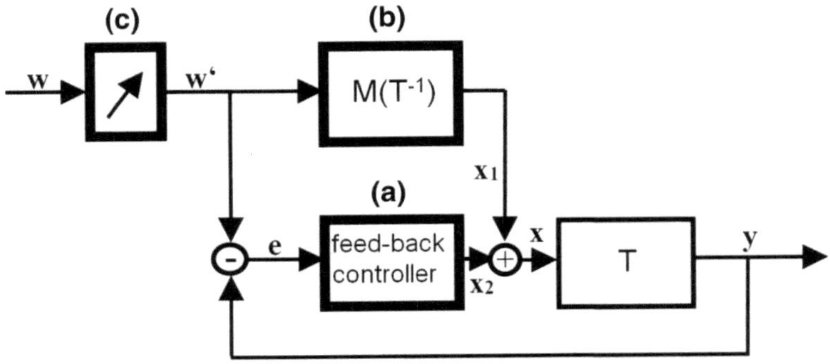

Fig. 1 Outline of a theoretical framework that combines open-loop and closed-loop control of the transformation T. Functions that are modulated by learning and that are potentially affected by age are marked by bold outlines: **a** the feed-back controller, **b** the internal model, **c** strategic corrections (Learning-related feedback that is used to modify the feedback controller, the internal model, and strategic corrections is not shown)

intentional strategic corrections. These three ways of dealing with a visuo-motor transformation will be described in more detail.

First, control of a tool transformation can be achieved via the online use of (visual) feed-back of the cursor movement. In this case, a continuous comparison between current cursor position and desired cursor position (movement goal) is used to generate control signals that on a moment-to-moment basis reduce the observed deviation of the current position from the intended position of the cursor. Learning can affect the efficiency with which the feed-back controller generates corrective control signals (Fig. 1a), and it might be subject to age-related changes. For example, age-related slowing could delay visual closed-loop control.

Second, an internal (inverse) model of the transformation, $M(T^{-1})$, can specify the movement parameters without relying on visual feed-back (Fig. 1b). This control mode is called feed-forward control. An internal model is a model of the tool, in this case of the mouse–cursor device, that mimics the input–output relation of the transformation implemented. Thus, it becomes possible to assign movement parameters directly to a desired movement goal based on previous motor experience. In terms of learning, feed-forward control improves as the internal model of the tool becomes more accurate. Changes of the internal model are conceived as reflecting adaptation proper, an implicit change that is not subject to conscious awareness and that cannot easily be modified intentionally.

Finally, there is a third process that can facilitate the adjustment to the transformation of a novel tool. This we call strategic adjustment. It results in strategic, intentional (and thus conscious) corrections of otherwise spontaneously executed movements (Fig. 1c). Examples of such corrections are "past pointing" or "side pointing". In the case of past pointing, having seen an undershoot of the target, one may deliberately aim at a position beyond the target at the next attempt. Strategic corrections are assumed to be based on explicit knowledge of the visuo-motor transformation. In terms of the two-route control scheme that combines feed-forward and feed-back control, as shown in Fig. 1, strategic corrections are deliberate changes of the intended movement goal that serves as an input to both the feed-back controller and the internal model.

Experimental Paradigm

For the experimental study of age-related variations of learning novel visuo-motor transformations, different paradigms have been developed that nevertheless follow the same logic. Typically, a movement of the hand is transformed into a motion of a cursor on a monitor. Transformations applied are mostly concerned with the amplitude or the direction of the movements (cf. Fig. 2). The focus on gain changes and rotations is related to the claim that movements are planned in terms of amplitude and direction (e.g., Vindras and Viviani 2002). In fact, the two types of adaptation have different behavioral characteristics and neural correlates. Only rarely more complex visuo-motor transformations such as that of a sliding first-order lever (cf. Fig. 2) have been studied. Such a transformation combines both, gain changes and rotations, which vary across different target locations.

In our experiments that we review in this chapter, the cursor was presented in a vertical plane on a monitor in about 1 m distance from the eyes of the seated participant, and movements of the right index finger, made in a horizontal plane,

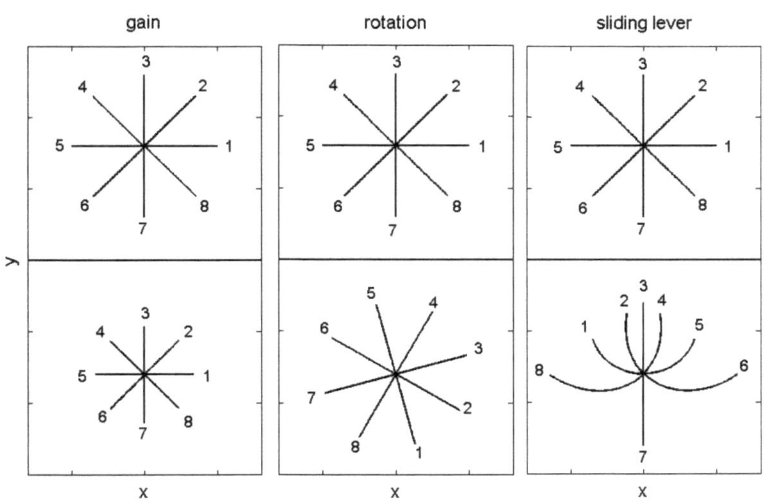

Fig. 2 Types of transformations studied. The *upper* half of each graph shows the typical target configuration with straight paths from the central start location to the targets. The *lower* half of each graph shows the associated movements of the hand. Numbers refer to the different target positions and the associated correct end positions of the hand

were recorded by means of a magnetic tracker. During prolonged periods of practice of the transformation (between 350 and 600 trials), movements to targets arranged on a circle around a central start location were produced. Mostly the cursor was visible during each movement, and the target had to be reached accurately. Other apparatuses and practice regimes have been used in other laboratories. For example, the cursor is often presented in the plane of the hand movements by means of a simple mirror arrangement, and instead of discrete accurate aiming movements, rapid reversal movements which require less accurate target acquisition are utilized during practice (e.g., Bock 2005). These variations, however, have astonishingly little effect on the findings.

The general experimental procedure we used comprised three parts: a pre-test phase, a practice phase, and a post-test phase. In the pre-test phase, baseline performance was measured before the participants experienced the novel tool transformation. In the practice phase, participants practiced a novel transformation (i.e., gain, rotation or complex transformation). In the post-test phase, participants were tested and results were compared against baseline performance. The test phases served to relate the observed learning and its age-related changes to the different mechanisms of learning a novel visuo-motor transformation (i.e., feed-back control, feed-forward control, and strategic corrections, cf. Fig. 1). Each of these mechanisms can be subject to age-related variations.

In the practice phase, visual feed-back of the output of the transformation, i.e., the position of the cursor on the screen, was continuously visible during movement in most experiments, so that all three processes could contribute to performance. However, a rough assessment of changes of visual closed-loop control can be made based on movement time in practice trials with concurrent visual feed-back. Of particular interest is the difference to movement time of approximately the same movements performed in the absence of visual feed-back. In the course of practice, however, movement time can also be reduced because feed-forward control in terms of the internal model becomes more accurate, and because explicit knowledge regarding the transformation becomes available for strategic corrections.

In order to identify age-related variations in the relative contributions of those processes, we used three different types of post-tests without visual feed-back (in the following referred to as open-loop tests because the visual closed loop is broken down by the withdrawal of visual feed-back). Due to the absence of visual feed-back, feed-back control of the tool should not be possible anymore, and thus not contribute to performance in the post-tests. Thereby we were able to assess feed-forward control related to an internal model and strategic corrections related to explicit knowledge without being confounded by feed-back control. In the following, the three tests are described in detail.

Test of Adaptive Shifts

Visual open-loop performance in this post-test reflects both the accuracy of an internal model of the transformation and strategic corrections. In our experiments we have used a measure called *adaptive shift*. For a visuo-motor rotation, for example, movement directions are measured in the pre-test in the absence of the visuo-motor rotation and before any practice with it. After practice, movement directions are measured again in the post-test. During this test participants are informed that the novel visuo-motor transformation is still in effect, even though they do not see the results of their hand movement in terms of cursor motion on the screen. *Adaptive shifts* in this example are the changes of movement direction from the pre-test to the post-test. With a visuo-motor rotation of 30°, for example, an adaptive shift of −30° would indicate perfect adaptation and an adaptive shift of 0° no adaptation at all.

Test of After-Effects

After-effects are considered as a measure of the implicit modification of an internal model that is largely unconfounded by strategic corrections of the movements. Similar to adaptive shifts, after-effects are changes of movement characteristics such as direction from a pre-test to a post-test. In this post-test, however, participants are informed that the visuo-motor transformation is no longer present. Because of the known absence of the transformation, there is no reason for the participants to apply strategic corrections. Thus, any changes that are still observed in this test reflect only the implicit modifications of an internal model. After-effects can also be seen as the residuals of adaptive shifts that remain even when the absence of the visuo-motor transformation is known.

Test of Explicit Shifts

Within our theoretical framework, strategic corrections are based on explicit knowledge of the transformation. Consequently, we routinely used a third type of test to assess the *explicit shift*. In this type of test, participants give verbal judgments of the correct hand movement required for the cursor to reach the target in the presence of the visuo-motor transformation. More specifically, again for the example of a visuo-motor rotation, a line is presented on the monitor with its one end in the start location; it can be rotated around this location, thus pointing in different directions. The participants instruct the experimenter to change the direction in which the line points until this direction matches the direction of the hand movement the participant thinks appropriate for the cursor to reach the target presented in that trial. These judgments are given in a pre-test where participants are informed about the absence of the visuo-motor transformation, and in a post-test where they are informed about the presence of that transformation. The changes of these judgments from pre-test to post-test are defined as explicit shifts.

Age-Related Variation of the Acquisition of Explicit Knowledge

The principal findings on the age-related variation of learning novel visuo-motor transformations consist of three elements:

1. an age-related reduction of the adaptive shift,
2. no age-related reduction of the after-effect,
3. an age-related reduction of explicit knowledge.

These elements had partly been observed in previous studies with somewhat different procedures (Bock 2005; Bock and Girgenrath 2006; Buch et al. 2003; McNay and Willingham 1998). In most of these studies the older participants were between 60 and 80 years of age. Our results revealed that similar changes are already observable at older working age (50–67 years) as compared to young adults (18–30 years). We shall illustrate the general pattern of results for adaptation to a visuo-motor rotation of 75° counter-clockwise (Heuer and Hegele 2008a).

Heuer and Hegele (2008a, Exp. 1a) had their participants practice movements to eight different targets for 480 trials in total. Targets were arranged around a central start position in a distance of 9 cm with angular separations of 45° (Fig. 2, rotation). During practice the initial direction error, measured 200 ms after the start of the movement, declined, and so did movement duration. In the older participants the initial direction error was larger throughout practice, and movement time was longer. The adaptive shifts, after-effects, and explicit shifts are shown in Fig. 3. First, the adaptive shifts were larger for the young participants than for the older ones. Even for the young participants, they were less than −75°, indicating that they did not fully compensate the visuo-motor rotation of 75°. Second, after-effects were considerably smaller than adaptive shifts, and for the young participants they were slightly (but not significantly) smaller than for the older ones. Third, explicit shifts were stronger again for the young participants than for the older ones.

The triad of age-related variation (reduced adaptive shifts, normal after-effects, and reduced explicit shifts) is a fairly robust phenomenon. It strongly suggests that the acquisition of explicit knowledge is impaired at older adult age, but not the acquisition of implicit knowledge (or of an internal model of the transformation). Sometimes the explicit shift in older adults is not even reliably different from zero. In some data sets, the stronger after-effects in older participants compared to

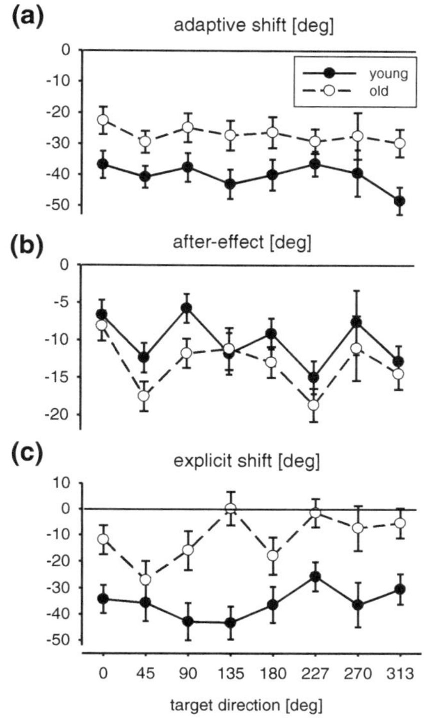

Fig. 3 The triad of age-related variations of the learning of novel visuo-motor transformations. **a** Adaptive shifts are reduced at older working age, **b** after-effects are not affected (or even stronger at older working age), **c** explicit knowledge is reduced (or even absent) in older learners. Error bars mark standard errors (after Heuer and Hegele 2008a)

young ones are statistically significant, which may be associated with an age-related increase of perseveration tendencies.

Even though the triad of age-related variation is a robust phenomenon, it is not found for all types of visuo-motor transformations. These differences are related to the complexity and the magnitude of the transformations. We will review the limiting conditions both for rotation and gain adaptation.

For visuo-motor rotations, the triad of age-related variation (or parts of it) has been found consistently for rotations of 60° and more (Bock 2005; Bock and Girgenrath 2006; Buch et al. 2003; McNay and Willingham 1998). However, with smaller rotations (45° and 30°) the results have been less consistent (Seidler 2006; Heuer and Hegele 2008a, Exp. 2b). Perhaps with these smaller visuo-motor rotations, age-related variations become apparent only beyond age 70. In a study of direction-dependent visuo-motor rotations between 0° and 30°, Heuer and Hegele (2008b) observed no age-related variations of adaptive shifts and after-effects, but of explicit shifts. In this experiment the target directions during practice were 0° (to the right), 45°, and 90° (forward) with the associated rotations of 30°, 16.5°, and 0°.

With respect to changes of the visuo-motor gain (Fig. 2, gain), the observations are similar to the findings for small visuo-motor rotations. Although there is some evidence of a general age-related impairment of gain adaptation at higher senior age (Seidler 2006), such an impairment is essentially absent at older working age (Heuer and Hegele 2007). This is true both for a constant gain change (1.5 instead of 1) and a variable gain change, which depended on the amplitude of the movement ($A_c = 3 \cdot \sqrt{A_h}$, with A_c and A_h as the amplitudes of cursor motions and hand movements, respectively). More specifically, target amplitudes for the cursor were 3, 6, and 9 cm, while the target amplitudes for the hand were 2, 4, and 6 cm with the constant gain and 1, 4, and 9 cm with the variable gain. Even though age-related variations of behavioral adaptation (adaptive shifts and after-effects) were absent under these conditions, there was some indication of a reduced explicit knowledge at older working age.

Whereas we have found no age-related variation of behavioral adaptation for a variable, amplitude-dependent gain change, we have found the full triad of age-related changes when the novel visuo-motor transformation was a direction-dependent change of the visuo-motor gain (Hegele and Heuer 2010a). A direction-dependent change of the visuo-motor gain had been shown to give rise to adaptation only with terminal visual feed-back during practice, but not with continuous visual feed-back (Heuer and Hegele 2008c). In contrast, adaptation to an amplitude-dependent change of the visuo-motor gain has been observed with both types of visual feed-back during practice (Heuer and Hegele 2008d). These findings suggest that learning of a direction-dependent visuo-motor gain might be particularly difficult. Such learning can be avoided as long as visual closed-loop control is possible to master the transformation, as is the case with continuous visual feed-back. However, with only terminal visual feed-back available during practice, performance can only be improved by increasing the accuracy of the internal model of the transformation and by generating explicit knowledge for strategic corrections. For this reason, implicit and explicit learning of the rather difficult direction-dependent gain seems to occur only with terminal visual feedback. Only with this difficult transformation the triad of age-related differences can be observed.

In addition to the basic visuo-motor transformations, we have also examined the complex visuo-motor transformation of a sliding first-order lever, which is similar to the transformation of laparoscopic instruments (cf. Heuer and Sülzenbrück 2009). This transformation is illustrated in Fig. 2 (sliding lever). There are large differences between the directions of the visual targets and the required hand movements. In contrast, the differences between the visual target amplitudes and the amplitudes of the required hand movements are rather small. Finally, the paths of the hand have to be curved in order to produce straight cursor motions. With this complex transformation, there was a reduction of the adaptive shifts of movement directions at older working age, but not of the after-effects (Heuer and Hegele 2009).

For the complex visuo-motor transformation of the sliding first-order lever, there was no test of explicit knowledge. However, there was some indirect evidence of an impairment of explicit knowledge at older working age. It is apparent from Fig. 2 that quite large adaptive shifts of direction are required for some targets (1, 5, 6, 8). For these targets, the observed adaptive shifts of the participants were clearly bimodal, with one mode around the perfect adaptive shift and the other mode around no adaptive shift at all. This all-or-none variation of adaptive shifts strongly suggests an influence of explicit knowledge: either the required large shift of direction is known and produced ("adapters") or not ("non-adapters"). With respect to the age-related variation of the acquired explicit knowledge, Heuer and Hegele (2009) found a smaller proportion of adapters among the older participants than among the young ones, suggesting an age-related decline.

Taken together, the existing evidence strongly suggests that the acquisition of explicit knowledge of novel visuo-motor transformations suffers at older working age. In general, this is associated with a reduced adaptive shift, that is, with a poorer behavioral adaptation. However, behavioral adaptation is reduced only when explicit knowledge is actually used for strategic corrections. It is not used for rather simple novel transformations, such as constant or amplitude-dependent gain changes or small visuo-motor rotations, for which poorer explicit knowledge is thus not associated with poorer behavioral adaptation. However, it is likely that these transformations are affected at older age above 70 years. The reason is that motor control becomes more tightly interwoven with cognitive processes at older age (e.g., Heuninckx et al. 2005). Therefore, the age-related decline of these cognitive processes becomes more likely to affect motor performance with novel visuo-motor transformations.

Explicit Knowledge and Strategic Adjustments

The triad of age-related variation of learning novel visuo-motor transformations includes a behavioral change (i.e., reduced adaptive shifts) and a change of explicit knowledge of the transformation. Adaptive shifts in turn reflect both implicit changes of an internal model of the transformation and strategic corrections based on explicit knowledge. Implicit and explicit adjustments to novel visuo-motor transformations seem to be functionally independent (Mazzoni and Krakauer 2006; Sülzenbrück and Heuer 2009; Taylor and Ivry 2011). Thus, in principle the age-related decline of adaptive shifts, that is, of behavioral adaptation, can be functionally independent of the age-related decline of the acquired explicit knowledge. On the other hand, explicit knowledge can be used for strategic corrections and thereby contribute to the observed adaptive shifts. Therefore, we explored the relation between both kinds of age-related changes, those of adaptive shifts and explicit knowledge, in some detail.

Altogether we have made three types of observation on the relation between age-related variations of adaptive shifts and explicit knowledge. According to the first type of observation, age-related variations of adaptive shifts can be attributed to changes of explicit knowledge. In some experiments we have subdivided the groups of young and older participants by their level of explicit knowledge. An example is shown in Fig. 4 (Heuer and Hegele 2008a, Exp. 1a), where the cutoff between groups with good and poor explicit knowledge was an explicit shift of

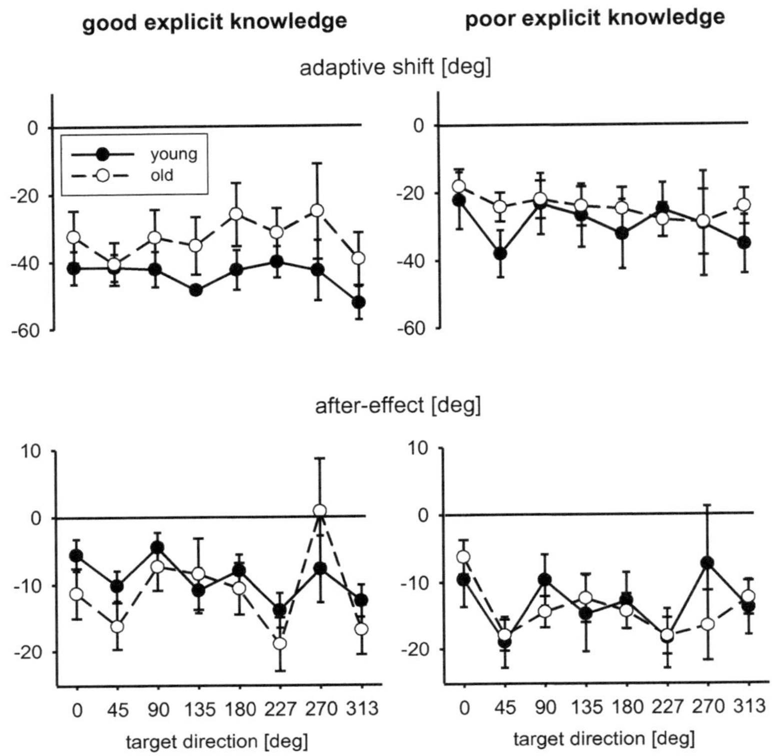

Fig. 4 Adaptive shifts and after effects in young and old participants with good and poor explicit knowledge of a visuo-motor rotation. Error bars mark standard errors (after Heuer and Hegele 2008a)

−25°. The mean explicit shifts were −45° and −37.3° in the young and older participants with a high level of explicit knowledge, and they were −7.5° and 1.4° in the young and older participants with poor explicit knowledge. As is evident from Fig. 4, participants with better explicit knowledge had larger adaptive shifts than participants with poorer explicit knowledge. At both levels of explicit knowledge, however, age-related variations of adaptive shifts were no longer present. In contrast to the adaptive shifts, after-effects did not co-vary with the level of explicit knowledge for both age groups. This finding also suggests that better explicit knowledge is not associated with poorer implicit knowledge, so that the smaller after-effects of younger participants which we observed in some experiments cannot be conceived as a consequence of their better explicit knowledge.

A second type of relation is an age-related reduction of explicit knowledge that is not accompanied by a reduction of behavioral adaptation. This pattern of results appeared with a small direction-dependent visuo-motor rotation (Heuer and Hegele 2008b) and an amplitude-dependent visuo-motor gain (Heuer and Hegele

2007). The third type of relation, finally, is an age-related decline of adaptive shifts that remains even when older and young participants are matched by their explicit knowledge. We have seen this pattern of results when explicit knowledge was particularly good both in the young and older participants, either because during practice only a single target was used (Heuer and Hegele 2008a, Exp. 1b) or because explicit knowledge had been acquired separately prior to the practice of a visuo-motor rotation (Hegele and Heuer, in press).

The findings on the relation between explicit knowledge and behavioral adaptation suggest the following conclusions. First, at older working age, the acquisition of explicit knowledge of visuo-motor transformations suffers, but not yet the acquisition of implicit knowledge, that is, of an internal model of the visuo-motor transformation. Second, when visuo-motor transformations are rather small or simple, strategic corrections based on explicit knowledge are not yet recruited, so that the age-related decline of explicit knowledge remains separated from behavioral adaptation. Third, for larger or more complex transformations, strategic corrections based on explicit knowledge become relevant. Thus, the age-related decline of behavioral adaptation becomes correlated with the age-related decline of explicit knowledge. Fourth, at older age it is not only the acquisition of explicit knowledge that suffers, but also its application for strategic corrections. This age-related change becomes apparent only when explicit knowledge is particularly good. Even though not statistically significant in that experiment, Fig. 4 illustrates the principle that the remaining age-related variation of adaptive shifts is stronger in those young and older participants with good explicit knowledge than in those with poor explicit knowledge. Therefore, in particular for high levels of explicit knowledge, stronger adaptive shifts in young than in older participants can be observed even when the age groups are matched in terms of explicit knowledge.

The Robustness of Age-Related Variations

The consistent finding of poorer acquisition of explicit knowledge of a novel visuo-motor transformation at older working age prompted us to explore simple counter-measures. Of course, the most straightforward and effective counter-measure is the avoidance of visuo-motor transformations, at least of the more complex ones (see chapter Ergonomic Design of Human–Computer Interfaces for Aging Users, Schlick et al.). If this is impossible, the obvious counter-measure is to apply a separate training until explicit knowledge of a novel visuo-motor transformation is more or less perfect, and only thereafter to practice movements with the transformation. In fact, recent findings of Hegele and Heuer (in press) suggest that separate training of explicit knowledge of a visuo-motor rotation could be a successful counter-measure, provided that only the older, but not the young participants, are trained. However, when training of both age groups is matched, the typical age-related variations remain. Some participants, and more among the older ones than among the young ones, even forget or unlearn the previously acquired explicit knowledge about the visuo-motor rotation during practice. A possible reason is that, in particular in older participants, the deviant direction of the hand movement relative to the motion of the cursor is not experienced during practice with the visuo-motor transformation. In fact, awareness of the position of one's own hand is rather limited in the presence of a transformation (e.g., Müsseler and Sutter 2009). Recent findings indicate that this tendency is indeed more prominent in older than in young adults (Rand et al., in press; Wang et al. 2012).

Slightly less direct than separate training of explicit knowledge of a visuo-motor transformation is the presentation of both the target for the cursor and the target for the hand in each trial. In the case of a visuo-motor rotation, the size of the rotation can actually be deduced by the participants based on the angle between the imaginary lines from the start position of the cursor to the two targets. In addition, the hand target enables an immediate explicit adjustment to the visuo-motor rotation, a kind of intentional "side pointing". With a quite similar procedure and a visuo-motor rotation of 45°, Mazzoni and Krakauer (2006) observed basically perfect immediate strategic corrections in young adults. In the course of practice, implicit adjustments developed on top of them, so that the visuo-motor rotation became gradually over-compensated. Taylor and Ivry (2011), in a quite similar task, observed that strategic corrections were reduced as the implicit adjustments grew larger, so that the over-compensation was gradually reduced in the later practice phase.

In our search for counter-measures against the age-related variation of learning novel visuo-motor transformations, we compared participants at young and older working age under conditions in which cursor targets and hand targets were presented concurrently (Hegele and Heuer 2010b). The visuo-motor rotation was −75°, i.e., in a clockwise direction. The initial direction error during practice is shown in Fig. 5a. The initial direction error was measured 200 ms after the start of the movements and is largely unaffected by closed-loop control. Its decline indicates behavioral adaptation to the visuo-motor rotation, similar to the adaptive shifts measured in the post-test. Standard practice conditions, in which only the cursor target was presented, were performed on the second day of the experiment with a visuo-motor rotation of 75° counter-clockwise. The initial direction error for the standard practice conditions is shown in Fig. 5b. As compared with the standard practice, the presentation of the hand target had essentially no effect on performance of the older participants, and young participants showed only a short-lived advantage (i.e., the first block of practice trials).

When two targets were presented concurrently, both groups of participants gave up to exploit the hand target for strategic corrections, the young participants slightly later than the older ones. Perhaps participants were over-burdened by the task to aim at both the hand and cursor targets concurrently in order to reach to the cursor target. As far as adaptive shifts, after-effects, and explicit shifts were concerned, the presentation of the hand targets had little effect. Only the adaptive shifts of the young participants were reduced to a level where they were no longer different from those of the older participants. According to these findings the

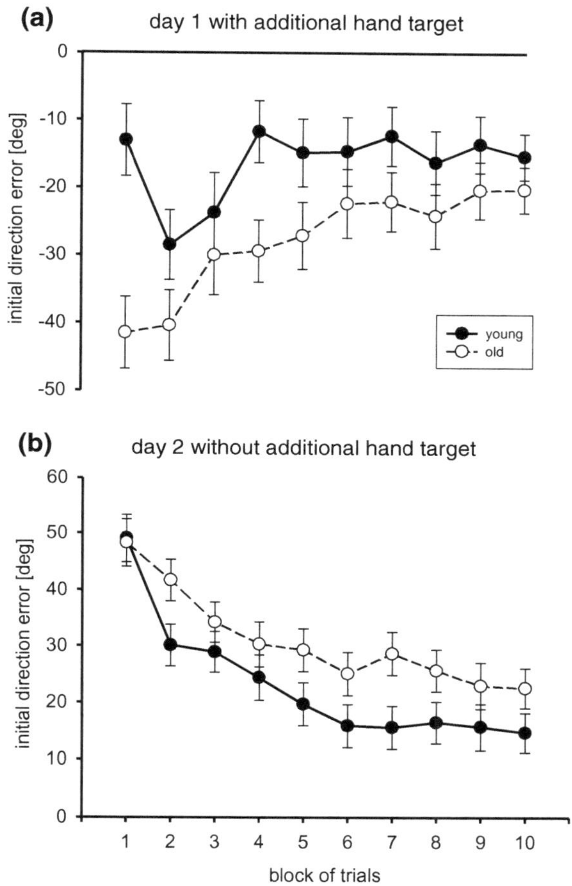

Fig. 5 Initial direction errors during practice with a visuo-motor rotation. **a** On day 1 practice was with a visuo-motor rotation of −75° (*clockwise*) and the concurrent presentation of targets for the cursor and for the hand; **b** on day 2 practice was with a visuo-motor rotation of 75° (*counter-clockwise*) and presentation of only the cursor target. Error bars mark standard errors (after Hegele and Heuer 2010b)

concurrent presentation of cursor targets and hand targets is not an effective counter-measure for the age-related variation of explicit knowledge and adaptive shifts.

The limited benefit of the concurrent presentation of cursor target and hand target not only in older adults, but also in young adults, contrasts with the findings of Mazzoni and Krakauer (2006) and Taylor and Ivry (2011). Most likely this difference is due to the larger rotation angle employed in our study as compared with previous studies. With an increasing angle between the two targets, sharing of attention between them might become more difficult.

Our third attempt to identify counter-measures for the age-related variations was concerned with the mechanical transparency of the visuo-motor transformation (Heuer and Hegele 2010). Humans have the capability of mechanical reasoning, and this capability can be utilized for the operation of mechanical tools, but perhaps not electronic tools. Even for electronic tools, however, visible or imaginable features can be used to create mechanical metaphors, that is, mechanical devices that mimic the transformation of the electronic tool. By such means, the mechanical transparency of the tool should be enhanced, and its use should be facilitated.

In our study we used a virtual sliding lever with the visuo-motor transformation illustrated in Fig. 2. With this virtual device, the dynamic transformation of a physical lever (cf. Heuer and Sülzenbrück 2009) is absent, but the relation between the movement of the finger and the motion of the cursor on the monitor is the same as if the motion of the cursor would correspond to the motion of the tip of a sliding lever. To enhance the mechanical transparency of the tool, its load arm was presented on the monitor in addition to the cursor that marked the tip of the virtual lever. The visible fulcrum was fixed at the bottom of the monitor, so that hand movements were accompanied by appropriate changes of the length and/or orientation of a line emanating from the fixed position. In a control group, however, only the cursor was presented during practice.

Wentink et al. (2002) had shown that the time needed to perform a simulated laparoscopic task was reduced when the shaft of the instrument was visible on the monitor. Even though our aiming movements were simpler than the task of Wentink et al. (2002), we observed faster movement times when the load arm was visible than in the control group, more so in the older than in the young participants. In addition, mean curvature of the cursor paths in both age groups was reduced compared to their respective control groups. However, the age-related variations of adaptive shifts were not affected. Thus, enhanced mechanical transparency facilitates visual closed-loop control, in particular at older working age, but it does not boost the learning of the visuo-motor transformation. To the contrary, at least under certain conditions, it can even result in poorer learning, suggesting a possible trade-off between the facilitation of closed-loop control and the acquisition of an internal model and/or explicit knowledge (Sülzenbrück and Heuer 2012).

Recently Bock and Hagemann (2010) drew attention to the fact that age-related changes observed in typical laboratory tasks may not always match changes observed in everyday tasks. For example, motivational conditions may differ. In typical laboratory tasks, accurate movements are a purpose in themselves, whereas they are typically embedded in a super-ordinate task in everyday life, and thus serve a super-ordinate purpose. Taking such considerations into account, our fourth attempt of identifying a counter-measure was to employ practice conditions in which movements were embedded into a super-ordinate task.

The super-ordinate task in the study (Heuer and Hegele, in preparation) was to hit a ball—presented as a circle on the monitor—so that it reached a virtual hole at a certain distance. For that purpose, participants had to perform rapid hand movements by which the position of a circular cursor on the monitor was controlled. Distances from the start position of the cursor to the ball and to the virtual hole were 3 and 12 cm, respectively; the diameters of the cursor, the ball, and the hole were 6, 12, and 30 mm, respectively. When the cursor hit the virtual ball, the direction of the ball motion was determined by the location of impact. The distance the ball travelled was determined by the velocity at impact, which initially was half the velocity of the cursor and gradually declined in the course of the motion because of simulated friction. In each trial, start position, virtual ball, and virtual hole were aligned. Thus, with a straight motion of the cursor along the line passing the centre of the virtual ball and the hole, the ball would move toward the hole. When a visuo-motor rotation was introduced, this task required accurate visual open-loop performance and straight movements. In spite of these special practice conditions with a super-ordinate purpose, the typical age-related variations were still present in tests of aimed movements performed after practice. Namely, there was an age-related decline of adaptive shifts and explicit knowledge, but no age-related decline of after-effects.

Secondary Effects of the Age-Related Variation of Explicit Knowledge

The age-related reduction of explicit knowledge of novel visuo-motor transformations should not only affect the behavioral adjustments to the transformation practiced, but should also produce secondary effects. We tested this for dual adaptation and transfer across the workspace. At older working age, the reduction or even absence of explicit knowledge in fact abolishes both these phenomena. We shall discuss dual adaptation and generalization in turn.

Dual adaptation refers to the concurrent learning of two different visuo-motor transformations. The problem with this kind of learning is to keep both explicit and implicit knowledge of both transformations separated, so that learning of the one transformation does not distort or abolish the acquired representation of the other one. There are several reports of successful dual adaptation, but also some reports of failures. According to Woolley et al. (2007), distinct motor requirements are a prerequisite of dual adaptation. In our experiment, we tested whether dual adaptation is possible for a task in which motor requirements are the same, but a visual separation of two cursor locations allows differentiating opposite visuo-motor rotations (Hegele and Heuer 2010c).

Participants performed aiming movements during practice of visuo-motor rotations. The start location of the hand was always the same, while the start locations of the cursor were displayed in the left or right half of the monitor. The visuo-motor rotations associated with the right and left visual start locations were 75° clockwise and counter-clockwise, respectively. After movements to the eight targets around the one start location, movements to the eight targets around the other start location were performed. This cycle was repeated throughout practice. In the tests performed after the concurrent practice, additional start locations were used which were midway between the centre of the monitor and the two eccentric start locations. The adaptive shifts, after-effects, and explicit shifts are shown in Fig. 6.

Fig. 6 Adaptive shifts (**a**), after-effects (**b**), and explicit shifts (**c**) after concurrent practice with visuo-motor rotations of 75° (*counter-clockwise*) and −75° (*clockwise*). The different rotations were associated with *left* (for *counter-clockwise*) and *right* (for *clockwise*) eccentric start locations on the monitor; the intermediate *left* and *right* start locations were used only in the test which was performed after concurrent practice. Error bars mark standard errors (after Hegele and Heuer 2010c)

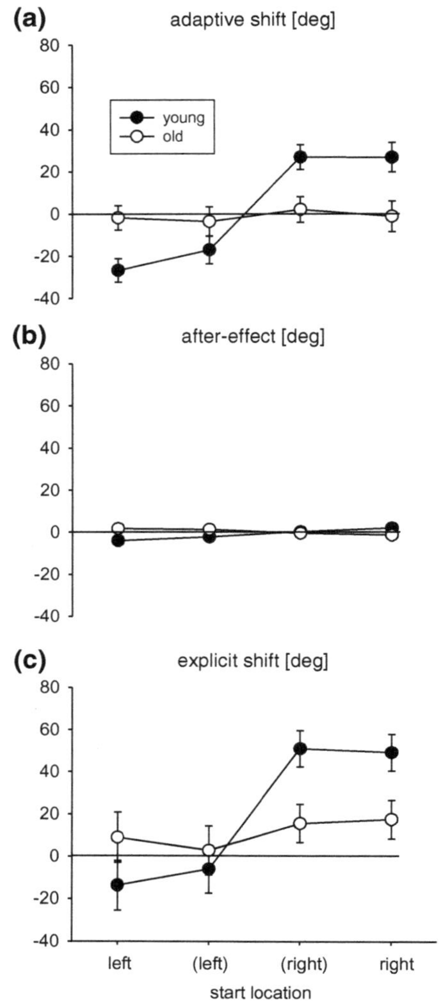

From Fig. 6, it is apparent that dual adaptation was found in the young participants, but not in the older ones. Young participants exhibited adaptive shifts in opposite directions depending on the visual start location on the monitor. With the new intermediate start locations in the tests, the adaptive shifts were essentially the same as with the more eccentric start locations presented during practice. The start-position-dependent adaptive shifts were accompanied by appropriate explicit shifts. The absence of after-effects strongly suggests that dual adaptation under the particular conditions studied was based on explicit knowledge only, whereas implicit knowledge was not acquired. The few young participants who did not acquire explicit knowledge were similar to the older participants with respect to the absence of adaptive shifts.

In the study of generalization of adaptation to visuo-motor transformations across the workspace (Heuer and Hegele 2011), movements were practiced with a particular arm configuration and start location in the right half of the workspace. In the tests performed after practice, a second arm configuration with the start location in the left half of the workspace was used in addition to that in the right half in order to examine the generalization across the workspace. On the monitor, however, the start location was always the same. Practice was carried out with only a single target. With such a practice protocol, implicit learning is known to be restricted to the target direction used during practice and adjacent target directions, whereas explicit learning generalizes across all target directions (cf. Heuer and Hegele 2008a). Consistent with the previous findings, young participants revealed explicit shifts that almost perfectly generalized across the workspace. More specifically, explicit shifts were present at all target directions with both the left and right start locations. In contrast, at older working age explicit shifts were essentially absent. After-effects were present in both age-groups, but only with the start location used during practice. According to these findings, it is only explicit learning that generalizes across the workspace, but not implicit learning. Adaptive shifts in the older participants were quite similar to after-effects, which reflected only implicit learning. Conversely, adaptive shifts in the young participants reflected both implicit and explicit learning.

Outlook

The findings reported in this chapter reveal a fairly robust age-related change of the plasticity of the human brain in learning novel visuo-motor transformations. This change is already observed at older working age, and is likely to become stronger after age 70. The main element of this change is the decline of explicit learning. Even though our attempts to identify practice conditions which counteract the age-related decline were not successful, it is clear that it can be compensated by way of special effort to train explicit knowledge in older learners. However, the same effort invested in the training of young learners is likely to have even larger benefits.

Perhaps the major question that remains is this: Why does the acquisition of explicit knowledge of novel visuo-motor transformations suffer at older working age? And why is the acquired explicit knowledge used only less efficiently for strategic corrections? Our tentative answer at present holds that, with increasing age, there is a progressive functional neglect of somatosensory information on the position of the arm. The acquisition of explicit knowledge of the visuo-motor transformation requires awareness of both the position of the cursor, the output of the transformation, and the position of the hand, the input of the transformation. With limited awareness of the position of the hand, not only the acquisition of explicit knowledge should suffer, but also the application of strategic corrections which requires intentionally "misdirected" hand movements.

As noted by Müsseler and Sutter (2009), the limited awareness of one's own hand position can be functional in the presence of a visuo-motor transformation. In fact, for mirror drawing better performance has been observed in a deafferented patient than in a healthy control group (Lajoie et al. 1992). Such superiority has also been observed when somatosensory information was degraded by repetitive transcranial magnetic stimulation over the contralateral somatosensory cortex (Balslev et al. 2004). At older age, the neglect of somatosensory information and the reliance on visual information might increase. Indeed, an increased reliance on visual feed-back at older age has been reported repeatedly for aiming movements (e. g., Yan et al. 1998). More closely related to the learning of visuo-motor transformations, some recent results (Rand et al., in press; Wang et al. 2012) suggest that older participants encounter more difficulties in judging differences between visual and somatosensory position information in the presence of visuo-motor rotations. Such difficulties found at older working age are perhaps partly caused by age-related deficits of sensory (visual and/or somatosensory) information processing, but partly by a stronger functional neglect of somatosensory information, accompanied by a stronger reliance on visual information.

Acknowledgments The research reported in this chapter has been supported by the DFG, grant He 1187/15. We are grateful to Nathalie Brackmann, Jan Delkus, Alexander Fölling, Eva Hanisch, Sandra Hardmann, Barbara Herbst, Marc Hülsey, Maia Iobidze, Thorsten Klar, Steven Li, Julia Neuhaus, Eckhard Rückemann, Helene Schneider, Jennifer Stube, and Aljessa Tschishova for their support in setting up and running the experiments.

References

Balslev D, Christensen LOD, Lee J-H, Law I, Paulson OB, Miall RC (2004) Enhanced accuracy in novel mirror drawing after repetitive transcranial magnetic stimulation-induced proprioceptive deafferentation. J Neurosci 24:9698–9702

Bock O (2005) Components of sensorimotor adaptation in young and elderly subjects. Exp Brain Res 160:259–263

Bock O, Girgenrath M (2006) Relationship between sensorimotor adaptation and cognitive functions in younger and older subjects. Exp Brain Res 169:400–406

Bock O, Hagemann A (2010) An experimental paradigm to compare motor performance under laboratory and under everyday-like conditions. J Neurosci Methods 193:24–28

Buch ER, Young S, Contreras-Vidal JL (2003) Visuomotor adaptation in normal aging. Learn Mem 10:55–63

Charness N, Holley P, Feddon J, Jastrzembski T (2004) Light pen use and practice minimize age and hand performance differences in pointing tasks. Hum Factors 46:373–384

Fouquet C, Petit GH, Auffret A, Gaillard E, Rovira C, Mariani J, Rondi–Reig L (2011) Early detection of age-related memory deficits in individual mice. Neurobiol Aging 32:1881–1895

Gallagher AG, McClure N, McGuigan J, Ritchie K, Sheehy NP (1998) An ergonomic analysis of the fulcrum effect in the acquisition of endoscopic skills. Endoscopy 30:617–620

Hegele M, Heuer H (2010a) Adaptation to a direction-dependent visuomotor gain in the young and elderly. Psychol Res 74:21–34

Hegele M, Heuer H (2010b) The impact of augmented information on visuo-motor adaptation in younger and older adults. PLoS ONE 5(8):e12071

Hegele M, Heuer H (2010c) Implicit and explicit components of dual adaptation to visuomotor rotations. Conscious Cogn 19:906–917
Hegele H, Heuer H (in press) Age-related variations of visuo-motor adaptation result from both the acquisition and the application of explicit knowledge. Psychol Aging
Heuer H, Hegele M (2007) Learning new visuo-motor gains at early and late working age. Ergonomics 50:979–1003
Heuer H, Hegele M (2008a) Adaptation to visuo-motor rotations in younger and older adults. Psychol Aging 23:190–202
Heuer H, Hegele M (2008b) Adaptation to direction-dependent visuo-motor rotations and its decay in younger and older adults. Acta Psychol 127:369–381
Heuer H, Hegele M (2008c) Constraints on visuo-motor adaptation depend on the type of visual feedback during practice. Exp Brain Res 185:101–110
Heuer H, Hegele M (2008d) Adaptation to a nonlinear visuo-motor amplitude transformation with continuous and terminal visual feedback. J Mot Behav 40:368–379
Heuer H, Hegele M (2009) Adjustment to a complex visuo-motor transformation at early and late working age. Ergonomics 92:1039–1054
Heuer H, Hegele M (2010) The effects of mechanical transparency on adjustment to a complex visuo-motor transformation at early and late working age. J Exp Psychol Appl 16:399–412
Heuer H, Hegele M (2011) Generalization of implicit and explicit adjustments to visuo-motor rotations across the workspace in younger and older adults. J Neurophysiol 106:2078–2085
Heuer H, Hegele M Age-related variation of visuo-motor adaptation during motor-skill learning (in preparation)
Heuer H, Massen C (2012) Motor control. In: Healy AF, Proctor RW (Vol. Eds.) Weiner IB (Ed.-in-Chief) Handbook of psychology, vol 4. Experimental psychology. Wiley, Hoboken, pp 320–354
Heuer H, Sülzenbrück S (2009) Trajectories in operating a hand-held tool. J Exp Psychol Hum Percept Perform 35:375–389
Heuninckx S, Wenderoth N, Debaere F, Peeters R, Swinnen SP (2005) Neural basis of aging: the penetration of cognition into action control. J Neurosci 25:6787–6796
Lajoie Y, Paillard J, Teasdale N, Bard C, Fleury M, Forget R, Lamarre Y (1992) Mirror drawing in a deafferented patient and normal subjects: visuoproprioceptive conflict. Neurology 42:1104–1106
Lee D, Kwon S, Chung MK (2012) Effects of user age and target-expansion methods on target-acquisition tasks using a mouse. Appl Ergonomics 43:166–175
Li SC, Lindenberger U, Sikström S (2001) Aging cognition: from neuromodulation to representation. Trends Cogn Sci 5:479–486
Mazzoni P, Krakauer JW (2006) An implicit plan overrides an explicit strategy during visuomotor adaptation. J Neurosci 26:3642–3645
McNay EC, Willingham DB (1998) Deficit in learning of a motor skill requiring strategy, but not of perceptuomotor recalibration, with aging. Learn Mem 4:411–420
Müsseler J, Sutter C (2009) Perceiving one's own movements when using a tool. Conscious Cogn 18:359–365
Myerson J, Hale S, Wagstaff D, Poon LW, Smith GA (1990) The information-loss model: a mathematical theory of age-related cognitive slowing. Psychol Rev 97:475–487
Neumayer LA, Gawande AA, Wang J, Giobbie-Hurder A, Itani KMF, Fitzgibbons Jr RJ, Reda D, Jonasson O, CSP #456 Investigators (2005) Proficiency of surgeons in inguinal hernia repair: effect of experience and age. Ann Surg 242:344–352
Proctor RW, Vu KL, Pick DF (2005) Aging and response selection in spatial choice tasks. Hum Factors 47:250–270
Rand MK, Wang L, Müsseler J, Heuer H Vision and proprioception in action monitoring by young and older adults (in press). Neurobiol Aging
Risucci D, Geiss A, Gellman L, Pinard B, Rosser J (2001) Surgeon-specific factors in the acquisition of laparoscopic surgical skills. Am J Surg 181:289–293

Sandfeld J, Jensen BR (2005) Effect of computer mouse gain and visual demand on mouse clicking performance and muscle activation in a young and elderly group of experienced computer users. Appl Ergonomics 36:547–555

Sawaki L, Yaseen Z, Kopylev L, Cohen L (2003) Age-dependent changes in the ability to encode a novel elementary motor memory. Ann Neurol 53:521–524

Seidler RD (2006) Differential effects of age on sequence learning and sensorimotor adaptation. Brain Res Bull 70:337–346

Seidler RD, Stelmach GE (1995) Reduction in sensorimotor control with age. Quest 47:386–394

Smith MW, Sharit J, Czaja SJ (1999) Aging, motor control, and the performance of computer mouse tasks. Hum Factors 41:389–396

Sülzenbrück S, Heuer H (2009) Functional independence of explicit and implicit motor adjustments. Conscious Cogn 18:145–159

Sülzenbrück S, Heuer H (2012) Enhanced mechanical transparency during practice impedes open-loop control of a complex tool. Exp Brain Res 218:283–294

Szafran J (1951) Changes with age and with exclusion of vision in performance at an aiming test. Quart J Exp Psychol 3:111–118

Taylor JA, Ivry RB (2011) Flexible cognitive strategies during motor learning. PLoS Comput Biol 7(3):e1001096

Teeken JC, Adam JJ, Paas FGWC, van Boxtel MPJ, Houx PJ, Jolles J (1996) Effects of age and gender on discrete and reciprocal aiming movements. Psychol Aging 11:195–198

Vindras P, Viviani P (2002) Altering the visuomotor gain. Evidence that motor plans deal with vector quantities. Exp Brain Res 147:280–295

Walker N, Millians J, Worden A (1996) Mouse accelerations and performance of older computer users. In: Proceedings of the human factors and ergonomics society 40th annual meeting, Human Factors and Ergonomics Society, Santa Monica, CA, pp 151–154

Walker N, Philbin DA, Fisk AD (1997) Age-related differences in movement control: adjusting submovement structure to optimize performance. J Gerontol: Psychol Sci 52B:P40–P52

Wang L, Sutter C, Müsseler J, Dangel RJZ, Disselhorst-Klug C (2012) Perceiving one's own limb movements with conflicting sensory feedback: the role of mode of movement control and age. Front Psychol 3:289

Welford AT (1981) Signal, noise, performance, and age. Hum Factors 23:97–109

Wentink M, Breedveld P, Stassen LPS, Oei IH, Wieringa PA (2002) A clearly visible endoscopic instrument shaft on the monitor facilitates hand-eye coordination. Surg Endosc 16:1533–1537

Woolley DG, Tresilian JR, Carson RG, Riek S (2007) Dual adaptation to two opposing visuomotor rotations when each is associated with different regions of workspace. Exp Brain Res 179:155–165

Yan JH, Thomas JR, Stelmach GE (1998) Aging and rapid aiming arm movement control. Exp Aging Res 24:155–168

Influence of Age and Expertise on Manual Dexterity in the Work Context: The Bremen-Hand-Study@Jacobs

Claudia Voelcker-Rehage, Eva-Maria Reuter, Solveig Vieluf and Ben Godde

Lessons Learned

Demographic change requires the promotion of horizontal careers, i.e., the training in and the change into different jobs with less physically demanding tasks. For this purpose, it is necessary to investigate which abilities are preserved or can be learned a new in older workers and thus which jobs and tasks are suitable for this target group. Tasks which require a high amount of manual dexterity like using precision tools (e.g., forceps, tweezers and screwdrivers), mounting of computer boards or soldering of electronic devices are prototypical tasks of the work context. Manual tasks require much less physical capacities than those involving the whole body, like lifting or carrying, and thus provide a potential work field for older employees. Although manual dexterity is a typical field where on average age-related decline can be observed, it has been shown that at the same time with adequate training this decline can be reduced or even prevented and that older adults remain able to learn such tasks. We investigated age-related differences in manual dexterity during the working life span with a special focus on precision

C. Voelcker-Rehage (✉) · B. Godde
Jacobs Center on Lifelong Learning and Institutional Development and AgeAct Research Center, Jacobs University, Bremen, Germany
e-mail: c.voelcker-rehage@jacobs-university.de

B. Godde
e-mail: b.godde@jacobs-university.de

E.-M. Reuter · S. Vieluf
Jacobs Center on Lifelong Learning and Institutional Development, Jacobs University, Bremen, Germany
e-mail: e.reuter@jacobs-university.de

S. Vieluf
e-mail: s.vieluf@jacobs-university.de

C. M. Schlick et al. (eds.), *Age-Differentiated Work Systems*,
DOI: 10.1007/978-3-642-35057-3_17, © Springer-Verlag Berlin Heidelberg 2013

grip control and tactile performance on a behavioral and neurophysiological level. Furthermore, we were interested whether work-related expertise due to continuous and elaborated use might postpone age-related changes. The core findings of our research project are:

- Manual dexterity declines already throughout the working life span. Also age-related differences in underlying neurophysiological correlates can be observed in middle-aged adults.
- Differentiating tactile stimuli on a comparable level to younger adults seems to require more frontal (i.e., cognitive) brain resources in older workers. This might have consequences for complex work situations requiring dual-tasking or task switching.
- Expertise seems to counteract age-related decline in finger force control and to postpone age effects for about ten years.
- Manual expertise, however, has highly specific effects in the domain of finger force control and only limited transfer to the domain of tactile perception can be observed. Expertise effects rather occur in complex than in simple tasks.
- Some differences in manual dexterity seem to be related to age rather than to reduced training or use as both experts and novices show age-related reductions in tactile perception and fine motor control.
- Especially older workers may benefit from short term interventions: age-related performance decline in tactile perception and underlying brain changes can be at least partly reversed whereas fine motor control of both young and older workers can be enhanced through practice.

Our results let us conclude that changes from physically demanding jobs into jobs requiring manual dexterity can be successfully implemented. However, systematic task-related training is required to counteract age-related decline in manual dexterity that might hinder such horizontal career tracks.

Introduction

Against the background of the demographic change (see Introduction and Overview, Schlick et al., this volume), continuously changing work demands and how they can be met by older employees require increasing attention.

With average absence rates between 3 and 6 % across 27 EU Member States and Norway, costs of disability and sickness benefits are estimated to be about 2.5 % of the GDP and thus 2.5 times those of unemployment (Edwards and Greasley 2010). In Germany, particularly in the production sector with predominantly physically demanding jobs, the number and duration of cases of illness show the second highest numbers after the service sector, where mental diseases are the main issue (Kraemer 2010). Besides impairments by skin and respiratory diseases or noise, injuries caused by lifting or carrying heavy weights contribute most to occupational diseases. As a consequence of the demographic development, a further increase of these numbers must be expected if older employees are exposed to physically demanding jobs for a longer time despite their decreasing physical abilities.

Statistical data for Germany show that in the year 2003 musculo-skeletal and cardiovascular diseases were responsible for about 37 % of all early retirement cases of men (women: 26.5 %; Gesundheitsberichterstattung des Bundes 2006). Moreover, labor unions and employers' mutual insurance associations expect that with increasing demands of the modern work environment aging and wearing processes will accelerate and therefore jobs cannot be fulfilled a whole working life any more. This will be particularly true for jobs with high physical demands.

Alternatives for older employees are provided by so-called adapted work places, which take account of the changed abilities or by vertical career trajectories into leading functions where the older employees can use their experiences and forward their knowledge to younger colleagues. However, the number of adaptive work places is limited, thus the probability for early retirement or long-term diseases increases. In addition, vertical careers are possible for part of the cohort of older workers only.

In terms of an age-differentiated career development strategy it is therefore necessary to promote horizontal careers, i.e. the training in and the change into different jobs with less physically demanding tasks also at higher age ("lifelong learning"). Therefore a career planning that allows early leave of physically highly demanding jobs and the adoption of new tasks might be promising and desirable. For this purpose, it is necessary to investigate which abilities are preserved or can be learned anew in older workers and thus which jobs and tasks are suitable for older workers.

Tasks which require a high amount of manual dexterity are prototypical tasks in a work context suitable for older workers. This includes the use of precision tools such as forceps, tweezers and screwdrivers, the mounting of computer boards as well as the soldering of electronic devices. Manual tasks require much less physical capacities than those involving the whole body like lifting or carrying and thus provide a potential work field for older employees. It might also be expected that a change from a physically demanding job to a technical position with fine motor demands is much more attractive to many older production employees than changing into jobs where mostly mental abilities are required. Moreover, it is known that older adults pay more attention to working accurately than working fast, as it is required in most manual jobs. Although manual dexterity is a typical field where on average age-related decline can be observed, it has been shown that at the same time with adequate training this decline can be reduced or even prevented and that older adults remain able to learn such tasks (Dinse et al. 2008; Voelcker-Rehage 2008).

Current State of Research

The manipulation of small objects with the hand (grasping, moving, and release) as demanded at various work places in production or manual jobs requires the dynamic and adaptive control of isometric forces as well as the exact haptic perception of those objects (Flanagan and Wing 1993). The precise adaptive control of fine finger forces allows the very fine modulation of small objects with very high precision. Herewith the actual grip force at each time point of the movement is only a small amount higher than the force required preventing the object from gliding through the hand. Too high grip forces might disturb this fine modulation or even lead to damage of sensitive objects. The amount of force required for object manipulation is related to object properties like mass, size, shape, and surface textures (Westling and Johansson 1987). The efficient regulation of grip force also depends on the sensitivity of the fingers. Cooling, anesthesia, or wearing gloves reduce haptic performance and lead to massive increase of grip forces (Monzée et al. 2003). Thus tactile discrimination is an essential requirement for the manipulation of objects. Here, not only spatial but also temporal discrimination is required to identify and distinguish shapes and textures of an object.

Recent studies on sensorimotor interactions also revealed that the excitability of cortical motor neurons can be influenced by somatosensory input, e.g. following peripheral nerve stimulation (Tecchio et al. 2006). Thus, a continuous information exchange between somatosensory cortical areas processing tactile and proprioceptive information and those brain areas involved in finger and hand movement control is essential for manual dexterity.

In our understanding, manual dexterity comprises the precise control and coordination of finger movements and fine finger forces and the related sensory and haptic abilities to perceive and manipulate objects (Johansson and Westling 1984). Common methods to investigate age-related changes in force control are so-called precision grip tasks. During these tasks, subjects apply fine grip forces between the tips of the index or middle finger and the thumb (Vaillancourt and Newell 2003). Generally, participants are provided with visual feedback regarding how closely their force output matches the target force level. Tactile perception typically is measured by threshold determination and discrimination tasks.

Other lines of research in this field focus on hand movements during reaching for objects or manipulating and moving objects and the adaptation to disturbances like altered force fields or visual feedback conditions (see Chap. Age-Related Variations in the Control of Electronic Tools, Heuer et al.). This is of particular interest since several work places require the use of precision tools under indirect visual feedback, as for instance surgeons do when performing an endoscopy.

Age-Related Differences in Manual Dexterity

A decline in manual dexterity is common in older adults and has been demonstrated to account for much of the observed impairment in everyday tasks, like pouring milk into a cup, preparing meals, or retrieving coins from a purse. Also work-related tasks, like tighten bolts, bending glasses or assembling printed circuit boards might be impaired by the decline of manual dexterity. Aiming at the understanding of the underlying mechanisms, the investigation of the regulation and coordination of fingertip forces has been given lot of attention during the last decades (Diermayr et al. 2011).

It has been repeatedly shown that movements of the upper extremities of older adults are slower and less precise than those of younger subjects (Cole 1991; Voelcker-Rehage and Alberts 2005; Vieluf et al. 2013). It is also known that older adults produce grip forces that are about two times higher than those of younger adults if it is required to grasp and hold objects of different weights and surface textures (Cole 1991). As possible causes, age-related changes on the perceptual, motor, and neuromuscular level are discussed.

Also tactile sensitivity is increasingly impaired with older age (Dinse et al. 2008; Reuter et al. 2012) and deficits in tactile sensitivity and perception and therefore in sensorimotor feedback loops play an important role for age-related decline in manual dexterity (Nowak et al. 2003). It has been shown, for example, that the judgment about the necessary force to grip an object is influenced by tactile cues which are transmitted by the contact with the object surface (Jones and Piateski 2006). Even if the temporal coordination between grip force and movement-induced load remains the same, wrong or imprecise sensory feedback may lead to unskillfulness in using the hands (Hermsdörfer and Blankenfeld 2008).

Only few studies addressed differences in fine motor and tactile abilities in middle-aged adults. Dinse and colleagues (2008) reported that the tactile 2-point discrimination threshold in a group of 47–55 years old adults was in between the thresholds of young adults (20–30 years of age) and adults 66 years and older. Lindberg and colleagues (2009) revealed that the control of the fine finger forces declines already in middle-aged adults. Moreover, to compensate these degenerative changes before a decline in manual dexterity becomes visible, those middle-aged adults modify their force generation strategy: they initiate the movement slower and focus more on accuracy and the prevention of overshooting forces.

Neurophysiological Correlates of Age-Related Differences in Manual Dexterity

The analysis of neurophysiological differences between age groups allows the characterization of processes underlying age-related decline in manual dexterity, the identification of the potentials of older adults as well as the development of possible training and intervention strategies.

Studies on neurophysiological correlates of age-related differences in force modulation and control of grasping movements are rare. Most studies on age-related decline in motor control have been done with finger tapping tasks and revealed more widespread activations in motor-related brain areas with increasing age (Calautti et al. 2001). However, this seems to be the case for only those older adults with performance levels similar to young adults (Reuter-Lorenz et al. 2000). Therefore such reorganization in cortical and subcortical networks might be interpreted as compensatory processes.

Ward and Frackowiak (2003) used functional magnetic resonance imaging (fMRI) to examine age-related differences in a kinetic grasping task. They concluded that in the course of the aging process the integration of visuo-spatial and sensorimotor processing becomes less effective and that an adaptive and plastic motor network is able to compensate these changes and allows performance comparable to young adults. Also Heuninckx and colleagues (2005) showed that older as compared to younger adults activate more sensorimotor and frontal brain areas and that these additional activations correlate with motor performance thus indicating compensation. Moreover, older adults showed stronger activations in brain areas related to the cognitive processing interpreted as a shift from automated to more cognitively controlled movement execution.

In daily life and also at the work place, often bimanual tasks that require the interaction of both hands have to be performed. Thereby not only the action performed by each hand can differ, but also the timing between the hands. Depending on the task, different amounts of motor attention and motor control are likely to be required, different patterns of brain activity can be expected and for instance can be measured with electroencephalography (EEG). It is known that during motor performance alpha power of EEG signals decreases, indicating changes in neural activity, and that alpha coherence of EEG signals, as indicator of the coupling between brain areas, increases (Manganotti et al. 1998). These changes are the stronger the more complex the task is. As revealed by Serrien and colleagues (2004), particularly, EEG coherence in the alpha (8–12 Hz) and beta band (13–30 Hz) is modulated during bimanual motor tasks. Task-related coherence—that is the difference in coherence between a rest condition and the motor task—in the alpha band mainly depends on task demands and processing load whereas task-related coherence in the beta band is related to motor processing itself (Serrien et al. 2004). It has also been shown that alpha band power and coherence are particularly age sensitive (Polich 1997). That is why we focused in our study on age- and expertise-related differences in the alpha band, with respect to brain activity and functional coupling.

Performance decline in the tactile domain is related to reduced sensitivity of mechanoreceptors in the skin and slower nerve conduction speed as well as to reduced cortical activation amplitudes and distortions of the topographic representations of the hand (Dinse et al. 2008; Godde et al. 2002). Dependent on the cognitive load of the respective task, differences are observed in the periphery (e.g., sensitivity), in the primary somatosensory cortex (simple spatial and temporal discrimination) or in higher association areas (pattern or object discrimination) (Dinse et al. 2008). EEG-studies on age-related differences during sensory processing revealed that predominantly the P3-component, a component of the event-related brain potential that is known to relate to attentional resource allocation during conscious signal processing, is sensitive to aging and is reduced particularly at parietal electrode sites (Polich 1997).

Role of Expertise for Performance of Older Adults

Studies on work-related expertise revealed that with increasing experience older workers acquire strategies to compensate for reduced reaction times and impaired memory performance. However, in the cognitive domain it seems that older experts are not able to outperform their younger colleagues by use of these strategies (Lindenberger et al. 1992). The effect of acquired expertise on manual dexterity is not well understood. On the motor (Furuya and Kinoshita 2008) as well as on the sensory level (Elbert et al. 1995) acquired expertise seems to positively influence baseline performance and learning outcome. Petrofsky and Lind (1975) showed that maximum hand grip force does not decline in the course of the working life if the work requires a high amount of grip force. Furthermore, neurophysiological correlates differed between experts and novices of the same age indicating a higher neural efficiency in experts (Del Percio et al. 2009). That is, experts reveal more focused and/or increased recruitment of specialized brain regions and less activation or inhibition of unspecific and irrelevant brain areas in expertise specific tasks (Babiloni et al. 2008), so that information can be processed faster and more accurately. For musicians and blind people, who reveal superior performance in tactile and auditory tasks as compared to non-musicians or sighted controls, it has been shown that their specific expertise goes along with an expansion in the respective cortical areas (Elbert et al. 1995). Movement expertise, on the contrary, is often related to more focal activations in specific motor cortical areas (Del Percio et al. 2009).

Aim of the Project

The aim of this project was to empirically examine the significance of age differences in fine motor control and tactile performance with regard to manual dexterity of older workers and to identify the mechanisms which underlie the decrease in manual dexterity during the working life. With study 1 we aimed to characterize in detail fine motor and haptic performance in different age groups. Electroencephalography (EEG) was used to examine age-related differences on the neurophysiological level in selected motor and tactile tasks. Experiments of study

1 were also designed to analyze the role of experience and expertise for the maintenance of manual dexterity in older workers. Moreover, they should reveal if preserved levels of performance are based on compensatory brain activation patterns rather than on "youth-like" brains, that is, only little changed cortical representations and activation patterns. In study 2 we examined the learning capacity of young and older workers. For this purpose we applied a motor short-term practice and a tactile stimulation intervention and asked if in older employees similar performance increases could be induced as shown for younger adults.

Methods

Sample

In total, 96 right handed subjects participated in two studies. All participants were healthy and had normal or corrected to normal vision. We excluded persons with any hobbies requiring the dexterous use of the hand or fingers, e.g. needle work or music making.

Participants were screened for demographic information that included their educational background, weekly working hours, hand usage regarding typing and handwriting, and health. Clinical manual dexterity was assessed using the Purdue Pegboard Test (Model 32020, Lafayette Instruments, Lafayette, IN, USA). As expected, clinical manual dexterity differed significantly between the age groups ($p < 0.05$). Older adults placed less pegs than younger adults.

Sample selection and prescreening were necessary to separate real age effects from those related to pathological alterations in the hand and nervous system and to distinguish effects of work-related expertise from those induced by hand use during leisure time.

In **study 1**, participants ($n = 72$) were of three age groups; 'young' (18–25 years of age), 'middle-aged' (34–46 years of age) and 'older' (55–65 years of age). Middle-aged and older participants were further subdivided depending on their occupational field, which was used as the criterion for expertise in manual dexterity. Service employees, for instance, office clerks, consultants and students, formed the group of *novices*. Opticians (performing also ophthalmic laboratory technical tasks), goldsmiths, dentists, dental technicians, watch makers, and hearing care professionals, who had at least 10 years of work-related experience, formed the group of *experts*. By use of a detailed questionnaire, experts were proven to use their hands more frequently than novices. We did not include experts in the young age group, because young subjects had not gained 10 years of work-related experience.

Consequently, we investigated five subsamples: 'novices young' (NY) ($n = 17$, 9 females, mean age $= 20.8$, $SD = 1.7$), 'novices middle-aged' (NM) ($n = 14$, 9 females, mean age $= 40.6$, $SD = 3.9$) and 'novices older' (NO) ($n = 15$, 8 females, mean age = 58.2, SD = 2.9) 'experts middle-aged' (EM) (n = 12, 7 females, mean age = 41, SD = 3.8), and 'experts older' (EO) (n = 14, 8 females, mean age = 57.9, SD = 2.9).

In **study 2**, 12 young (25–35 years, 6 female, M = 29.67, SD = 2.87) and 12 older (54–65 years, 7 female, M = 57.75, SD = 3.65) participants (all novices) were assessed.

Measure of Isometric Force Modulation

A six degree of freedom force transducer (Mini-40 Model, ATI Industrial Automation, Garner, NC, USA; Fig. 1) that was affixed to a table in front of the participant in a comfortable position was used for measuring grip force during a force-tracking motor task. The target force level and the actual grip force produced by the subjects were displayed on a screen.

This device was used to assess performance parameters in different precision grip tasks. To assess maximum voluntary contraction, participants used a precision grip (i.e. thumb and index finger only) of their dominant (right) hand to exert an isometric force against the force transducer.

Sinusoidal tracking tasks. Participants were instructed to match their grip force to the target as accurately as possible. Four different target sine waves were selected as target force profiles in Experiment 1. They differed in two parameters: the force range and the speed of the target sine waves. Force levels were either relative to the individual maximum voluntary contraction (MVC) (10–20 %), or fixed (2–12 N). As previous studies have shown, these force levels can be maintained relatively easily and they evoke no fatigue (Voelcker-Rehage and Alberts 2005). Two different frequencies were used: a commonly used slower frequency of 0.2 Hz (Voelcker-Rehage and Alberts 2005) and 1 Hz as a faster condition. Consequently, the four conditions were: fast/fixed, fast/relative, slow/fixed, slow/relative.

Bimanual tracking tasks. To study the effects of task complexity on brain potentials (EEG), participants performed two bimanual force modulation tasks as described above (force range: 2–12 N, frequency: 0.5 Hz, trial duration: 30 s, number of trials: 7) presented independently for both hands. The tasks differed in the starting point of the target force for the right hand. Either both profiles started at the minimum and consequently moved symmetrically, or, in the asymmetric condition, the left hand started at the minimum and the right hand at the maximum, so that the profiles run contradictory.

Measures of Tactile and Haptic Performance

Six experimental tactile tasks and one haptic task were applied to the participants. In the tactile tasks the hand was held out passively and different stimuli were presented on the tip of the right or left index finger (Fig. 1). In the haptic task stimuli were actively explored with the right hand. In all the tasks described in the following, the subjects' sight was shielded by a curtain so that they were not able to see their hand or the applied stimulus.

Touch detection Threshold (TDT) was defined by the use of custom made, calibrated von-Frey-filaments, which are thin nylon threads of different thickness with plastic handles. These filaments are a standardized and common tool to measure the lightest detectable touch. Each filament represents a pressure force ranging on a logarithmic scale from 18 to 6500 mg, 0.177 from 63.743 mN, respectively.

A *frequency discrimination task* (FDT) was conducted with a piezoelectric waver (piezo: TeleSensory, MountainView, Ca; casing and controller: metec AG, Stuttgart, Germany). The waver contained eight plastic pins in two rows which could be moved individually. Controlled by the software Presentation (Neurobehavioral Systems, Albany, Ca) all pins together were driven to vibrate at a given frequency against the fingertip of the participant. Each trial consisted of a pair of vibrations of either a 20 Hz reference frequency or a test frequency between 10 and 19.5 Hz. After each trial the subject had to decide as quickly and accurately as possible, whether the first or second stimulus was faster and thus had a higher frequency (Voelcker-Rehage and Godde 2010). In all tasks where the waver was used, the response of the subject was indicated by a button press with the index or middle finger, respectively, of the non-stimulated hand.

In order to define the spatial discrimination ability on the fingertip, hemispherical plastic domes (JVP Domes, Stoelting, Wood Dale, IL, USA) were used for a *grating orientation task* (GOT). The surface was interrupted by a grating made of bars and grooves of equal width varying in their spacing. The experimenter applied the domes perpendicularly to the subject's finger. The participant had to indicate whether the gratings were parallel or orthogonal to the axis of the finger.

Whereas the GOT was used in study 1, a different orientation discrimination task was performed with the piezoelectric waver in study 2 (*Piezoelectric orientation task,* POT). Participants had to discriminate if two pins of the piezoelectric waver applied to the tip of the left index finger were oriented along the finger axis or perpendicular to it. Stimuli were applied either to a trained area at the distal part of the fingertip or to a control area more proximal at the same fingertip.

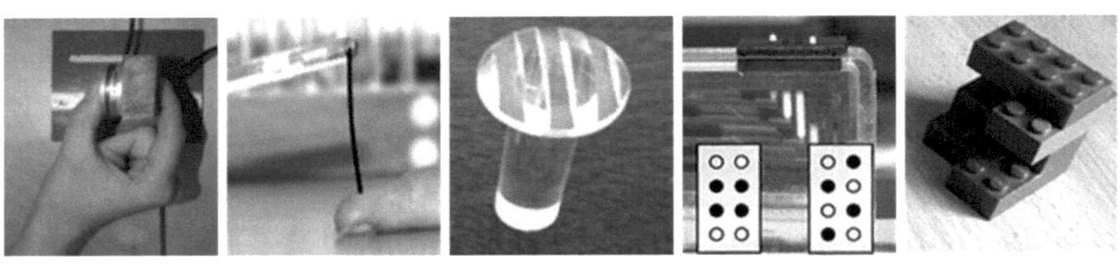

Fig. 1 Experimental devices, from *left* to *right*: Force transducer grasped in precision grip, von-Frey-filaments, hemispherical plastic domes, piezoelectric waver with two exemplary patterns and cubic objects made of LEGO™ bricks

For the *pattern discrimination task* (PDT) six different patterns were built with the pins of the piezoelectric waver. All patterns consisted of four of the eight pins. The task was to decide whether two patterns were the same or different.

A *variation of the PDT* was combined with EEG. Subjects performed a tactile two-choice discrimination task with unequal occurrence ratio of the stimuli provided with the piezoelectric waver to the tip of their left index finger. Either a straight line consisting of four pins or a zigzag line was presented. In about 80 % of the trials the same stimulus was presented which are referred to as standard trials. In 20 % of the trials the respective other stimulus was presented and these trials are referred to as deviant trials.

Cubic objects made from four LEGO™ bricks were used for an *object discrimination task* (ODT). The objects differed in one brick of the object. The subject received two objects consecutively and had to decide whether these two objects were the same.

Training Paradigms

To analyze the effect of short-term practice on *motor performance*, participants exercised a sine wave tracking task (5–25 % of MVC, 0.3 Hz) over 10 practice trials of 30 s duration. Performance in the last trial was compared to the first trial to indicate any practice effects.

In the *tactile domain*, we applied 20 min of tactile high-frequency stimulation (tHFS) with the piezoelectric waver on the distal part of the left fingertip. tHFS consists of 20 single pulses within 1 s stimuli trains and with 5 s intervals between each train. This stimulation paradigm has been shown to be very effective in inducing somatosensory cortical reorganization and changes in tactile discrimination performance (Voelcker-Rehage and Godde 2010). To assess effects of tHFS on discrimination performance, d', as the difference between the z-transforms of the hit rate and the false alarm rate, in the POT for stimuli at the tHFS area was compared to d' at a control area more proximal at the finger tip. d' is a standard measure for sensory discrimination that takes into account not only the sensitivity for the target but also the ability to prevent false alarms.

Behavioral Data Analysis

Precision grip tasks. The primary motor outcome variable for the force-tracking task was the relative root mean square error (rRMSE) as indicator for deviations from the target force.

To assess different causes for movement deviations in sine wave tracking, e.g., deviations in speed or force level, we applied a curve fit. A curve fit is a specific form of a regression analysis and calculates the curve that fits best to the applied forces. It reveals information about different sine wave parameters and allows identifying mechanisms responsible for deviations from the target force as indicated by the rRMSE. Thus we calculated the amplitude (A), the period length or frequency (F), the phase shift (S), and the intercept (Y) of the curve fitted to the applied forces. The individual parameters were subtracted from the target sine wave parameters and these difference values were further analyzed (dA, dF, dS, dY).

Tactile tasks. TDT was determined by calculating the mean of the weights at the six points of return. To determine discrimination thresholds in the FDT and the GOT we interpolated from the observed data the frequency or grating differences that would have given 75 % correct responses (Voelcker-Rehage and Godde 2010). Regarding the PDT and the ODT, error rates were analyzed. Moreover, in the tasks that required a button press, i.e., FDT and PDT, reaction times (RT) of correct responses were also analyzed. For the PDT in combination with EEG and for the POT, d' was calculated.

Electroencephalographic Recordings

During selected tasks we used a 32 channel EEG system (Active 2, Biosemi) to record brain potentials from 32 electrodes positioned after the international 10–20 system.

In line with Serrien et al. (2004) as a neurophysiological correlate of motor attention power of the alpha frequency band (8–12 Hz) was calculated at the electrode positions Fz and Cz and compared between rest condition and motor tasks, called *task-related power* (TRPow). As a correlate of movement planning and control we analyzed *task-related coherence* (TRCoh; difference in coherence between rest and task condition) for the alpha-band within the mesial structure, including the connections between Fz and Cz and their functional coupling to both hemispheres (electrodes: F3/4, C3/4 and P3/4).

For analysis of EEG data related to tactile discrimination we focused on the amplitude and the topography of the late *P3 component* (also known as *P3b*) of the event-related potential (ERP). As tactile discrimination does not only demand sensory encoding but also higher order cognitive processes, we assume the P3 to be a good indicator of age-related differences in tactile discrimination. ERPs were obtained by averaging time segments from -100 until 700 ms relative to the stimulus onset. P3-peaks were automatically identified at the electrode positions FZ, CZ, and PZ within a time windows of 200–550 ms after stimulus onset and visually controlled. These electrodes were chosen because P3 amplitudes are maximal over the midline electrodes and known to be largely symmetrically distributed (Fabiani et al. 2000). Peak amplitudes and latencies were exported for further statistical analysis.

Statistical Analysis

In general, analysis of variance (ANOVA) was performed for the variables of interest with between factors age and expertise and tested behavioral and neurophysiological variables (study 1) or test repetition (study 2) as within factors.

All analyses reported are Greenhouse-Geisser corrected, if appropriate. Follow-up contrasts were corrected for multiple comparisons and results are reported as significant (with $p < 0.05$) or marginal significant ($p < 0.1$) effects.

Results

Study 1: Effects of Age and Expertise on Manual Dexterity

Research Question 1.1: Effects of Age and Expertise on Precision Grip Performance

Our first research question was if and how force modulation performance is influenced by age in adults throughout their working life. Further we were interested in the effect of work-related expertise, i.e. extensive use of the fingers and long-term practice in manual dexterity tasks, on force modulation performance in the different age-groups. For this purpose, participants of the five age and expertise groups (cf. sample description) performed the four sinusoidal tracking tasks described above.

For all four sine wave task conditions, the ANOVA revealed a significant effect of age (only marginal significant for slow/fixed condition). Older participants performed on a lower level as compared to middle-aged and younger adults. Post-hoc contrasts always revealed a significant difference between young and older adults, whereas the performance of the middle-aged subjects neither significantly differed from the performance of either younger or older adults (for slow/relative condition a marginally significant difference between middle-aged and older adults was found). The middle-aged adults' performance was on a level between the young adults and the older adults (cf. Fig. 2).

Regarding expertise effects, except for the slow/fixed conditions, in all conditions experts performed significantly better than novices. As displayed in Fig. 2, expertise seems to compensate for age-related decline in a way that older experts reached performance levels comparable to middle-aged novices.

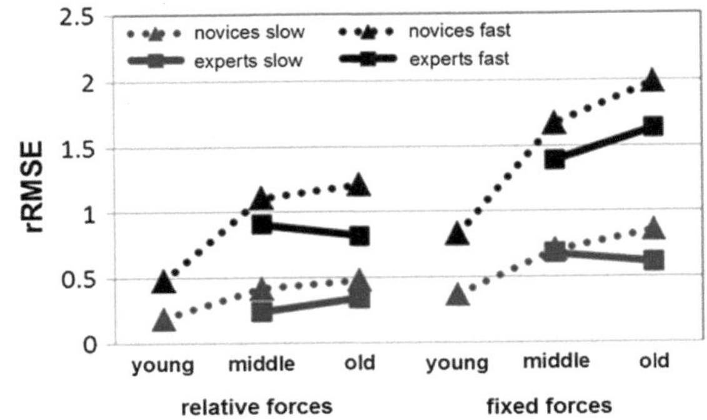

Fig. 2 Mean performance (rRMSE) for relative and fixed forces with high (*fast*) and low (*slow*) frequencies for young, middle-aged and older novices and middle-aged and older experts

Research Question 1.2: Influence of Age on Specific Force Tracking Characteristics

To analyze the influence of age on specific force tracking characteristics, young and old novices (cf. sample description) performed the four sinusoidal tracking tasks described in the methods section. A curve fit was calculated that allowed identifying mechanisms responsible for deviations from the target force.

Deviation of the approximated curve from the target sine wave was always higher for the group of older participants (cf. Fig. 3 for example curve fits; significant effects for dA, dS, dY and marginally significant effect for dF). Comparing the four different sine wave tasks, we found a main effect of speed and force range for dA, dF, dY, and a speed by force range interaction for dA and dY. The interaction of age and force range was marginally significant for the curve fit parameters dA and dF showing higher differences in dA and smaller differences in dF between the age-groups in the fixed force conditions than in the relative force condition.

Combined, the error occurring in different tasks seems to be based on different mechanisms, shown by different adjustments of curve fit parameters. Generally, older adults seem to perform the tasks with lower modulation amplitudes and

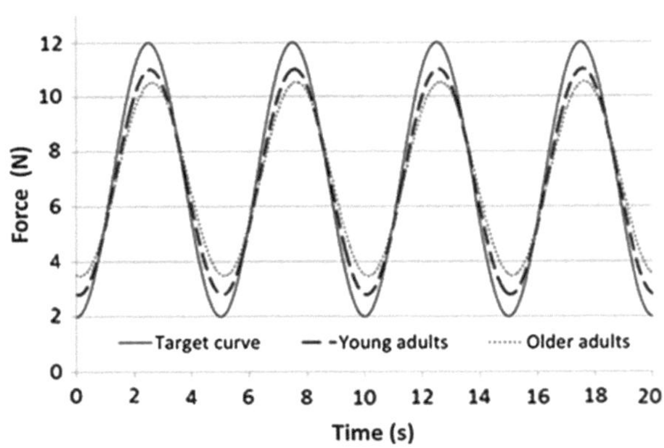

Fig. 3 Fitted curves for young (*dashed*) and older adults (*dotted*) in the condition slow/fixed. The target sine wave is shown as solid line

baseline force levels (intercept) as well as a larger phase shift as compared to younger adults.

Research Question 1.3: Neurophysiological Correlates of the Effects of Age and Expertise on Precision Grip Performance

To study performance differences and its neurophysiological correlates over the working lifespan, participants performed two bimanual tracking tasks of different difficulty as described above.

47 subjects (NM: 11; NA: 12; EM: 10; EA: 14) participated in this part of the study. 13 young novices served as a comparison group. Six subjects had to be excluded from the analysis because they were not able to perform the task sufficiently well.

We found significant main effects of age and expertise. The tracking variability was higher for older in comparison to middle-aged adults. Experts performed both tracking tasks with lower variability of force production. Comparing both tasks, tracking variability was higher in the asymmetric condition than in the symmetric one, indicating a higher difficulty of the asymmetric task. This effect interacted with age (marginally significant). The difference between the tracking variability in the symmetric condition and the asymmetric condition was larger for the older as compared to the middle-aged adults (cf. Fig. 4).

Analysis of task-related power (TRPow) of the alpha band revealed no age or expertise effect. For the whole group TRPow decreased for the symmetric as well as for the asymmetric task, but this TRPow decline did not differ between the tasks. Thus one might assume that the same amount of motor attention was paid to both tasks, indicating that despite variation of the starting point the motor requirements remained the same. That is, the same amount of force had to be generated with different demands on timing and movement control (cf. Fig. 5, *left panel*).

For task-related coherence (TRCoh), results showed marginal significant main effects for expertise as well as for the interaction of age and expertise (cf. Fig. 5, *middle and right panel*). While TRCoh of middle-aged novices and experts did not differ, in the group of older adults experts had higher TRCoh than novices,

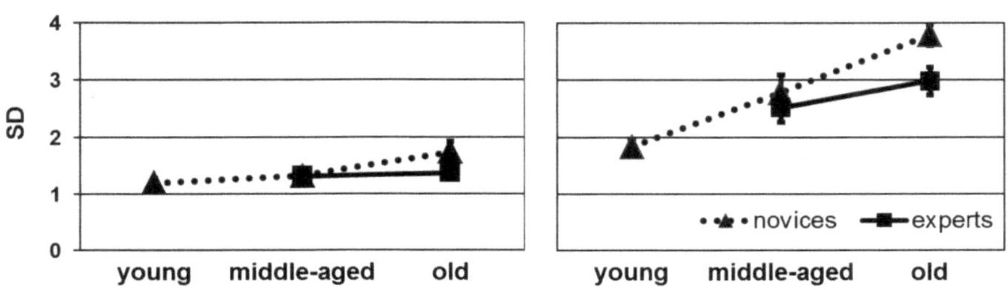

Fig. 4 Tracking variability (SD), mean value for the left and the right hand, in the symmetric (*left*) and the asymmetric (*right*) condition

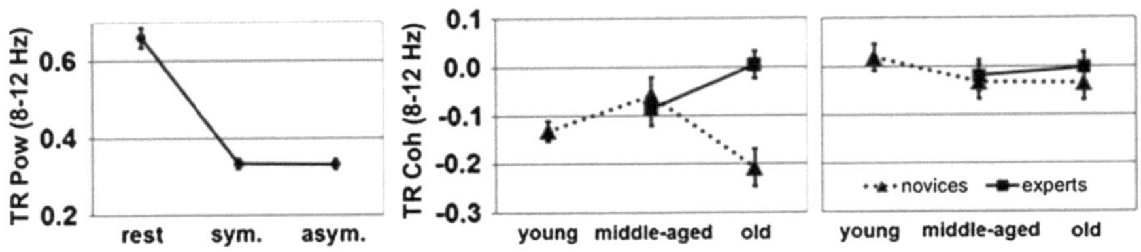

Fig. 5 *Left* Task-related power in the alpha band during periods of rest, symmetric and asymmetric motor performance averaged for the four groups. *Middle* and *right* Task-related coherence in the alpha band during periods of symmetric (*middle*) and asymmetric (*right*) motor performance

especially in the symmetric condition (significant interaction of expertise, age and task). Further TRCoh in the asymmetric condition was higher than the TRCoh in the symmetric condition.

Research Question 1.4: Effects of Age and Expertise on Touch Perception

Next we investigated the decline in touch perception throughout the working life. Again, also the influence of work-related expertise on tactile and haptic perception as well as on age-related decline was examined. Participants performed five tactile tasks (TDT, FDT, GOT, PDT, ODT; cf. Methods) in the same order in two individual sessions. The ODT was split into four runs of 15 trials each, which were interleaved with the other tests.

Influence of age. The age groups differed regarding their performance in the five tactile tasks (cf. Fig. 6). The young and middle-aged participants performed better than the old participants in all tasks (main effect of age for all tasks). They had lower thresholds and made less errors than the older subjects.

Tukeys' post hoc test confirmed that young adults performed significantly better than older adults in all these tests. Middle-aged subjects' performance was either on the level of the young subjects or in between the performance of the young and the older group, depending on the task. Younger adults were proven to not differ significantly from middle-aged participants in all of the tests. However, middle-aged subjects differed significantly from older adults in GOT and FDT.

Only the RT of young subjects in the PDT and the FDT was shorter than the RT of middle-aged and older subjects. This difference, however, was not statistically significant. In the haptic ODT a trend towards a significant age effect was revealed. Older subjects tended to perform less accurate than middle-aged subjects.

Influence of expertise and its interaction with age. ANOVA for the factors age and expertise did not reveal significant effects of expertise or expertise by age interactions in any of the conducted tasks (cf. Fig. 6). Only the PDT revealed a trend for an expertise effect with experts performing better than novices.

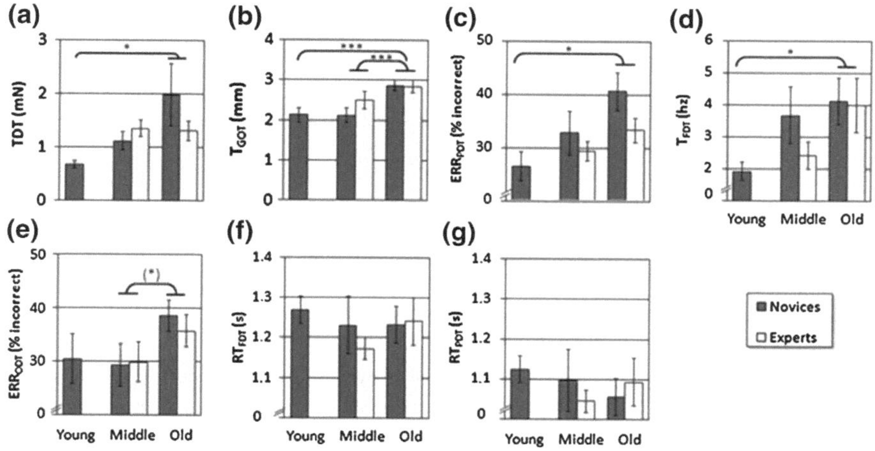

Fig. 6 Mean performance of subsamples: **a** Touch detection threshold. **b** Grating orientation threshold. **c** Frequency discrimination threshold. **d** Pattern discrimination error rate. **e** Object discrimination error rate. **f** Reaction time frequency discrimination. **g** Reaction time pattern discrimination (modified from Reuter et al. 2012)

Research Question 1.5: Neurophysiological Correlates of the Effects of Age on Tactile Performance

Age-related decline in tactile discrimination performance is likely to result from changes on a peripheral and on a central level. In this study we investigated the relationship between tactile performance and neurophysiological correlates of conscious processing and the allocation of cognitive resources as indicated by the P3 component of the ERP.

Only the novices from the three age groups were analyzed in the tactile EEG experiments in order to reveal age-related differences regardless of expertise effects. After exclusion of some subjects due to bad EEG signals or performing at chance levels, 28 subjects remained in the three age groups (12 young, 8 middle-aged, 8 older adults). A variation of the PDT was performed with EEG as tactile two-choice discrimination task (cf. Methods).

For the deviant stimuli, ANOVA revealed a significant age effect for d' as an indicator of discrimination performance. Older adults performed on a lower level than young adults. Performance of middle-aged subjects was in between the performance levels of the other two groups but did not differ significantly (Fig. 7).

ANOVA for the P3-Amplitudes with the within factor electrode position and the between factor age group revealed differences in the distribution of the maximal P3 amplitudes after presentation of deviant stimuli with increasing age (significant position × age interaction). For young and middle-aged adults P3 amplitudes differed between all three electrode positions and showed a fronto-parietal gradient with maximal P3 amplitudes over the parietal cortex. Older adults had reduced amplitudes at Pz in comparison to the younger subjects resulting in an

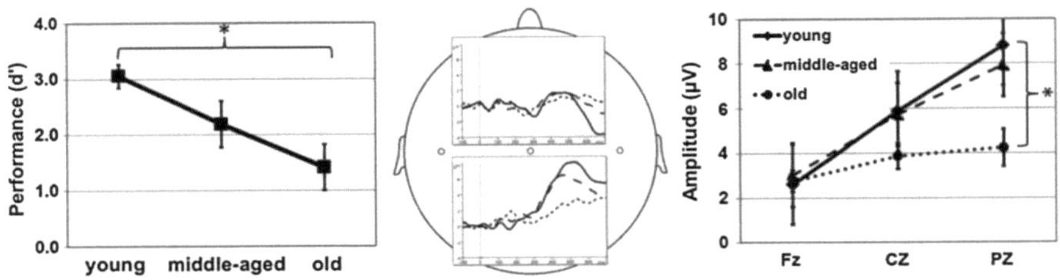

Fig. 7 *Left* Tactile discrimination performance measured as d' for deviant stimuli. *Middle* Grand average ERPs from young, middle-aged and old subjects at the electrode positions Fz (*top*) and Pz (*bottom*). *Right* P3 peak amplitudes at the electrode positions Fz, Cz and Pz

equipotential distribution of P3 amplitudes between frontal, central, and parietal brain regions (cf. Fig. 7). P3 latencies were prolonged with increasing age (main effect of age).

Interrelation between P3 peak amplitude topography and tactile discrimination performance. To learn about the significance of P3 peak amplitudes for behavioral outcomes we conducted a regression analysis with the amplitudes at the electrode positions Fz and Pz as predictors for performance separately for all age groups. Results revealed that in both, young and old, but not in middle-aged adults, 81 % of the variance could be explained by the P3 peak amplitudes. Interestingly, in young adults only the peak amplitude at PZ significantly contributed significantly to explain the variance in performance whereas the peak amplitude at FZ did not. The opposite was true for the older subjects. Here, peak amplitude at FZ was a significant predictor for performance but not PZ amplitude. Higher amplitudes at the respective positions were associated with better discrimination accuracy.

Study 2: Practice Effects on Manual Dexterity

Research Question 2.1: Effects of Short-Term Motor Practice

To examine age effects on the ability to improve the motor abilities, young and older workers exercised a sine wave tracking task (5–25 % of MVC, 0.3 Hz) over 10 practice trials of 30 s duration. Generally, older participants performed significantly worse than younger adults. However, the practice effect was similar for both age groups with a performance increase of about 24 % (cf. Fig. 8).

Research Question 2.2: Effects of Tactile High-Frequency Stimulation

To investigate age differences in plasticity in the tactile domain, performance in a tactile orientation discrimination task on the part of the finger that was stimulated for 20 min with tHFS was compared to performance with a not stimulated part of

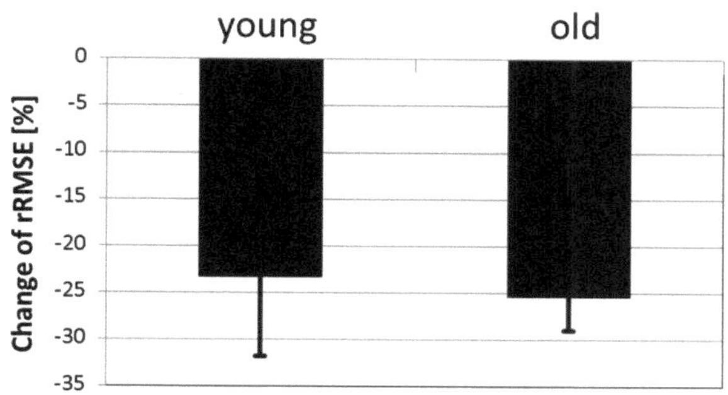

Fig. 8 Mean change in performance (rRMSE) in the precision grip task from first to last practice trial for both age groups

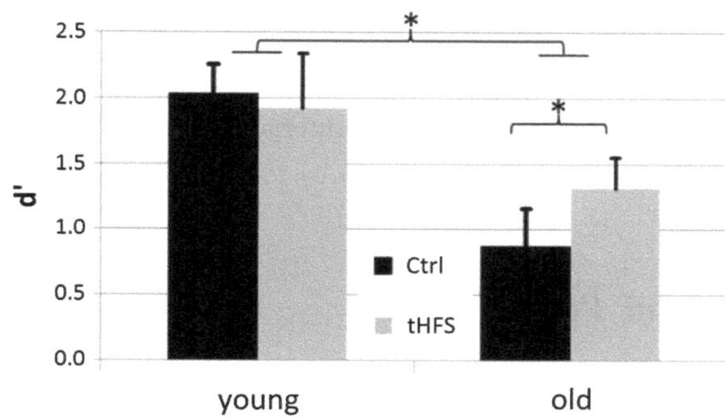

Fig. 9 Tactile orientation discrimination performance (d′) at the control and the stimulation site for both age groups

the finger. Again, performance levels of older adults were generally lower as compared to young adults. However, on contrast to the motor domain, only older adults could profit from the stimulation intervention. Whereas performance of the young adults was similar at the stimulation and the control site, sensitivity of older adults was about 52 % higher at the stimulation site as compared to the control site. Thus, whereas young adults performed on a very high level even at the control site, tHFS seemed to be sufficient to reduce the age-related decline in tactile performance in the group of older adults (cf. Fig. 9).

Discussion

As expected, in all motor and sensory tasks older workers revealed lower performance levels as compared to younger workers. Herewith we confirmed earlier studies showing age-related decline in precision grip performance (e.g. Cole 1991) and tactile perception (Dinse et al. 2008; Reuter et al. 2012) and contributed to the research gap on middle-aged adults.

The time course of the age-related decline differed between the motor and tactile domain. In the motor domain, significant differences could be found between young and older workers but not between the middle-aged and either of the other age groups. Thus, one might conclude that in contrast to our assumption, middle-aged workers were not significantly affected by age-related decline as compared to young workers. However, motor performance of middle-aged workers was always in between the performance of the young and older workers and the linear trend observed (cf. Fig. 2) indicated a continuous performance decline during the adult lifespan. Thus, middle-aged subjects might already show a subliminal decline in motor tasks.

In the tactile domain, significant group differences were found between middle-aged and older workers for selected tasks (GOT, ODT) but not between young and middle-aged workers. Thus, for tactile performance age-related changes seem to occur more stepwise and to get relatively severe from the age of 45 years onwards.

Overall our data suggest that the aging process of manual dexterity seems to start slowly but continuously goes on during the working lifespan.

Expertise effects were mainly seen for the motor tasks, whereas for the tactile tasks expertise neither counteracted those losses nor enhanced general touch perception.

Motor Performance

In line with a study by Lindberg and colleagues (2009) the decline in precision grip performance was prominent already in middle adulthood. Since all participants were in the active work force, an inactive lifestyle, as often associated with age-related decline, is not a possible explanation. Even the argument of reduced strength in older ages does not hold for our study, as MVCs were not significantly different for the age groups. Thus, other factors such as decreased somatosensory functioning (Reuter et al. 2012) and less efficient muscle recruitment, respectively reduced intra- and intermuscular coordination, might be responsible for the age-related differences in hand functioning and performance in force modulation tasks.

Interaction between motor task characteristics and age. The curve fit revealed that older adults showed higher deviations from the target sine wave in amplitude, phase, and intercept than younger adults, irrespectively of task. Thus, age-related differences in force modulation tasks seem to be relatively stable and independent of task characteristics. Age-related differences are only slightly stronger under the fast tracking conditions as compared to the slow ones.

The higher deviations from the target sine wave in amplitude, phase and intercept indicate that older adults performed on a lower force level in general, especially under the fixed conditions, so that the minimum was reached correctly and the amount of force produced was as much as possible in the given time (frequency was only marginally affected by age). The peak of the sine wave was not reached by the older participants. Thus, we conclude that older adults performed the task in a smaller scale, but with an appropriate timing. As the amount of force produced in the given time is reduced in spite of unchanged MVC the age-related change can be assumed as slowing (Jagacinski et al. 1995).

The influence of expertise. In line with our previous findings (Vieluf et al. 2012) Expertise seems to have the potential to compensate for age-related decline, particularly in older adults. As shown in Fig. 2, with increasing task requirements the influence of expertise became more important: In the fast force tracking conditions experts showed much better performances as compared to novices. In all conditions expertise was able to postpone age-related decline by at least 10 years. In all task conditions older experts performed on the level of middle-aged novices or even below.

Neurophysiological correlates of motor performance. For both, the symmetrical and asymmetrical tracking tasks, we found the expected decrease in alpha power during motor performance in comparison to the rest condition. However, there was no effect of task complexity, that is, we found no difference between the symmetrical and asymmetrical condition. Thus, it might be assumed that, although the cognitive demands might be different for both conditions, both conditions required the same amount of specific motor-related attention. Basically the tasks for each hand were the same and only required different timing and integration of motor patterns of both hands. Regarding coherence in the alpha-band, results revealed increased TRCoh during the asymmetric compared to the symmetric condition. Increased task-related alpha coherence during motor tasks has been associated with more need for movement planning and control. This was also reported by Serrien and colleagues for a comparable task combining force maintenance with a force modulation task (Serrien et al. 2004). Thus the very low or even negative TRCoh values during the symmetric condition could be an indicator for the integration of both hands into one movement pattern, which can be controlled mainly by low level sensorimotor programs, whereas two different tasks require more planning and control. This assumption is supported by the higher task-related coherence under the asymmetric condition, indicating more difficulties to integrate two movement patterns that differ in their temporal constraints. Further analysis will show if coherence in the beta frequency range, related to motor processing itself (Serrien et al. 2004), was differently affected by these motor tasks.

Tactile Performance

We showed that with increasing age touch perception decreases throughout ones working life and that expertise neither counteracts those losses nor enhances general touch perception measured in different tactile and haptic tasks. However, it remains open whether work-related expertise in manual dexterity tasks may result in improved sensory processing abilities in tasks more closely related to the work setting. Our results are in line with the conclusion made by Grant et al. (2000) that tactile and haptic expertise has a finite limit in terms of transferability. This is why future studies should focus more on haptic perception. Such studies should also include the left hand because work-related tasks are mostly executed bimanually which may increase the differences between experts and novices with regards to their sensorimotor experience.

Neurophysiological correlates of tactile performance. In line with previous studies, our results revealed a reduced P3 amplitude over the parietal cortex and a more equipotent scalp distribution (Fjell and Walhovd 2001) in older in comparison to middle-aged and younger adults. These differences in P3 topography indicate that older adults seem to be less able to allocate parietal resources and therefore seem to rely more on frontal activation as a compensatory attempt (Getzmann and Falkenstein 2011). This interpretation is supported by our findings of a strong positive relationship between frontal P3 amplitudes and performance levels in the group of older subjects: while younger adults' performance does only depend on the parietal activation, older adults need to recruit frontal brain areas in order to perform on a relatively high level.

Training Results

Results from our motor and tactile practice/training experiments confirm that older adults preserve their abilities to improve their performance in tasks requiring manual dexterity already by short term motor practice or tactile stimulation. This is in line with recent studies from the motor (Voelcker-Rehage and Alberts 2005) and tactile (Dinse et al. 2008) domains. Surprisingly, in the tactile training experiment only the older adults improved their performance. This might be explained by an already very high performance level of the younger adults at baseline (ceiling effect). Moreover, it shows that already impaired persons can especially profit from this kind of intervention.

Summary of Results

Our results confirmed age-related differences in basic components of manual dexterity, finger force control and tactile perception, even already during middle adulthood. However, in the motor domain these age effects could be at least partly compensated by expertise. Neurophysiological data showed that better performance in older adults was related to a reorganization in the sensorimotor and frontal cortical areas indicating compensatory activation patterns in older adults. However, although older adults generally performed at a lower baseline performance level, they were able to improve motor and tactile functioning by short term practice or stimulation interventions. Particularly in the tactile domain such an intervention was well suited to attenuate age-related decline.

Outlook

Our study results offer starting points for work-related measures like on-the-job training, opportunities for horizontal career changes, or rehabilitation after injuries or diseases. Often, older workers are hesitant to change their jobs and to learn new tasks, because it is widely assumed that older adults are not as able to learn as younger people. We could show that performance improvements can be induced in both, younger and older adults. Both short-term (intervention) and long-term (expertise) effects were revealed. This indicates that training measures for older workers even in jobs demanding manual dexterity might be successful and promising. Nevertheless, future research must show if not only basic components of manual dexterity are trainable, but also if abilities in these components can be transferred to work-related tasks and if such work-related tasks can be trained as well. The same is true for the transferability to everyday tasks, because manual dexterity plays a pivotal role not only in the work context but also in everyday life. Decline in manual dexterity has much more impact on independent living and the need for social support than vision or hearing impairments. Thus, maintenance and improvement of manual dexterity is important not only with respect to horizontal careers but also for the life phase after retirement and therefore, regular training of hand and fingers through permanent use in daily life might also be effective in attenuating age-related decline in older adults.

Although expertise was shown to be highly specific in the cognitive domain (Krampe and Charness 2006), expertise had strong effects on such arbitrary measures of manual dexterity that on first sight might not be directly linked to every-day activities. However, the fact that differences regarding these measures could be found in experts as compared to novices indicates the significance of these measures. Otherwise they would not occur as the result of intensive training during the working life.

Further, longer term training protocols should be tested to investigate the potentials and limits of workers in different age groups and with respect to different preconditions. As shown for older workers in our tactile training paradigm, one might expect that novices profit to a higher degree from training programs since they start at a lower level. However, it might also be the case that experts show higher learning gains due to more optimal preconditions due to long term use and available cognitive strategies.

More detailed analyses of the neurophysiological correlates of motor and tactile performance seem to be promising for the understanding of age-related changes in manual dexterity and the effects of training and expertise. It is of particular interest if learning capacity and learning mechanisms are different between young and older workers as shown for our tactile intervention paradigm. Such findings would have direct impact on the design of training interventions for older workers.

Acknowledgments We would like to thank all subjects participating in the study, the students and interns assisting in carrying out the experiments and Britta Molter and Dirk Hormann for carefully proofreading the manuscript (DFG VO 1432/7-1).

References

Babiloni C, Del Percio C, Iacoboni M, Infarinato F, Lizio R, Marzano N (2008) Golf putt outcomes are predicted by sensorimotor cerebral EEG rhythms. J Physiol 586:131–139

Calautti C, Serrati C, Baron JC (2001) Effects of age on brain activation during auditory-cued thumb-to-index opposition: a positron emission tomography study. Stroke 32:139–146

Cole KJ (1991) Grasp force control in older adults. J Motor Behav 23:251–258

Del Percio C, Babiloni C, Marzano N, Iacoboni M, Infarinato F, Vecchio F et al (2009) "Neural efficiency" of athletes' brain for upright standing: a high-resolution EEG study. Brain Res Bull 79(3–4):193–200

Diermayr G, McIsaac TL, Gordon AM (2011) Finger force coordination underlying object manipulation in the Elderly—a mini-review. Gerontol 57(3):217–227

Dinse HR, Wilimzig C, Kalisch T (2008) Learning effects in haptic perception. In: Grunwald M (ed.) Human haptic perception–basics and applications. Birkhäuser, Basel, pp 165–182

Edwards P, Greasley K (2010) Absence from work. Eurofound, Dublin. Available at http://www.eurofound.europa.eu/docs/ewco/tn0911039s/tn0911039s.pdf (10 May 2010)

Elbert T, Pantev C, Wienbruch C, Rockstroh B, Taub E (1995) Increased cortical representation of the fingers of the left hand in string players. Science 270(5234):305–307

Fabiani M, Gratton G, Coles MGH (2000) Event-related brain potentials. In: Cacioppo JT, Tassinary LG, Berntson GG (eds) Handbook of psychophysiology. Cambridge University Press, New York, pp 53–84

Fjell AM, Walhovd KB (2001) P300 and neuropsychological tests as measures of aging: scalp topography and cognitive changes. Brain Topogr 14(1):25–40

Flanagan JR, Wing AM (1993) Modulation of grip force with load force during point-to-point arm movements. Exp Brain Res 95:131–143

Furuya S, Kinoshita H (2008) Expertise-dependent modulation of muscular and non-muscular torques in multi-joint arm movements during piano keystroke. Neuroscience 156(2):390–402

Gesundheitsberichterstattung des Bundes (2006) Heft 30: Gesundheitsbedingte Frühberentung, Berlin

Getzmann S, Falkenstein M (2011) Understanding of spoken language under challenging listening conditions in younger and older listeners: a combined behavioural and electrophysiological study. Brain Research, pp 8–22

Godde B, Berkefeld T, David-Jürgens M, Dinse HR (2002) Age-related changes in primary somatosensory cortex of rats: evidence for parallel degenerative and plastic-adaptive processes. Neurosci Biobehav Rev 26:743–752

Grant AC, Thiagarajah MC, Sathian K (2000) Tactile perception in blind Braille readers: a psychophysical study of acuity and hyperacuity using gratings and dot patterns. Percep Psychophysics 62:301–312

Hermsdörfer J, Blankenfeld H (2008) Grip force control of predictable external loads. Exp Brain Res 185(4):719–728

Heuninckx S, Wenderoth N, Debaere F, Peeters R, Swinnen SP (2005) Neural basis of aging: the penetration of cognition into action control. J Neurosci 25(29):6787–6796

Jagacinski RJ, Liao MJ, Fayyad EA (1995) Generalized slowing in sinusoidal tracking by older adults. Psychol Aging 10(1):8–19

Johansson RS, Westling G (1984) Roles of glabrous skin receptors and sensorimotor memory in automatic control of precision grip when lifting rougher or more slippery objects. Exp Brain Res 56:550–564

Jones LA, Piateski E (2006) Contribution of tactile feedback from the hand to the perception of force. Exp Brain Res 168:298–302

Kraemer B (2010) Absence from work—Germany. Eurofound, Dublin. Available at http://www.eurofound.europa.eu/ewco/studies/tn0911039s/de0911039q.htm (10 May 2012)

Krampe RT, Charness N (2006) Ageing and Expertise. In: Ericsson KA, Charness N, Feltovich PJ, Hoffmann RR (eds) The Cambridge handbook of expertise and expert performance. Cambridge University Press, Cambridge, pp 471–488

Lindberg P, Ody C, Feydy A, Maier MA (2009) Precision in isometric precision grip force is reduced in middle-aged adults. Exp Brain Res 193(2):213–224

Lindenberger U, Kliegl R, Baltes PB (1992) Professional expertise does not eliminate negative age differences in imagery-based memory performance during adulthood. Psychol Aging 7:585–593

Manganotti P, Gerloff C, Toro C, Katsuta H, Sadato N, Zhuang P, Leocani L, Hallett M (1998) Task-related coherence and task-related spectral power changes during sequential finger movements. Electroencephalogr Clin Neurophysiol 109(1):50–62

Monzée J, Lamarre Y, Smith AM (2003) The effects of digital anesthesia on force control using a precision grip. J Neurophysiol 89(2):672–683

Nowak DA, Glasauer S, Hermsdörfer J (2003) How predictive is grip force control in the complete absence of somatosensory feedback? Brain 127:182–193

Petrofsky JS, Lind AR (1975) Aging, isometric strength and endurance, and cardiovascular responses to static effort. J Appl Physioliology 38(1):91–95

Polich J (1997) EEG and ERP assessment of normal aging. Electroencephalogr Clin Neurophysiol 104(3):244–256

Reuter-Lorenz PA, Jonides J, Smith EE, Hartley A, Miller A, Marshuetz C et al (2000) Age differences in the frontal lateralization of verbal and spatial working memory revealed by PET. J Cogn Neurosci 12(1):174–187

Reuter E, Voelcker-Rehage C, Vieluf S, Godde B (2012) Touch perception throughout working life: effects of age and expertise. Exp Brain Res 216:287–297

Serrien DJ, Pogosyan AH, Brown P (2004) Cortico-cortical coupling patterns during dual task performance. Exp Brain Res 157:79–84

Tecchio F, Zappasodi F, Melgari JM, Porcaro C, Cassetta E, Rossini PM (2006) Sensory-motor interaction in primary hand cortical areas: a magnetoencephalography assessment. Neurosci 141(1):533–552

Vaillancourt DE, Newell KM (2003) Aging and the time and frequency structure of force output variability. J Appl Physiol 94:903–912

Vieluf S, Mahmoodi J, Godde B, Reuter E, Voelcker-Rehage C (2012) The influence of age and work-related expertise on fine motor control. GeroPsych 25(4):199–206

Vieluf S, Godde B, Reuter EM, Voelcker-Rehage C (2013) Age-related differences in finger force control are characterized by reduced force production. Exp Brain Res

Voelcker-Rehage C, Alberts JL (2005) Age-related changes in grasping force modulation. Exp Brain Res 166:61–70

Voelcker-Rehage C (2008) Motor-skill learning in older adults—a review of studies on age-related differences. Eur RevAging Phys Act 5(1):5–16

Voelcker-Rehage C, Godde B (2010) High frequency stimulation improves tactile but not motor performance in older adults. Mot Control 14:460–477

Ward NS, Frackowiak RSJ (2003) Age-related changes in neural correlates of motor performance. Brain 126:873–888

Westling G, Johansson RS (1987) Responses in glabrous skin mechanoreceptors during precision grip in humans. Exp Brain Res 66:128–140

Physiological Responses of Two Male Age Groups to Working in Deep Cold and Subjectively Experienced Stress and Strain

Karsten Kluth, Mario Penzkofer and Helmut Strasser

Lessons Learned

Recommendations for tolerability of different cold exposures and the design of cold-protective clothes seem to be useful and very exact. They can be found in standards, rules and regulations, as well as in the requirements of the Occupational Health and Safety Act. However, there is a controversial discussion about these recommendations and standards in the scientific literature of occupational medicine and ergonomics since present knowledge about the impact of such extreme stress on strain is not secured. It rather lies in a grey area between scientific knowledge and assumptions. This deficiency needs to be ameliorated, primarily in order to create preventive occupational health and safety and, ultimately, also to increase the efficiency of the work process.

Actually, a well-regulated organization of working times and breaks has not been established, and no consistent knowledge exists as to whether an additional age-differentiated organization is necessary.

Therefore, in the context of an age-differentiated employment of the workforce, the physiological responses of heart rate, blood pressure, core temperature and skin surface temperature, as well as the energy expenditure were analyzed in a field study carried out in an industrial deep cold-storage depot. In whole workday tests, possible age-dependent effects of order-picking during cold exposures to

K. Kluth (✉) · M. Penzkofer · H. Strasser
Department of Mechanical Engineering, Siegen University, Siegen, Germany
e-mail: kluth@ergonomie.uni-siegen.de

M. Penzkofer
e-mail: penzkofer@ergonomie.uni-siegen.de

H. Strasser
e-mail: h.strasser@aws.mb.uni-siegen.de

C. M. Schlick et al. (eds.), *Age-Differentiated Work Systems*,
DOI: 10.1007/978-3-642-35057-3_18, © Springer-Verlag Berlin Heidelberg 2013

approximately −24 °C on the strain of 30 male Subjects (Ss) were measured. The Ss were classified into two age groups (20–35 and 40–65 year-olds).

The main results of the study are as listed below:

- Order-picking in the cold store at approximately −24 °C was associated with work-related increases of heart rate by 35 bpm and more for both groups.
- Due to lowered maximum heart rate and capacity utilization with advanced age on the one hand, and an almost identical cardiovascular activation of younger and older Ss during working in the extreme cold on the other hand, may inhere a health risk for the older Ss in the long run.
- While working continuously for 120 min in the cold store at −24 °C, the core temperature, in comparison to the value at the outset, decreased by 1.5 K (for the 20–35 year-olds) and by 2.2 K (for the 40–65 year-olds).
- In both age groups, a complete warming-up during the breaks with a duration of 20 min at temperatures of approximately +20 °C, was not possible for all individuals.

Based on the findings, the following recommendations can be derived to improve occupational safety and health for working in the cold:

- Due to the strong core temperature decreases in the cold store, improvements in parts of the cold-protective clothes seem to be necessary.
- A re-organization of the warming-up breaks with a duration longer than the 20 min granted at present is advisable.

Introduction

The history of frozen food goes back to the year 1920. Clarence Birdseye created the worldwide first perfect quick freezing process which allowed for conservation of food. Only 10 years later, the first frozen groceries were offered for sale in Springfield, Massachusetts. In the year 1950, already 64 % of the grocery stores in the US had frozen food cabinets, and the sales already exceeded $ 1 billion. In the year 2009, the assortment of frozen and refrigerated groceries was one of the reflationary markets in the food sector, with the US as the world market leader. Due to a constantly improving quality and product variety of frozen food, this trend will continue in the future and, accordingly, will also lead to an increasing employment for working in the cold.

An effective protection against perishable influences on refrigerated or frozen food is only provided if a continuous cold chain exists from the manufacturer, through intermediate storage and the trade to the consumer. Order-picking systems in deep cold-storage depots—used as intermediate storages—represent a key element in the distribution of refrigerated or frozen food. Due to daily updated assortment of goods in the stores, order quantities for each of the products are small and the order is diversified. Loading units existing just of one type of article become more and more unusual even on the level of the wholesale or for the manufacturers of finished goods.

Despite rather successful automation in logistics, order-picking frozen food characterizes the main work in the cold and has largely remained "manual work." Employees must stay at temperatures of approximately −24 °C for a considerable length of working time, which is interrupted by warming-up breaks. The transfer and stockpile of the goods—with a weight up to 15 kg—for the order-pickers alone represents an already physically high demanding task which is aggravated by the exposure to the extremely low temperatures. On the one hand, the working environment with temperatures down to −24 °C leads to an unusual load. On the other hand, a physical "zero load" also would rather increase the cold strain, however. This is caused due to a reduced "heating effect" of the physically active musculature as an important source of warmth. Therefore, special attention should be paid to the protection of the employees of deep cold-storage depots, particularly the group of workers who are permanently exposed to these temperatures. With some restrictions, this is also true for the employees who have to work in the chill room with temperatures of approximately +3 °C.

If sufficient protection from cold is missing and if extremely long exposures would occur, potential health risk cannot be excluded. At least, for especially susceptible individuals, possibly short term diseases such as infections of the respiratory tract, colds and, in the long run, hypertension cannot always be avoided (see e.g. Hassi et al. 2005).

Therefore, an optimal protection against cold is demanded for workers of every age, and in the context of the research project the following hypotheses had to be proven.

- The regulation of working time and warming-up breaks which are currently used in practice leads both to an objectively measurable higher cold-strain for older people and to subjectively increased sensations of cold. The question has to be answered, whether job design measures with the objective of preventive health protection take into account the influence of age on physical strain.
- An empirically ascertainable age-related maximum cold exposure exists, up to which no negative impacts on the physical capacity and the subjective sensations of cold can be found.

Current State of Research

Aside from some older studies (Adolph and Molnar 1946; Budd and Warhaft 1966; Hellstrøm et al. 1970; Kleinöder 1988)—except to Tochihara et al. (1995)—there have rarely been any studies that examined the effects of working in the cold on the human body during real tasks. Thus, neither suitable and humane regimes of work and rest periods nor instructions with respect to the ergonomic design of workplaces in cold-storage depots in accordance with work-physiological, organizational, and legal requirements are based on established work-scientific knowledge. Therefore, this work-physiological field study is an attempt to objectify the physiological strain of work in deep cold environments and focus on aging as an important ergonomic issue.

The thermoregulation occurs autonomously via physiological mechanisms (i.e., more or less pronounced changes of heart rate, blood pressure, body core temperature, skin surface temperature and energy conversion). Despite protective clothing and peripheral vasoconstriction, certain heat losses in the body, which have to be compensated for, always occur (cp. Strasser and Kluth 2006). Those heat losses can only be managed by considerably increased energy expenditure.

Even slight deviations from the comfortable thermic zone impede well-being as well as the ability to react, concentrate and perform (Mäkinen 2007); furthermore, this deviation entails temperature regulation mechanisms that put stress on the body. The cold-induced reduction of the blood circulation of skin and limbs causes cold sensations as well as limitations of movement, sensitivity and agility. Last but not least, the scientifically proven age-related decrease of the energy metabolism and physical efficiency inevitably leads to a deterioration of the human capacity for protection against hypothermia with advancing age. Physical illnesses could be caused or exacerbated. Interestingly, due to its "heating effect", the physically strenuous work of manual order-picking, which from an ergonomics point of view, at a first glance, seems necessary to be reduced, must be seen rather as helpful and necessary than harmful. To monitor a more or less balanced thermoregulation, it seems advisable to focus on "core temperature" and "skin temperature" as well as "heart rate" and "blood pressure".

The body core temperature of a healthy human being—generally taken rectally—is a function of work stress, surrounding temperature and time of day, the latter causing a variation of about 1 K. In the case of work stress, the rectal temperature increases linearly with oxygen intake. An increase of oxygen intake of 1 L (i.e. workload somewhat above the endurance level) can be due to an increase of the temperature in the range of 0.5–0.8 K. The surrounding temperature, however, does not affect core temperature as much as work-induced temperature changes. When lowering the surrounding temperature from 10 to 0 °C, Hellstrøm et al. (1970) found out that the rectal temperature dropped by only 0.5 K after a 1-h exposure to the cold, but other scientific sources have reported a slight increase of the core temperature when carrying out physically strenuous work in the cold (for details see Forsthoff 1983). This author found small core temperature increases in the range between 0.2 and 1.0 °C even at a surrounding temperature of −30 °C. These results were based on physiological responses to order-picking with physical activities close to the limit of the endurance level of not specially trained males and were also confirmed by Kleinöder (1988) in laboratory tests simulating activities at −28 °C.

According to a German standard (DIN 33403 2000), core temperature can drop to 35 °C without any health risk, but any further cooling down would be life-threatening. Yet, a sufficient number of scientific study results (cp. e.g., Boennemark et al. 1969; Forsthoff 1983) show that substantial drops in core temperature—even at extremely low temperatures—can be prevented through an increase of work performance and/or through a sufficiently adapted increase in the insulation of the cold-protective clothes.

Skin temperature, as expected, shows significantly higher variations under different climatic conditions than does core temperature. Skin temperature changes of some degrees, e.g. 2–3 K, compared to equivalent changes in core temperature are harmless. Nevertheless, body surface (skin) temperature is also of decisive importance because the heat transmission from the body core to the environment via the skin surface is mostly controlled through the variation of the blood circulation (vasoconstriction and vascular dilatation). Up to 95 % of the entire heat exchange with the surroundings takes place in the skin.

Experiments carried out in the laboratory under perfectly controlled conditions by Müller-Arnecke and Hold (1999) led to significant skin temperature profiles, especially at the hands and feet. They were characterized by substantial decreasing values which did not rise to their outset values even in the case of a moderately continuous exposure to the cold at 0 °C for 3 times 50 min, each, followed by warming-up pauses of 10 min (at 21 °C). One of the reasons for this phenomenon was that despite the wearing of protective clothes during the laboratory experiments (tracking tests and controlling tasks in a sitting position to simulate the operation of machines), not enough body heat was produced to avoid the cooling down. An energy expenditure during work of approximately 4 kJ/min—as assumed in the tests—corresponded just to light manual work, light physical sorting tasks in a sitting position or office work.

Other laboratory tests (e.g. Kleinöder 1988) showed that an air temperature of −30 °C led to a decrease of the foot temperature of up to 10 °C, even when working on a treadmill at 3 km/h. In field studies, mentioned already above, Forsthoff (1983) measured a considerable temperature drop while working at −28 °C which was associated with (negative) cold sensations in the extremities that could only be compensated with increased insulation of the boots.

The tasks of order picking, characterized by a great number of changes of body postures and movements of arms and legs, seem to be rather beneficial to the blood circulation down to the toes and the finger tips than do uniform operations, which hitherto have been simulated in laboratory studies. This conclusion is confirmed by studies of Imamura et al. (1998).

As a consequence of insufficient blood circulation of the body periphery, a loss of mobility, sensitivity and agility occurs, especially in the hands and feet, a phenomenon that is often significant when working in the cold (cp. Wenzel and Piekarski 1980). At a hand skin temperature of 10 °C, which already causes significant pain (cp. Kleinöder 1988), the fine-coordinative efficiency is reduced to only 35 % of that at normal skin temperature (cp. Wenzel and Piekarski 1980). At hand temperatures of 20 °C, which can also cause cold sensations, the loss of efficiency is around 25 %. Apart from the temperature itself, the cooling speed affects cold sensations, too. When lowering the temperature by approximately 1 K per minute, a strong cold sensation is caused in the area of still quite comfortable skin temperatures (28–31 °C).

Heart rate indicates physical strain of the whole body and remains more or less constant at a higher level ("steady state") if the stress does not lead to fatigue. Only a continuous increase marks physical fatigue with stress above the endurance level (cp. Strasser and Müller-Limmroth 1983). Heat strain leads to a sometimes drastic increase of heart rate (in case of vasodilatation in the body periphery) for reasons of thermoregulation. The process is characterized by an increased convective heat transport away from the body core or the musculature towards the skin as a cooling surface through the blood in which case the heart is used as a circulating pump. Therefore, statements on the strain of the circulatory system when working at low temperatures are inconsistent. Adolph and Molnar (1946) observed an increase, whereas Budd and Warhaft (1966) registered a decrease of heart rate. O'Hanlon and Horvath (1970), in contrast, did not note any statistically significant alterations.

In the rare experimental tests that have been carried out so far and are comparable to the cold exposure of this field study (especially Forsthoff 1983), increases as well as decreases of heart rate have been measured. In the study carried out by Müller-Arnecke and Hold (1999), heart rate decreased significantly during test time at a constant cold exposure of 3 × 50 min. This decrease, however, cannot be interpreted as a consequence of the exposure, but can rather be explained with a certain decline of the "intentional basic tension" of the test subjects known from laboratory tests for decades (cp. Strasser 1981). Yet, the authors mentioned above did not interpret the specific climatic issues a systematic effect, which occurs, at least, when comparing the heart rate profiles of the two tested extreme values during the exposure to the cold, i.e. −5 °C and +21 °C.

Besides heart rate, blood pressure is another strain parameter for the assessment of the activity state of the cardiovascular system. Starting from the basal blood pressure (the lowest value in the morning) and the intermittently recorded blood pressure, physiological reactions in the form of substantial increases, indeed, occur under static and dynamic stress in addition to circadian fluctuations. According to Tochihara et al. (1995) and Ozaki et al. (2001), due to peripheral vasoconstriction, pronounced increases in blood pressure could be measured as physiological responses to cold exposures. Also measurements by Forsthoff (1983) in resting as well as in working exhibited increases in blood pressure of up to 20 mmHg (systolic) and 10 mmHg (diastolic). Thus, local and sudden cold stimuli caused a pronounced increase in blood pressure (cp. Hines and Brown 1975). However, persons who have been acclimatized to the cold seem to respond considerably weaker.

Materials and Methods

As described above, most studies which investigated the effects of cold exposures on humans were carried out under laboratory conditions and, therefore, are not comparable to real tasks in industry.

To comply with real working conditions, the investigations in this field study ("Study I") were all carried out in the cold-storage depot of an industrial company. With the aim of assessing whether order-picking in deep cold is sustainable at any age in the long term, and of objectifying protection against the cold, an age-related analysis of the effects of working in the cold was carried out. For comparison purposes, in addition, also subjectively experienced stress and strain were determined by means of a standardized questionnaire which was designed in cooperation with the participating company ("Study II").

Study I

Study Design and Procedure

To make the stay in the cold store bearable, the employees had to wear cold-protective clothing consisting of thermo underwear, a pullover, a pair of trousers, and, more importantly, a cold-protective suit. The latter consisted of a thick jacket and long trousers, a thick hat, special thermo gloves—normally made of leather—and cold-insulating boots.

The independent test variables at a temperature in the cold store of −24 °C were the cold exposure duration of 80, 100, and 120 min per shift. To determine age-related maximum length of cold exposure for working in chill rooms and cold stores, the duration of the working phases was varied. Furthermore, it was assumed that an extended duration of exposure increases the physiological responses of the body and perhaps the endurance level would be exceeded. Identical warming-up phases of 20 min, each, at approximately +20 °C and workload adapted to the real job should have guaranteed real-life physiological responses to working in the cold (at approximately +3 °C and −24 °C) as dependent parameters.

A test layout was developed (cp. Fig. 1) which enabled the measurement of the physical strain associated with work in the deep cold-storage depot. Before each cold exposure, a standard physical working capacity $(PWC)_{130}$ test (cp. HVBG 2002) was carried out to enable statements about all individuals' physical capacities of the cardiovascular system in the submaximal area. Moreover, a short case history with details on the physical conditions of the participants was obtained by means of interviews (cp. Table 1).

The PWC_{130} test—taken on a bicycle ergometer—started with a load of 50 W and a speed of 60 rpm. The load was increased every 2 min by 25 W until the predefined target value, a heart rate of 130 beats per minute (bpm) was reached. At least three load stages had to be passed through by the Ss, and the total test duration had to be less than 12 min. After a following recovery phase and the instrumentalization of the Ss for the data acquisition, the Ss had to work for 80, 100, and 120 min in the cold store at −24 °C on the first day. During the three working phases, the Ss were required to order-pick prepared pallets with a defined total weight of up to 3.2 tons. On average, the order-pickers moved 227 items with

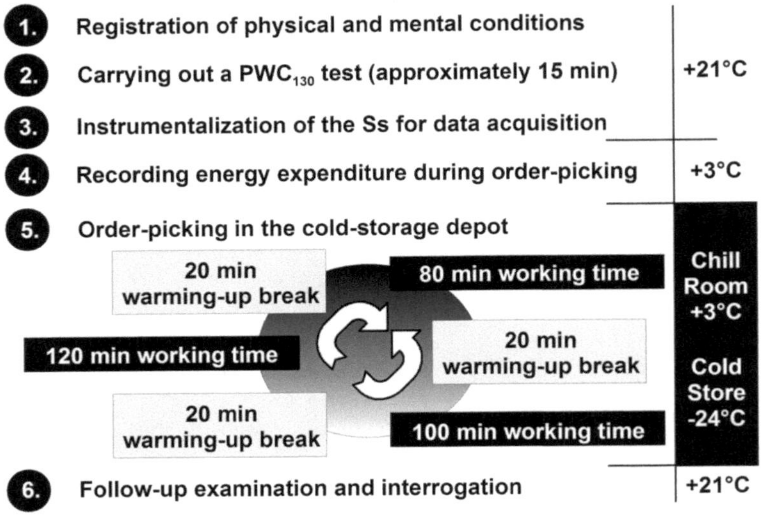

Fig. 1 Diagram of the test procedure

a mean total weight of approximately 1.6 t/h (cp. Kluth and Strasser 2006). On the second experimental day, they had to carry out the same work in the chill room at +3 °C. To avoid cross-over-effects, the sequence of the working phases and the experimental days (cold store, chill room) was varied.

Measurement Parameters

For the objectification of the physical strain associated with working in the cold, physiologically important parameters could be registered with the measurement systems shown in Fig. 2. "Heart rate" and "skin temperature", e.g., were measured continuously and every 15 min "blood pressure" and "body core temperature" were recorded discretely. Moreover, for the identification of the intensity of the physical work and heat production, energy expenditure was quantified with the mobile spirometry system Cortex® MetaMax 3B for a duration of 15 min, each, in the chill room and the cold store. The metabolic exercise testing system measured the oxygen and carbon dioxide content of the breathing air and all other physical respiratory parameters using the breath-by-breath or mixing chamber methodology.

Heart rate via "beat-to-beat" measurements was determined using the Polar® S810i system, which is optimized for scientific investigations and performance diagnostic examinations.

The systolic and diastolic blood pressure was discontinuously measured every 15 min via a commercially available, full-automatic 24-h long-term blood pressure monitor that uses the oscillometric method and was certified according to an European Directive (93/42/EEC 1993). In the same time interval the tympanum temperature was measured by an infrared ear thermometer.

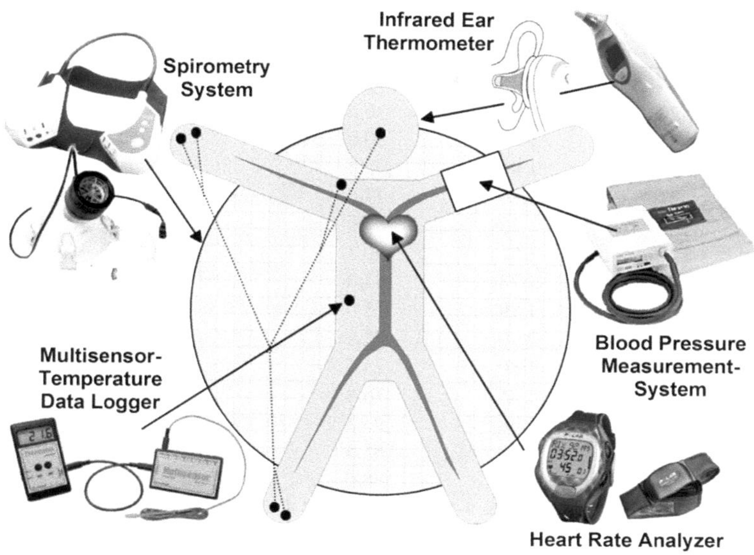

Fig. 2 Measurement system for the determination of stress and strain of the subjects in the cold-storage depot

Furthermore, temperature sensors, which were attached to the body for the recording of the skin temperature, and data loggers were used; both allowed the employees to move freely without being hooked up to a stationary measuring device. Skin temperature was taken at the nose and at 6 additional measuring points on the right part of the body, i.e., the tip of the index finger, the proximal phalanx of the ring finger, the dorsal part of the big toe, the sole of the foot, the kidney area, and the shoulder.

These skin temperature measuring points were chosen based on the work of Müller-Arnecke and Hold (1999), who worked out first approaches for a weight of skin temperature measurements from which mean skin temperature values could be derived. Due to reduced blood circulation of skin and extremities, the risk of cooling of hands and fingers as well as feet and toes is particularly high while working in cold (cp. Brajkovic and Ducharme 2006). Here the mean skin temperature should not drop to less than 30 °C and the local temperature of fingers and toes should not drop to less than 12 °C.

Participants

Fifteen male subjects (Ss), in two age groups (20–35 and 40–65 year-olds; cp. Table 1), had to carry out whole workday tests under predetermined, realistic working conditions.

Table 1 Characteristics of the group of 15 younger (01–15) and 15 older subjects (16–30)

Ss	Age [Years]	Height [m]	Weight [kg]	BMI	rPWC130 [W/kg]
01	24	1.75	53.9	17.6	1.45
02	26	1.82	86.3	26.1	1.70
03	28	1.85	80.0	23.4	1.41
04	28	1.95	99.2	26.1	1.97
05	21	1.80	77.8	24.0	1.86
06	20	1.75	64.6	21.1	1.72
07	29	1.82	90.3	27.3	1.26
08	21	1.78	84.4	26.6	1.74
09	31	1.79	75.0	23.4	1.04
10	29	2.00	75.8	19.0	1.40
11	29	1.80	75.4	23.3	1.18
12	25	1.85	76.8	22.4	1.02
13	29	1.86	95.3	27.5	1.21
14	24	1.80	61.5	19.0	2.03
15	24	1.88	88.6	25.1	1.92
x ± Sd	**25.9 ± 3.5**	**1.83 ± 0.1**	**79.0 ± 12.4**	**23.5 ± 3.2**	**1.53 ± 0.3**
16	43	1.85	73.9	21.6	1.47
17	47	1.87	100.7	28.8	0.85
18	46	1.76	65.1	21.0	1.87
19	59	1.84	125.4	37.0	1.37
20	62	1.78	116.8	36.9	1.25
21	57	1.83	89.7	26.8	1.34
22	63	1.78	99.6	31.4	1.05
23	53	1.76	81.2	26.2	1.52
24	62	1.75	73.6	24.0	1.66
25	63	1.79	90.0	28.1	1.25
26	57	1.86	94.1	27.2	1.86
27	56	1.87	105.8	30.3	1.98
28	55	1.71	82.0	28.0	1.46
29	64	1.72	63.5	21.5	1.83
30	48	1.80	89.2	27.5	1.48
x ± Sd	**55.7 ± 6.9**	**1.80 ± 0.1**	**90.0 ± 17.8**	**27.8 ± 4.9**	**1.48 ± 0.3**

Study II

Procedure

Subjectively experienced stress and strain associated with working in the cold was determined via systematic face-to-face interviews of two age groups of professional male and female order-pickers (63 younger and 65 older employees). The questionnaire which comprised 57 items was not confined to the basic work in the chill room and the cold store and its associated subjective feelings, such as muscle and joint complaints as well as cold sensations or frostbite-symptoms, but also covered questions on the working conditions, the cold-protective clothing, the work equipment, the work-rest regulation and motivational aspects. Assessments concerning the physical strain and the entire workplace had to be carried out in some final topics.

Similar question forms used in this case had already been proven in previous studies (cp. Kluth and Strasser 2003 or Strasser and Kluth 2008). The investigation was carried out in form of interviews under the supervision of a person who was familiar with the topic. Arising questions and uncertainties could be clarified immediately and a return rate of 100 % could be guaranteed.

Participants

For taking part in the investigation, a work experience of order-picking in the cold store of more than a year was a prerequisite. At the time of the investigation, the age of the subjects was between 19 and 49 years. Two groups of 19–35-year olds and 36–49 year-olds, respectively, somewhat below and above the average age of the German workforce, were built. These two samples allowed not only a gender-specific analysis of the subjective assessments as was done by Penzkofer et al. (2010) but also to draw age-related conclusions. Table 2 gives a survey of additional characteristic data of the two samples.

Unfortunately, the number of males and females, however, varied within the two age groups (43 males and 20 females in the group of the younger Ss and 19 males and 46 females in the group of the older Ss). In consideration of these facts, a classification of the 128 interrogated order-pickers into homogeneous groups suitable for a more easy comparison was not possible. The group of younger males was accompanied by a rather great number of older females. Additionally, due to the low average age of the Ss and the age difference between the two age-groups of just 10 years, it could not be expected that strong "ageing effects" would be found.

Table 2 Specific data of the sample of the 128 employees of cold-storage depots

Characteristics	Younger Ss (N = 63)	Older Ss (N = 65)	
Age	29.5 ± 4.1	40.9 ± 3.5	[Years]
Height	176.0 ± 10.1	169.3 ± 8.5	[cm]
Weight	76.1 ± 12.9	71.6 ± 11.8	[kg]
BMI	24.5 ± 3.2	25.0 ± 4.0	
Working experience	5.1 ± 2.7	8.0 ± 3.1	[Years]
Order-picking in cold-storage depots	4.0 ± 4.6	6.5 ± 2.8	[Years]
Daily working time	6.5 ± 4.6	4.7 ± 1.2	[h]

Results

Study I

Physiological Responses of Heart Rate

Heart rate responses to the cold exposures—allowing a rating of whole body strain—of the two age groups are shown as 5-min-mean values in Fig. 2.

The heart rate values (work pulses) for the younger Ss in the cold store (cp. black area in top chart of Fig. 3) mark a high physical strain. Increases of 35 bpm and more above the resting level with an average of 70 bpm were no rarity during the phases of work. Due to increases over time, the endurance level sometimes was exceeded. Individual increases of up to 60 bpm were partly reached. For approximately half of the Ss, the base resting level of heart rate in the following warming-up period was not reached.

The work pulses registered for the older Ss were also in the range of 30–35 bpm above the resting level with an average of 77 bpm. Between the two age-groups there were just small differences in heart rate increases measured in each of the three working phases, which nevertheless are highly significant. This statistical significance of differences between the two age groups was verified with the two-sided t test (cp. Sachs 1974). Work-related increases in heart rate of approximately 30–35 bpm for the older Ss nearly match the average for the group of younger Ss. In the working phases with a duration of 80 and 100 min, the heart rate values of the 40–65 year-olds were not substantially, but were significantly lower than those of the 20–35 year-olds (80 min: $p < 0.0001$; 100 min: $p = 0.001$). In the longest phase with a duration of 120 min, there were no significant age-related differences between the two age groups ($p = 1$).

For the identification of the individual resting level (70 bpm on average for the younger Ss vs. 77 bpm on average for the older Ss) 2-min-mean-values of heart rate were calculated above the whole workday and their minimum was detected. It was assumed that the minimum of heart rate appeared during the assessment of physical and mental conditions in the morning, the follow-up examination in the evening or in one of the warming-up breaks. In each of these phases the Ss were in a sedentary position. Therefore, 35 bpm were regarded to limit the endurance level (see among others Strasser 1981 and Hettinger 1970).

In the chill room at approximately +3 °C—carrying out the same work, with regard to the extent of physical load handled and the course of time—substantially and highly significant lower levels of heart rate for the two age groups were registered (cp. bottom chart of Fig. 3). On average, the values were 5–10 bpm under the work pulses in the cold store. Probably this is due to both, lower cold stress and to wearing less heavy protective clothing (2 kg vs. 5 kg). In individual cases, heart rates of approximately 20 bpm below the work pulses in the cold store were measured in the chill room. With regard to age, also only moderate and again

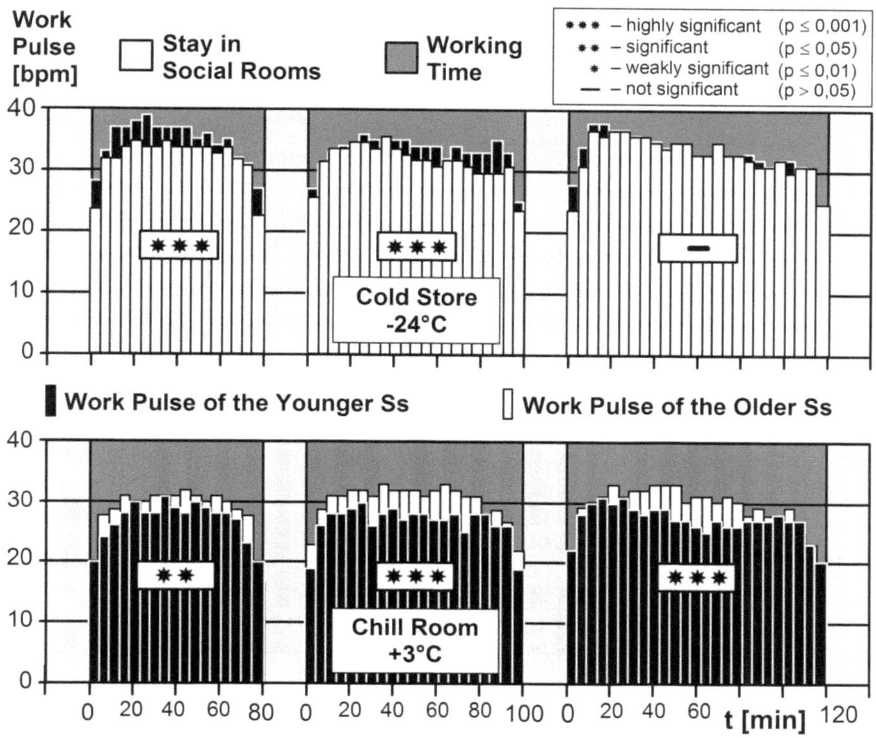

Fig. 3 "Work Pulse" -time diagram averaged over 15 male Ss at the age of 20–35 years (*height of black columns*) and 15 male Ss at the age of 40–65 years (*height of white columns*). 5-min-mean values during three working periods over 80, 100 and 120 min in the cold store at −24 °C (*top chart*) and during the same three working periods in the chill room at +3 °C (*bottom chart*). Symbolic labeling of statistically significant differences in heart rate increases between the two age groups of the two-sided t-test (–: $p \geq 0.05$, *: $p < 0.05$, **: $p < 0.01$, ***: $p < 0.001$)

not substantial differences of 5 bpm between the two groups were found. The heart rate of the younger Ss was lower than those of the older Ss, in the 100 and 120 min working phase the differences were even highly significant ($p < 0.0001$).

Blood Pressure Associated with Working in the Cold

According to classification criteria of the World Health Organization (WHO), blood pressure values of up to 140 mmHg (systolic) and 90 mmHg (diastolic) are referred to as normotonia, whereas values of 140–160 mmHg (systolic) and 90–95 mmHg (diastolic) are called borderline hypertension. Systolic values in excess of 160 mmHg and diastolic values greater than 95 mmHg are considered as pathological hypertension.

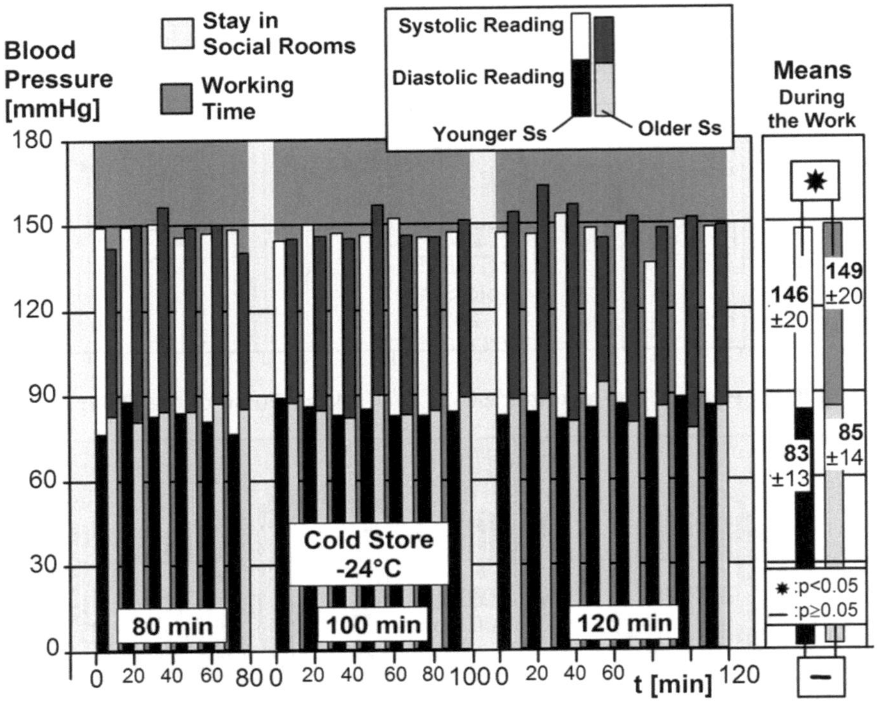

Fig. 4 Discretely measured mean blood pressure values (pair of columns in the very *right* part) and values during the cold exposures with means of the systolic and diastolic blood pressures for the two age groups with symbolic labeling of statistically significant differences (–: $p \geq 0.05$, *: $p < 0.05$, **: $p < 0.01$, ***: $p < 0.001$)

The blood pressure values of the younger and older Ss shown in Fig. 4 were still within the upper range of what is considered normal or, in some cases, borderline hypertension.

Averaged over the subjects of the respective age-group, maximum values of 146 ± 20 and 83 ± 13 mmHg were measured for the younger Ss as well as 149 ± 20 and 85 ± 14 mmHg for the older Ss. Age-related differences could not be verified. In the chill room statistically significant lower blood pressure values (in the range of 140 mmHg systolic and 80 mmHg diastolic) were detected for both age groups ($p < 0.005$).

Age-Related Skin Temperature Decreases Associated with Cold Exposure

As was expected, the temperature sensors for the skin temperature located under the thick cold protective clothing for the cold store exhibited no substantial temperature changes in the area of the shoulder and the kidney. The skin surface temperature dropped only to an insignificant amount. The warming-up process during the breaks was finished after a short time.

Figure 5, however, shows that a considerable temperature decrease was recorded at the nose, which was exposed to the cold without any protection. As quickly as the nose got cold, however, it was also warmed up again during the breaks—partly due to a more intensive blood circulation in the face after leaving the cold area.

On average the surface temperature dropped to +15 °C, causing a distinct feeling of cold for the working persons. The same was true for the fingers, the surface temperatures of which also reached just +16 °C, on average. In this case, however, the fingers got warmed up during the breaks. A somewhat smaller but continuous decrease of temperature during an increasing working time was recorded at the toes. They only completely warmed up again within a minimum of 20 min after leaving the cold store, and only if the boots were taken off during the break. The temperature curve of the measuring position under the sole of the foot was inconspicuous. The temperature of the foot, during continuously walking when working in the cold, was even a bit higher than during resting during the break. The mean temperature at the finger tips was as low as in the case of the nose.

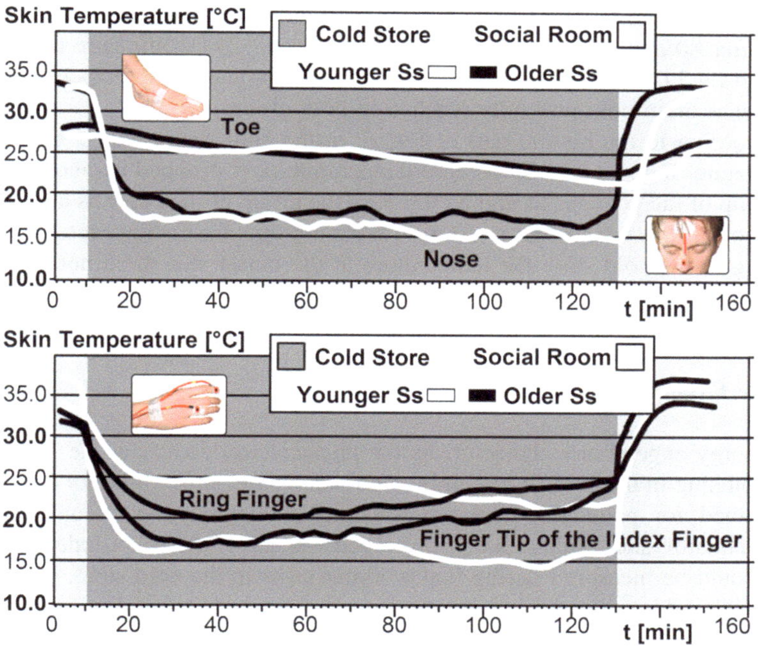

Fig. 5 Skin temperature [°C] at nose and toe (*top chart*) as well as ring finger and tip of the index finger (*bottom chart*) of male order-pickers prior to, during, and after the cold exposure of 120 min at approximately −24 °C in the cold store. Means of 15 subjects (Ss) in two age groups

It was nearly impossible to differentiate skin temperatures between the two age groups, yet, for the ring finger, a slightly age-related difference could be observed. Otherwise, the results for the younger test persons without restrictions were comparable to the older group. For the other 6 measuring positions, no age-related differences in skin temperature were recorded.

Working in the chill room could be considered as almost unproblematic. The skin surface temperatures were only decreased by a few Kelvin. Even for working in the cold store, the measured deviations from values under normal environmental temperatures were not remarkable. During the breaks, the initial values were easily reached again. As shown for core temperature, no age-related differences could be measured.

Age-Related Tympanum Temperature Decreases Associated with the Cold Exposure

Figure 6 shows that core temperature taken with an ear thermometer at the tympanum every 15 min at a surrounding temperature of −24 °C differed by 1.3 (after the 80-min exposure to the cold) to 1.5 K (after the 120-min exposure) in the 20–35 year-olds compared to 36.6 °C at the outset. Therefore, the temperature taken at the tympanum, which on principle is lower by approximately 2 K than the temperature taken in the rectum, shows a substantial decrease. In the 40–65 year-olds, after 80 and 120-min working in the cold at −24 °C, the core temperature dropped by 2.0–2.2 K. After a 20-min warming-up phase, the temperature at the outset was often not quite fully reached in both groups.

According to the bottom part of Fig. 6, in the chill room, with a surrounding temperature of about +3 °C, the core temperature only dropped by about 0.6 K in the group of the younger Ss and by 0.9 K in the group of the older Ss at the end of each cold exposure. After a 20 min warming-up phase, however, similar to working in the cold store the temperature at the outset was often not quite fully reached.

Energy Expenditure

The energy expenditure characterizes the physical workload, i.e. the intensity of order-picking in the cold. While in literature an energy expenditure of 360 W was determined for palletizing and order-picking moderately heavy units in cold (cp. Gebhardt and Müller 2003), averaged values of 520 W—independent of age—could be measured during real working tasks in the cold store. Despite the same physical workload for younger and older Ss, order-picking in the cold, as will be shown in the discussion, leads to an increasing physical strain with advancing age.

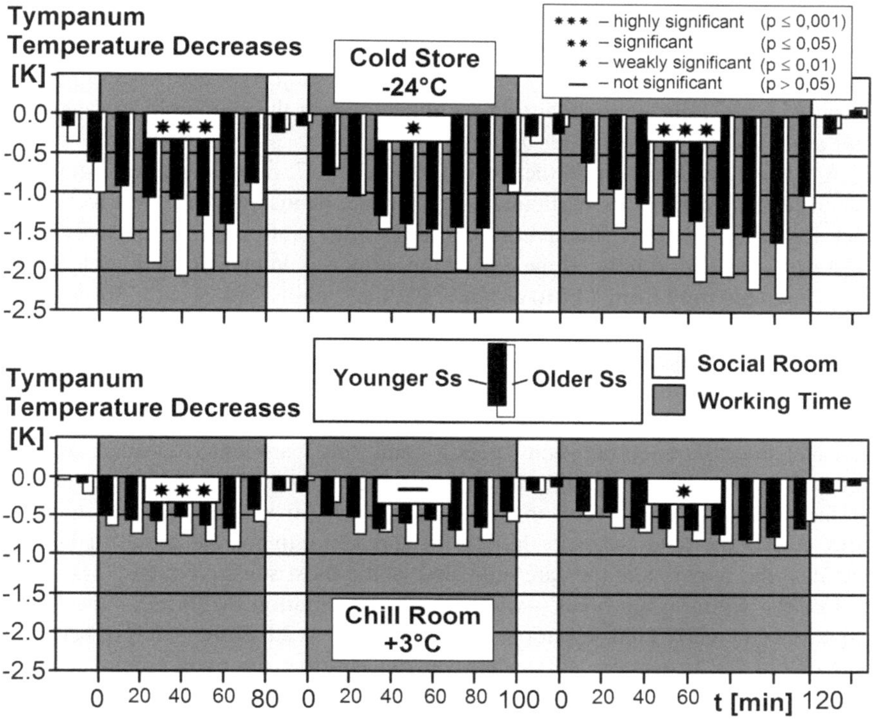

Fig. 6 Tympanum (*core*) temperature decreases [K] of male order-pickers prior to, during and after cold exposures of 80, 100, and 120 min. Means of 15 subjects (Ss) in two age groups, each, with symbolic labeling of statistically significant differences of the two-sided t test (–: $p \geq 0.05$, *: $p < 0.05$, **: $p < 0.01$, ***: $p < 0.001$)

Study II

Subjective Cold Experiences and Assessment of Physical Complaints

In order to identify possible health risks due to the cold exposures, the employees were also asked for their cold sensations. A special hazard of the cold exposure results from the cooling off of several sensitive body regions in the course of the working time. Based on experience, the acra are especially prone to such cooling off, which can be accompanied by negative impacts on the ability to work if the hands are affected (Imamura et al. 1998). If the affected body areas are not completely re-warmed in the warming-up period, the skin temperature decreases further in the following work phase.

To gain a deeper insight into the subjective cold experiences during work in a deep cold-storage depot, the order-pickers were asked to quantify the time-dependent cold sensations in various body areas, such as forehead, nose, mouth and ears, in the upper part of the body (chest/back, arms, hands and fingers) as well as in the lower body half (abdomen/buttocks, legs, feet and toes). In a procedure already applied by Strasser and Kluth (2006), these subjective cold sensations were recorded in 30-min intervals during the time spent in the deep cold-storage depot and at the end of the warming-up phase.

As expected and shown in the left column of Fig. 7, remarkably cold sensations occurred in the cold store particularly at the nose, hands and fingers as well as the feet and toes. However, the averaged ratings after prefixed time intervals of the cold exposure and, finally, at the end of the work (cp. values of the middle part of Fig. 7) ranged only from "1" to around "2", i.e., "chilly" to "cold". Surprisingly, in all problematic regions, cold sensations were felt more frequently by the younger order-pickers. While the difference is negligible for the fingers with 56 % affected in the group of the younger and 52 % in the group of the older Ss. The younger Ss felt cold sensations in the other regions more often. At the nose, hands and feet the difference between "young" and "old" amounted even to 14 %. In contrast, consistent age-related differences could not be identified in the intensity of the cold sensations. With maximal ratings of 2.6 and 2.8 ("cold" to "very cold") after the cold exposure lasting 90 and 120 min for the group of younger females, the fingers and toes are indicated as the most sensitive areas. Also at the end of the warming-up break—with an average duration of 20 min—the Ss still experienced clearly cold sensations with ratings up to 2.0 at the feet (cp. very right part of Fig. 7). Especially noticeable was the fact that the male employees more frequently indicated cold sensations than their female colleagues.

Working in a cold store with temperatures of approximately −24 °C as shown in the results of Study I requires considerably higher resources of an order-picker's physical performance capacity than a comparably high physical work at a normal surrounding temperature.

Since order-pickers normally move approximately 230 items with a mean total weight of 1.6 t/h and weight units up to 15 kg, it seemed advisable to also collect data associated with complaints in muscles and joints. For the investigation into the physical condition of the order-pickers, special question forms were used which are based on a compatible visualization of all body parts, where possible complaints in muscle parts and joints could be indicated. Also statements concerning the "frequency", "duration" and "intensity" of possible complaints had to be made on a 4-step scale from 0 (never, none) to 4 (daily, very strong, beyond the working day). The upper part of Fig. 8 shows the percentage of employees who had experienced complaints in the most affected body areas while working in the cold store. In the other, not visualized, body areas such as hands and fingers, arms or legs, the frequency of occurrence had not exceeded 20 % and, therefore, due to low importance had not regarded been anymore.

Apparently, the rear part of the left and right shoulder area, the area of the neck, the upper back and the lower back (lumbar vertebra) were affected by a substantial percentage of the Ss (up to 48 %). Age-related differences, however, could only be quantified at the back. While over 40 % of the group of the younger Ss felt complaints in this area, surprisingly only 22 % of the older Ss were concerned.

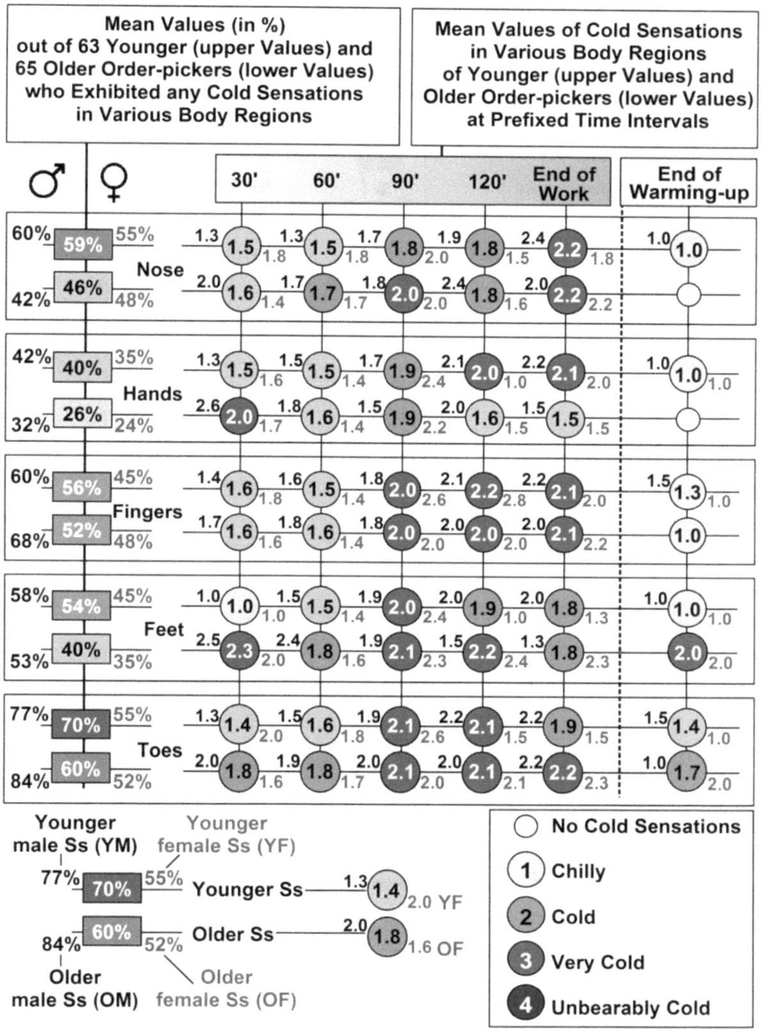

Fig. 7 Course of cold sensations in various body regions over a work period of 120 min at −24 °C and successive warming-up phases. Mean values of younger order-pickers (*upper numbers*) and older order-pickers (*lower numbers*) who exhibited any cold sensations on a scale from "0" (*no negative sensation*) to "4" (*unbearably cold*)

Due to the lower part of Fig. 8, on average, with ratings for the "frequency" between "2" and "3" the complaints were experienced, at least, once a week. With ratings for the "duration" close to "4", they lasted till the end or beyond the workday. However, with ratings of a maximum of "2" for the "intensity" on a 4-step scale, they were classified as no more intensive than "rather strong". Again, age-related differences could only be identified regarding the back area. Although

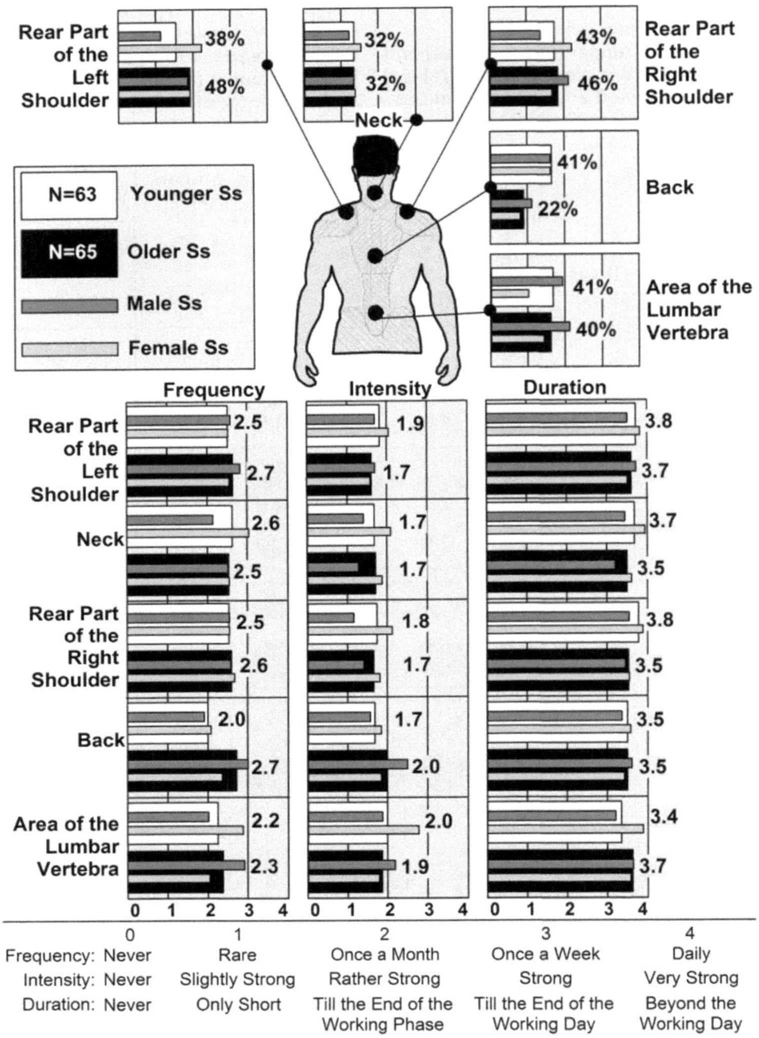

Fig. 8 Subjective evaluation of physical complaints. Percentages of order-pickers who reported any complaints in the visualized body regions (*upper part*). The *lower part* shows frequency, intensity and duration of experienced complaints with numbers representing mean values averaged over the group of males and females

complaints in this area were felt more often by younger order-pickers, they were experienced more frequently and more intensively by the older employees. Similar results were yielded in the chill room. Since order-picking in the two storage areas is the same, except for the prevailing temperature, substantial differences could not be expected here.

A final question referred to the overall experienced physical body strain. The two age groups exhibited no substantial differences for both the chill room and the cold store. With ratings of -1.3 ± 1.8 for the younger and -1.3 ± 1.9 for the older males which means "slightly demanding", and ratings of -2.0 ± 1.4 for the younger and -1.8 ± 1.6 for the older females which means already "quite demanding", differences between males and females are more clearly.

For further information about, e.g. the assessments of the working environment, the protective clothing, the work-rest regulation or the structure of the questionnaire, see Penzkofer et al. (2011).

Discussion

Work-Related Heart Rate Increases

Based on established knowledge of physiological mechanisms, a useful vasoconstriction in the body's periphery, and a possibly resulting specific blood-pressure increasing effect, as a response to working in the cold, can hypothetically be expected. Such an effect may also cause the heart rate to increase. The effects of the protective clothing may be involved, too. On the one hand, while working in the cold store, almost all body areas are covered by the protective clothing, so thermoregulatory responses to the cold may only partly take effect. On the other hand, the additional weight of the protective clothing of approximately 5 kg can also have an impact on activation of the cardiovascular system.

Heavy heat protection clothing (e.g. a flame-retarding smelter suit with an aluminum backed smelter coat), weighing approximately 5.5 kg, has, as shown in prior research (cp. Hertting et al. 1984), a clear effect on work-related heart rates. Increases of approximately 10 bpm due to the weight of the clothing alone have been recorded. To a somewhat less extent due to less heavy weight, this also seems to apply for wearing cold-protective clothing.

Although, to some extent, even highly significant work-related heart rate increases with respect to age were indicated by the two-sided t-test. Based on expected higher physical strain for the older Ss, substantial differences between the two age groups—with differences in their mean values merely of 5 bpm—did not occur. Nevertheless, it has to be considered that heart rate responses to a predefined load are dependent on several influencing variables, such as individual fitness, whereby strict limitations in heart rate with advanced age exists. Attention should be paid to the fact that, for every human, the maximum heart rate (HR_{max}) is lowered per year of life, according to the following formula: $HR_{max} = 208 - 0.7 \times \text{age}$ (cp. Tanaka et al. 2001). Therefore, the capacity utilization of heart rate increases is reduced with advancing age, so a work pulse profile with nearly identical increases above the resting level—and considering the higher resting levels of the older age group—is more problematic for older than for younger employees. According to the results shown in Fig. 3, an at least 10 % higher capacity utilization of possible heart rate increases during all working phases must be considered as substantially higher strain for the group of older Ss.

Regarding maximum heart rate decreasing with age and reduced capacity utilization, correspondingly adapted workday break regimes have to be provided for older employees. Due to lower thermal stress of order-picking at temperatures of approximately +3 °C, another possibility would be the assignment of older personnel to the chill room. Despite its substantial physical stress, order-picking work itself is—due to the heat production—more of a benefit than an additional stress as long as the work takes place within the order-picker's physiological performance capacity and thus within the endurance limit. Therefore, we cannot recommend to reduce the intensity of the physical work in the cold store.

Blood Pressure

The blood pressure values, which were approximately 10 mmHg and 5 mmHg higher while working in the cold store with −24 °C, were primarily due to two reasons. Firstly, the additional weight of the cold-protective clothing can stimulate the body's blood circulation. Secondly, the extreme climatic stress—despite the suitable protective clothing—poses higher demands on the thermo-physiological regulation mechanisms. Laboratory studies carried out in the past (cp. Ozaki et al. 2001) investigated that the blood pressure rises substantially with cold stress. This is due to the peripheral vasoconstriction in order to minimize the heat loss in the body's periphery.

The initial values of the blood pressure (measured in the morning) were already increased—compared to the standard values of 120/80 mmHg—with values of 140 mmHg systolic and 80 mmHg diastolic for the younger Ss and 140/85 mmHg for the older Ss. Due to this fact there were no clear increases in blood pressure visible at the beginning of the cold exposure and, furthermore, substantial increases while working up to 2 h in the chill room (+3 °C) or cold store (−24 °C) could no longer be expected. However, no decrease of the blood pressure values to the normal values—especially the systolic value—could be registered during the 20-min warming-up breaks.

The high initial values of the blood pressure may be derived from the unusual situation as a "guinea pig" and the associated "white coat syndrome" (cp. Dolan et al. 2004). To asses this theory and, in addition, to show the influence of the cold exposure on the blood pressure in more detail, 12 male subjects between 21 and 27 years were asked to carry out further tests in the dry storage (+18 °C) and in the cold store (−24 °C). In order to register normal daily values of the blood pressure, the Ss were further instructed to measure their blood pressure at home. In comparison to their daily values, an increase of 15 mmHg systolic and 6 mmHg diastolic during working in the cold store was measured. Furthermore, working in the cold led to blood pressure increases of 5 mmHg systolic and diastolic in comparison to the blood pressure measurement in the dry storage.

The combination of the cold protection clothing—in general, satisfactory—and the manual material handling of goods showed a physiological strain, increasing with age, that leads to an improved blood circulation due to the many changes of body posture and the movement of the arms and legs, the positive effect of which can also be felt in the toes. This effect, however, is not sufficient compensation for the heat lost during working in the extreme cold while picking up and carrying goods weighing up to 15 kg.

Core and Skin Temperature

Drops of the body core temperature up to 2.2 K show, that cold stress cannot be completely compensated by the combination of a suitable cold-protective clothing and the physical strenuous work of order-picking. The significantly higher temperature decrease of the older Ss cannot be interpreted without considering the registered skin surface temperatures. The skin surface temperature at the toes and fingers of the older Ss, in contrast to the younger Ss, show very inconsistent temperature profiles, which was especially true for the longest working phase (120 min). This may indicate a thermoregulatory disorder, as the body is not able to keep the blood at core temperature in the periphery. Rather the body core loses warm blood—documented through the greatest decrease of the body core temperature in this working phase—into the outer extremities which is caused by limited vasoconstriction.

It has to be concluded that a 20-min break is not long enough for a complete rewarming of the body for all Ss when working in the cold store. The investigations showed that, independent of age, the warming-up break should be at least 30 min. No age-related regulation of working time and warming-up breaks is advisable. The breaks in the cold store, however, need to be sufficiently long. The cold-protective clothing is largely adequate, but despite heat production during high demanding physical work, individual body parts are highly affected by the cold. Therefore, the clothing still needs to be improved.

Cold Sensations and Physical Complaints Associated with Working in Deep Cold

The subjectively experienced cold sensations yielded surprising results. Despite clear physiological disadvantages of elderly people—particularly with respect to heat generation—negative cold sensations more frequently occurred in the group of the younger order-pickers. This probably is due to the fact that a certain acclimatization to the extreme cold over years takes place. In some body parts cold sensations had not subsided completely at the end of the warming-up break, lasting approximately 20 min on average. Therefore, longer warming-up breaks should be provided for both, younger and older employees.

Order-picking in the cold—regardless of age—leads to complaints (no more often than once a week) especially in the upper part of the body, which were felt to be "rather strong". These experienced physical complaints are surprisingly similar, in terms of frequency, intensity, and duration with working in a normal environment. Working in the cold and in a normal environment besides the temperature differed, however, by wearing cold-protective clothing. Whereas working in a normal surrounding temperature can be facilitated by tools, such as packing and lifting aids, this is not possible in the cold store for the foreseeable future.

Outlook

The test design of these studies allowed statements about the maximum cold exposure up to which no negative impact on the physical capacity and the subjective sensations of cold can be measured. Moreover, there is still a lack of clarity how long the warming-up break should be, to guarantee a riskless task fulfilment for every age of the workers and to optimize the personal proficiency level in the long run. This should be clarified using a similar test design, certainly with a variable length of warmin-up breaks, in the future.

But there are more reasons to carry out further examinations in the cold. The working population—due to the demographic changes—will become older and even more females will be present in the workforce. In particular, physically demanding work, which a few decades ago was performed only by males, is increasingly assigned also to female workers. Therefore, in the future more older female workers will have to perform physically demanding work under particularly difficult environmental conditions. This effect can also be seen in the company participating in the field studies. The cold-storage depots were built in the late 90s, when the company increased the offer of refrigerated and frozen food. Based on the knowledge of the physiological characteristics concerning the heat generation of young and old males and females, respectively, at that time mostly younger males were employed, especially in the extreme cold environment with temperatures of approximately −24 °C. At the time when the investigation was conducted, however, order-picking in the cold was carried out by a surprisingly high proportion of females which amounted to an extraordinary percentage (52 %).

Over time, the workplace of order-picking—due to the working-time, rather high wages and quick acquisition of skills—has become more attractive for females with regard to a re-entry into the job, e.g., after giving birth to children.

Regarding the growing number of female order-pickers and in knowledge of their physiological characteristics, it seems to be necessary to carry out the same work-physiological investigations—as shown in this study—with females.

Under thermal stress the human being is subject to the laws of thermodynamics, whereby males seem to have better anatomical and physiological properties to protect them from cold. In order to keep the core temperature at a normal level in case of an increasing temperature gradient, the energy expenditure in the organism has to be intensified substantially. However, the energy expenditure of females is lower than that for the males, which leads to a significantly lower heat production (cp. Boothby et al. 1936). An important source of warmth is—besides the liver—the physically active musculature. So, when considering that females can dispose of only 70 % of the muscular strength compared to males (cp. Hettinger 1968) and that their muscle mass amounts to only 36 % of their overall body weight compared to 42 % for males, the females' ability to develop heat is probably lower. The female body, however, consists of 10 % more fat, which seems to be advantageous when trying to retain warmth; yet, an almost 15 % thinner female skin probably leads to less effective insulation against loss of warmth. The ratio between body surface and body volume is expected to determine the degree of thermal storage and, therefore, maybe also of importance for the tolerability to endure cold exposures. Compared to females, males often have a substantially larger volume with respect to their skin surface and, therefore, are more able to produce and accumulate vital body warmth.

All the above-mentioned characteristics (differences in morphology and metabolism) give rise to the question whether females are necessarily at a disadvantage when exposed to short-term or medium-term cold stress or whether they might be able to resist the cold due to a gender-specific, vegetative and hormonal control system, which increases the metabolic productivity in the musculature leading to an increasing production of warmth.

First results of a study concerning the physiological responses of younger and older females while order-picking in deep cold were published by Kluth et al. (2012) and Baldus et al. (2012).

Acknowledgments We are thankful for the cooperation of the employees, and are grateful for the support provided by the management of ALDI GmbH and Co. KG. We would also like to thank all subjects participating in the study, the students assisting in carrying out the field studies and the associates of the Ergonomics Division at the University of Siegen involved in the research project (DFG grant no. STR 392/5-1).

References

Adolph EP, Molnar GW (1946) Exchanges in heat and tolerances to cold in men exposed to outdoor weather. Amer J Physiol 146:507–537

Baldus S, Kluth K, Strasser H (2012) Order-picking in deep cold—physiological responses of younger and older females. Part 2: body core temperature and skin surface temperature. Work J Prev Assess Rehabil 41(1):3010–3017

Boennemark B, Johansson L, Keiding J, Magnusson H, Palmgren B, Söderqvist A (1969) Arbete i fryshus, Stockholm: Rapport A-141.69, Arbetsmedicinska Institutet

Boothby WM, Berkson J, Dunn HL (1936) Studies of the energy of metabolism of normal individuals, a standard for basal metabolism, with a nomogram for clinical application. Amer J Physiol 116:468

Brajkovic D, Ducharme MB (2006) Facial cold-induced vasodilation and skin temperature during exposure to cold wind. Eur J Appl Physiol 96(6):711–721

Budd GM, Warhaft N (1966) Body temperature, shivering, blood pressure and heart rate during standard cold stress in Australia and Antarctica. J Physiol 186:216–232

Dolan E, Stanton A, Atkins N, Den Hond E, Thijs L, McCormack P, Staessen J, O'Brien E (2004) Determinants of white-coat hypertension. Blood Pressure Monit 9:307–309

Forsthoff A (1983) Arbeit in −28 °C, Arbeitsphysiologische Untersuchung zur klimatischen Belastung bei Körperarbeit in extrem tiefen Umgebungstemperaturen unter besonderer Berücksichtigung der Kühlhausarbeit. Dokumentation Arbeitswissenschaft, Band 9

Gebhardt H, Müller BH (2003) Ergonomische Gestaltung von Kältearbeitsplätzen, Technik 32. Bundesanstalt für Arbeitsschutz und Arbeitsmedizin, Dortmund

Hassi J, Rytkönen M, Kotaniemi J, Rintamäki H (2005) Impacts of cold climate on human heat balance, performance and health in circumpolar areas. Int J Circumpolar Health 64:459–467

Hellstrøm B, Berg K, Vogt Lorentzen F (1970) Human peripheral warming during exercise in cold. J Appl Physiol 29:191–199

Hertting R, Hettinger T, Eissing G (1984) Einfluss von Schutzkleidung auf die Beanspruchung des Menschen bei Arbeit unter Wärmestrahlungsexposition. Arbeitsmed Sozialmed Präventivmed. 19(1):9–14

Hettinger T (1968) Isometrisches Muskeltraining, 3rd edn. Thieme, Stuttgart

Hettinger T (1970) Angewandte Ergonomie. Bartmann, Frechen

Hines EA, Brown GE (1975) Proceedings of the staff meeting mayo clinic 7, 322 1932. In: Leblanc J (ed) Man in the cold. Charles C. Thomas Publisher, Springer, USA

HVBG (2002) Leitfaden für die Ergometrie bei arbeitsmedizinischen Vorsorgeuntersuchungen nach berufsgenossenschaftlichen Grundsätzen (Anhang 2). In: Berufsgenossenschaftliche Grundsätze für arbeitsmedizinische Vorsorgeuntersuchungen, Gentner, Stuttgart

Imamura R, Rissanen S, Kinnunen M, Rintamäki H (1998) Manual performance in cold conditions while wearing NBC clothing. Ergonomics 41(10):1421–1432

Kleinöder R (1988) Ergonomische Gestaltung von Kältearbeit bei −30 °C in Kühl- und Gefrierhäusern, Schriftenreihe der Bundesanstalt für Arbeitsschutz, Forschung, Wirtschaftsverlag NW, Dortmund

Kluth K, Strasser H (2003) Subjective evaluation of a newly developed scanner checkout in comparison with a conventional cash register system via standardized working tests. In: Strasser H, Kluth K, Rausch H, Bubb H (eds) Quality of work and products in enterprises of the future. Ergonomia, Stuttgart, pp 275–278

Kluth K, Strasser H (2006) Heart rate and blood pressure responses to working in the cold In: Pikaar RN, Koningsveld EAP, Settels PJM (eds) Proceedings of the 16th Triennial Congress of the International Ergonomics Association, Maastricht, The Netherlands, Proceedings-CD, 6 pages

Kluth K, Baldus S, Strasser H (2012) Order-picking in deep cold—Physiological responses of younger and older females. Part 1: heart rate. Work J Prev Assess Rehabil 41(1):3002–3009

Mäkinen TM (2007) Human cold exposure, adaptation, and performance in high latitude environments. Am J Hum Biol 19:155–164

Müller-Arnecke HW, Hold U (1999) Ergonomische Gestaltung von Kältearbeit im Bereich von 0 °C, Schriftenreihe der Bundesanstalt für Arbeitsschutz und Arbeitsmedizin, Fb 845. Wirtschaftsverlag NW, Dortmund

O'Hanlon JF, Horvath SM (1970) Changing physiological relationships in men under acute cold stress. Can J Physiol Pharmacol 48:1–10

Ozaki H, Nagai Y, Tochihara Y (2001) Physiological responses and manual performance in humans following repeated exposure to severe cold at night. Eur J Appl Physiol 84:343–349

Penzkofer M, Kluth K, Strasser H (2010) Order-picking in deep cold—a gender-related analysis of subjectively-assessed effects. In: Vink P, Kantola J (eds) Advances in occupational, social, and organizational ergonomics. CRC Press, Taylor and Francis, Boca Raton, pp 168–177

Penzkofer M, Kluth K, Strasser H (2011) Subjectively assessed age-related stress and strain associated with working in the cold. Theor Issues Ergon Sci. doi:10.1080/1463922X.2011.617114

Sachs L (1974) Angewandte Statistik, Planung und Auswertung. Methoden und Modell. Springer, Heidelberg, p 209–216

Strasser H (1981) Arbeitswissenschaftliche Methoden der Beanspruchungsermittlung - Beanspruchungsprofile unter dem Aspekt der Ausführbarkeit und Erträglichkeit menschlicher Arbeit, Band 69 der Schriftenreihe Arbeitsmed Sozialmed Präventivmed, Gentner, Stuttgart

Strasser H, Kluth K (2006) Sensations of cold and physiological responses to groceries handling in cold-storage depots. In: Pikaar RN, Koningsveld EAP, Settels PJM (eds) Proceedings of the 16th Triennial Congress of the International Ergonomics Association, Maastricht, The Netherlands, Proceedings-CD, p 6

Strasser H, Kluth K (2008) Physiological responses of core and skin temperature to groceries handling in cold storage depots and sensations of cold. Z Arb wiss 62(4):249–256

Strasser H, Müller-Limmroth W (1983) Ergonomie an der Kasse - aber wie? Arbeitswissenschaftliche Schriftenreihe des Bayrischen Staatsministeriums für Arbeit und Sozialordnung, München

Tanaka H, Monahan K, Seals D (2001) Age-predicted maximal heart rate revisited. J Am Coll Cardiol 37:153–159

Tochihara Y, Ohkubo C, Uchiyama I, Komine H (1995) Physiological reaction and manual performance during work in cold storages. Appl Hum Sci 14(2):73–77

Wenzel HG, Piekarski C (1980) Klima und Arbeit, Bayerisches Staatsministerium für Arbeit und Sozialordnung, München

Standards and Legal Requirements

DIN 33403 (2000). Climate at the workplace and in its environments—Part 2: Effect of the climate on the heat balance of human beings, Beuth, Berlin

93/42/EEC (1993) European Council Directive concerning medical devices. Official J Eur Commun No L 169

Index

C. M. Schlick et al. (eds.), *Age-Differentiated Work Systems*,
DOI: 10.1007/978-3-642-35057-3, © Springer-Verlag Berlin Heidelberg 2013

S (*cont.*)

T

U

V

W

Y

Z

MIX
Papier aus verantwortungsvollen Quellen
Paper from responsible sources
FSC® C105338

If you have any concerns about our products,
you can contact us on
ProductSafety@springernature.com

In case Publisher is established outside the EU,
the EU authorized representative is:
Springer Nature Customer Service Center GmbH
Europaplatz 3, 69115 Heidelberg, Germany

Printed by Libri Plureos GmbH
in Hamburg, Germany